Antenna and Wave Propagation

About the Authors

Ashish Mathur is a senior faculty in the department of Electronics and Communication at JIET School of Engineering and Technology for Girls, Jodhpur under Poornima JIET Foundation. He is working since 6 years and has also worked with SMCET, Jaipur and BMIT, Jaipur for two years respectively. He has completed his Diploma in Electronics and Communication from Technical Board, Jodhpur in 2002 from Government Polytechnic College, Bikaner and his B.E. in the same branch from Rajasthan University in 2005 from Sri Balaji College of Engineering and Technology, Jaipur. Presently he is pursuing his Master's degree in technical field and his areas of interest are Digital and Analog Communication and Antenna.

Ranjana Trivedi is a senior faculty in the department of Electronics and Communication at Jodhpur Institute of Engineering and Technology, Jodhpur under Poornima JIET Foundation. She is working since 6 years and has also worked with JNIT, Nathdwara and Arya Group of Colleges. She has completed her B.E. in Electronics and Communication from Rajasthan University in 200[illegible] from Engineering College, Bikaner. Presently she is pursuing her Master's degree in technical field and her areas of interest are Microwave and Antenna and wave Propagation.

Geetika Bhati Mathur is a senior faculty in the department of Electronics and Communication at Jodhpur Institute of Engineering and Technology, Jodhpur under Poornima-JIET Foundation. She is working since 6 years and has also worked with Poornima Group of Colleges, Jaipur for more than one year. She has completed her B.E. in Electronics and Communication from Rajasthan University in 2005 from Modi Institute of Engineering and Technology, Laxmangarh. Presently she is pursuing her Master's degree in technical field and her areas of interest are Antenna and Wave Propagation, Radar and Audio Video Systems.

Antenna and Wave Propagation

Ashish Mathur
Ranjana Trivedi
Geetika Bhati Mathur

Department of Electronics & Communication Engineering,
JIET Group of Institutions,
Jodhpur (Rajasthan)

Published by:
Scientific Publishers (India)
5-A, New Pali Road, P.O. Box 91,
Jodhpur – 342 001 (India)

E-mail: info@scientificpub.com
www.scientificpub.com

ISBN: 978-81-7233-692-9

Lasertype set : Rajesh Ojha

Printed in India

Dedicated

to

"MAA SARASWATI"

Bramhanandam Paramsukhadam Kevalam Gyaanmurtim,
Dwandateetam Gagansadrasam Tatvamasyaadilakshyam,
Ekam Nityam Vimalamachalam Sarvdhisakshibhutam,
bhavateetam trigunrahitam Sadgurum tam namami.

to all

our GURU'S,

our parents and to

Dhruv and Pradnya

PREFACE

Today, in the day to day life, when communication systems have become the basic part of one's life, for increasing the business, for security, for multimedia or more than that say for the purpose of mere entertainment; the need of antenna has increased in the communiqué market. Thus, as antennas are the basic parts of these systems therefore no one can ever imagine any data transfer via wireless, mobile, optical or satellite communications without them or even chatting with our friends sitting at the other part of the earth. These antennas work along the transmitter and receiver systems to instantaneously transfer the important information across the globe in few seconds.

Generally, on an outing everyone have seen many kinds of antenna mounted on the roof of a building, on the towers, on back of vehicles, on Wi-Fi sets on military tankers or even on the cordless mobile phone sets carried by all. The antennas are designed according to the frequency and the application for which it has to be used. For last many years the antennas are developed according to one's personal need. Therefore, these antennas are of many types, Yagi-Uda antennas used for T.V. Reception, Dish (Lens) antennas for Digital T.V. reception purpose, Parabolic reflectors along with the helical antennas for satellite communication, whip antenna for FM reception on the mobile vehicles, Horn antenna are mounted on the surface of the aircraft for receiving and transmitting various frequencies, last but not the most commonly used antennas are the microstrip antennas which are made up of only a patch of a conductive material on the mobile phone sets. Recently this technology is advancing towards the development of "SMART ANTENNAS".

The motivation of writing this text came when one our author during her regular classroom teaching felt the absence of a suitable text for this topic. This book is to fill the void and is designed for the final year undergraduate or to assists the first year post graduate students of electronics and communication engineering and allied subjects. It may be difficult to cover the entire text in one semester. Depending on other courses offered and the emphasis given in a programme, a teacher may omit one or two topics.

Book is a perfect blend of theoritical and practical aspects of the topic. Concepts and analytical mathematical treatment are given only to the extent required. Worked out examples are given where considered necessary. To cover the practical topics and by keeping in mind the need of dissertation project for the post graduate students a brief description of an antenna designing software IE3D is introduced as a unit of this book.

UNIT 1: Antenna Fundamentals

This unit is the introductory preface for understanding the working of an antenna system. It includes all the basic parameters which play an important role in understanding how an antenna works or the parameters to be kept in mind during the designing of an antenna.

UNIT 2: Types of antenna

This part of the book covers all the types of antenna used today for different types of communication techniques. Here all the emphasis is on how a particular antenna works, where these designed antennas can be used, what are the suitable frequency range for the same and much more.

UNIT 3: Antenna Arrays

Here in this unit the emphasis is on the theory of using more than one antenna of the same type as an array; so as to increase the gain of the system. Further, we have discussed various techniques like the old method and the Binomial array technique for the designing of an array.

UNIT 4: Radio wave propagation-I

The text given under this unit is providing the information related to all the three modes of propagation, i.e. the ground wave, sky wave and the space wave propagation. Along with this it gives the relevant structures of the ionosphere which changes according to day and night. It is also covered that how this change effects the transmission during both these different timings.

UNIT 5: Radio wave propagation-II

This unit is a continuous sequence of the 4 unit. Here we discuss the basic parameters included during different modes of propagations, according to which the height of antenna, distance between two stations are been decided and much more.

UNIT 6: Advance antenna Topics

This part of the book is written by keeping in mind the need of our post graduate students where they are briefed about the advance applications of antenna such as- Smart Antenna, antennas for biomedical applications, antennas for monitoring RF radiation, IR antennas, phased array.

UNIT 7: IE3D Antenna designing software

This unit full fills the entire basic requirement for a basic learner of this software used for the purpose of designing of own antenna.

We set out to write this book with an aim of giving a broad, yet fairly in depth, and up-to-date coverage of Antenna and Wave Propagation. How far we have succeeded in this aim is for the readers to judge. We would be grateful for comments from the readers, especially students, teachers and practising professionals.

Ashish Mathur
Ranjana Trivedi
Geetika Bhati Mathur

ACKNOWLEDGEMENT

Many have contributed to the successful preparation of this text. We would like to place on record our grateful thanks to all of them.

We started writing this book after we were inspired and motivated to do so by our honourable Director Er. Navneet Agarwal Sir (JIET Group of Institution, Jodhpur).

We are also thankful to Mr. Sunil Trivedi for his kind help as and when required regarding surfing and browsing. We give our especial thanks to Er. Ankit Bhandari to design the layout for our book.

All our administrator staff and few other colleagues at the institute have in some way contributed to this process.

We thank, Mr. Pawan Kumar and Tanay Sharma, Directors, Scientific Publisher (India), Jodhpur, meticulously processed the manuscript with remarkable speed, both during the editorial and production stages, and made valuable improvements.

Mr. Rajesh Ojha, for his skilful services, in word processing the handwritten manuscript. He did his job so efficiently and delightfully that we had no hesitation in revising, modifying and correcting the computerized manuscript many times.

A number of persons in our family especially our parents and children have been well wishers of this activity. Those were the unspoken words by many of them then "Look at us", which were then been ignored by us. We feel sorry for that time and say thanks to them to understand our condition.

We are overwhelmed, when we think of the fact that there are so many who have worked to make this effort a success although there is only our name printed in the column of authors but we are indebted to each of them.

It is our experience that both science and religion have their roles to play in one's life. While science has helped me to think and reason rationally, religion has carried us beyond the realm of thought and reasoning. A great seer of India has blessed this effort. Immense pleasure is felt by us to write this last page on the land of Tannot Check post near Jaisalmer, with the blessing of Goddess Tanot Rai Mata ji. It is with this great humility that we offer this text at the feet of the Supreme Being.

Ashish Mathur
Ranjana Trivedi
Geetika Bhati Mathur

CONTENTS

UNIT 1
ANTENNA FUNDAMENTALS

CONTENTS

ANTENNA FUNDAMENTALS

1.1 ANTENNA THEORY HISTORY

What was the origin of antenna?

1. To begin with the topic, let us discuss with the invention of a compass, as they in some sense (receives) magnetic field (but it was not an electromagnetic field).
2. The human eye of course receives high frequency electromagnetic waves (light) of particular frequency. Technically the eye could be classified as an antenna; however, since it can't transmit wave, it is really a sensor. The colours that we see (red, green, and blue) are each waves of different frequencies that the eyes can detect.
3. Ben Franklin's kite experiment was not quite an antenna, as that captured lighting discharge, which was a direct current path where the energy was not transferred independent of the medium it travels.
4. The first experiments that involved both the coupling of electricity and magnetism along with its definitive relationship were that done by Faraday somewhere around the 1830s.

He slid a magnet in the coil of a wire attached to a galvanometer. In moving the magnet, he was creating a time varying magnetic field which as a result (from Maxwell's Equations) had a time-varying electric field.

This coil acted as a *loop antenna* and received the electromagnetic radiation, which was (received) detected by the galvanometer that is, the work of an antenna. Interestingly, the concept of electromagnetic waves was not even thought up by anyone at this point.

5. Later in 1886, Heinrich Hertz developed a wireless communication system in which he forced an electrical spark to occur in the gap of a *dipole antenna.* He used a *loop antenna* as a receiver, and observed a similar disturbance.
6. By 1901, Marconi send information across the Atlantic, for this he used several vertical wires attached to the ground for transmitting the signal, across the Atlantic, the receiver antenna was a 200 meter wire held up by a kite.

Figure 1.1 A painting of great scientist Michael Faraday experimenting in his chemistry lab.

7. In 1906, Columbia University had an experimental wireless station where they used a transmitting aerial cage which was made up of wires and suspended in air.
8. In early antenna literature waveguides were said as ``hollow metal pipes". Major antennas and their discovery years (fabrication)-

   (a) Yagi Uda Antenna - 1920

(b) Horn Antenna - 1939

(c) Antenna Arrays - 1940

(d) Parabolic Reflectors - (Late 1940 - early 1950)

(e) Patch Antennas - 1970

(f) PIFA - 1980

9. Latest research on antenna is going on with **meta-materials** (materials that have engineered dielectric and magnetic constants)

This research focuses on making antennas smaller, particularly in communications for personal wireless communication devices (e.g. cell phones).

A lot of work is being performed on numerical modeling of antennas, so that their properties can be predicted before they are built and tested.

1.2 INTRODUCTION TO ANTENNA

Defination 1 - A metallic device for radiating or receiving radio waves.

Defination 2 - Antenna or aerial as the means for radiating or receiving waves.

Defination 3 - Antenna is transitional structure between free space and a guiding device. (The guiding device or transmission line may be coaxial line or hollow pipe (waveguide) and it is used to transport electromagnetic energy from the transmitting source, to antenna or from antenna to the receiver.

Defination 4 - Antenna or Aerial is a system of elevated conductors which couples or matches the transmitter or receiver to free space.

Defination 5 - Antenna is the structure associated with the region of transition between a guided wave and a free space wave (or vice-versa).

Defination 6 – Transmitting antenna is connected to transmitter by a transmission line, to force electromagnetic waves into free space which travels in space with the velocity of light.

Defination 7 – Receiving antenna is connected to radio receiver, to receive or to intercepts a portion of electromagnetic waves travelling through space.

Basics of Antenna:

(Explanation 1)

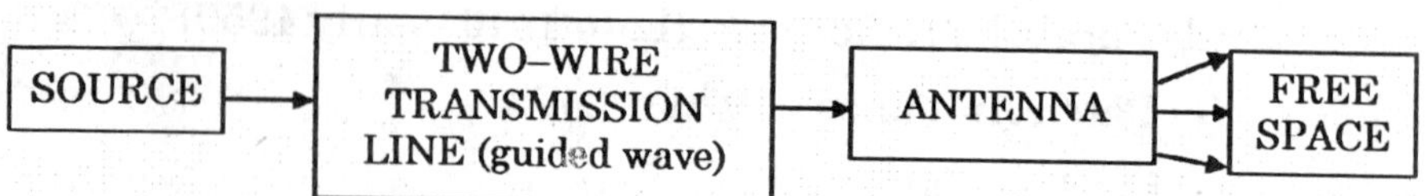

Figure 1.2 Basic Block diagram of an antenna system.

Transmission Line

It is a wired path, which transfers or guides radio frequency energy from source to destination, with minimum attenuation and other radiation losses.

- During this transfer of energy through transmission lines, the energy is confined within the vicinity of the line and also the guided waves which follow along the line, that is, one dimensional does not spread out into space.
- Now, to understand how the antenna Radiates it is important to know here that, if an **Infinite lossless transmission line** is connected to a source or generator a uniform travelling wave (or incident wave) appears along the line in the forward direction from source towards load and maxi-mum energy is said to be transferred from source to load.
- Similarly same condition can be achieved if we use a **finite length transmission line,** which is terminated by a load equal to **characteristic impedance of the line (Z_o)** at far end from the source end.
- Again, if the same transmission line is terminated by a load other than the `Z_o' (whether high or low) mismatch occurs and a backward wave (or reflected wave) appears on the line in the backward direction from load to source end, also along with a forward wave.
- Now, due to interference in both these waves (forward and backwards), standing waves occur which are undesirable and is a important reason of energy loss.
- **Conclusion:** If the line is lossless and of infinite length or a finite transmission line is terminated with characteristic Impedance (Z_o), then **all the energy sent through forward or incident waves are absorbed by the load and hence no backward or reflected waves, thereby,**

no loss across the line, that is, all the energy is transferred from source to load without any loss.

Let us now look at the figure given below-

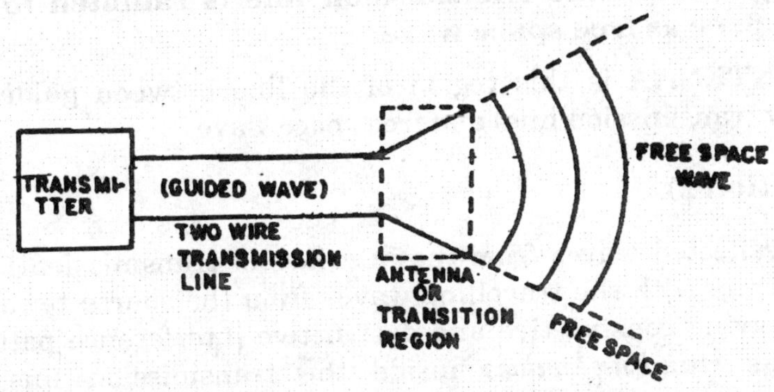

Figure 1.3

- Two wire transmission lines are connected with a transmitter.
- It is also assumed that the transmission line is perfectly matched so that only forward wave (incident wave) travels in forward direction that is, guided wave, guided by transmission line is travelling along the line.

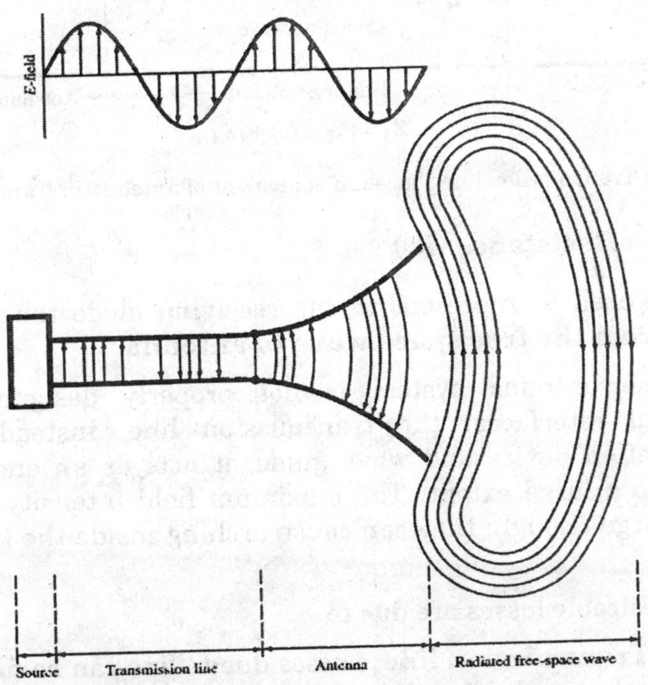

Figure 1.4 Antenna as a transition device.

Now, if this transmission line spreads out gradually till the distance between the two lines are several wavelengths. Then, in this region, the wave which is guided by the transmission line is radiated into free space as free space wave.

∴ **ANTENNA** is this region of the line between guided wave (guided by transmission line) and free space wave.

(Explanation 2)

The reflected waves from the interface of transmission line and antenna along with the travelling waves from the source towards the antenna having constructive and destructive interference patterns is referred as *standing waves* inside the transmission line which represent *packets of energy concentration and storage, such or in* ***resonant devices***. A wave pattern is shown in Figure is used to represent antenna in receiving mode.

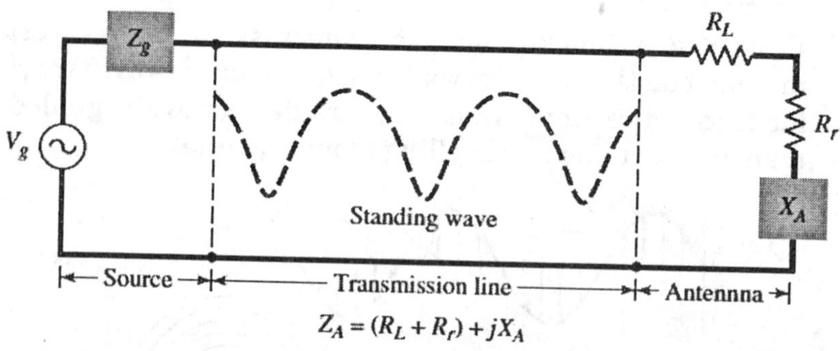

Figure 1.5 Transmission line Thevenin equivalent of antenna in transmitting mode

Radiation Resistance (R_r)

It is used to represent in the receiving mode the **transfer of energy from the free space wave to antenna.**

If the antenna system is not properly designed (that is, unmatched interfaces) the transmission line, instead of energy transportation device and wave guide, it acts as an energy storage element to a large extent. The maximum field intensity of standing wave is large enough; they can cause arching inside the transmission line.

Undesirable losses are due to

(1) **Transmission line:** Losses due to line can be minimized by selecting low loss lines.

(2) **Antenna:** Losses due to antenna can be decreased by reducing the loss resistance represented by R_l.

(3) **Standing waves:** Losses due to the standing waves can be reduced, (that is, the energy storage capacity of the line can be minimized), by matching the impedance of the antenna (load) to characteristic impedance of line. This is same as matching loads to transmission lines (where the load here is the antenna).

In addition to receiving or transmitting energy, **an antenna in an advanced wireless system is usually required to optimize or accentuate the radiation in some directions and suppress it in others.** (Hence the antenna must also serve as a directional device in addition to a *probing device*). It can be now well said that the antenna serves to a communication system the same purpose that eyes and eyeglasses serve to the human.

1.3 WHY DO ANTENNA RADIATES?

To understand why antennas radiate, is the prime necessity in understanding the fundamentals of antennas. So in this chapter, in I[st] phase we introduce a low key explanation with no mathematics, on how and why an antenna radiates electro-magnetic fields. While in the second phase we carry out mathematical derivation of radiation mechanism through single wire, two wires and in dipoles.

To start with phase first, of discussion, let us start with some basic physics. There is an electric charge which is a quantity of nature such as mass, weight or density which is possessed by every object. We all are most likely electrically neutral we don't have a net charge that is positive or negative (that is, protons and electrons) respectively).

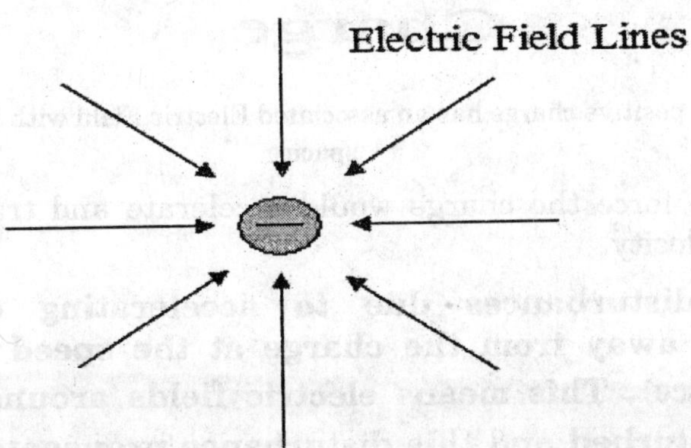

Figure 1.6 A negative charge has an associated Electric Field with it, everywhere in space.

Whereas, some materials (like metals) are electrically conductive as they have loosely bound electrons. Therefore, when a voltage is applied across a metal, the electron travel around a circuit, this flow of electrons is **electric current.**

Now for the let us talk about the charge. Let us say that for some reasons, there is a **negatively charged particle** sitting somewhere in space. All the charged particles will have an associated electric field with them as illustrated in the given figure-

So this negatively charged particle produces an electric field around it, everywhere in space. The electric field is a vector quantity; it has a magnitude (how strong the field strength is) and a direction (in which direction does the field point). The field strength dies off (becomes smaller in magnitude) as we move away from the charge. Further the magnitude of the E-field depends on how much charge exists. If the **charge is positive,** the E-field lines point away from the charge.

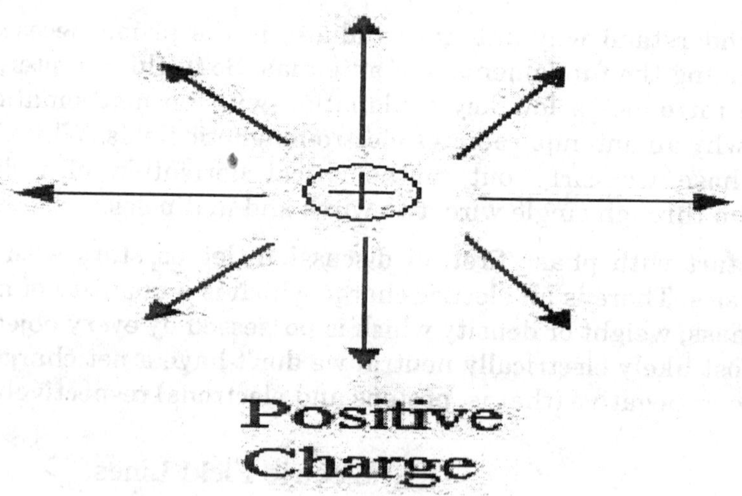

Figure 1.7 A positive charge has an associated Electric Field with it, everywhere in space.

Due to force the charge would accelerate and travel away at a constant velocity.

The disturbances due to accelerating charges will propagate away from the charge at the speed of light (c = 3×10^8 m/sec). This means electric fields around the charge will be disturbed and this disturbance propagates away from the charge with speed of light as in given figure-

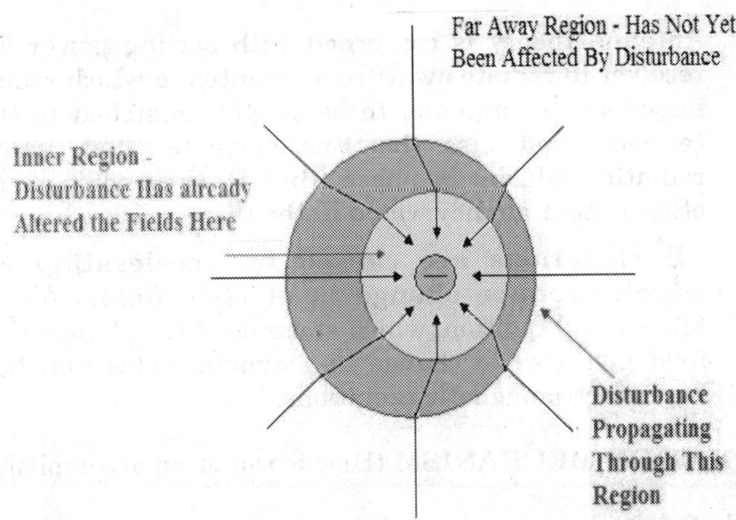

Figure 1.8 The E-fields when the charge is accelerated.

Given figure explains that, once the charge is accelerated, the field regains its original state. However, the fields can only respond to events at the speed of light. Hence, if a point is very far away from the charge it will take time for the **disturbance (or change in electric fields)** to propagate to the point.

The above figure, illustrates three regions-

(1) Innermost Region – In this the field is close to the charge. It is said the *Region of disturbance*)

(2) Middle Region – It is called as the *Region of varying field* from old magnitude and direction to their new magnitude and directions.

(3) Outermost Region – It is the *Undisturbed Region.* It has same magnitude and direction as the charges are nearly constants.

We have already laid the fundamental reason for radiation. The fields changes due to moving charges. If a moving set of charges (electric current) is generated then there will be radiations for sure.

As we know that, cables and wires used for different purposes due to, accelerating electric charge in them radiates energy, like antennas, but very poor antennas. It is because of the fact that these transmission medium carry current from one terminal to other in circuit or a loop. The radiation in the wired loop is cancelled by current flowing in the adjacent wire (that is travelling the current in opposite directions).

Summary

(1) Antenna theory is concerned with getting power from the receiver to radiate away from an antenna which requires the impedance of antenna to be roughly matched to that of a receiver, and also that the currents, that, causes the radiation add up in phase (that is, they don't cancel each other output as they would in the transmission line)

(2) **All radiations are caused by accelerating charges which produce change in electric fields.** Also due to Maxwell's Equation, which states that the change in electric field give rise to change in magnetic fields and hence we have electromagnetic radiations.

1.4 RADIATION MECHANISM (How is radiation accomplished?)

A. Single Wire:

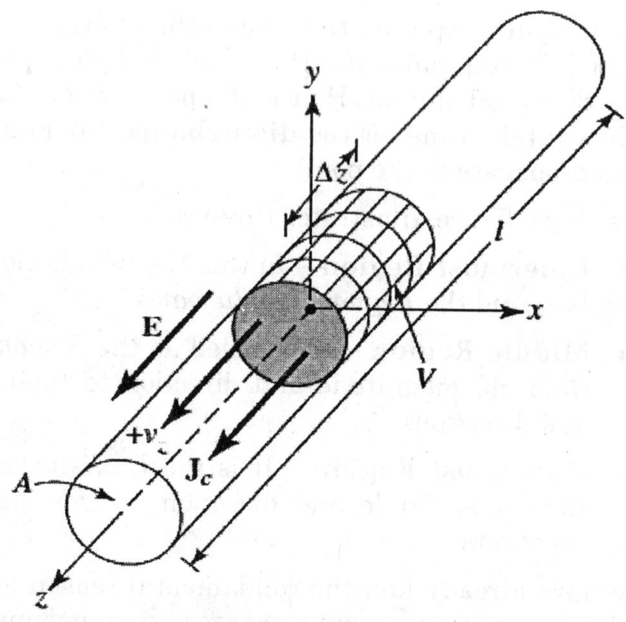

Figure 1.9 Charge uniformly distributed in a circular cross section cylinder wire.

(1) Conducting wires are material whose prominent characteristic is the motion of electric charges and the creation of current flow.

(2) Let us assume –

An electric volume charge density be 'q_v'(Coulombs/m^3),

Distributed uniformly in a circular wire

With cross sectional area `A' and volume `V'

The total charge is `Q' within the volume `V', moving in the `Z' direction with a uniform velocity `v_z' (meter/sec).

(3) Therefore, current density `Jz' (Ampere/m^2) over cross section of the wire is-

$$J_z = q_v\, v_z \tag{1}$$

As $J = \dfrac{I}{A}$ i.e current per area, and

$$J = \frac{Q}{t.A} \qquad \text{as} \qquad v_z = \frac{d}{t} \quad \therefore t = \frac{d}{v_z}$$

Hence, $J = \dfrac{Q.v_z}{d.A}$ as d. A= V

$$\therefore J = \frac{Q.v_z}{V}$$

Now, as $q_v = \dfrac{Q}{V}$, therefore, J= $q_v . v_z$

(4) If the wire is made up of, an **ideal electric conductor**, the current density `Jz' (Amperes/m^2) resides on the surface of the wire and it is given by-

$$J_s = q_s.\ v_z \tag{2}$$

Where, q_s (coulombs/m^3) = Surface charge density

(5) If the **wire is very thin** (ideally zero radius) then the current in the wire can be represented by-

$$I_z = q_l\ .\ v_z \tag{3}$$

Where, q_l (Coulombs / m^3) is charge per unit length.

(6) Now examining the current density of a very thin wire (**Conclusion**: can be applied to all the different conditions mentioned above). **If the current is time varying then the current are-**

$$\frac{dI_z}{dt} = q_l\ \frac{dv_z}{dt} = q_l\ .\ a_z \tag{4}$$

Where, $\frac{dv_z}{dt} = a_z \left(m/s^2\right) = acceleration$

(7) If the wire is of the length `l', then

$$l \cdot \frac{dJ_z}{dt} = q_l \cdot l \frac{dv_z}{dt} = q_l \cdot l \cdot a_z \qquad (5)$$

This is the basic **relation between current and charge.** It serves as the fundamental **relation of electromagnetic radiations.**

It states (very important)-

``To create radiation, there must be a time - varying current or an acceleration (or deceleration) of charge".

(8) We usually refer to-

Currents in time - harmonic application and

Charge is most often mentioned in transients.

(9) (a) To create charge acceleration (or deceleration) the wire must be-

- Curved
- Bent
- Discontinuous
- terminated

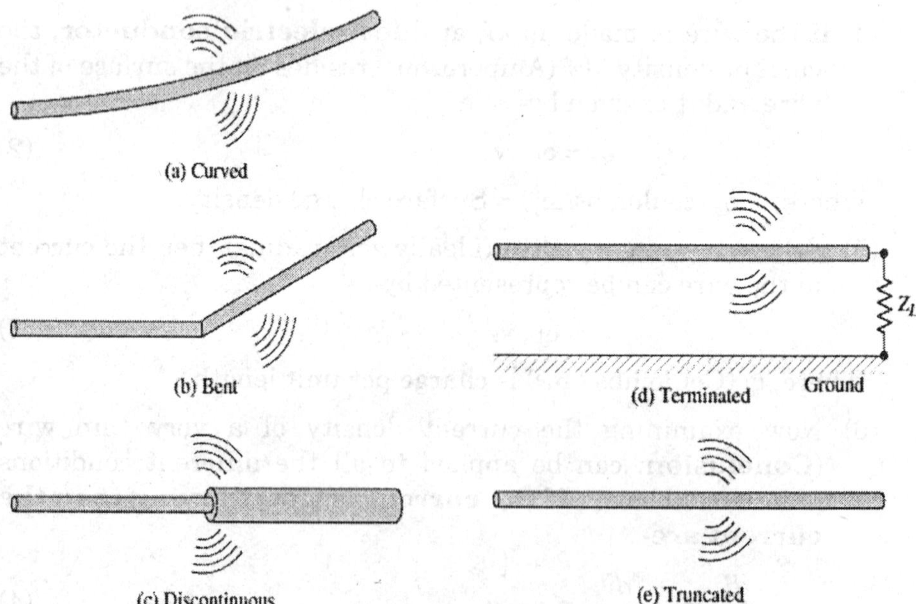

Figure 1.10 Wire configurations for radiation.

(b) Periodic charge acceleration (or deceleration) or time varying current is also created when charge is oscillating in a time harmonic motion, for a $\lambda/2$ dipole.

(10) A qualitative understanding of the radiation mechanisms may be obtained by considering a pulse source attached to an open ended conducting wire, which may be connected to the ground through a discrete load at its open end.

Summary (very important):

(1) If a *charge is not moving*, current is not created and there is **no RADIATION.**

(2) If a charge is moving with *uniform velocity*.

(a) There is no radiation if the wire is *straight and infinite* in extent.

(b) There is **radiation** if the wire is *curved bent, discontinuous, terminated or truncated.*

(3) If a *charge is oscillating* in a time-motion, it radiates even if the wire is straight.

(11) When the wire is initially energized, the charges (free electron) in the wire are set in motion by the electrical lines of force created by the source.

(12) When charges are accelerated in the source-end of the wire and decelerated (negative acceleration with respect to original motion); during reflection from its end, it is suggested that the radiated fields are produced at each end and along the remaining part of the wire.

(13) **Stronger radiation with a more broad frequency spectrum occurs if the pulses are of shorter or more compact duration. While continuous time harmonic oscillating charge produces ideally, radiation of single frequency determined by the frequency of oscillation.**

(14) The **acceleration** of the charges is accomplished by the external source in which forces set the charges in motion and produce the associated field radiated.

(15) The **deceleration** of the charges at the end of the wire is accomplished by the internal (self) forces associated with the induced field due to the buildup of charge concentration at the ends of the wire.

(16) The internal forces receive energy from the charge buildup as its velocity is reduced to zero at the ends of wire.

(17) Therefore, **charge acceleration due to an exciting electric field and deceleration due to impedance discontinuities or smooth curves of the wire are mechanisms responsible for electromagnetic radiation.**

B. Two Wire:

(1) Let us consider a voltage source connected to a two - conductor transmission line which is connected to an antenna. Applying a *voltage* across the two-conductor transmission - line *creates an electric field* between the conductors.

(2) The electric field has associated with it electric lines of force which are tangent to the electric field at each point and their strength is proportional to the electric field intensity.

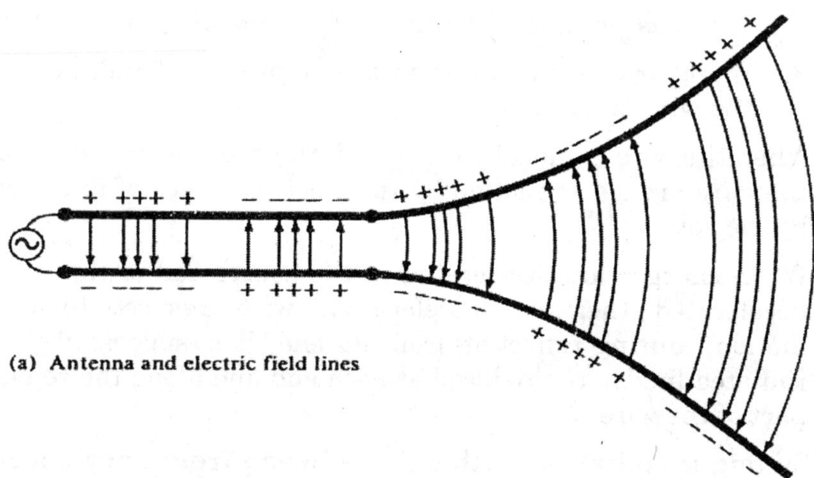

Figure 1.11 Antenna and electric field lines

(3) The electric lines of force have a tendency to act on the free electrons (detachable from atoms) associated with each conductor and force them to be displaced.

(4) The movement of the charges creates a current that in turn creates magnetic field intensity.

(5) Associated with the magnetic field intensity are magnetic lines of force which are tangent to the magnetic field.

(6) **Electric field lines start on positive charges and end on negative charges.** They also can start on a positive charge and end

at infinity, start at infinity and end on negative charge, or form closed loops neither starting nor ending on any charge.

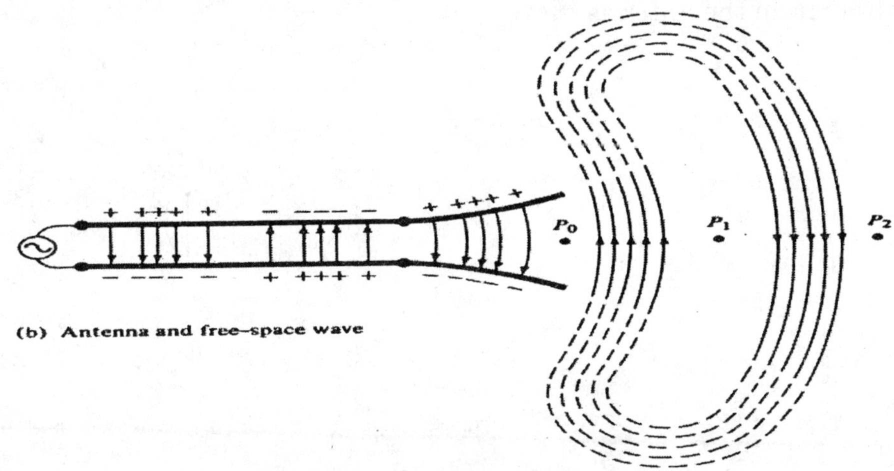

Figure 1.12 Antenna and free space wave

(7) *Magnetic field lines always form closed loops encircling current carrying conductors because there are no magnetic charges.*

(8) The electric field lines drawn between the two conductors, helps to exhibit the distribution of charges. *(If we assume that the voltage source is sinusoidal, we expect the electric field between the conductors to also be sinusoidal with a period equal to that of the applied source).*

(9) **The relative magnitude of electric field intensity is indicated by the density (bunching) of the lines of force with the arrows showing the relative direction (positive or negative).**

(10) **The creation of time varying electric and magnetic fields between the conductors forms electromagnetic waves which travel along the transmission line**, as in the above Figure 1.4.4.

(11) **The electromagnetic waves enter the antenna and have associated with them electric charges and corresponding currents. If we remove the part of the antenna structure, *free space waves* can be formed by "connecting" the open ends of electric lines** (dashed lines) as in figure 2.

(12) The free space waves are also periodic but a constant phase point moves outwardly with the spread of light and travels a distance of $\lambda/2$ (to P_1) in the time of **one half of a period.**

(13) Close to antenna the constant phase point 'P_o' ,moves faster than the speed of height but approaches the speed of light at points far away from the antenna.

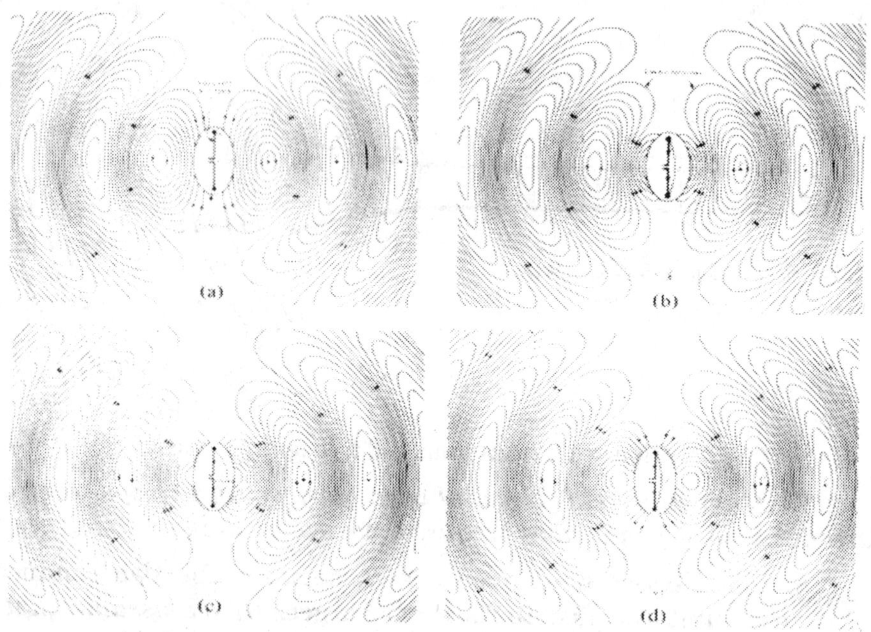

Figure 1.13 Displays the creation and travel, of free space waves

Where λ, = wavelength.

(It is similar like dropping a pebble in calm water, the way the waves are created and are processed outwards; the same is the case with electromagnetic waves / free space waves)

That is, the initial electric disturbance by the source is of short duration; the created electromagnetic waves travel, inside the transmission line, then into the antenna, and finally are radiated as free space waves, even if the electric source has ceased to exist.

When electromagnetic waves are within the transmission line and antenna, their existence is associated with the presence of the charges inside the conductors.

When the waves are radiated, they form closed loops and there are no charges to sustain their existence.

Conclusion

Electric charges are required to excite the fields but are not needed to sustain them and may exist in their absence. This is in direct analogy with water waves.

C. Dipole:

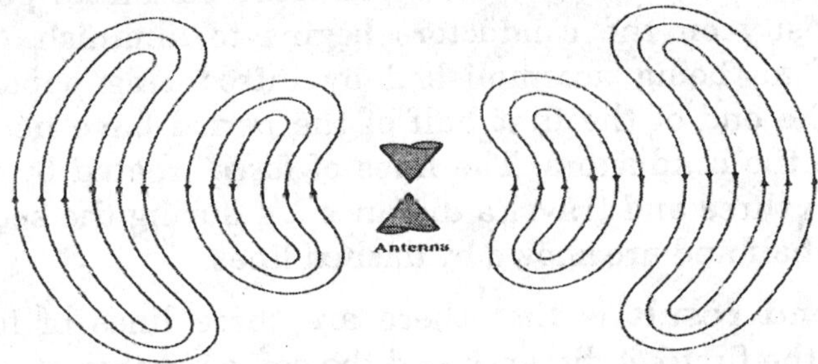

Figure 1.14 Electric field lines of free space wave for biconical antenna

(a)

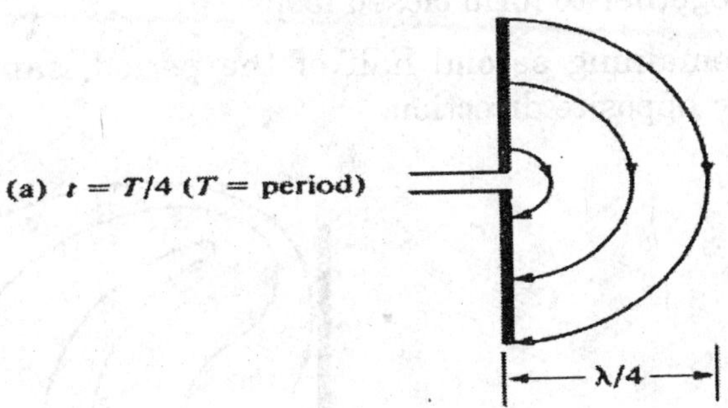

Figure 1.15

It displays the lines of force created between the arms of a small center-fed dipole in the first quarter of the period during which time the charge has reached its maximum value (assuming sinusoidal time variation and the lines have travelled outwardly a radial distance $\lambda/4$). Let us assume that the numbers of lines formed are three.

(b)

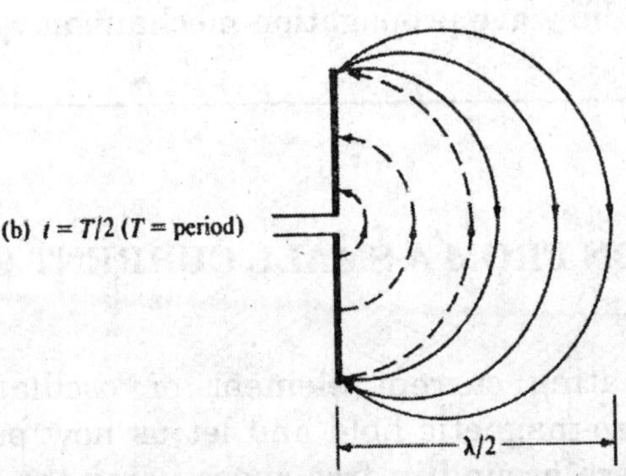

Figure 1.16

During the next quarter of the period, the original three lines travel an additional $\lambda/4$ (a total of $\lambda/2$ from the initial point) and the charge density on the conductors begins to diminish. This can be thought of an being accomplished by introducing opposite charges which at the end of the first half of the period have neutralized the charges on the conductors. The lines of force created by the opposite charges are three and travel a distance $\lambda/4$ during the second quarter of the first half and are shown by dashed lines.

The **end result** is that there are three lines of force pointed upward in the first $\lambda/4$ distance and the same number of lines directed downward in the second $\lambda/4$.

Since there is no net charge on the antenna, then the lines of force must have been forced to detach themselves from the conductors and to unite together to form closed loops.

In the remaining second half of the period, same procedure is followed but in opposite direction.

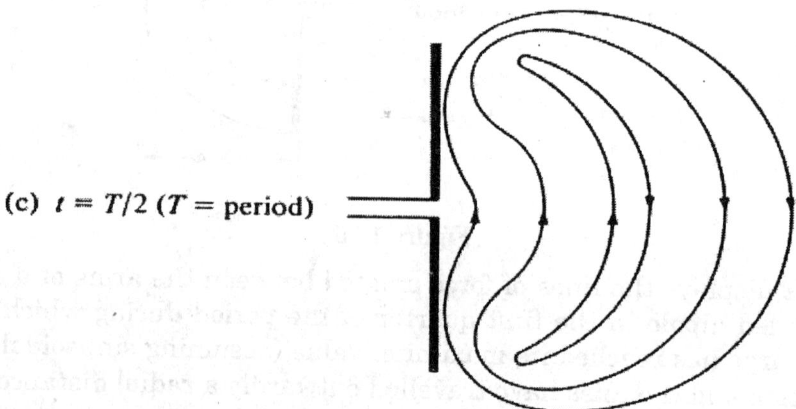

Figure 1.17 Formation and detachment of electric field lines for short dipole.

> 1. Explain radio wave propagation mechanism with examples.
> **(R.T.U. 2009)**

1.5 RADIATION FROM A SMALL CURRENT ELEMENT
(Phase second)

An alternating current element or oscillating electric dipole possesses electro-magnetic field and let us now proceed to find these fields everywhere around in free space using the concept of retarded vector potential.

Let the elemental length (dl) of the wire is placed at the origin of the spherical co-ordinate and `I' be current flowing through it. The length is so short that current is constant along the length. The current element (I.dl) position is indicated in the given figure-

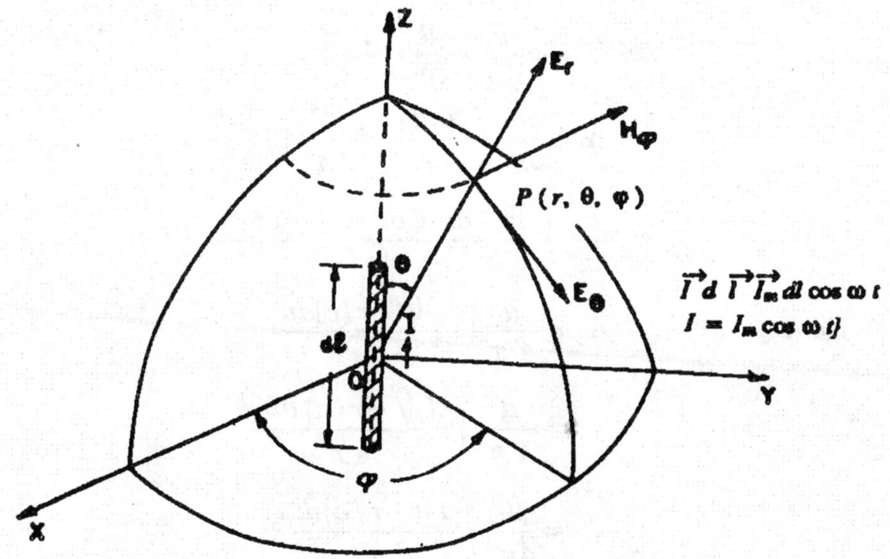

Figure 1.18

(In fact, this isolated current element concept seems to be very much unreal but still any physical circuit or antenna system carrying current may be considered to consists of a large number of such current element joined **in series** that is, end to end, and hence if the electromagnetic fields of this building block is known; the electromagnetic field of any physical antenna having specified current distribution can be calculated.

1.5.1 Magnetic Field Component

To find the EMF at any arbitrary point P (r, θ, ϕ) and for this purpose firstly the vector potential `A' is to be calculated. The general expression for magnetic vector potential is given by-

$$A(r) = \frac{\mu}{4\,\pi} \int \frac{J\left(t - r/c\right)}{r} dv$$

Where, A = vector potential acting along `z' direction

(∴ it have only `z' component eg. 'A_z' retarded in time by (r/c) second).

As this element is excited by the current.

$$I = I_m \text{ Cos wt.}$$

$$\therefore \int J.dv$$

Is replaced by- *I.dl*

$$as \Rightarrow J = \frac{I}{dA} \times \frac{dl}{dl} = \frac{J.\,dl}{dv}$$

$$\therefore A = \frac{\mu}{4\pi}\int \frac{J.dv}{r} = \frac{\mu}{4\pi}\int \frac{I\,.\,dl}{r}$$

$$A_z \frac{\mu}{4\pi}\int \frac{\text{Im}.\,dl.\,Cos\ w\left(t - r/c\right)}{r}$$

Hence,
$$A_z = \frac{\mu}{4\,\pi}\int \frac{J.\left(t - r/c\right)dv}{r}$$

$$= \frac{\mu}{4\,\pi}\int \frac{J\left(t - r/c\right)ds\ dl}{r}$$

$$= \frac{\mu}{4\pi}\int \frac{I\left(t - r/c\right)dl}{r}$$

as
$$I = \int J\ .\,ds.$$

$$A_z = \frac{\mu}{4\,\pi}\int \frac{I_m\ Sin\,w\left(t - r/c\right)dl.}{r}$$

$$= \frac{\mu}{4\,\pi}\,.\,\frac{I_m}{r}\int Sin\,w\left(t - \frac{r}{c}\right).\,dl$$

$$= \frac{\mu}{4\,\pi}\,.\,\frac{I_m}{r}\ Cos\ w\left(t - r/c\right)dl$$

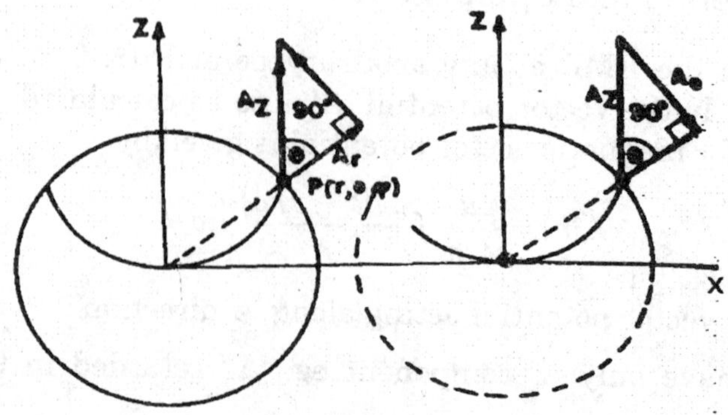

Figure 1.19 Resolution of vector potential at point P (r, θ, ϕ) of previous given figure.

Now the magnetic field intensity `H' is obtained from the magnetic vector potential-

$$B = \nabla \times A$$

$$= \mu H$$

Due to spherical symmetry in the problem in x-y plane the field is symmetrical and so $\frac{\partial}{\partial \theta} = 0.$

From the previous and this figure, it is seen that components of A are-

$$\frac{\partial}{\partial \phi} = 0$$

$$A_\phi = 0$$

$$A_r = A_z \; Cos\; \theta$$

$$A_\theta = -A_z \; Sin\; \theta$$

Where $A_z = \frac{\mu}{4\pi} \int \frac{I_m \; dl \; Cos\; w\,(t - r/c)}{r}$ (d)

Now from $(\nabla \times A)$-

$$(\nabla \times A)_r = \frac{1}{r\; Sin\theta}\left[\frac{\partial}{\partial \theta}(A_\phi \sin\theta) - \frac{\partial A_\theta}{\partial_\phi}\right] \quad (a)$$

$$(\nabla \times A)_\theta = \frac{1}{r}\left[\frac{1}{\sin\theta}\frac{\partial A_r}{\partial \phi} - \frac{\partial}{\partial r}(A\phi\, r)\right] \quad (b)$$

$$(\nabla \times A)_\theta = \frac{1}{r}\left[\frac{r}{\partial r}\left\{(A_\phi\, r) - \frac{\partial A_r}{\partial \theta}\right\}\right] \quad (c)$$

and (d) we have-

$$(\nabla \times A)_r = \mu H_r = 0 \qquad \therefore H_r = 0$$

$$(\nabla \times A)_\theta = \mu H_\theta = 0 \qquad \therefore H_\theta = 0$$

and $$(\nabla \times A)_\phi = \frac{1}{r}\left[\frac{\partial}{\partial_r}\left(A_\theta\, r - \frac{\partial\, A_r}{\partial \theta}\right)\right] = \mu H_\phi$$

or $$\mu\, H_\phi = \frac{1}{r}\left[\frac{\partial}{\partial r}\{-A_z\; Sin\;\theta,\, r\} - \frac{\partial}{\partial \theta}\{A_z\; Cos\;\theta\}\right]$$

Foot note 1

For $\left[\frac{\partial}{\partial_r}(A_\theta r)\right]$

As $A_\theta = -A_z \sin\theta = \frac{-\mu I_m\, dl\, Cos\, w(t - r/c)}{4\pi r}.\, Sin\,\theta$

or $r\, A_\theta = \frac{-\mu I_m\, dl\, Cos\, w(t - r/c).\, Sin\,\theta}{4\pi}$

or $\frac{\partial}{\partial\theta}(r\, A_\theta) = \frac{-\partial}{\partial r}\left[\frac{\mu I_m\, dl\, Cos\, w(t - r/c) - Sin\,\theta}{4\pi}\right]$

$$= \frac{+\mu I_m\, dl\, Sin\,\theta}{4\pi}.\frac{\partial}{\partial r}[Cos\, w(t - r/c)]$$

$$= \frac{\mu I_m\, dl \sin\theta}{4\pi}.[- Sin\, w(t - r/c).(-w/c)]$$

$$\frac{\partial}{\partial r}(r\, A_\theta) = \frac{-\mu I_m\, dl\, Sin\,\theta.\, w.\, Sin\, w(t - r/c)}{4\pi c}$$

Foot note 2

For $\left[\left(\frac{\partial A_r}{\partial\theta}\right)\right]$

$$A_r = A_z\, Cos\,\theta = \frac{\mu I_m\, dl\, Cos\, w(t - r/c)\, Cos\,\theta}{4\pi r}$$

$$\therefore \frac{\partial}{\partial\theta}(A_r) = \frac{\partial}{\partial\theta}\left[\frac{\mu I_m\, dl\, Cos\, w(t - r/c)\, Cos\,\theta}{4\pi r}\right]$$

$$= \frac{\mu I_m\, dl\, Cos\, w\, t_1}{4\pi r}.\frac{\partial}{\partial\theta} Cos\,\theta$$

$$= \frac{\mu I_m\, dl\, Cos\, wt_1}{4\pi r}.(- Sin\,\theta)$$

$$\therefore \frac{\partial'}{r\theta}(A_r) = \frac{-\mu I_m\, dl\,.\, Cos\, wt_1 (Sin\,\theta)}{4\pi r}$$

where $t_1 = t - r/c$

$$|H| = H_\phi = \frac{1}{\mu r}\left[\frac{\partial}{\partial r}\{-A_z \ Sin\ \theta, r\} - \frac{\partial}{\partial_\theta}(A_z \ Cos\ \theta)\right]$$

From Foot Note 1 and 2

$$|H| = \frac{1}{\mu r}\left[-\frac{\mu\ I_m\ dl\ w}{4\ \pi\ c}(Sin\ \theta\ Sin\ wt_1) - \left(-\frac{\mu\ I_m\ dl}{4\ \pi r}.Sin\ \theta.Cos\ wt_1\right)\right]$$

$$\therefore |H| = H_\phi = \frac{I_m\ dl}{4\ \pi}\ Sin\ \theta\left[-w\frac{Sin\ wt_1}{c\,r} + \frac{Cos\ wt_1}{r^2}\right]$$

1.5.2 Electric Field Component:

Now let us calculate the components of E (E_r, E_θ, E_ϕ) which can be done by using Maxwell's equation -

$$\nabla \times H = \frac{\partial D}{\partial t} = \varepsilon\frac{\partial E}{\partial t}$$

$\because D = \varepsilon E$, writing above equation in its components form we get-

$$\varepsilon\frac{\partial E_r}{\partial t} = (\nabla \times H)_r = \frac{1}{r \sin\theta}\left[\frac{\partial}{\partial\theta}(H_\phi \sin\theta) - \frac{\partial}{\partial\theta}(H_\theta)\right] \quad \text{(a)}$$

$$\varepsilon\frac{\partial E_\theta}{\partial t} = (\nabla \times H)_\theta = \frac{1}{r}\left[\frac{1}{\sin\theta}\left(\frac{\partial H_r}{\partial\theta}\right) - \frac{\partial}{\partial r}(H_\phi\ r)\right] \quad \text{(b)}$$

$$\varepsilon\frac{\partial E_\phi}{\partial t} = (\nabla \times H)_\phi = \frac{1}{r}\left[\frac{\partial}{\partial r}(H_\theta r) - \frac{\partial H_r}{\partial\theta}\right] \quad \text{(c)}$$

Putting,

$H_r = 0\ and\ H_\theta = 0$, from the magnetic field component in the above equation (a, b, c.) we get-

(a) $$\varepsilon\frac{\partial E_r}{\partial t} = (\nabla \times H)_r = \frac{1}{r Sin\,\theta}\left[\frac{\partial}{\partial\theta}(H_\phi\ Sin\ \theta)\right] \quad \text{(d)}$$

Foot Note 3

For $\varepsilon\frac{\partial E_r}{\partial t}$

$$\varepsilon \frac{\partial E_r}{\partial t} = \frac{1}{r \, Sin \, \theta} \left[\frac{\partial}{\partial \theta} \left\{ \frac{I_m \, dl \, Sin^2 \theta}{4 \pi} \left(- \frac{w \, Sin \, wt_1}{cr} + \frac{Cos \, wt_1}{r^2} \right) \right\} \right]$$

$$= \frac{1}{r \, Sin \, \theta} \left[\frac{I_m \, dl}{4\pi} \left\{ - \frac{w \, Sin \, wt_1}{cr} + \frac{Cos \, wt_1}{r^2} \right\} \cdot \frac{\partial}{r\theta} \left(Sin^2 \theta \right) \right]$$

$$= \frac{1}{r \, Sin \, \theta} \left[\frac{I_m \, dl}{4\pi} \left\{ - \frac{w \, Sin \, wt_1}{cr} + \frac{Cos \, wt_1}{r^2} \right\} 2 \, Sin \, \theta \, . \, Cos \, \theta \right]$$

$$\frac{\partial E_r}{\partial t} = \frac{2 \, I_m \, dl}{4 \, \pi \, E} \left[- \frac{w \, Sin \, Wt_1}{cr^2} + \frac{Cos \, wt_1}{r^3} \right] . \, Cos \, \theta$$

$$\int \partial E_r = \frac{2 \, I_m \, dl \, Cos \, \theta}{4 \, \pi \, \varepsilon} \int \left[- \frac{w \, Sin \, wt_1}{cr^2} + \frac{Cos \, wt_1}{r^3} \right] dt$$

$$E_r = \frac{2 \, I_m \, dl \, Cos \, \theta}{4 \, \pi \, \varepsilon} \left[\frac{w \, Cos \, wt_1}{cr^2 \, w} + \frac{Sin \, wt_1}{r^3 w} \right]$$

$$E_r = \frac{2 \, I_m \, dl \, Cos \, \theta}{4 \, \pi \, \varepsilon} \left[\frac{Cos \, wt_1}{cr^2} + \frac{Sin \, wt_1}{w \, r^3} \right]$$

(b) $$\varepsilon \frac{\partial E_\theta}{\partial t} = (\nabla \times H)_\theta = \frac{1}{r} \left[- \frac{\partial}{\partial r} \left(H_\phi \, r \right) \right] \quad \text{(e)}$$

Foot note No. 4

$$\frac{\varepsilon \, \partial \, E_\theta}{\partial \, t} = - \frac{1}{r} \frac{\partial}{\partial \, r} \left[\frac{I_m \, dl \, Sin \, \theta}{4 \, \pi} \left(- \frac{w \, Sin \, wt_1}{cr} + \frac{Cos \, wt_1}{r^2} \right) r \right]$$

$$= \frac{1}{r} \frac{\partial}{\partial \, r} \left[\frac{I_m \, dl \, Sin \, \theta}{4\pi} \left(+ \frac{w \, Sin \, wt_1}{c} - \frac{Cos \, wt_1}{r} \right) \right]$$

$$= \frac{I_m \, dl \, Sin \, \theta}{4 \, \pi \, r} \left[\frac{\partial}{\partial \, r} \left(\frac{w \, Sin \, wt_1}{c} \right) - \frac{\partial}{\partial r} \left(\frac{Cos \, wt_1}{r} \right) \right]$$

as $t_1 = \left(t - \frac{r}{c} \right)$

$$\frac{\varepsilon \partial E_\theta}{\partial t} = \frac{I_m \, dl \, Sin\,\theta}{4\pi r}\left[\frac{w}{c} Cos\, wt_1\left(\frac{-w}{c}\right) - \frac{r\frac{\partial}{\partial r}Cos\, wt_1 - Cos\, wt_1}{r^2}\right]$$

$$\frac{\partial E_\theta}{\partial t} = \frac{I_m \, dl \, Sin\,\theta}{4\pi\varepsilon . r}\left[\frac{-w^2}{c^2} Cos\, wt_1 - \left\{\frac{-r\, Sin\, wt_1 (-w/c) - Cos\, wt_1}{r^2}\right\}\right]$$

$$\int \partial E_\theta = \frac{I_m \, dl \, Sin\,\theta}{4\pi\varepsilon . r}\int\left[\frac{-w^2}{c^2} Cos\, wt_1 + \frac{Sin\, wt_1(-w/c)}{r} + \frac{Cos\, wt_1}{r^2}\right]dt$$

$$E_\theta = \frac{I_m \, dl \, Sin\,\theta}{4\pi\varepsilon . r}\left[\frac{-w^2\, Sin\, wt_1}{c^2\, w\, r} + \frac{w}{cr^2}\frac{Cos\, wt_1}{w} + \frac{Sin\, wt_1}{r^3\, w}\right]$$

(c) $\varepsilon \dfrac{\partial E_\phi}{\partial t} = (\nabla \times H)_\phi = 0$

or $E_\phi = 0$

From Foot Note 3 and 4

$$E_\phi = 0$$

$$E_\theta = \frac{2\, I_m \, dl \, Cos\,\theta}{4\,\pi\,\varepsilon}\left[\frac{Cos\, wt_1}{c\,r^2} + \frac{Sin\, wt_1}{w\,r^3}\right]?$$

$$E_\phi = \frac{I_m \, dl \, Sin\,\phi}{4\,\pi\,\varepsilon}\left[-\frac{w\, Sin\, wt_1}{c^2\, r} + \frac{Cos\, wt_1}{c\,r^2} + \frac{Sin\, wt_1}{wr^3}\right]$$

$$H_\phi = \frac{I_m \, dl \, Sin\,\theta}{4\pi}\left[\frac{Cos\, wt_1}{r^2} - \frac{w\, Sin\, wt_1}{cr}\right]$$

$$E_r = \frac{2\, I_m \, dl \, Cos\,\theta}{4\pi\varepsilon}\left[\frac{Cos\, wt_1}{cr^2} + \frac{Sin\, wt_1}{w\, r^3}\right]$$

$$E_\theta = \frac{I_m \, dl \, Sin\,\theta}{4\pi\varepsilon}\left[-\frac{w\, Sin\, wt_1}{c^2\, r} + \frac{Cos\, wt_1}{cr^2} + \frac{Sin\, wt_1}{w\, r^3}\right]$$

Conclusion:

Out of six components of Electromagnetic field-

H_r $H\phi$ H_θ and E_r E_θ E_ϕ, only three components E_r, E_θ, H_ϕ exist in the current element of Hertzian dipole and rest components are everywhere zero.

1. The radiation field components at a distance `r' due to a current element of length `l' in the free space and carrying current (I_m Sin wt) are given by :-

$$E_\theta = \frac{I_o\ w\ dl\ \sin\theta}{4\ \pi\ \varepsilon\ r\ c^2}\ .\cos w\left(t - \frac{r}{c}\right)$$

$$H_\phi = \frac{I_o\ w\ dl\ \sin\theta}{4\ \pi\ c\ r}\cos w\left(t - \frac{r}{c}\right)$$

Where ε, permittivity of free space

c, velocity of light **(Raj. Univ. 2005)**

2. Also show that, (i) $\left|\frac{E\theta}{H\phi}\right|$ is equal to impedance of free space.

(Raj. Univ. 2005)

1.5.3 Induction (near) field and Radiation (far) field

Current element so simple gives complicated field like those which are given by the equation-

$$H_\phi = \frac{I_m\ dl\ \text{Sin}\ \theta}{4\pi}\left[+\frac{\text{Cos wt}_1}{r^2} - \frac{\text{w Sin wt}_1}{cr}\right]$$

Where Amplitude $= \frac{I_m\ dl\ \text{Sin}\ \theta}{4\pi}$

Induction (near) field $= \frac{\text{Cos wt}_1}{r^2}$

Radiation (far) field $= \frac{\text{w Sin wt}_1}{cr}$

(i) Induction Field:

- It is proportional to $\frac{1}{r^2}$
- This field predominate at points close to the current element (where r = small). *This field is more effective in the vicinity of the current element only.*

- **It represents the energy stored in the magnetic field surrounding the current element or conductor.**
- This energy is alternatively stored in the field and returned to source current element during each half cycle.

(ii) Radiation field

- It is proportional to $\frac{1}{r}$
- It accounts for the radiation of EM waves from the conductor under suitable conditions.
- The radiation terms are an additional term in comparison to steady currents because **in steady current analysis no account was taken for the time of propagation.**
- The radiation component of magnetic field is produced by the alternating electric field and the electric radiation component arises from the alternating magnetic field.
- The flow of current or the movement of charge, in the conductor thus establishes the local induction fields, whereas the radiation fields exist as a consequence of the changing induction fields.
- **Just near the conductor, magnetic field is in phase with the current in the conductor, whereas the electric field varies in phase with the change on either end of the conductor element.**
- **Therefore, in this region the H-field and E-field have, a phase difference or π/2 radian and are at right angle to each other in space that is, E_θ and H_ϕ are in phase in the far field.**

$$E_r \; \alpha \, 1/r^2 \text{ (induction term)}$$

$$E_\theta \; \alpha \frac{1}{r^2} \text{ (induction term)} + \frac{1}{r} \text{ (radiation term)}$$

- Besides these fields both (E_r and E_θ) contains a term varying with $\frac{1}{r^3}$. **This is called electrostatic field or just electric field terms due to its similarity with components of field of an electrostatic dipole. It is very important near the current element.**

- The two terms in **E**r, belongs to induction field and are negligible at a distance where r >> λ. **Hence, no radial E component is present in the radiation field.**
- The last two terms of E_θ, also belongs to induction field and becomes negligible at a distance where r >> λ. Hence, the radiation field of E_θ where r >> λ, is given by

$$E_\theta = \frac{I_m \, Sin\,\theta \, dl}{4\,\pi\,\varepsilon}\left[-\frac{w\,Sin\,wt_1}{c^2 r}\right] \text{ where } \left(\frac{1}{r^2} \;\&\; \frac{1}{r^3} \text{ neglected}\right)$$

$$= -\frac{w\,I_m \; dl\;Sin\,\theta\;Sin\,wt_1}{4\,\pi\,\varepsilon\,c\,.\,cr}$$

$$= \frac{2\,\pi\,f\,I_m\;dl\;Sin\,\theta\;Sin\,wt_1}{4\pi\,\varepsilon\,.\,\frac{1}{\sqrt{\mu\,\varepsilon}}\,.\,cr}$$

as

$$c = \frac{1}{\sqrt{\mu\,\varepsilon}}$$

$$\frac{f}{c} = \frac{1}{\lambda}$$

$$w = 2\pi\,f$$

$$= \frac{c/\lambda\;I_m\;dl\;Sin\,\theta\;.\;Sin\,wt_1}{2\,\varepsilon\,.\,\frac{1}{\sqrt{\varepsilon}\,\sqrt{\mu}}\,.\,cr}$$

$$= \frac{I_m\;dl\;Sin\,\theta\;.\;Sin\,wt_1}{2\,\frac{\sqrt{\varepsilon}}{\sqrt{\mu}}\,.\,r\lambda}$$

as

$$\eta = \sqrt{\frac{\mu}{\varepsilon}} = 120\,\pi$$

$$= \frac{-\,\eta\;I_m\;dl\;Sin\,\theta\;Sin\,wt_1}{2\,\lambda\,r}$$

$$\therefore E_\theta = \frac{-\,60\,\pi\,I_m\;dl\;Sin\,\theta\;Sin\,w\,(t - r/c)}{\lambda\,r}$$

$$|E\theta| = \frac{60\,\pi\, I_m\, dl}{\lambda r}$$

As maximum value of Sine is unity when $[\theta = 90^\circ]$

Similarly, at a distance r $>>\lambda$

$$H_\phi = \frac{I_m\ dl\ Sin\ \theta}{4\pi}\left[-\frac{w\ Sin\ wt_1}{cr}\right]\left(as\ \frac{1}{r^2} = neglect\right)$$

$$= \frac{-2\pi f\ .\ I_m\ dl\ \sin\theta\ \sin wt_1}{4\ \pi\ cr}$$

$$H_\phi = \frac{-I_m\ dl\ Sin\theta\ Sin\ w\left(t - r/c\right)}{2\lambda r}$$

$$|H_\phi| = \frac{I_m\ dl}{2\ \lambda r}$$ [as maximum value of sine is unity when $\theta = 90^\circ$]

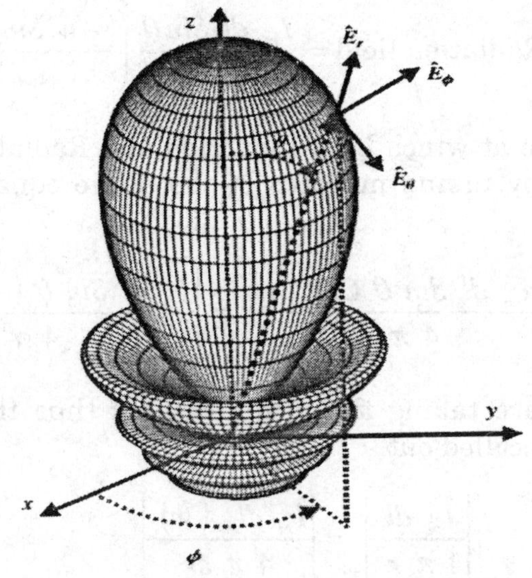

Figure 1.20

Analysis:

$$|E_\theta| = 120\ \pi\ |H_\phi|$$

As H_ϕ is tangent to parallel latitude on a sphere and Eθ is tangent to the meridian.

1.5.4 Distance at which Induction & Radiation Fields are equal

As the distance from the current element increases, both induction and radiation fields emerge and start decreasing. Since induction field varies inversely as square of distance (that is, $1/r^2$) so diminishes rapidly and hence is important near the conductor but the radiation field which varies inversely as distance (that is, $1/r$) diminishes relatively less and is in position to propagate to a larger distance.

However, a distance reaches from the conductor at which both induction and radiation fields becomes equal and the particular distance depends on the wavelength (or frequency) used. The two fields will thus, have equal amplitude at that particular distance [For H_φ]

$$\text{Induction field} = \frac{I_m\ dl\ Sin\,\theta}{4\,\pi}\left(\frac{Cos\ wt_1}{r^2}\right)$$

$$\text{Radiation field} = \frac{I_m\ dl\ Sin\,\theta}{4\,\pi}\left(\frac{-\ w\ Sin\ wt_1}{c\,r}\right)$$

Distance at which both Induction and Radiation fields are equal is obtained by taking modulus of the above equations and equating them-

$$\left|\frac{I_m\ dl\ Sin\ \theta\ Cos\ wt_1}{4\ \pi\ r^2}\right| = \left|\frac{I_m\ dl\ Sin\ \theta\ (-\ w)\ Sin\ wt_1}{4\ \pi\ cr}\right|$$

As we are taking the modulus, only thus the Cosine and Sine term gets cancelled out.

$$\left|\frac{I_m\ dl}{4\ \pi\ r^2}\right| = \left|\frac{I_m\ dl\ (w)}{4\ \pi\ cr}\right|$$

$$\left|\frac{1}{4\ \pi\ r}\right| = \left|\frac{2\pi f}{4\ \pi\ c}\right|$$

$$or\quad \frac{1}{r} = \frac{w}{c}$$

$$\frac{1}{r} = \frac{2\pi}{\lambda}\quad or\ r = \frac{\lambda}{2\pi} = \frac{\lambda}{6}$$

Current element at this distance is serving for both the fields.

$$r = 0.159\ \lambda$$

After this distance, beyond l/6 Induction field vanishes while Radiation field contribution to Radio wave Propagation (that is, radio communication).

> 1. Show that the distance at which both the induction and radiation fields are equal, is given by-
>
> $r = \frac{\lambda}{6}$ *where* λ = *wavelength.* **(Raj. Univ. 2006)**

1.5.5 Radiation Pattern of Current Element / Dipole:

(i) E_θ & H_ϕ , are in time phase in far field as it has got the dimension of pure resistance

$$\left(Hence \quad \eta = 120\,\pi = 377\ \Omega\right)$$

(ii) Both E_θ *and* H_ϕ *are proportional to* $\sin\theta$.

(iii) Pattern is independent of ϕ. Hence this gives conclusion that space pattern is **doughnut shape** being figure of 8 about the axis of the dipole.

as $\sin\theta = 1$ *at* 90° *and* $\sin\theta = 0$ *at* 0° (*along the axis*).

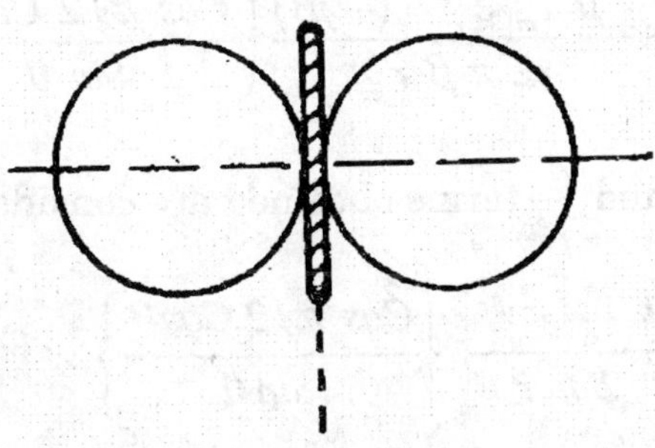

Figure 1.21

> 1. Derive the expression of magnetic field and electric field component for small current element. **(Raj. Univ. 2008)**

1.6 RADIATION FROM A HALF-WAVE DIPOLE

Half wavelength / half wave dipole (λ/2) antenna, is one of the simplest antenna. It has complex directional system, e.g., antenna arrays etc. A λ/2 antenna is the fundamental **radio antenna** of metal rod or tubing or thin wire with physical length of half wavelength in free space at the frequency of operation.

A λ/2 antenna is also called as *Hertz* antenna or sometimes called as half wave doublet.

Dipole or Doublet means the same thing as long as electrical dimension of the two are same.

A dipole antenna may be defined a "**symmetrical antenna in which the two ends are at equal potential relative to the midpoint.**"

Formula used-

(1) $$A_z = \frac{\mu I_m e^{-jBr}}{2\pi Br}\left[\frac{Cos\ \pi/2\ Cos\ \theta}{Sin^2\ \theta}\right]$$

(2) $$\mu H_\phi = -Sin\theta \frac{\partial A_z}{\partial r}$$

Now substituting the value of `A_z' from (1) in (2) we have-

$$\mu H_\phi = -Sin\ \theta \frac{\partial}{\partial r}\left[\frac{\mu I_m e^{-jBr}}{2\pi \beta r}\left(\frac{Cos\ \pi/2\ Cos\ \theta}{Sin^2\ \theta}\right)\right]$$

$$= \frac{-\mu I_m e^{-jBr}(-jB)}{2\pi\beta r}\left(\frac{Cos\ \pi/2\ Cos\ \theta}{Sin\ \theta}\right)$$

Here, the $\frac{1}{r^2}$ and $\frac{1}{r^3}$ terms obtained are considered negligible.

$$H_\phi = \frac{j I_m e^{-jBr}}{2\pi r}\left\{\frac{Cos\ \pi/2.Cos\theta}{Sin\theta}\right\}$$

$$|H_\phi| = \left|\frac{I_m e^{-jBr}}{2\pi r}\left(\frac{Cos\ (\pi/2\ Cos\ \theta)}{Sin\ \theta}\right)\right|$$

$$|H_\phi| = \frac{I_m}{2\pi r}\ \frac{Cos\ (\pi/2\ .\ Cos\ \theta)}{Sin\ \theta}\ Amp/m^2$$

This is the required magnetic field Intensity expression for half wave dipole. The electric field expression for the radiation field can be achieved from-

$$\frac{E_\theta}{H_\phi} = \eta = 120\,\pi$$

$$|E_\theta| = \eta\,|H_\phi|$$

$$= 120\,\pi\,.\,\frac{I_m}{2\pi r}\frac{Cos\,(\pi/2\;Cos\;\theta)}{Sin\;\theta}$$

$$|E_\theta| = \frac{60\,I_m}{r}\left\{\frac{Cos\;(\pi/2\;Cos\;\theta)}{Sin\;\theta}\right\}\;volt/m$$

This is the expression for the amplitude, the electric field intensity for the radiation field of a λ/2 antenna or a λ/4 antenna (monopole). Since E_θ & H_ϕ are in time phase ∴ maximum value in time of the Poynting Vector is just the product of the peak values of E_θ & H_ϕ *i.e.*-

$$P_{max} = |E_\theta|\,|H_\theta|$$

Average value

$$P_{av} = \frac{E_\theta}{\sqrt{2}}.\frac{H_\phi}{\sqrt{2}}$$

$$= \frac{E_\theta\,H_\phi}{2}$$

$$P_{av} = \frac{P_{max}}{2}$$

$$P_{av} = \frac{E_\theta\,H\phi}{2} = \frac{15\,I_m^{\,2}}{\pi\,r^2}\left\{\frac{Cos\;\pi/2\;Cos\;\theta}{Sin\;\theta}\right\}^2$$

In practice, r. m. s. current is of importance for measurement purposes etc. So

$$\because\;\frac{I_m}{\sqrt{2}} = I_{ms}$$

$$or\quad I_m = \sqrt{2}\,I_{ms}$$

$$\therefore P_{av} = \frac{15\left(\sqrt{2}\, I_{rms}\right)^2}{\pi\, r^2} \left\{\frac{Cos\,(\pi/2\ Cos\ \theta)}{Sin\ \theta}\right\}^2$$

$$P_{av} = \frac{30\ I_{rms}^{\ 2}}{\pi\ r^2} \left\{\frac{Cos\,(\pi/2\ Cos\,\theta)}{Sin\,\theta}\right\}^2 W/m^2$$

This is the expression for average power in terms of r. m. s. current.

1.7 POWER RADIATION BY A HALF WAVE DIPOLE (that is, λ/2 antenna) AND ITS RADIATION RESISTANCE

For the calculation of total power radiation by a λ /2 antenna and its radiation resistance can be found in the same manner as has been adopted in case of current element power calculations that is, by using Poynting Vector method. The elemental area of the spherical shell is given by-

$$ds = 2\,\pi r^2 \sin\theta\, d\,\theta$$

∴ Total power radiated from λ /2 antennas-

$$W = \oint P_{av}\ ds$$

$$= \int_0^{\pi} \frac{30\ I^2\ rms}{\pi\ r^2} \left\{\frac{Cos^2\,(\pi/2 Cos\ \theta)}{Sin^2\ \theta}\right\} 2\pi r^2\ Sin\theta.d\,\theta$$

Where r = radius of hemisphere

$$W = 60\ I^2_{\ rms} \int_0^{\pi} \frac{Cos^2\,(\pi/2\ Cos\,\theta)}{Sin\ \theta}\, d\theta$$

As, 2 $Cos^2\ \theta = 1 + Cos\ 2\theta$

$$W = 60\ I^2_{\ rms} \int_0^{\pi} \frac{1}{2}\left[\frac{1 + \cos\,(\pi\ \cos\ \theta)}{\sin\ \theta}\right]$$

$$W = 60\quad I^2_{rms}\ .\ I$$

$$\text{Where}\quad I = \int_0^{\pi} 1/2 \left[\frac{1 + Cos\,(\pi\ Cos\ \theta)}{Sin\ \theta}\right] d\theta$$

$$= 1.219$$

∴ Total power radiated ⇒ $W\ = 60\ I^2\ rms\ .\ 1.219$

$$W = 73.140\ \ I^2_{rms}$$

Therefore, total power radiated by half wave dipole in free space remote from the ground, is given by-

$$W_r = I^2_{rms} . R_r$$

Radiation resistance for centre fed half wave dipole is given by-

$$R_r = 73.14 \approx \mathbf{73\ \Omega}$$

And *for quarter wave antenna* it is given by-

$$R_r = 73.14/2 = \mathbf{36.57\ \Omega}$$

Radiation Resistance is the function of, height above the ground with range for half dipole antenna equals to (60 – 90) Ω

NOTE:

1. **The $\lambda/2$ antennas are called as HERTZ antennas. They have spherical radiation pattern.**
2. **The $\lambda/4$ antennas are called as MARCONI antennas. They have semi-spherical (hemisphere) radiation pattern.**

1. Explain radiation from a current element in free space for half wave antenna. **(R.T.U. 2009)**

1.8 TYPES OF ANTENNAS

Now, before switching over to the fundamental parameters of antenna, let us take a quick look on various types of antenna and their general application; so that a overview is set about antenna system in our daily routine. Thus, here in this section we are going to discuss a broad *classification of antennas*.

1.8.1 Wire Antennas

They can be broadly classified as-

(1) Straight wire (Dipole) (2) Loop (3) Helix

(1) Dipole

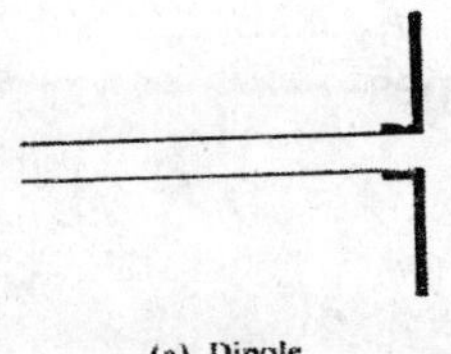

(a) Dipole

Figure 1.22

(2) Helix

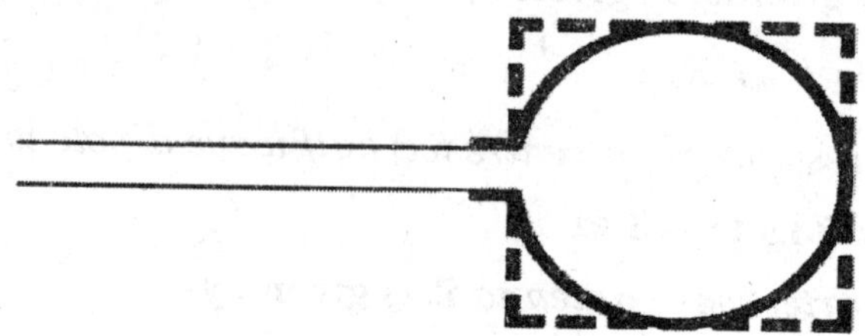

(b) Circular (square) loop

Figure 1.23

(3) Loop (Circular/Square)

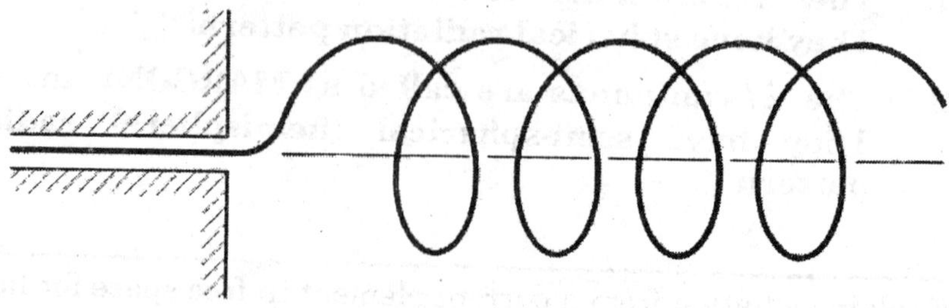

(c) Helix

Figure 1.24

Applications:

These antennas are generally used in the fields of-

- Automobiles
- Buildings
- Ships
- Aircraft
- Space-Craft

1.8.2. Aperture Antenna:

They are of many shapes-

(1) Pyramidal

(2) Conical

(3) Rectangular

(1)Pyramidal Horn

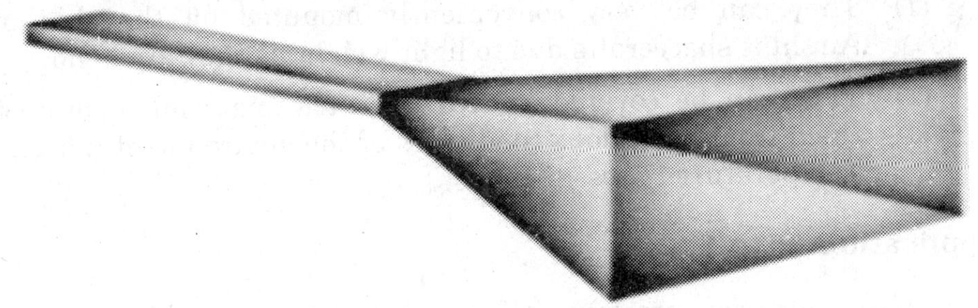

(a) Pyramidal horn

Figure 1.25

(2) Conical Horn

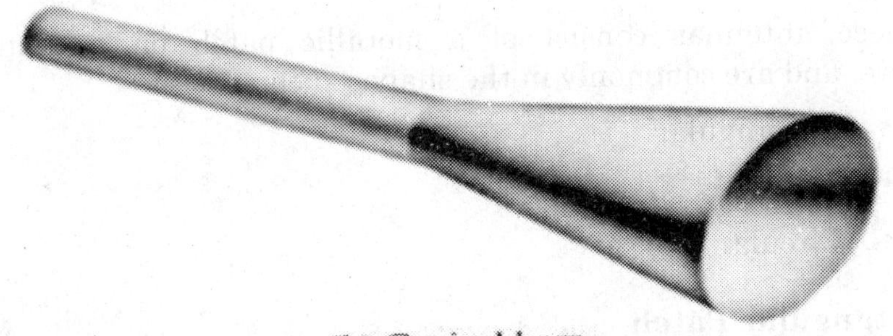

(b) Conical horn

Figure 1.26

(3) Rectangular waveguide

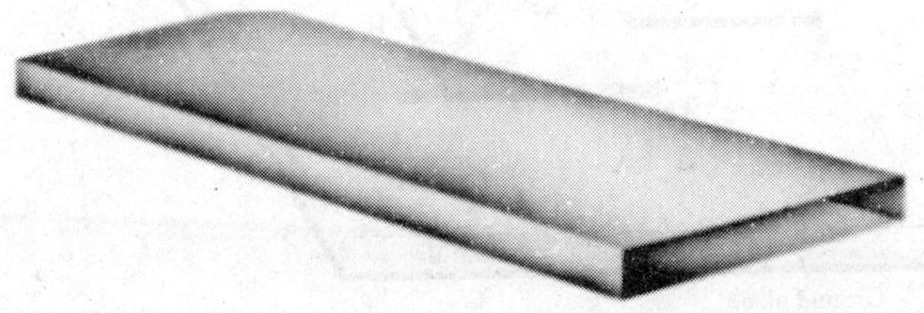

(c) Rectangular waveguide

Figure 1.27

Advantages

(1) They can be very conveniently mounted on the skin of Airships spacecrafts due to light weight as they are hollow.

(2) They can be covered with a dielectric material to protect them from hazardous conditions of the environment (for e.g. Temperature / pressure).

Applications

These antennas are generally used at the higher frequencies (that is, at the microwave ranges) in -

- Aircraft
- Spacecraft

1.8.3 Microstrip Antennas

These antennas consist of a metallic patch on a grounded substrate, and are commonly in the shapes such as-

- Rectangular
- Square
- Circular

(1) Rectangular Patch

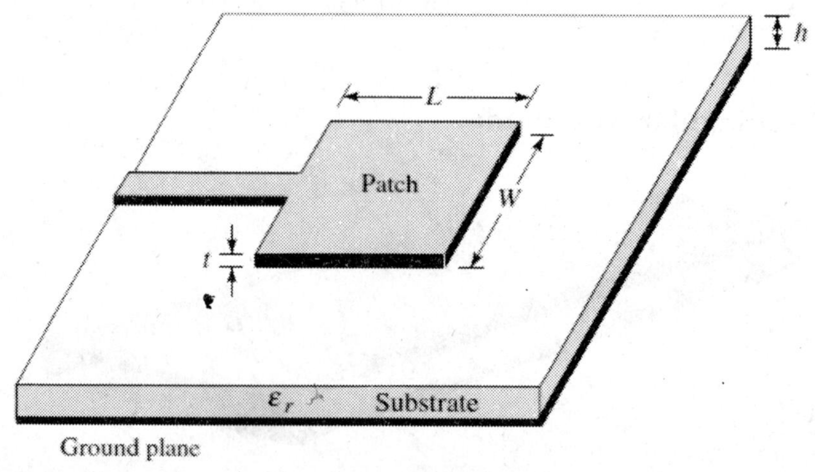

(a) Rectangular

Figure 1.28

(2) Circular Patch

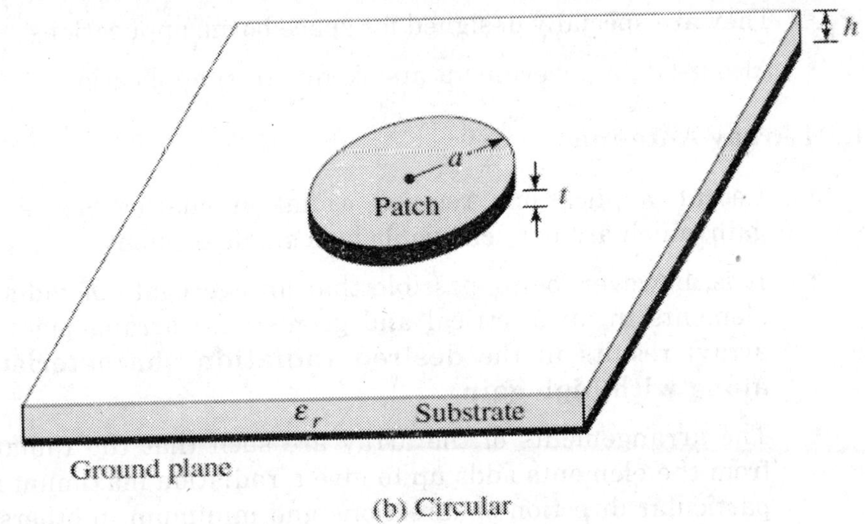

(b) Circular

Figure 1.29

Advantages

These are most commonly used antennas of the modern world, due to-

1. Ease of fabrication
2. Attractive radiation characteristics (especially low cross-polarization radiation)
3. Ease of analyses
4. Low profile
5. Comfortable to planar and non-planar surface
6. Simple
7. In-expensive to fabricate using modern printed circuit technology
8. Mechanically Robust when mounted on rigid surface
9. Compatible with MMIC designs
10. Versatile in terms of resonant frequency, Polarization, Pattern, Impedance.
11. It can be easily mounted on the surface of High performance aircraft, Spacecraft, Satellite, Missiles, Cars, Mobile Telephones etc.

Applications

- They are specially designed for space borne applications.
- Also used for government and commercial applications.

1.8.4 Array Antennas

- Recent applications require radiation characteristics and gain which are not achievable by a single element.
- It is, however, being possible that an aggregate of radiating elements in an electrical and geometrical arrangement (an array) results in the **desired radiation characteristics, along with high gain.**
- The arrangements of the array are such that the radiation from the elements adds up to give a radiation maximum in a particular direction or directions and minimum in others, or as desired.

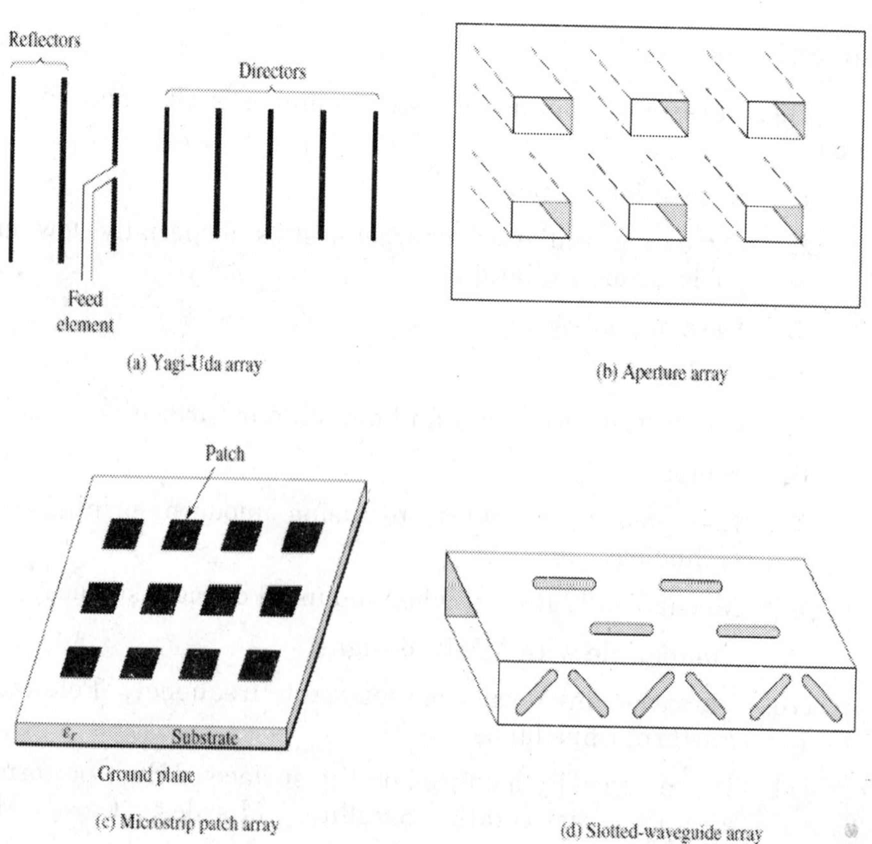

Figure 1.30

1.8.5 Reflector Antennas

The success in the exploration of outer space is mostly due to the reflector antennas. Due to the need to communicate over great distances; we required sophisticated antennas in order to transmit and to receive signals that had to travel millions of miles. For such applications we switch to parabolic reflectors.

These antennas are built with diameters as large as 305 m. Such large dimensions are needed to achieve the high gain required to transmit or receive signals after millions of miles of travel.

Types of Reflectors

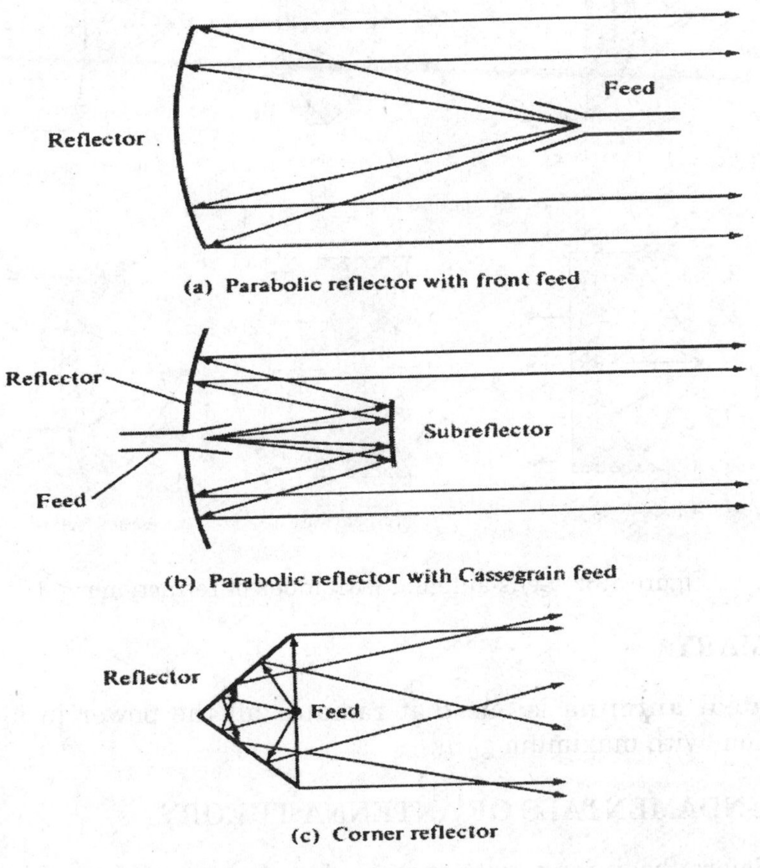

Figure 1.31

1.8.6 Lens Antennas:

Lenses are primarily used to-

- **Collimate incident divergence energy to prevent it from spreading in undesired direction.**

- By properly **shaping** the geometrical configuration and choosing **appropriate material** of the lenses, they can transform various forms of divergent energy into plane waves.
- They are used in same application as that of parabolic reflectors.
- **Their dimensions and weight become exceedingly large at lower frequencies.** Therefore, these antennas are used in high frequency applications.

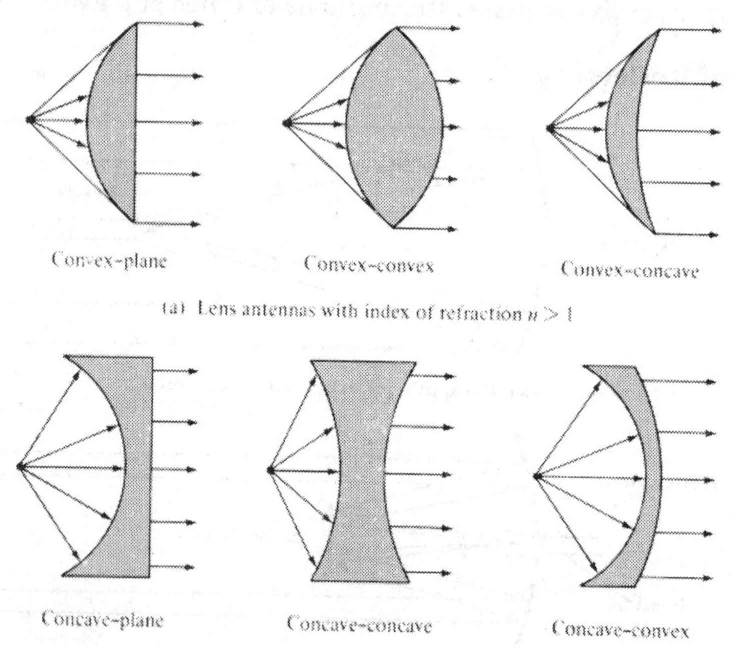

Figure 1.32 Lens antenna with index of refraction n < 1.

SUMMARY

Ideal antenna is one that radiates all the power in a desired direction, with maximum gain.

1.9 FUNDAMENTALS OF ANTENNA THEORY

Before beginning with types of antenna, their working and designing of an antenna, let us first understand the fundamental parameters like Radiation pattern, Field Regions, Directivity, Efficiency, Gain, Bandwidth, Polarization etc., that characterizes an antenna.

So let us learn that basic of antenna. We'll start with frequency and then will deal with other parameters one by one.

1.9.1 Frequency

Frequency is one of the most important concepts for the working of antenna and fortunately, it isn't too complicated.

Basic of frequency: Antenna's function is to transmit or receive EM wave (Electromagnet waves). EM wave includes the light from the sun, the waves received by ccll phone or radio, etc.

All electromagnetic waves propagate at the same speed in air or in space. This speed (speed of light) is roughly 671 million miles per hour (1 billion km per hour). This is roughly a million times faster than the speed of sound (which is about 761 miles per hour at sea level). The speed of light is denoted by \`c'.

According to \`S.I.\` units in science-

$$c = 3,00,000,000 \frac{\text{meters}}{\text{second}}$$

$$= 3\times10^{8} \text{ m/sec.}$$

Now before defining frequency, let us define what is an "electromagnetic wave"?

This is an electric field that travels away from some source (say antenna, sun, radio tower, etc.). A travelling electric field has an associated magnetic field with it. Both these fields are perpendicular to each other and both are also perpendicular to the direction of wave propagation.

Generally these waves are of sinusoidal in nature-

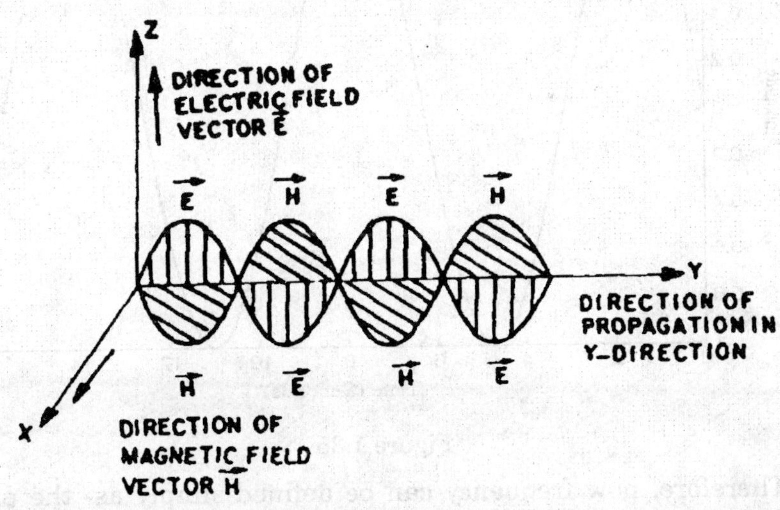

Figure 1.33

EM waves vary with space (position) and time. The **spatial variation** can be explained by the following diagram-

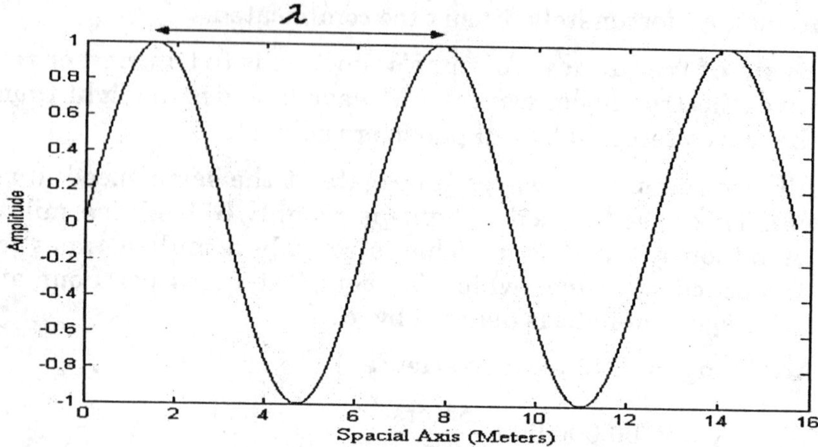

Figure 1.34

When these waves are, plotted as a function of space, waves repeats after every `λ' meters which is called as the **wavelength.**

The **temporal variation** can be explained by the following diagram-

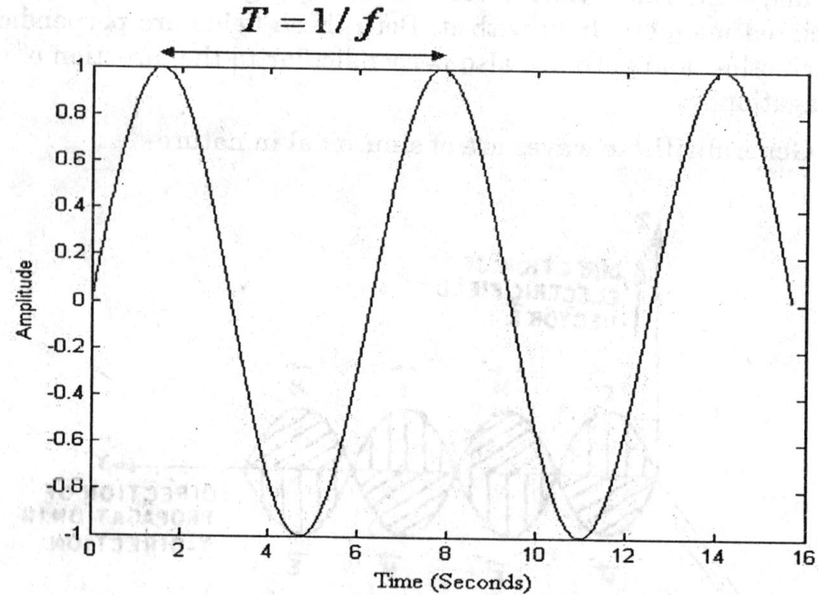

Figure 1.35

Therefore, now frequency can be defined simply as- the number of complete cycles, the wave completes in one second (two hundred cycles per second is written as 200 Hz).

Mathematically it is given by-

$$f = \frac{1}{T}$$

To understand in layman language:

How fast someone walks depends on the steps they take **(wavelength)** multiplied by the rate at which they take steps **(frequency)**.

Similarly the speed that the waves travel is how fast the waves are oscillating in time (t), multiplied by the size of the step the waves are taken per period (λ). The equation that relates frequency, wavelength and the speed of light is given by the formula-

$$c = f\,\lambda$$

Basically, the frequency is just a measure of how fast the wave is oscillating.

And since all Electromagnetic waves travel at the same speed, **the faster it oscillates the shorter the wavelength and longer wavelength implies a slower frequency.**

1.9.2 Frequency Advanced Concepts

Frequency is very fundamental parameter of antenna. To understand this, we introduce here the mathematical, "Fourier Analysis". To understand it, let us begin with finding out the frequency of the following waveform-

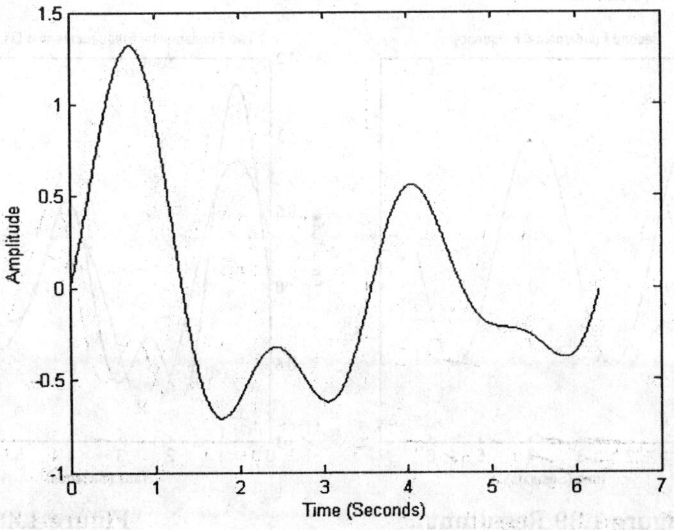

Figure 1.36

The first answer, of the above question will be that as, the period of the given waveform is not constant; in other words the wave given is non-periodic over the plotted region. But this is not the right answer. The answer is here is a **Fundamental Secret.**

Let us break the given waveform; into its "Building blocks" or its frequencies. This decomposition is done with a **Fourier Transform** (or Fourier series for periodic waveforms).

The first component is a sinusoidal wave with period T=6.28 = (2 π) and amplitude 0.3 as shown in given figure-

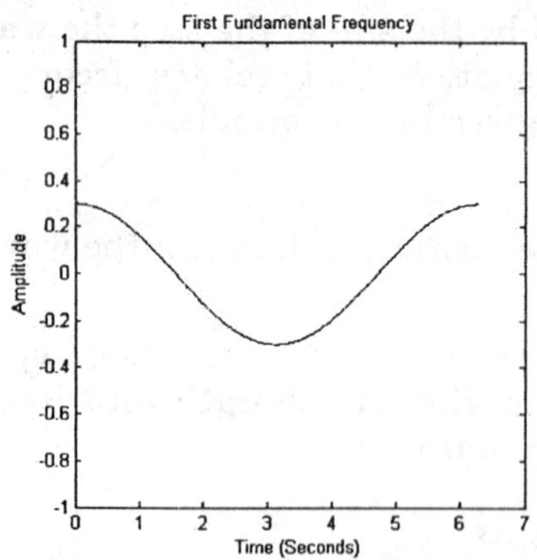

Figure 1.37 Resultant

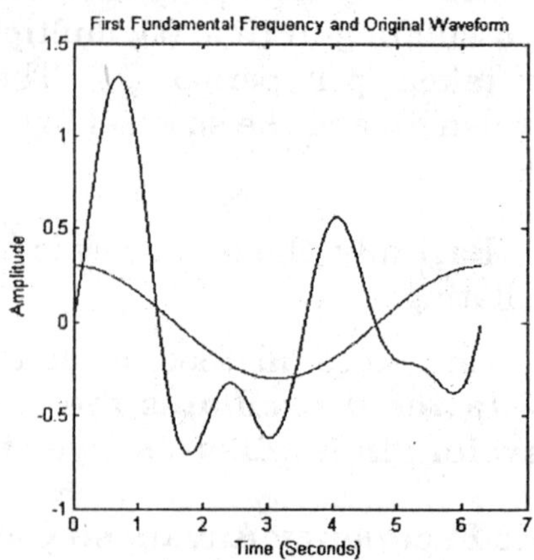

Figures 1.38

The second frequency will have a period half as long as the first (twice the frequency). The second component as given in the figure is given, along with the sum of first two frequencies compared to the original waveform.

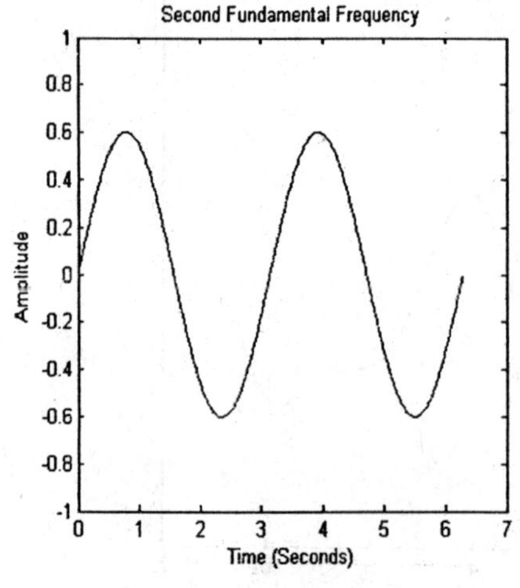

Figure 1.39 Resultant

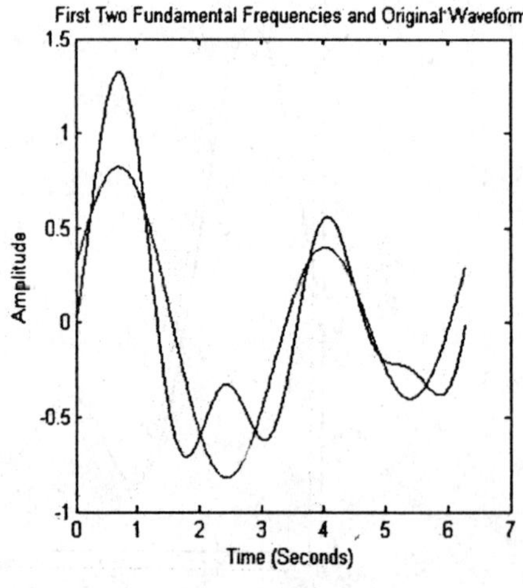

Figure 1.40.

(We see that the sum of the first two frequencies is starting to look like the original waveform).

The third frequency component is 3 times the frequency as the first. The sum of the first three components is shown in the next figure-

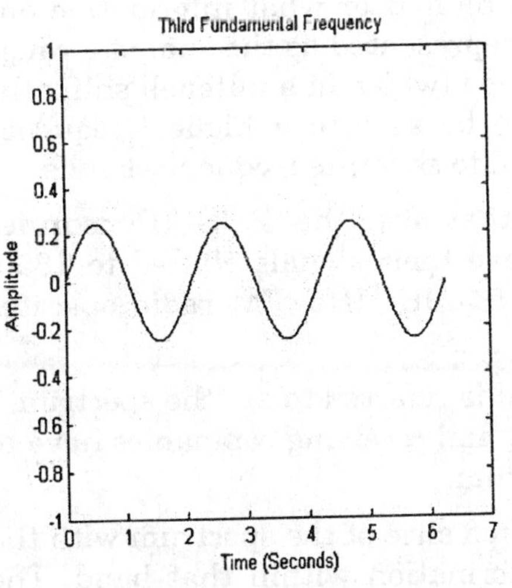

Figure 1.41 Resultant

First Three Fundamental Frequencies and Original Waveform

Figure 1.42

Finally adding the fourth frequency component, we get the original waveform, shown in Figure 1.43.

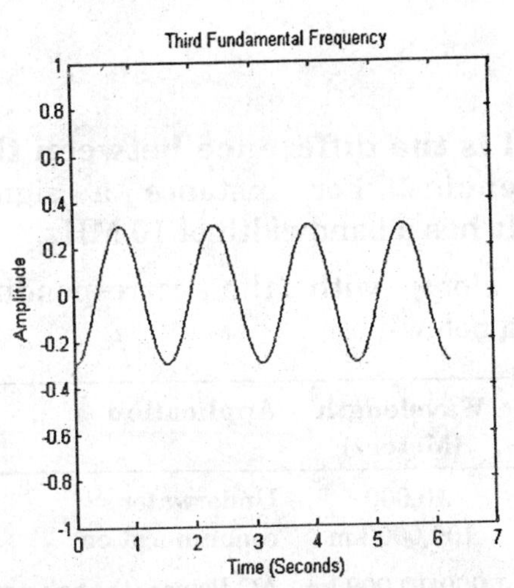

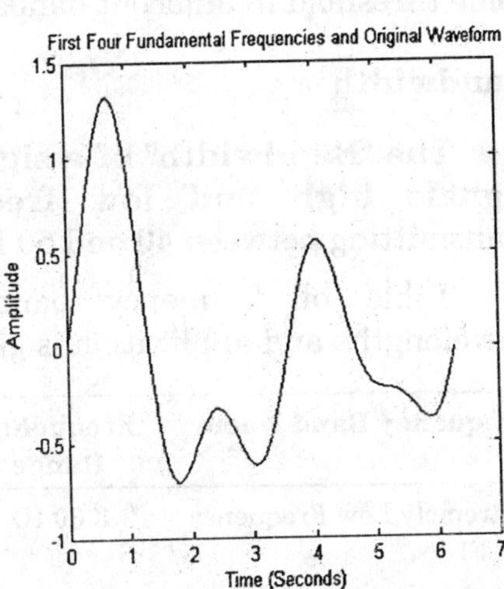

Figure 1.43

It is same for TV signals, cell phone signals, second waves that travel when one speak. In general, waveforms are not made up of a discrete number of frequencies, but rather a continuous range of frequencies, that is,, frequency bands.

Frequency Bands

How do a cell phone and the television work, simultaneously? Both these devices uses antenna to receive information from electromagnetic waves, so why is not there a problem of interference in the working of device at the same time?

The answer to this question is, no matter what information one wants to send, the waveform can be represented as the sum of a range of frequencies. By the use of modulation (which in a nutshell shifts the frequency range of the waveform to be sent to a higher frequency band), the waveforms can be relocated to separate frequency bands.

As an example, cell phones that use the **PCS (Personnel Communication Service)** band have their signals shifted to 1850-1900 MHz. Television broadcast at 54-216 MHz. FM radio operates between 87.5 -108 MHz.

The set of all frequencies in total is referred to as "the spectrum". For all the devices, data transmitting and receiving companies have to pay, to get access to part of the spectrum.

For instance, AT&T has to bid on a slice of the spectrum with the FCC, for the "right" to transmit information within that band. The transmission of Electromagnetic energy is greatly regulated. When AT&T is sold a slice of the spectrum, they cannot transmit energy at any other band (technically, the amount transmitted must be below some threshold in adjacent bands).

Bandwidth

The "Bandwidth" of a signal is the difference between the signals high and low frequencies. For instance a signal transmitting between 40 and 50 MHz has a bandwidth of 10 MHz.

Table of frequency bands along with the corresponding wavelengths and application is given below-

Frequency Band Name	Frequency Range	Wavelength (Meters)	Application
Extremely Low Frequency (ELF)	3-30 Hz	10,000-100,000 km	Underwater communication
Super Low Frequency (SLF)	30-300 Hz	1,000-10,000 km	AC Power (though not a transmitted waves)
Ultra Low Frequency (ULF)	300-3000 Hz	100-1,000 km	
Very Low Frequency (VLF)	3-30 kHz	10-100 km	Navigational Beacons
Low Frequency (LF)	30-300 kHz	1-10 km	AM Radio

Medium Frequency (MF)	300-3000 kHz	100-1,000 m	Aviation and AM Radio
High Frequency (HF)	3-30 MHz	10-100 m	Shortwave Radio
Very High Frequency (VHF)	30-300 MHz	1-10 m	FM Radio
Ultra High Frequency (UHF)	300-3000 MHz	10-100 cm	Television, Mobile Phones, GPS
Super High Frequency (SHF)	3-30 GHz	1-10 cm	Satellite Links, Wire less Communication
Extremely High Frequency (EHF)	30-300 GHz	1-10 mm	Astronomy, Remote Sensing
Visible Spectrum	400-790 THz (4*10^14-7.9*10^14)	380- 750 nm (nanometers)	Human Eye

1.10 FUNDAMENTALS PARAMETERS OF ANTENNAS

1.10.1 Radiation Pattern

Defination

(1) A **radiation pattern** can be defined as the variation of the power radiated by an antenna as a function of the direction away from the antenna.

(2) A mathematical function or a graphical representation (Graphical representation is represented as by the function of directional coordinates) of the radiation properties of the antenna as a function of space coordinates.

- Radiation pattern (power variation as a function of arrival angle) in most of the cases is determined in far-field region.
- Radiation properties include-

  (1) Power

  (2) Flux

  (3) Density

  (4) Radiation Intensity

  (5) Field Strength

  (6) Directivity

  (7) Phase or Polarization.
- Radiation property is 2 or 3-Dimensional spatial distribution of radiated energy as a function (of the observers position along a path or surface of constant radius).

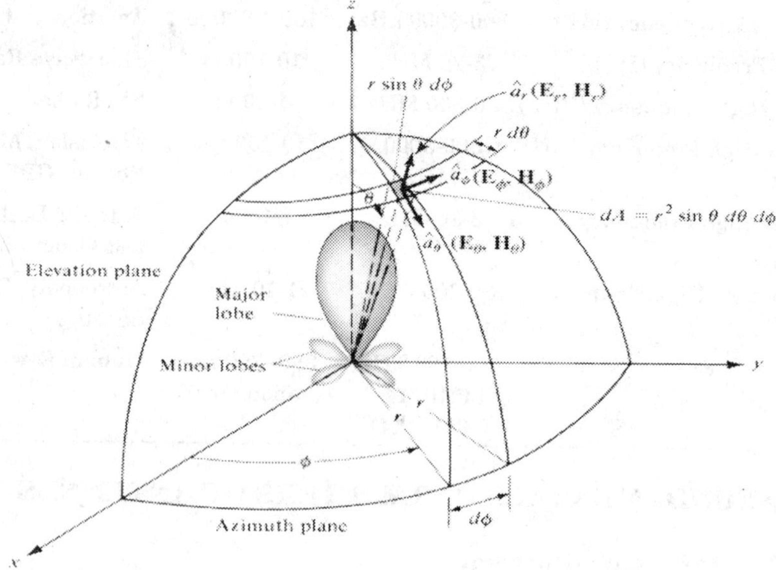

Figure.1.44

Power-pattern: *Trace of received power* at a constant radius in called as power pattern.

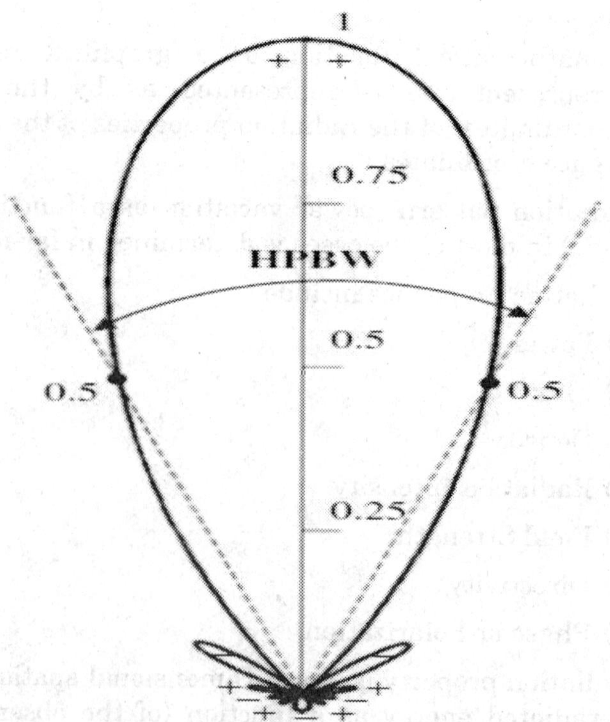

Figure.1.45 Power Pattern (in linear scale)

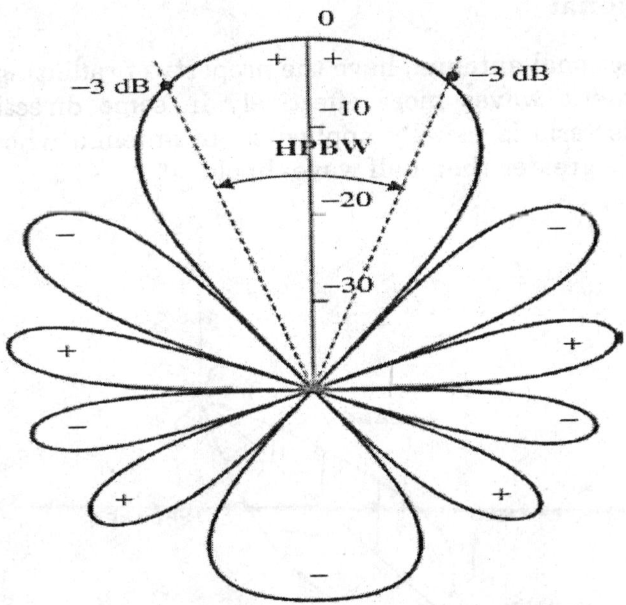

Figure 1.45(b). Power Pattern (in dB)

Field pattern: *Graph of* spatial variation of the electric (or magnetic) *field* along a constant radius is called an **amplitude field pattern.**

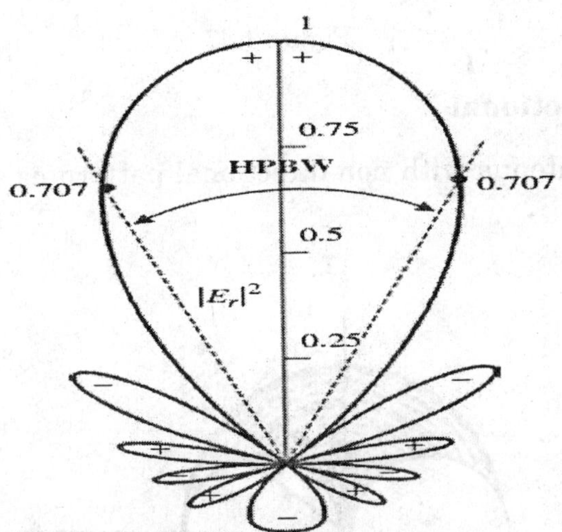

Figure 1.46 Field Pattern (in linear scale)

NOTE: In practice 3-D pattern is measured and recorded in a series of 2-D pattern.

Antennas can be categorized according to their radiation patterns-

(a) Directional

A directional antenna, have the property of radiating or receiving electromagnetic waves more effectively in some directions than in others. This term is usually applied to an antenna whose maximum directivity is greater than half wave-dipole.

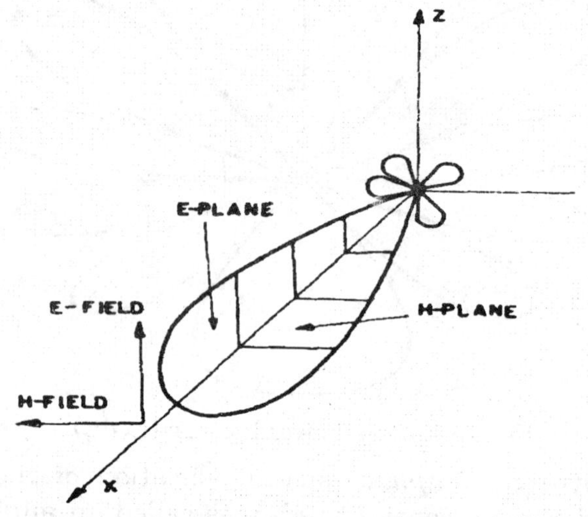

(b) E-plane and H-plane principal patte rns..

Figure 1.47.

(b) Omni-directional

It is an antenna with non-directional pattern as shown in Figure below-

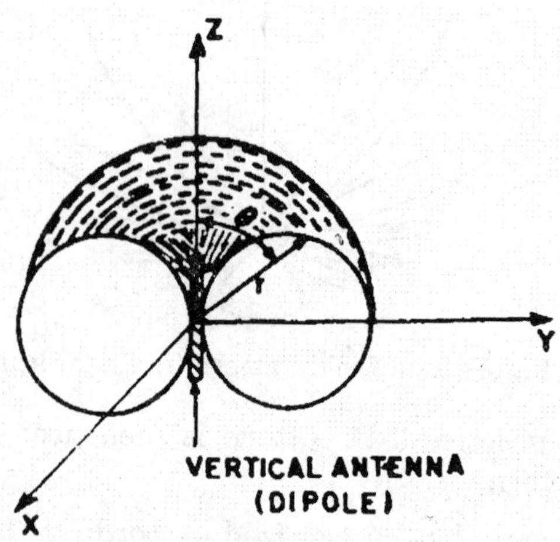

Figure 1.48

It is defined as one antenna, having an essentially non-directional pattern in a given plane (azimuth) and a directional pattern in orthogonal plane (elevation).

(c) Isotropic

A hypothetical lossless antenna having equal radiation in all directions is considered as an ideal antenna.

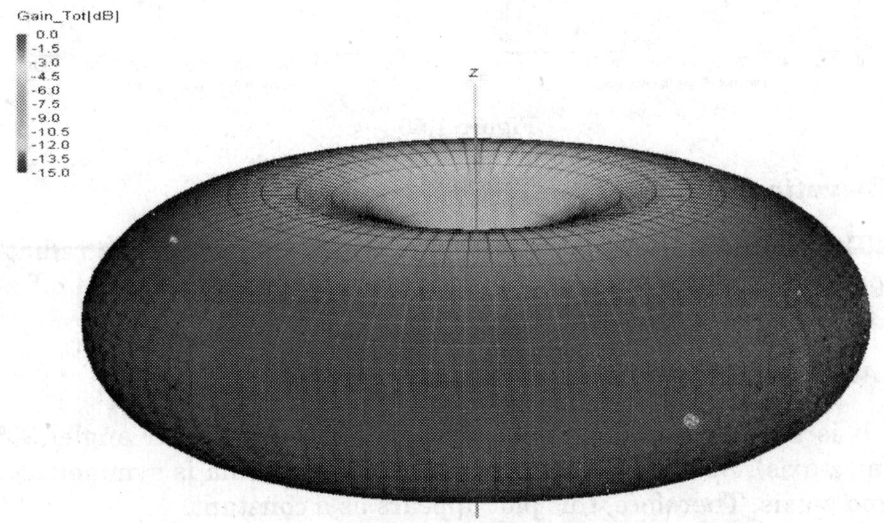

Figure 1.49 Radiation pattern of an isotropic antenna placed at the origin.

This is an example of a **donut shaped** or **toroidal pattern**. In this case, along the z-axis, corresponds to the radiation directly overhead the antenna, where a very little power is transmitted.

In the x-y plane (perpendicular to the z-axis), the radiation is maximum. **These plots are useful for visualizing as to in which directions the antenna radiates.**

The given pattern is minimum at 0° and 180° and becomes maximum at 90° off the axis (like in Broadside antenna).

2-D Pattern

It is simpler, to plot the radiation patterns in 2-D; which are given as "slices" through the 3-D plane. Standard spherical coordinates are used, where-

θ = angle measured off the z-axis.

ϕ = angle measured counter clockwise off the x-axis.

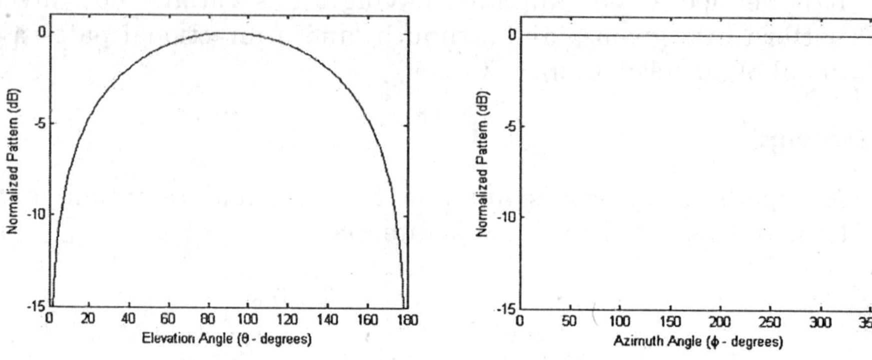

Figure 1.50

(a) Elevation Pattern

The elevation (height) pattern is in which the plot of the radiation pattern is represented as a function of the angle, measured off z-axis (for fixed azimuth angle).

(b) Azimuthal Pattern

It is the function of azimuthal angle for a fixed polar angle (90° off the z-axis). As the 3-D pattern of isotropic antenna is symmetrical around z-axis, Therefore, this plot appears as a constant.

Isotropic Antenna

A pattern is "isotropic" if the radiation pattern is same in all the directions. These antennas don't exist in practice but are sometimes discussed, as means of comparison with real antennas. Some antennas may also be described as "Omni directional", which actually means, that it is isotropic in a single plane (like in 3-D figure).

1. Isotropic radiator is only a theoritical concept it cannot be designed practically, why? How is this concept used for defining the power gain and directivity of an antenna? **(Raj. Univ. 2003)**
2. Explain the following terms- i) Isotropic Radiation **(Raj. Univ. 2004)**

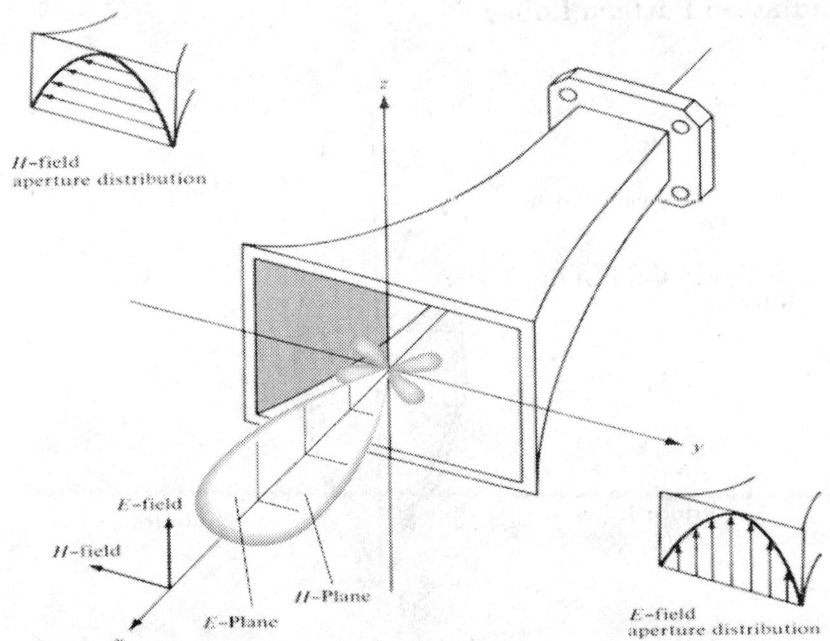

Figure 1.51.

Principal Patterns

For **linearly polarized antenna**, performance is often described in terms of its principal E- plane pattern and H-plane pattern.

E-plane

The plane which contains the electric-field vector along with the direction of maximum radiation is called as E-plane.

H-plane

The plane which contains the magnetic-field vector along with the direction of maximum radiation is called as H-plane.

Principal E and H-plane patterns for a pyramidal horn antenna

The basic aim of the principal pattern is to orient most antennas, so that at least one of the principal plane patterns coincides with one of the geometrical principal planes.

Therefore, in above figure-

x-z plane (elevation plane $\phi = 0$) principal E-plane

x-y plane (azimuthal plane $\theta = \pi/2$) principal H-plane.

Radiation Pattern Lobes

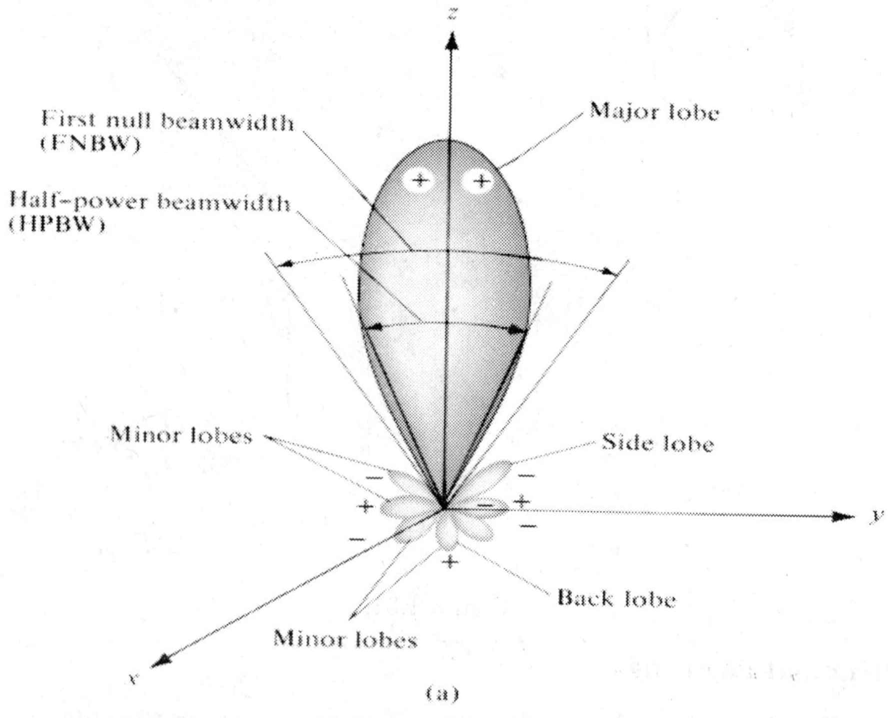

Figure.1.52

The above figure, demonstrates a symmetrical 3-D polar pattern with a number of radiation lobes. Some are of greater radiation intensity than others, but all are categorized as lobes.

Radiation lobe

-Portion of the radiation pattern bounded by the regions of relatively weak radiation intensity.

Lobes

Various parts of a radiation pattern are referred as lobes. They are

- Major/Main Lobe
- Minor Lobes
- Side Lobes
- Back Lobes

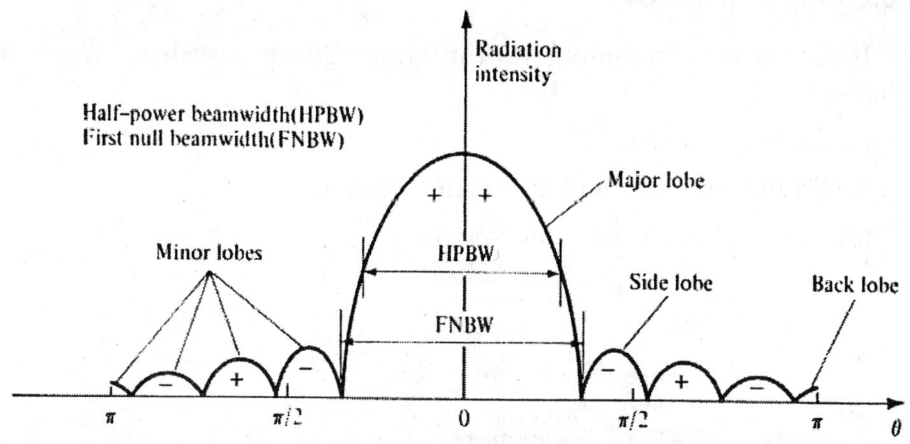

Figure 1.53

(1) Major Lobe: It is the radiation lobe, containing the direction of maximum ($\theta = 0$). (In some antennas, like split beam antenna, there exists more than one major lobe).

(2) Minor Lobes: These lobes are other than the major lobe.

(3) Side Lobes: A radiation lobe in any direction other than the intended lobe (usually side lobe is adjacent to the main lobe and occupies the hemisphere in the direction of the main beam).

(4) Back Lobes: A radiation lobe whose axis makes an angle of approximate 180° with respect to the main beam of antenna. (that is minor lobe which is opposite to that of major lobe)

Minor lobes usually represent radiation in undesired directions, hence they should be minimized. Side lobe < –20 dB (undesirable) (This requires very careful design and construction.)

$$\text{Level of minor lobes} = \frac{\text{Power density in lobe}}{\text{Power density in major lobe}}$$

It is also called as **side lobe ratio/level.**

NOTE: In most Radar system low **side lobe ratio** are very important to minimize, **false target indication** through the side lobes.

1. Define the following terms regarding Antenna:

   (i) Radiation pattern **(Raj. Univ. 2006, 2008)**

1.10.2 FIELD REGIONS

It is space surrounding the antenna. There are three types of regions-

(1) Reactive near field.

(2) Radiating near field of Fresnel Region.

(3) Far-field or Fraunhofer Region.

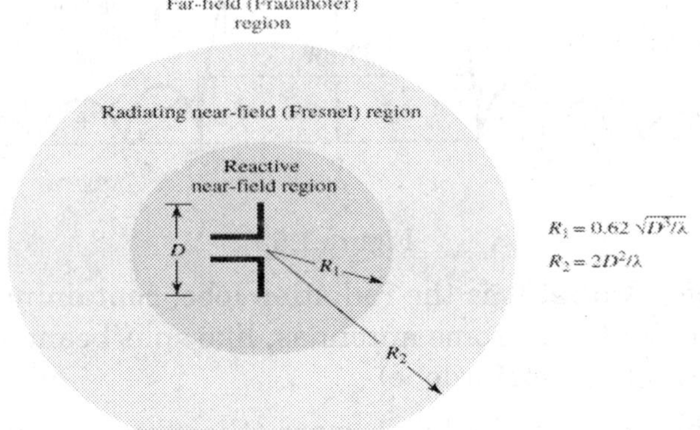

Figure 1.54 Field regions of an antenna

Regions can be given by-

$$0.62\sqrt{\frac{D^3}{\lambda}} < R < \frac{2D^2}{\lambda}$$

(1) Reactive near-field region

That portion of the near-field region immediately surrounding the antenna where in the reactive field predominates. Reactive fields, means that the E-fields and H-fields are out of phase by 90° to each other.

It is $R_1 < 0.62\sqrt{\frac{D^3}{\lambda}}$ from the antenna surface

where λ = wavelength

D = largest dimension of Antenna

(2) Radiating near-field region

This region is also called as **Fresnel Region.** This field region, of an antenna is between the reactive near field region and far field region wherein radiation fields predominate and wherein angular field distribution, is dependent upon the distance from the antenna (means the shape of radiation patterns may vary with distance).

If the antenna has a maximum dimension that is not large compared to the wavelength this region may not exist.

Or else it is-

$$R_L < \frac{2D^2}{\lambda}$$

In this region, the field pattern is generally, a function of the radial distance and the radial field component.

(3) Far - Field Region

This region is also called as **Fraunhofer region (optical term).** It is region, of the field, of an antenna, where the angular field distribution is essentially independent of the distance from the antenna. It means the radiation pattern does not change shape with distance. The inner boundary is taken to be at the radial distance-

$$R_L \text{ is greater than } \frac{2D^2}{\lambda}$$

In this region, the field components (that is, E-fields and H-fields) are essentially transverse orthogonal to each other and in phase and angular distribution is independent of the radial distance where the measurements are made.

$R = 2\,D^2/\lambda$ and the outer one at infinity.

This region is most important as it determines the antennas radiation pattern. Also as the antennas are used to communication wirelessly from long distances, so this is the region of operation for such antennas.

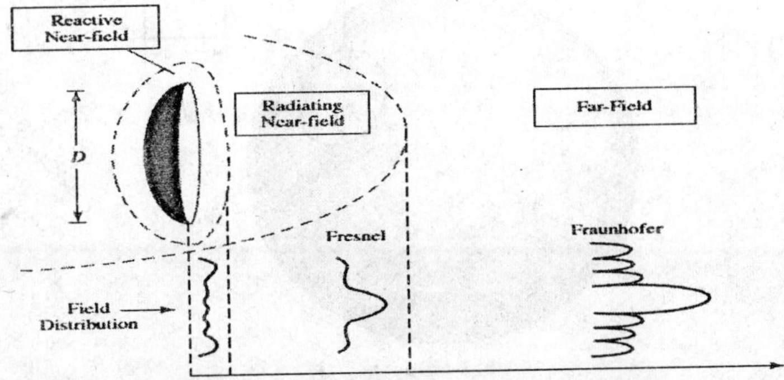

Figure 1.55 Typical changes of antenna amplitude pattern shape from reactive near field towards the far field.

1.10.3 RADIAN AND STERADIAN

(1) Radian

- **Measure of a plane angle is radian.**

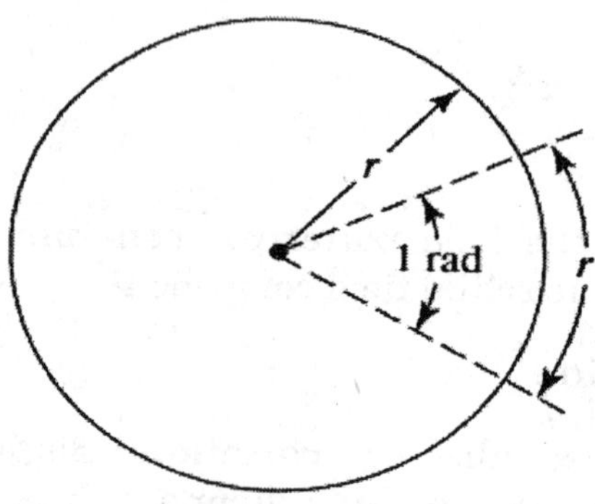

(a) Radian

Figure 1.56

- **One radian** is defined as the plane angle with its vertex at the centre of a circle of radius `r' that is subtended by an arc whose length is `r'

Now, as the circumference $= 2\pi r$, therefore it is $= \dfrac{circumference}{arc}$

$$= \frac{2\pi r}{r} = 2\pi \text{ radians (in a full circle).}$$

(2) Steradian

- **Measure of a solid angle is Steradian.**

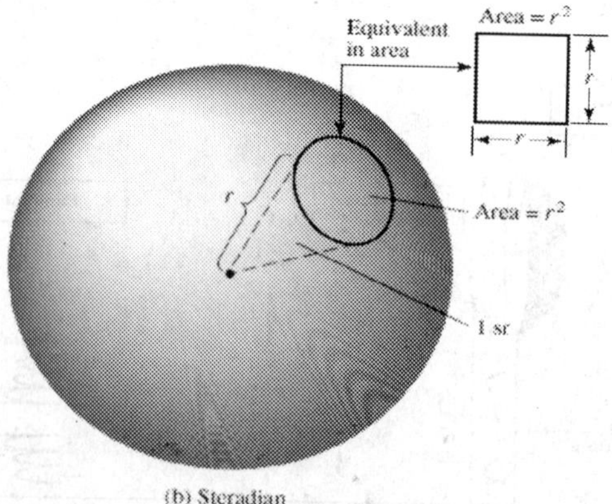

(b) Steradian

Figure 1.57

- One steradian is defined as the solid angle with its vertex at the centre of a sphere of radius 'r' that is subtended by a spherical surface area equal to that of a square with each side of length `r'.

As the area of sphere = $4\pi r^2$, therefore it is $= \dfrac{Area}{arc(3D)} =$

$\dfrac{4\pi r^2}{r^2} = 4\pi$ steradian (in a full sphere).

Example: Let there be a sphere with radius = r and Infinitesimal area = dA, is given by-

$$dA = r^2 \sin\theta \, d\theta \, d\phi \, (m^2)$$

Therefore element of solid angle d Ω of sphere ⇒ $d\,\Omega = \dfrac{dA}{r^2}$

$$= \sin\theta \, d\theta \, d\phi$$

1.10.4 BEAMWIDTH AND SIDE LOBE LEVELS

Antennas are characterized by their beam widths and side lobe levels. The Radiation pattern is given by-

$$R(\theta) = Sin\,\theta \, \frac{Sin\left(4\left(\theta - \frac{\pi}{2}\right)\right)}{4\left(\theta - \frac{\pi}{2}\right)}$$

The 3-D view of this radiation pattern is given as-

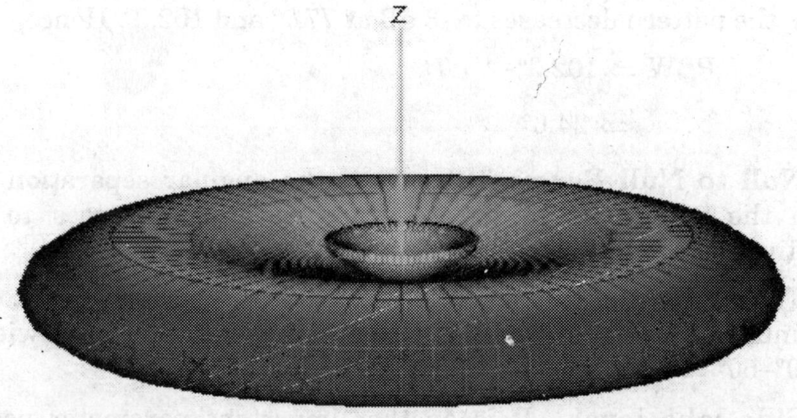

Figure 1.58 3D Radiation Pattern

The polar (polar angle measured off of z-axis) plot is given by-

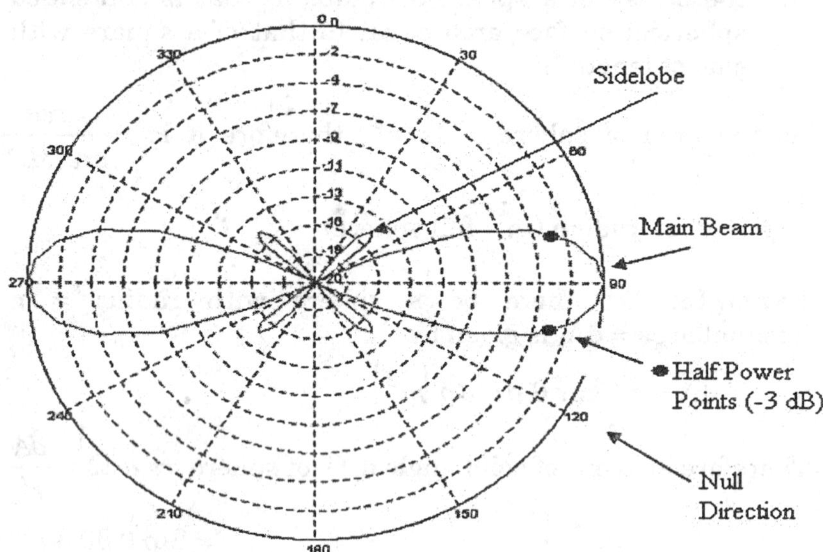

Figure 1.59 Polar Radiation Pattern.

The **main beam** is the region as round the direction of maximum radiation (usually the region that is within 3 dB of the peak of the main beam). In the figure main beam is centered at 90°.

The **side lobes** are smaller beams that are away from the main beam. These side lobes are usually radiation in undesired directions which can never be completely eliminated. The sidelobes in figure occurs roughly at 45° and 135°

Therefore, **the half power Beamwidth (HPBW)** is the angular separation in which the magnitude of the radiation pattern decreases by 50% (or -3dB) from the peak of the main beam. From the above figure, the pattern decreases to -3 dB at 77.7° and 102.3°. Hence,

$$\text{HPBW} \Rightarrow 102.3° - 77.7°$$

$$\Rightarrow 24.6°$$

Null to Null Beamwidth - It is the angular separation from which the magnitude of the radiation pattern decreases to zero (negative infinity dB) away from the main beam.

Therefore, from the above given figure, the pattern goes to zero (or minus infinity) at 60° and 120°. Hence, the Null-Null Beamwidth is $\Rightarrow 120° - 60° \Rightarrow 60°$

Side lobe level - It is another important parameter used to characteristic radiation patterns. The sidelobe level is the maximum

value of the sidelobes (away from the main beam). From the 2 figure, sidelobe level **(SLL)** is - 14.5 dB.

1.10.5 ANTENNA BEAMWIDTH

(Measure of the directivity of an antenna)

Defination 1: Antenna beam-width is an angular width in degrees, measured on the radiation - pattern (major lobe) between points where the radiated power has fallen to half its maximum value.

Defination 2: In a plane containing the direction of the maximum of a beam, the angle between the two directions in which the **radiation intensity is one half the maximum value of the beam.**

Defination 3: Beamwidth is used to describe 3-dB beamwidth.

This is called as beam-width - between half power point or half power beam width (HPBW) because the power at half power points is just half = $3-dB$ **beamwidth,** because at half power points, the power is 50% of the maximum power value of major lobe.

At these half power points - the field intensity (that is, voltage) equals $1/\sqrt{2}$ *or* 0.707 times its maximum value or 3-dB down from maximum value.

Radiation Pattern itself is a function of direction.

$$P_1 P_2 \rightarrow \text{Half power (3 dB) points}$$

$$\therefore P_1 OP_2 \rightarrow \text{Angle} = \text{Beam width.}$$

It is also given by-

Angular width (in degrees) of major lobe between the two directions at which the radiated or received power is one half of the maximum power.

Beamwidth decreases, side lobes increase and *vice -versa.*

The Beamwidth of an antenna **is used to describe resolution capabilities of antenna** to distinguish between two adjacent radiating sources or Radar targets.

$$\frac{FNBW}{2} = HPBW$$

(If separation is smaller, then the antenna will tend to smooth the angular separation distance).

1.10.6 HALF POWER BEAMWIDTH

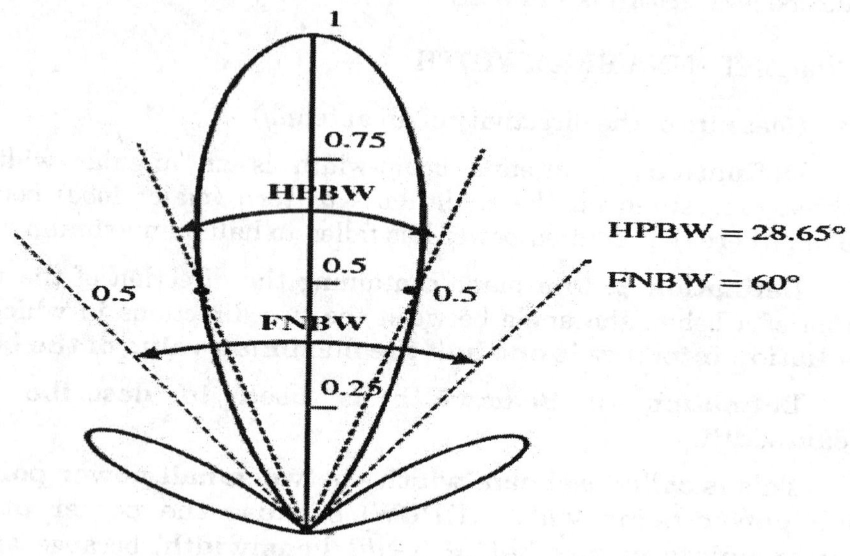

Figure 1.60

0 dB = main lobe.

3 dB = half power points

Sometimes radiation pattern is also described in terms of **angular width between first nulls or first side lobes, known as beamwidth between first nulls** and is abbreviated as **(BWFN)** or beamwidth - **10 dB down** from the pattern maximum.

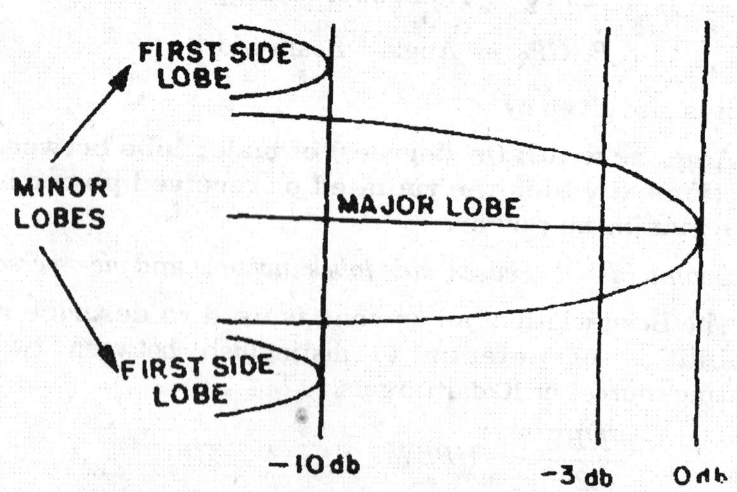

(c) Beam width on rectangular co-ordinates on logarithmic scale (decibel).

Figure 1.61

$$D = \frac{4\pi}{\Omega_A} = \frac{4\pi}{B}$$

Where D – *Directivity*

Ω_A – *solid Angle*

B - Beam Area

B ≈ HPBW in horizontal plane * HPBW in vertical plane

B ≈ HPBW in E- plane * HPBW in H- plane

(B is in Square radians)

$$B \approx \theta_E \times \theta_H$$

$$\therefore D = \frac{4\pi}{\theta_E \times \theta_H} \text{ in radians}$$

$$D = \frac{4\pi\,(57.3)^2}{\theta_E \times \theta_H}$$

$$D = \frac{41{,}257}{\theta_E{}^\circ \times \theta_H^\circ}$$

Factors affecting the beamwidth-

(1) Shape of radiation pattern

(2) Wavelength

(3) Dimensions (e.g. radius of aperture)

As, D is proportional to $\frac{1}{B}$

NOTE: Narrow beam is desirable and accuracy of direction finding as D is inversely proportional to beam width.

1. Explain following: **i. Half Power Beam width. (R.T.U. 2009)**
2. Explain any four terms of antenna- **(a) Antenna beam width (Raj. Univ. 2007)**
3. Explain the following terms- **(i) Beam width of an antenna (Raj. Univ. 2004)**

1.10.7 BANDWIDTH

Defination 1 - Bandwidth describes the range of frequencies over which the antenna can properly radiate or receive energy.

Defination 2 - The range of frequencies within which the performance of the antenna. With respect to, some characteristics, conforms to a specified standard.

Defination 3 - It can be considered as the range of frequencies on either side of a center frequency (Resonant frequency) where the antenna characteristics (input impedance, pattern, beamwidth, gain, direction, and radiation efficiency) are within an acceptable value of those at the centre frequency.

Bandwidth may be specified in many different ways-

(a) B.W. over which the **gain** is higher than same acceptable value.

(b) B.W. over which at least a given **front to back ratio** is achieved.

(c) B.W. over which the **SWR** on the transmission line can be maintained below.

Therefore, now the defination of antenna bandwidth becomes that- it is a width (range) of frequency over which the antenna maintains certain required characteristics like gain, front to back ratio or **SWR** pattern (Shape or direction, Polarization and impedance).

NOTE: These requirements may change with operation, ∴ specification are set to meet the needs of a particular application.

With increase in side -lobe levels; Gain decreases and there is change in impedance value; pattern and polarization characteristics. One of these factors, determines the *low frequency limit* and other factors (e.g. change of pattern shape or direction) the *high frequency limit.*

The band width within which the antenna maintains a given set of specifications; it depends on two characteristics –

(1) Impedance bandwidth - (input impedance, and radiation efficiency)

(2) Pattern bandwidth - (gain, side lobe level, beamwidth, polarization and beam direction)

At low frequency ($\lambda/2$ or less)

It is determined by impedance variation because the pattern characteristics are insensitive to frequency that is, pattern changes less rapidly. ∴ Due to this antenna performance is limited by impedance variation and Bandwidth depends on its Reciprocal `θ'.

$$\Delta w = w_2 - w_1 = \frac{w_r}{Q} = Band\ width.$$

or $$\Delta w = \frac{w_r}{Q} \qquad \therefore w_r = 2\pi f_r$$

$$\therefore \Delta 2\pi f = \frac{2\pi f_r}{Q}$$

$$= \Delta f = \frac{f_r}{Q}$$

Or $$\Delta f\ \alpha\ \frac{1}{Q}$$

Where, f_r - centre / Resonant/ designed frequency and

$$Q = \frac{2\pi\,(Total\ energy\ stored\ by\ antenna)}{(Energy\ discipated\ or\ radiated\ per\ cycle)}$$

Thus, it can be concluded here that as the **'Q' decreases the bandwidth increases.**

Frequency independent antenna

Like in, Log-periodic antenna etc., having unlimited bandwidth where lower and upper frequencies limits are specified independently. In such, cases, the band width is represented by a ratio of higher to lowest operating frequency.

Log Periodic ⇒ 20 :1

100: 1

And bandwidth percentage is given by-

$$B.W.\ \% = \frac{Operating\ Range}{Centre\ frequency} \times 100$$

Example

If minimum frequency = 98 MHz and Maximum frequency = 102 MHz is given for any antenna then calculate its bandwidth?

Solution: $$B.W. = \frac{102-98}{(102-98)/2} = \frac{4}{100} = 0.04$$

$$B.W.\% = 0.04 \times 100$$
$$= 4\%$$

Ans.

The desired Bandwidth is one of the determining parameters used to decide upon an antenna. For instance many antenna types have very narrow bandwidths and cannot be used for wide band operation.

Broadband antenna $\Rightarrow \dfrac{Upper\ \ frequency}{Lower\ \ frequency} = \dfrac{10}{1}$

∴ 10 - upper frequency is 10 times greater than lower frequency.

Narrowband antenna $\Rightarrow \dfrac{Upper - Lower}{Centre\ \ freq.\ of\ \ Bandwidth} \%$

= say 5% - It means that the frequency difference of acceptable operation is 5% of centre frequency of bandwidth.

Fractional Bandwidth (FBW)

The Bandwidth is often specified in terms of its Fractional Bandwidth (FBW). The antenna θ also relates to Bandwidth.

It is defined as the measure of how wideband the antenna is. If the antenna operates at the centre frequency f_c between lower frequency and the upper frequency f_2 (where, $f_c = (f_1 + f_2)/2$), then the fractional bandwidth FBW is given by:

$$\text{FBW} = \frac{f_2 - f_1}{f_c}$$

The fractional bandwidth varies between 0 to 2, and is often quoted as a percentage (between 0% and 200%). As the percentage increase, the bandwidth becomes wider.

Wideband antennas typically have a Fractional Bandwidth of 20% or more. Antennas with a FBW of greater than 50% are referred to as ultra- wideband antennas.

1.10.8 ANTENNA BEAM EFFICENCY: (B_E)

It is used to judge the quality of transmitting and receiving antennas. For an antenna with its major lobe coincident with z-axis (θ = 0), the beam efficiency is defined as-

$$B_E = \frac{Power\ transmitted\,/\,received\ within\ cone\ angle\ \theta_1}{Power\ transmitted\,/\,received\ by the\ antenna},$$

θ_1 = half angle of cone (within which the percentage of total power is found) also called as angle where first null or minimum occurs.

$$B_E = \frac{\int_0^{2\pi}\int_0^{\theta_1} \Phi(\theta, \phi) \sin\theta \, d\theta \, d\phi}{\int_0^{2\pi}\int_0^{\pi} \Phi(\theta, \phi) \sin\theta \, d\theta \, d\phi}$$

Beam efficiency indicates the amount of power in major lobe compared to total power.

$$B_E \text{ or } \varepsilon_M = \frac{\Omega_M}{\Omega_A} = \frac{\textit{Main beam area}}{\textit{Total beam area}}$$

$$\Omega_A = \Omega_M + \Omega_m$$

Where, Ω_m - minor lobe area

Divide Ω_A to both the sides-

$$1 = \frac{\Omega_M}{\Omega_A} + \frac{\Omega_m}{\Omega_A}$$

$$1 = \varepsilon_M + \varepsilon_m$$

Main Beam Efficiency-

$$\varepsilon_M \rightarrow \frac{\Omega_M}{\Omega_A}$$

Stray Factor-

$$\varepsilon_m = \frac{\Omega_m}{\Omega_A}$$

1.10.9 ANTENNA EFFICIENCY: (η)

"The efficiency of an antenna relates the power delivered to the antenna and the power radiated or dissipated within the antenna". The efficiency can be written as the ratio of the radiated power to the input power of the antenna:

$$\varepsilon = \frac{P_{radiated}}{P_{input}}$$

A high efficiency antenna has most of the power radiated away at the antenna's input.

A low efficiency antenna has most of the power absorbed as losses within the antenna.

The losses associated with in an antenna are typically –

(1) Conduction losses (due to finite conductivity of the antenna)

(2) Dielectric losses (due to conduction within a dielectric which may be present within an antenna.

The efficiency can also be defined as the ratio- $\dfrac{\textit{Power Radiated}}{\textit{Total Input power}}$

$$\eta = \frac{W_r}{W_T} = \frac{W_r}{W_r + W_l}$$

$$\eta = \frac{W_r}{W_r + W_l} = \frac{W_r}{W_T} \times \frac{4\pi\Phi(\theta,\phi)}{4\pi\Phi(\theta,\phi)}$$

$$= \frac{4\pi\Phi(\theta,\phi)}{W_T} \times \frac{W_r}{4\pi\Phi(\theta,\phi)}$$

$$= G_p \cdot \frac{1}{G_d}$$

$$\eta = \frac{G_p}{G_d}$$

Where, W_r = Power radiated

W_l = Ohmic losses.

If current flowing in the antenna is I, then

$$\eta = \frac{I^2 R_r}{I^2 (R_r + R_l)} \quad \text{or}$$

$$\eta\% = \left(\frac{R_r}{R_r + R_l}\right) \times 100$$

Where, R_r = Radiation Resistance

R_l = Ohmic loss Resistance of antenna conductor

$R_r + R_l$ = Total effective resistance

Loss Resistance

(1) Ohmic loss in the antenna conductor

(2) Dielectric loss

(3) I^2 R loss in antenna and ground system

(4) Los in earth connections

(5) Leakage loss in insulation

NOTE : η – Fraction of total energy supplied to the antenna which is converted into electromagnetic waves.

1. Define the following terms regarding Antenna- (i) Antenna Efficiency **(Raj. Univ. 2006, 2008)**
2. Explain any four terms of antenna- (a) Antenna efficiency. **(Raj. Univ. 2007)**

1.10.10 RADIATION INTENSITY

Defination1 - The power radiated from an antenna per unit solid angle.

It is a far - field parameter and can be obtained by simply multiplying the radiation density by the square of the distance.

$$U = r^2 \cdot W_{rad}.$$

where W - Instantaneous Poynting vector (W/m^2)

ε - Instantaneous Electric field Intensity (V/m)

H - Instantaneous Magnetic Field Intensity (A/m)

Radiation Power density

Electromagnetic waves are used to transport information through a wireless medium or a guiding structure, from one point to other. It can be then assume that power and energy are associated with electromagnetic fields. The quantity used to describe the above is instantaneous Poynting Vector defined as- $W = \varepsilon \times H$

∴ Poynting Vector = Power density = W_{rad} = Power radiated

$= R_e$ (E X H)

∴ U = Radiation Intensity (W/unit solid angle)

W_{rad} = Radiation Density (W/m²)

1.10.11 DIRECTIVITY

Directivity is a fundamental antenna parameter. It is a measure of how ``directional" antenna's radiation pattern is. An antenna that radiates equally in all directions would have effectively zero directionality, and **the directivity of non-directional antenna would be 1 (or 0 dB)**

Directivity is a function of angle, but the angular variation is described by its radiation pattern.

Defination 1: The *ratio* of the radiation intensity in a given direction from the antenna to the radiation intensity averaged in over all directions.

$$D = \frac{U}{U_o}$$

Defination 2: The *average radiation intensity* is equal to the total power radiated by the antenna divided by 4π.

$$\therefore D = \frac{U}{P_{rad}\ /4\pi}$$

$$\therefore D = \frac{4\pi . U}{P_{rad}}$$

Defination 3: The **directivity of non-isotropic source** is equal to the ratio of its radiation intensity in a given direction over that of an isotropic source.

$$D = \frac{U}{U_o}$$

Defination 4: If the direction is not specified, it implies the direction of maximum radiation intensity (maximum directivity)

$$D_{max} = D_o = \frac{U_{max}}{U_o} = \frac{U_{max} . 4\pi}{P_{rad}}$$

Where, D = Directivity (dimensionless)

D_o = Max. Directivity (dimensionless)

U = Radiation Intensity (W/unit solid angle)

U_{max} = Maximum Radiation Intensity (W/unit solid angle)

U_o = Radiation Intensity of isotropic source (W/unit solid angle)

P_{rad} = Total Radiated Power (W)

More on Directivity

As for antennas normalized radiation pattern can be written as a function in spherical coordinates: $F(\theta, \phi)$

Radiation pattern is normalized; peak value of F over complete range of angle is 1. Mathematically, the formula for directivity (D) is written as :-

$$D = \frac{1}{\frac{1}{4\pi}\int_0^{2\pi}\int_0^{\pi}|F(\theta, \phi)|^2 \sin\theta d\theta d\phi}$$

Example: Let us consider, one antenna with radiation pattern - $F(\theta, \phi) = \sqrt{\sin(\theta)}$ and the other antenna with radiation pattern - $F(\theta, \phi) = (\sin\ \theta)^5$.

Patterns are plotted for both the antennas as given in figure (a) and figure (b) respectively. Here patterns are functions of the polar angle `θ', and not a function of azimuth angle.

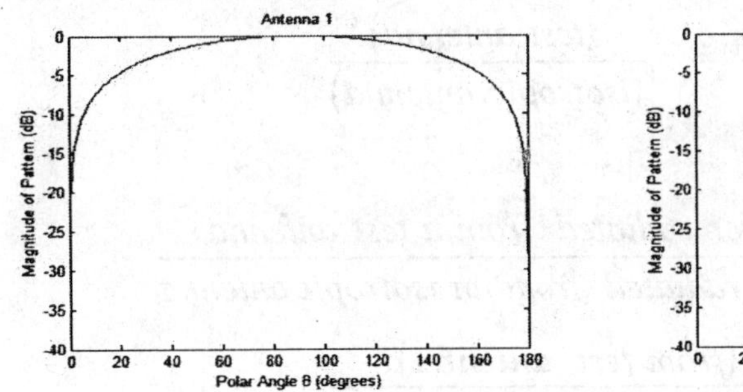

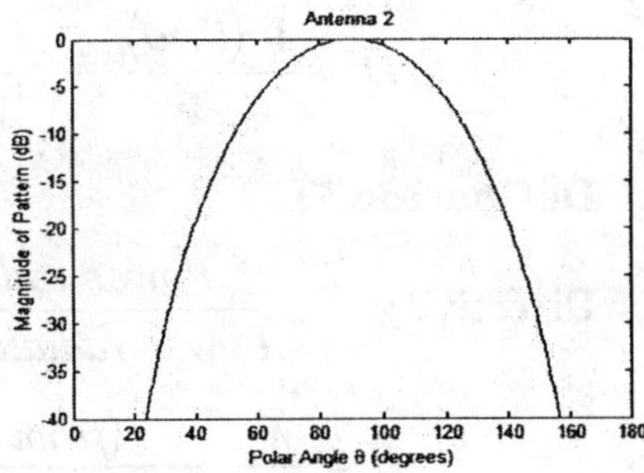

Figure 1.62. Radiation pattern - 2 D

Radiation pattern for antenna 1 is less directional then the 2 antenna.

Antenna 1 - Directivity 1.273 (1.05 dB)

Antenna 2 - Directivity 2.707 (4.32 dB)

Increased directivity implies a more `focused' antenna. (∴ Now we can say that, Antenna 2 receives 2.707 times more power in its peak direction than an isotropic antenna would receive.)

NOTE: Antennas for mobile phones should have *a low directivity* because the signal can come from any direction, and antenna should receive them

NOTE: Antennas for **satellite dish** have a very *high directivity* as they have to receive signals from a fixed director along with a greater distance.

Antenna Type	Typical Directivity	Typical Directivity (dB)	
Short Dipole	1.5	1.76	Non-directional
Half Wave Dipole	1.64	2.15	Less-directional
Patch (Microstrip) Antenna	3.2-6.3	5-8	Directional
Horn Antenna	10-100	10-20	Directional
Dish Antenna	10-10,000	10-40	Highly Directional

The maximum directive gain is called as directivity.

Defination 5: Ratio of maximum intensity to average radiation intensity-

$$D = \frac{\Phi(\theta, \phi)_{max}}{\Phi_{av}} \quad (both\ of\ test\ antenna)$$

Defination 6: Ratio of maximum radiation intensity of the subject antenna to the radiation intensity of an isotropic or reference antenna radiating the same total power

$$D = \frac{\Phi(\theta. \phi)_{max}}{\Phi_o} \quad \frac{(test\ antenna)}{(isotropic\ antenna)}$$

Defination 7:

$$\text{Directivity} = \frac{Power\ radiated\ from\ a\ test\ antenna}{Power\ radiated\ from\ an\ isotropic\ antenna}$$

$$D = \frac{W'}{W''} \quad \frac{(from\ test\ antenna)}{(from\ an\ isotropic\ antenna)}$$

$$\because \text{Average radiation intensity } (\Phi_{av}) = \frac{W}{4\pi}$$

$$\therefore D = \frac{\Phi(\theta, \phi)_{max}}{W/4\pi} = \frac{4\pi\Phi(\theta, \phi)_{max}}{W}$$

$$\therefore D = \frac{4\pi \text{ maximum radiation intensity}}{\text{Total radiated power}}$$

$$D(\theta, \phi) = \frac{\Phi_{max}}{\Phi}.D$$

ϕ_{max} = maximum Radiation Intensity

ϕ = Radiation intensity in direction of (θ, ϕ)

Directivity in decibels = D (dB) = 10 $\log_{10}$ D

$$D = \frac{4\pi}{B.A} \quad \text{and} \quad D = \frac{4\pi}{\Omega_A}$$

> 1. Explain the following terms-
>
> (i) Directivity of an antenna. **(Raj. Univ. 2004)**

1.10.12 GAIN

(It can be also well said as figure of merit)

Defination 1: The ability of an antenna or antenna system to concentrate the radiated power in a given direction or conversely to absorb effectively the incident power from the direction is specified by-

1. Antenna Gain
2. Directive Gain
3. Directivity
4. Simply Gain
5 Power Gain

As the antenna is a passive element, thus gain of an antenna is

not $\neq \dfrac{Output}{Input}$

Passive - System or device which functions only when acted upon by an external signal.

Active - System or device which generates its own signal.

$\therefore$ **Gain is a relative term in which actual antenna is compared with the reference antenna.** (Hypothetical / Isotropic antennas, radiates in all the directions).

Defination 2: It is closely related to the directivity and is a measure that takes into account the efficiency of the antenna as well as its directional capabilities.

(1) **Absolute gain** - The ratio of intensity in a given direction to the radiation intensity that would be obtained if the power accepted by the antenna were radiated isotropically.

The radiation intensity corresponding to the isotropically radiated power is equal to the power accepted (input) by the antenna divided by 4 π.

$$Gain = 4\,\pi \frac{\text{radiation intensity}}{\text{total input (accepted) power}}$$

$$Gain = 4\,\pi \frac{U(\theta, \phi)}{P_{in}} \quad \text{Dimensionless}$$

(2) **Relative gain** - The Ratio of the power gain in a given direction to the power gain of a reference antenna in its referenced direction. Power input should be same for both antennas.

Reference antennas are those whose gain is known, generally lossless isotropic antennas. (E.g. Dipole/ Horn)

$$Gain = \frac{4\,\pi\, U(\theta, \phi)}{P_{in}} \quad (\text{dimension less})$$

(Lossless isotropic source)

NOTE: When direction is not defined - power gain is used.

Defination 3. This Defination is favored for microwave antenna

The Ratio of maximum radiation Intensity in a given direction to the maximum radiation intensity from a reference antenna produced in the same direction with same power input.

$$\text{G} = \text{Gain} = \frac{\text{Maximum Intensity from Test antenna}}{\text{Maximum Intensity from Reference antenna (with same power input)}}$$

Gain of subject antenna (G_o) and is known as Gain with respect to, isotropic antenna-

$$\therefore G_o = \frac{\phi'_m}{\phi_o}$$

ϕ'_m = Maximum Radiation from test antenna

ϕ_o = Radiation from lossless / Reference antenna

NOTE: (Defination 4)

1) **As gain denotes concentration of energy, the high values of gain are associated with narrow beam width.** $\therefore$ Gain is closely related to Directivity (D)

2) When antenna efficiency is 100%, that is, antenna is without any internal losses, or when efficiency is cent percent then gain (G) and Directivity (D) are equal/or can be interchangeably used.

   Gain = Directivity

3) Although **gain** can be given in any direction from antenna but generally it **is in the direction of maximum gain**.

$$G_o\ (\theta, \phi) \cdot \frac{\phi_m}{\phi} = G_o$$

The gain in a direction for which Radiation Intensity 'φ' is not maximum may be designated by specifying the angle 'ϕ' at which it is measured. In general G_o (θ, ϕ) is utilized for this purpose.

where θ_m = Maximum Radiation Intensity

φ = Radiation Intensity in the direction of θ and ϕ.

Angle θ = Angle of Co-latitude

Angle ϕ = Polar angle.

Defination 5: Power radiated by antenna is more in one direction than other. For the same antenna direction of maximum radiation is greater than, that produced by an isotropic radiator emitting the same total power. Directivity is represented by the shape of its radiation pattern or polar diagram, while *the increased signal in a particular direction indicates that the practical antenna has a gain in that particular direction.* For all the power available is deflected and radiated in that particular direction and thus, there is a gain in that direction.

∴ Signal power received by a receiver at a distant point in the direction of maximum radiation, for any antenna the gain can be defined as-

$$\text{Gain (G)} = \frac{\textit{Max. Power received from given antenna}\ (P_1)}{\textit{Max. Power received from ref. antenna}\ (P_2)}$$

(for same input power).

$$G = \frac{P_1}{P_2}$$

Where, the signal power is proportional to the square of the radiation field at the point.

Defination 6: Let certain amount of power is supplied to an isotropic antenna, which say can produce a certain field strength (E) at a given distance from antenna.

If this isotropic antenna is replaced by any practical antenna under the same condition, then practical antenna will also radiate the same power. But now intensity of radiations in many directions are smaller than the field strength produced by isotropic antenna at the same distance and vice versa.

$$\therefore\ G_0 = \frac{E_1}{E_2} = \frac{\text{field strength at given direction in front}}{\text{field strength at given front direction of isotropic antenna}}$$

Gain in dB = 10 $\log_{10}$ G

(a) DIRECTIVE GAIN

Defination 1: All antennas concentrate its radiated energy to more or less extent in certain preferred directions. **The extent to which practical antenna concentrates its radiated energy relative to that of some standard antenna is termed as directive gain.**

Defination 2: It is also given by the Ratio of Radiation Intensity in a given direction to the average radiated power.

Where,

$\Phi(\theta, \phi)$ = Radiation intensity in a particular direction

Φ_w = Average Radiation Intensity in that direction = $\dfrac{W_r}{4\pi}$

$$\therefore\ \textit{Directive gain} = \frac{\textit{Radiation Intensity in particular direction}}{\textit{Average Radiated power}}$$

$$G_d(\theta, \phi) = \frac{\Phi(\theta, \phi)}{\Phi_{av}} = \frac{\Phi(\theta, \phi)}{W_r/4\pi}$$

$$= \frac{4\pi\,\Phi(\theta, \phi)}{W_r} \text{ or}$$

$$G_d(\theta, \phi) = \frac{4\pi\,\Phi(\theta, \phi)}{\int \Phi\, d\Omega}$$

$$G_d(in\ dB) = 10 \log_{10} G_d(\theta, \phi)$$

Defination 3: It is qualitative measure of the, **extent to which the total power radiated can be concentrated in one direction.**

NOTE: It does not depend on the power input to antenna, antenna loss or power consumed in terminating resistance.

Defination 4: The Ratio of power density (that is, Poynting vector) in particular direction at a given distance to the power density that would be radiated at some distance by isotropic antenna, radiating the same total power.

1. Define the following terms regarding Antenna:
i) Directive Gain **(Raj. Univ. 2006, 2008)**

(b) POWER GAIN

Defination 1: It compares two power densities of actual and isotropic antenna on the assumption that both are radiating *at the same total power;* or on second assumption that both antenna have *same input power.*

Therefore, it can be given by-

$$G_p = \frac{\text{Power density radiated in particular direction by subject antenna}}{\text{Power density radiated in that direction by an isotropic antenna}}$$

$$G_p = G_d$$

$$where\ \eta = Efficiency\ \ factor\ lies\ \ between\ 0\ to\ 1$$

$$where\ \eta = 1$$

Means the power gain is equal to dielectric gain at VHF and UHF, and is given by-

$$G_P = \frac{\text{Radiation intensity in a given direction}}{\text{Average total input power}}$$

$$= \frac{\Phi(\theta, \phi)}{\frac{W_T}{4\pi}}$$

$W_T = W_r + W_l$ = Total input power

$$G_p = \frac{4\pi\,\Phi(\theta, \phi)}{W_T}$$

On the other hand power gain can also be given as-

$$G_p = \frac{\text{Power input supplied to subject antenna in the direction of maximum radiation}}{\text{Power input supplied to reference antenna (for same field strength at the same given point)}}$$

$$G_p = \left(\frac{V_1}{V_2}\right)^2$$

$$G_p\ (dB) = 10 \log_{10} G_p$$

$$\text{Or } G_p\ (dB) = 10 \log_{10} \left(\frac{V_1}{V_2}\right)^2 = 10 \times 2 \log_{10} \left(\frac{V_1}{V_2}\right)$$

$$G_p\ (dB) = 20 \log_{10} \left(\frac{V_1}{V_2}\right)$$

$$G_p\ (dB) = 10 \log_{10} \frac{P_1}{P_2}$$

Power gain depends on-

(1) Sharpness of lobe

(2) Volume of solid radiation pattern

NOTE: A gain of 3dB means that the power received far from the antenna will be 3dB (twice as much) higher than what would be received from a lossless isotropic antenna with the same input power.

Directive Gain or Power Gain:

$$G_d = \frac{4\pi\,\Phi(\theta, \phi)}{W_r}$$

and

$$G_p = \frac{4\pi\,\Phi(\theta, \phi)}{W_T}$$

Where, $W_T = W_r + W_l$

$$\frac{G_d}{G_p} = \frac{W_T}{W_r}$$

Relation between

Directive Gain and Directivity can be given by-

$$G_d = \frac{4\pi\,\Phi(\theta, \phi)}{W_r}$$

$$D = \frac{\Phi(\theta, \phi)_{max}}{\Phi_{av}} \frac{(test\ antenna)}{(test\ /\ isotropic\ antenna)}$$

Again as we know that, for Loss less isotropic antenna -

Directive Gain = Directivity

$$G_d = D$$

1.10.13 RADIATION RESISTANCE

The antenna is a radiating device, in which the power (that is,, energy per unit time) is radiated into space in the form of electromagnetic waves. Power dissipation is given as- $W' = I^2R$

If it is assumed that all this power appears as Electromagnetic Wave or radio wave, then the power (W') can be divided by square of the current (I^2) that is- $R_r = \frac{W'}{I^2}$

At the point where it is fed to the antenna and obtain a **fictious resistance** called as **radiation resistance**; it is denoted by R_r or R_a or R_o. It represents a relation between total energy radiated from a transmitting antenna and current flowing in antenna.

The radiation resistance (R_r) can be thus defined as – "The fictious resistance which, when substituted with the antenna, will consume the some power as is actually radiated".

Energy supplied to an antenna is dissipating in the form of electromagnetic waves, as ohmic losses (dielectrics, insulators, ground and other surrounding objects).

Though the radiation from antenna is desirable & useful phenomena but still as far as antenna is concerned, it represents nothing but a loss just as the energy used in heating the wire of the antenna (that is, ohmic loss) is a loss.

Dissipated power = I^2R.

Total power loss = Ohmic loss + Radiation loss

$$W = W' + W''$$

$$W = I^2 R_r + I^2 R_l \qquad \text{Or } W = I^2 (R_r + R_l)$$

If $R = R_r + R_l$, then $W = I^2 R$.

However, for half wavelength antenna the power (W'') lost as heat is very less, only few percent of the total supplied power to the antenna; because radiation resistance of antenna is very less.

The value of radiation resistance depends on-

(1) Configuration of antenna.

(2) Point where R_r is considered

(3) Location of antenna with respect to the ground & other objects.

(4) **Corona discharge** - luminous discharge round the surface of antenna due to ionization of air etc.

1.10.14 RECIPROCITY THEOREM

This theorem was given by, Rayleigh Helmholtz and was generalized by, J. R. Carson. Thus it is also called as **Rayleigh Reciprocity theorem.**

Reciprocity is one of the most useful (and fortunate) property of antennas. Reciprocity states that the received and transmitted properties of an antenna are identical. Hence, antennas do not have distinct transmit and receive radiation patterns.

Statement

If an ***e.m.f.*** is applied to the terminals of an antenna no. 1 and the current measured at the terminals of another antenna no. 2 then an equal current both in amplitude and phase will be obtained at the terminals of antenna no. 1, if the same emf is applied to the terminals of antenna no. 2.

OR

If a ***current 'I'***, at the terminals of antenna no. 1 induces an emf E_{21} at the open terminals of antenna No. 2 induces an emf E_{12} at the open terminals of antenna no. 1, then-

$$E_{12} = E_{21}, \text{ provided that } I_1 = I_2$$

NOTE: Reciprocity only applies for the antennas operating in a linear medium made of linear materials (magnetic materials that exhibit hysteresis are not linear).

Assumption

(1) **E. m. f.** is of same frequency

(2) **Medium** between the two antennas are linear, passive and isotropic.

(3) **Generator** producing emf and the ammeter for measuring the current have zero impedance or if not, then both the generator and ammeter impedance are equal.

Explanation

(1) A transmitter with frequency `f' and zero impedance is connected to the terminals of antenna no. 2, generating a current I_2 and inducing an emf 'E_{12} 'at the open terminals of antenna no. 1.

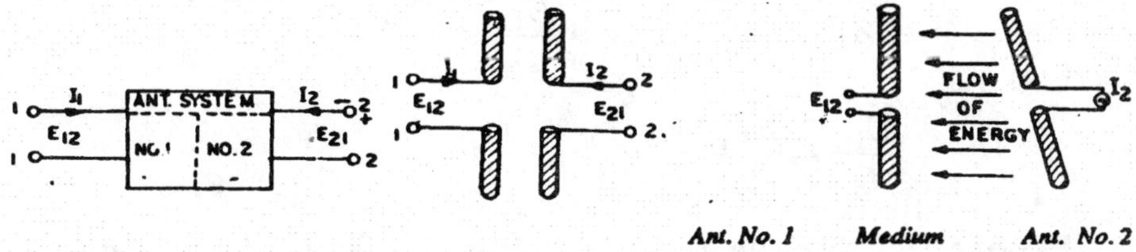

Figure 1.63 Current I_2 inducing and emf E_{12} in antenna no. 1

(2) Now the same transmitter is transferred to antenna no. 1 which is generating a current 'I' and inducing a voltage E_{21} at the open terminals of antenna no. 2.

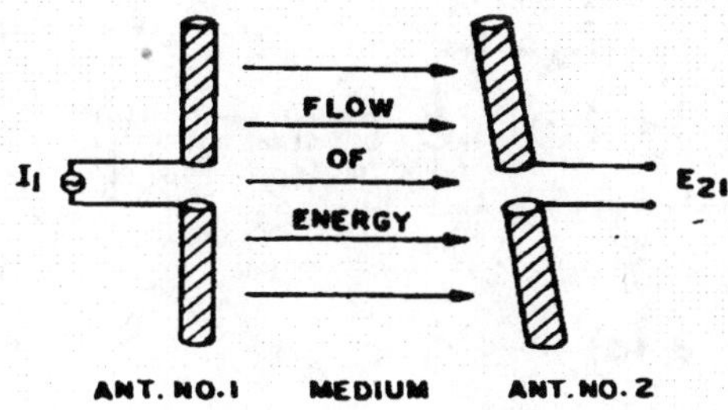

Figure 1.64 Current I_1 inducing and emf E_{21} in antenna no. 2

Thus according to the statement of Reciprocity Theorem provided-

$I_1 = I_2$ Provided E_{12} = E21

∵ Ratio of emf to current is impedance-

$$\therefore \frac{E_{12}}{I_2} = Z_{12} \quad \text{- Transfer Impedance (Case I)}$$

and $$\frac{E_{21}}{I_1} = Z_{21} \quad \text{- Transfer Impedance (Case II)}$$

The ratio of voltage (E_1) of one circuit to the current I_2 in the second circuit is defined as transfer impedance Z_T or Z_{12} that is, $Z_T = Z_{12} = \frac{E_1}{I_2}$

Thus from the reciprocity it follows that the two ratios that is, two impedance are equal and is given by- $E_{12} = E_{21}$

This, of course, is nothing but **mutual impedance (Z_m) between the two antennas.**

$$\therefore Z_m = Z_{12} = Z_{21} = \frac{E_{12}}{I_2} = \frac{E_{21}}{I_1}$$

or
$$\frac{E_{12}}{I_2} = \frac{E_{21}}{I_1}$$

Proof : To prove the Reciprocity theorem for antennas, the space between the antenna no. 1 and antenna no. 2 are replaced by the given network of linear, passive and bilateral impedances are shown in given figures below-

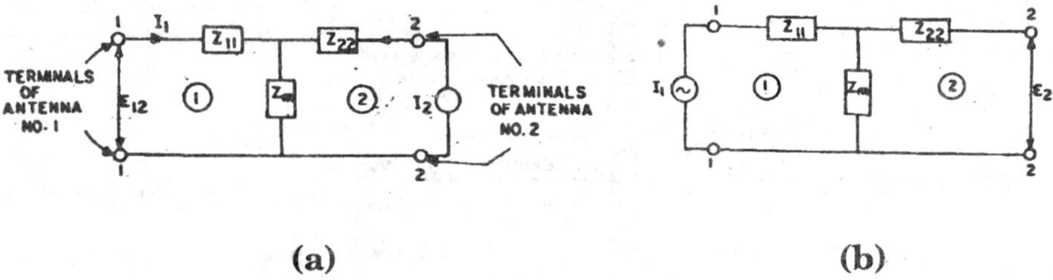

(a) (b)

Figure 1.65. Current I_1 inducing and emf E_{21} in antenna no. 2

Where, Z_{11}, Z_{12} = self impedance of antenna no. 1 and 2 respectively

Z_m = mutual impedance between two antennas

1, 1 = Terminals of antenna no. 1

2, 2 = Terminals of antenna no. 2

Now by applying Kirchhoff's mesh law to (4) figure (a)-

From Loop (2)

$$(Z_{22} + Z_m) I_2 - Z_m I_1 = 0 \quad \text{or} \quad I_2 = I_1 \frac{Z_m}{Z_{22} + Z_m} \qquad (1)$$

From Loop 1

$$(Z_{11} + Z_m) I_1 - Z_m I_2 = E_{12}$$

From Eq^n (1)

$$(Z_{11} + Z_m) I_1 - Z_m \left(\frac{Z_m I_1}{Z_m + Z_{22}} \right) = E_{12}$$

$$I_1\left[\left(Z_{11}+Z_m-\frac{Z_m^2}{Z_m+Z_{22}}\right)\right]=E_{12}$$

$$I_1\left[\frac{Z_{11}Z_m+Z_m^2+Z_{11}Z_{22}+Z_mZ_{22}-Z_m^2}{Z_m+Z_{22}}\right]=E_{12}$$

$$\text{or } I_1\left[\frac{Z_{11}Z_{22}+Z_m(Z_{11}Z_{22})}{Z_m+Z_{22}}\right]=E_{12}$$

$$I_1=\frac{E_{12}(Z_m+Z_{22})}{Z_{11}Z_{22}+Z_m(Z_{11}+Z_{22})} \tag{2}$$

Now on substituting the value of I_1 from equation (2) in equation (1) we have-

$$I_2=\frac{E_{12}(Z_m+Z_{22})}{Z_{11}Z_{22}+Z_m(Z_{11}+Z_{22})}\cdot\frac{Z_m}{(Z_m+Z_{22})}$$

$$\therefore I_2=\frac{E_{12}(Z_m)}{Z_{11}Z_{22}+Z_m(Z_{11}+Z_{22})} \tag{3}$$

Similarly from the figure (b), the value for I_1 obtained is-

$$I_1=\frac{E_{21}(Z_m)}{Z_{22}Z_{11}+Z_m(Z_{22}+Z_{11})} \tag{4}$$

Now on comparing the equation (3) and (4), we find that all the values are same except E_{12} and E_{21} and I_1 and I_2.

Now according to the Reciprocity theorem if $E_{12} = E_{21}$, then $I_1 = I_2$, and by applying the same condition, that is-

$I_1 = I_2$

We get $\dfrac{E_{12}(Z_m)}{Z_{11}Z_{22}+Z_m(Z_{11}+Z_{22})}=\dfrac{E_{21}(Z_m)}{Z_{11}Z_{22}+Z_m(Z_{11}+Z_{22})}$

After all cancellation, we are left with the term- $E_{12} = E_{21}$

Hence theorem is proved.

Advantage

From the Reciprocity theorem-

(1) Directivity Pattern

(2) Aperture

(3) Terminal Impedance of antenna can be proved same whether it is transmitting antenna or receiving antenna.

NOTE: However the current distribution in receiving antenna is not same as in transmitting antenna. Further Reciprocity theorem is equally applicable to, two separate antennas and also for two point, of the same antenna.

Limitations

(1) Although the **Rayleigh-Carson theorem (Reciprocity)** is applicable to Radio communication but it *fails to be true only when the propagation of Radio wave is effected by the presence of Earth's magnetic field.*

(2) It holds well for all practical radio work but *it fails for long distance communication through ionosphere.*

Voltage and currents Relation

The circuit theory concept of Rayleigh Carson theory can be used directly to deduce few terminal properties of antennas. Let us consider the figure 1 (a) and 1 (b) in which an antenna system has two ``terminal pairs" 1-1 and 2-2. As shown in figure 1 (b) the two terminals pairs are of two dipole antennas.

Now from the circuit theory, the voltage and currents, for linear system, are given by the simultaneous equation as-

$$E_1 = Z_{11} I_1 + Z_{12} I_2 \tag{5}$$

$$E_2 = Z_{21} I_1 + Z_{22} I_2 \tag{6}$$

Where, Z11, Z_{22} - self impedances of antennas or terminals 11 and 22

Z_{12}, Z_{21} - Mutual Impedances between antennas

I_1, I_2 - currents flowing in antenna no. 1 and 2.

Now according to the Reciprocity theorem, an emf E_{12} is induced at open terminals of antenna no. 1 (that is, $I_1 = 0$) by current I_2 of antenna no. 2. Applying this condition to equation (5)

$$E_1 = E_{12} = Z_{11}.\ 0 + Z_{12}.\ I_2$$

$$Z_{12} = \frac{E_{12}}{I_2}$$

Similarly, the second condition is that an emf E_{21} is induced at the open terminals of antenna no. 2 (that is, $I_2 = 0$) by the current I_1

$$E_2 = E_{21} = Z_{21} I_1 + Z_{22} . 0 \qquad \text{or} \qquad Z_{21} = \frac{E_{21}}{I_1}$$

But as –

$$Z_{12} = Z_{21}$$

$$\frac{E_{21}}{I_1} = \frac{E_{12}}{I_2}$$

Thus, **currents and voltages at the terminals of two antennas satisfy the Reciprocity Theorem.**

Application of Reciprocity Theorem

The reciprocity theorem may be used to derive the following very important properties of transmitting and receiving antennas.

(1) Equality of Directional Pattern

The directional patterns of transmitting and receiving antennas are identical if all the media are linear, passive, and isotropic.

(2) Equality of Directivities

(3) Equality of Effective lengths

(4) Equality of Antenna Impedance

1. Give the statement of Reciprocity Theorem and prove it with diagrams. **(R.T.U. 2009)**

1. Define and prove reciprocity theorem and also relation between voltage and current. **(Raj. Univ. 2008)**
2. Explain and prove reciprocity theorem and application of the theorem. **(Raj. Univ. 2007)**
3. What is reciprocity theorem? What is practical application of this theorem? What is difference between antenna impedance and radiation resistance? **(Raj. Univ. 2003)**

1.10.15 EFFECTIVE AREA / EFFECTIVE APERTURE / CAPTURE AREA

A transmitting antenna transmits electromagnetic waves and receiving antenna receives a fraction of the same. It is a useful parameter for calculating the receive power of an antenna.

Defination (1): An effective area or aperture over which it extracts electromagnetic energy from travelling electromagnetic waves.

Defination (2): Effective aperture parameter describes how much power is captured from a plane wave.

Defination (3): The ratio of power received at the antenna load terminal to the Poynting Vector or power density (W/m²) of the incident wave.

$$\text{Effective Area} = \frac{\text{Power Received}}{\text{Poynting vector of incident wave}}$$

$$A_e = \frac{W}{P} \approx A_e \quad \text{or} \quad W = P A_e$$

where W = Received power, (Watts)

P = Poynting vector of incident plane wave, (W/m^2)

A = Effective Area, (m^2)

This area effects in the losses intrinsic to the antenna (ohmic loss, dielectric loss etc.)

Let a receiving antenna be placed in the field of plane polarized travelling waves, having an effective area 'A' and the receiving antenna (dipole) , terminated at a load impedance, Z_L.

where $Z_L = R_L + j\ X_L$

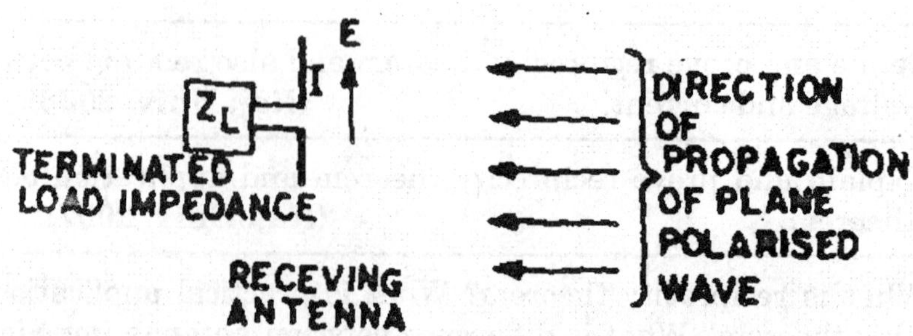

Figure 1.66 Receiving antenna in the field of plane polarized wave.

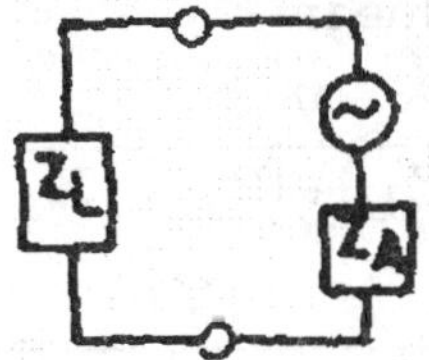

Figure 1.67 Equivalent circuit of the previous figure.

If, I = Terminal current, then Received Power,

$$= W = I_{rms}^2 R_L$$

Where, W = Received power.

R_L = Load resistance

I_{rms} = Terminal r.m.s. current.

$$A = \frac{W}{P} = \frac{I_{rms}^2 R_L}{P}$$

∴ the antenna extracts energy from incident electromagnetic waves, delivers the same terminated load impedance 'Z_L' and power flowing per m² or Poynting Vector is P (W/m²). This entire circuit can be replaced by an equivalent circuit as above figure, according to Thevenin's theorem.

V = Equivalent Thevenin's voltage.

Z_A = Equivalent Thevenin's impedance.

The voltage 'V' is induced by passing electromagnetic waves it produces current I_{rms} through terminal load impedance 'Z_L'.

$$I_{rms} = \frac{\text{Equivalent voltage}}{\text{Equivalent impedance}}$$

$$I_{rms} = \frac{V}{Z_L + Z_A} \text{ Ampere}$$

Where, $Z_A = R_A + j\,X_A$ = complex antenna impedance and

$$R_A = R_r + R_l = R_r$$

Where, R_r= Radiation Resistance

R_l=Loss Resistance

If R_L=0 (assumed)

Now putting these value in-

$$I_{rms} = \frac{V}{(R_L + jX_L) + (R_A + jX_A)}$$

$$|I_{rms}| = \frac{|V|}{\sqrt{(R_A + R_L)^2 + j(X_A + X_L)^2}}$$

$$|I_{rms}| = \frac{|V|}{\sqrt{(R_L + R_r + R_l)^2 + j\,(X_L + X_A)^2}}$$

Where, X_L = Load Resistance

X_A = Antenna Reactance

As the power received by terminal load impedance is-

$$W = I_{rms}^2 \cdot R_L$$

Or $$W = \frac{V^2 . R_L}{(R_L + R_A)^2 + (X_L + X_A)^2}$$

This is the **power delivered by an antenna at the terminating load** impedance'Z_L'. Now by the defination of effective aperture 'A_e'-

$$A_e = \frac{\text{Power received (W)}}{\text{Poynting vector (P)}} \qquad m^2 \text{ or } \lambda^2$$

$$A_e = \frac{V^2 R_L}{[(R_L + R_A)^2 + (X_L + X_A)^2]P} \qquad m^2 \text{ or } l^2$$

or, $$A_e = \frac{V^2 R_L}{[(R_r + R_L + R_l)^2 + (X_L + X_A)^2]P}$$

This expression is for effective area or aperture or capture area in terms of power density (Poynting vector), induced voltage (V) antenna impedance (Z_A) and terminating load impedance (Z_L).

The induced voltage 'V' is maximum when the antenna is oriented for maximum response and; the antenna and the incident wave both have the same polarization.

$$R_A = R_l + R_r$$

Where, R_l = mismatch between antenna.

Z_L = Antenna losses.

According to **Maximum Power Transfer** theorem, maximum power will be transferred from antenna to the antenna terminating load if , and only if-

$$X_L = -X_A$$

and $R_L = R_A = R_r + R_l$

or $R_L = R_r$ if $R_L = 0$

Thus maximum power received in Antenna, terminating load impedance 'Z_L' can be obtained if-

$$R_L = R_r \quad \text{and} \quad X_L = -X_A$$

Substitute these values in the given equation-

$$W = \frac{V^2 R_L}{(R_L + R_A)^2 + (X_L + X_A)^2}$$

$$\therefore W = \frac{V^2 R_L}{[R_L + (R_l + R_r)]^2 + [X_L + (-X_L)]^2}$$

$$= \frac{V^2 R_L}{(R_L + R_L)^2 + 0}$$

$$= \frac{V^2 R_L}{4 R_L^2} = \frac{V^2}{4 R_L}$$

$$W_{max} = \frac{V^2}{4 R_L} = \frac{V^2}{4 R_r}$$

This is the maximum power received in antenna terminating load impedance 'Z_L'; under the condition of maximum power transfer and without antenna loss and corresponding effective aperture is known as maximum effective aperture.

∴ Maximum Effective Aperture can be given as-

$$(A_e)_{max} = \frac{\text{Maximum received power}}{\text{Power density of incident wave}} = \frac{W_{max}}{\rho}$$

$$(A_e)_{max} = \frac{V^2}{4 R_r . P} \quad m^2 \text{ or } \lambda^2$$

1. Prove maximum effective aperture is- $(A_e)_{max.} = \dfrac{V^2}{4P\,R_r}$ $[\lambda^2 \text{ or } m^2]$ **(Raj. Univ. 2008)**
2. Define the following terms regarding Antenna- (i) Effective Area **(Raj. Univ. 2006)**

Effective Ratio: The ratio of effective area and maximum effective area (α)

$$\alpha = \frac{A_e}{A_e\,(\text{max.})}$$

$0 < \alpha < 1$ [This value is the best percent of efficiency]

The terminated load impedance is generally, not at the antenna terminal, but the load is in the receiver which itself is connected with a transmission line. In such a case the load impedance 'Z_L' is the equivalent impedance which appears across the terminals of the antenna. Thus the power delivered to receiver and load impedance 'Z_L' is same if transmission line is loss less and less if the Transmission line is lossy by an amount equal to loss in the time.

A general relation for the effective aperture in terms of peak gain (G) of any antenna is given by -

$$A_e = \frac{\lambda^2}{4\,\pi}\,.\,G$$

Now, that we understood what is effective aperture, *we will use this concept for calculating received power from a plane wave. For this we have to discuss Friis transmission formula.*

1.10.16 FRIIS TRANSMISSION FORMULA

It is very interesting to understand this concept. Consider two antennas in free space (that is, there is no obstruction nearby) separated by a distance R.

Assume that `$P_T$' watts of total power are delivered to the transmitting antenna. At any instance, assume that the transmit antenna is omnidirectional, lossless and that the receiving antenna is in the far field of transmit antenna. Then the power `P' of the plane wave incident on the receiving antenna at a distance 'R' from the transmitting antenna is given by-

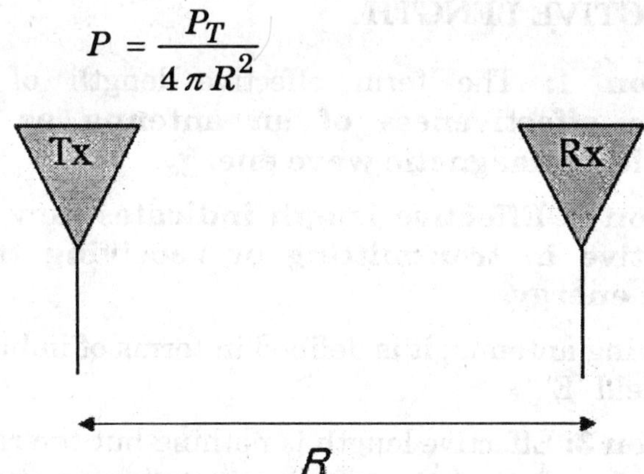

Figure 1.68 Transmit (Tx.) and Receive (Rx.) Antennas separated by R.

It the transmit antenna has a gain in the direction of the receiving antenna given by G_T, then the power equation above becomes-

$$P = \frac{P_T}{4\pi R^2} \cdot G_T$$

The gain term factors in the directionality and losses of a real antenna. Assume now that the receive antenna has an effective aperture given by A_{ER}. Then the power received by this antenna (P_R) is given by-

$$P_R = \frac{P_T}{4\ \pi R^2}\ G_T \cdot A_{ER}$$

Since the effective aperture for any antenna can also be expressed as-

$$A_e = \frac{\lambda^2}{4\pi} \cdot G$$

The resulting received power can be written as-

$$P_R = \frac{P_T\ G_T\ G_R\ \lambda^2}{(4\pi R)^2}$$

This is known as the Friis Transmission Formula.

It relates the free space path loss, antenna gains & wavelength to the received and transmits powers. This is one of the fundamental equations in antenna theory.

Finally, if the antennas are not polarization matched, the above received power could be multiplied by the **Polarization Loss Factor (PLF)** to properly account for this mismatch.

1.10.17 EFFECTIVE LENGTH

Defination 1: The term effective length of an antenna represents the **effectiveness of an antenna as radiator or collector of electromagnetic wave energy.**

Defination 2: Effective length indicates how for an antenna is effective in transmitting or receiving the electromagnetic wave energy.

For receiving antenna; it is defined in terms of induced voltage 'V' and incident field `E'.

Defination 3: Effective length is nothing but the ratio of induced voltage at the terminal of the **receiving antenna** under open circuited condition to the incident Electric field Intensity (or strength) E.

$$\therefore \textit{Effective length} = \frac{\textit{Open circuited voltage}}{\textit{Incident field strength (electric)}}$$

$$l_e = \frac{V}{E} \quad \text{m} \quad \text{or wavelength}$$

units V = volts.

E = volts/m.

l_e = m or wavelength

-Where V is induced voltage under open circuited condition at the antenna terminals due to E (uniform exciting field).

As the induced voltage `V' also depends on the effective aperture and hence effective length and effective aperture of an antenna are related to each other as deduced here.

As we know that-

$$A_e = \frac{V^2 R_L}{\left[(R_A + R_L)^2 + (X_A + X_L)^2\right] P} \qquad \because P = \frac{E^2}{2}$$

$$V^2 = \frac{A_e \cdot \left[(R_A + R_L)^2 + (X_A + X_L)^2\right] P}{R_L}$$

$$V = \frac{\sqrt{A_e \cdot \left[(R_A + R_L)^2 + (X_A + X_L)^2\right] E^2}}{\sqrt{Z \ R_L}}$$

$$\text{And as } l_e = \frac{V}{E} = \frac{\sqrt{A_e \cdot \left[(R_A + R_L)^2 + (X_A + X_L)^2\right] E^2}}{\sqrt{Z \ R_L}}$$

Under the conditions for maximum effective aperture, when-

$X_A = -X_L$ and $R_A = R_r + R_l = R_L$

or $R_A = R_r = R_L$ if $R_L = 0$ then,

$$\therefore l_e = \frac{\sqrt{(A_e)_m (2R_r)^2}}{\sqrt{Z R_r}}$$

$$l_e = 2\sqrt{\frac{(A_e)_m R_r}{Z}} \qquad m \text{ or } \lambda$$

$$\therefore (A_e)_{max} = \frac{l_e^2 Z}{4 R_r}$$

This gives relation between `$A_e$' and `l_e'.

Now for transmitting antenna - The effective length is that length of an equivalent linear antenna that has the same current I(c), at all the point along its length and that radiation the same field intensity 'E' as the actual antenna.

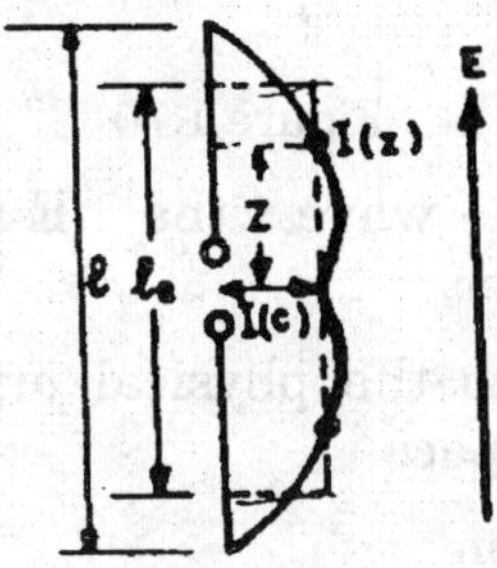

Figure 1.69

Where I (c) = current at the terminals of actual antenna

I (z) = Current at any point Z of the antenna

l_e = Effective length

l = Actual length

$$I(c)\, l_{et} = \int_{-l/2}^{+l/2} I(z)\, dz \qquad I_{let} = \frac{1}{I(c)} \int_{-l/2}^{+l/2} I(z)\, dz$$

$$I_{let} = \frac{2}{I(c)} \int_{o}^{l/2} I(z)\, dz$$

1. Explain any four terms of antenna-
   (a) Antenna effective length **(Raj. Univ. 2007)**

1.10.18 POLARIZATION (in terms of `E')

Defination: Polarization or plane of polarization can be defined by the direction in which the electric vector `E' is aligned during the passage of at least one full cycle.

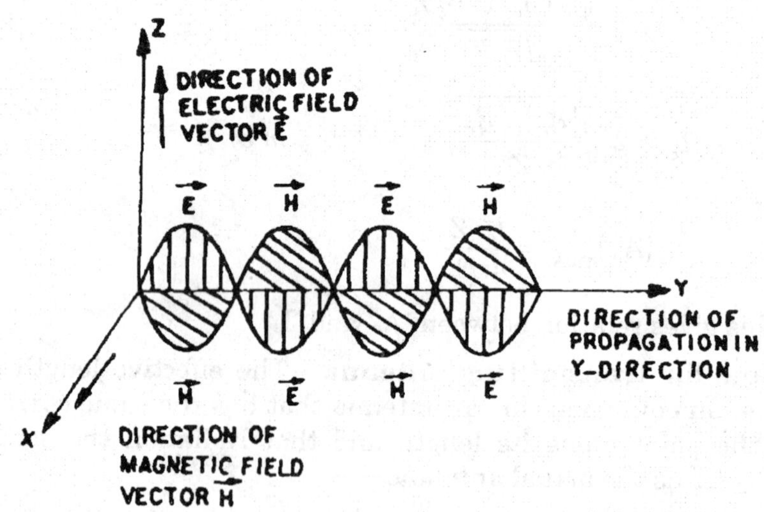

Figure 1.70

In electromagnetic wave the E-field and H-field are perpendicular to each other.

Polarization refers to the physical orientation of the radiated electromagnetic wave in space-

(1) Linear Polarization

(2) Circular Polarization

(3) Crossed Polarization

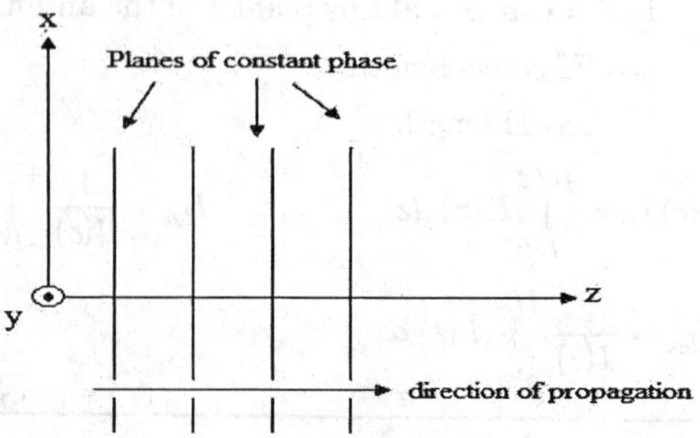

Figure 1.71 Graphical Representation of E-field travelling +z-direction

(1) Linear Polarization

Electromagnetic wave is called so, if they all have same alignment in space. They can once again be characterized as-

(a) **Vertical linear polarization** - If the E-field is vertical or lies in vertical plane, the wave is said to be vertically polarized.

(b) **Horizontal linear polarization** - If the E-field is horizontal or lies in horizontal plane, the wave is said to be horizontal polarized.

(When θ = 0, means angle is null).

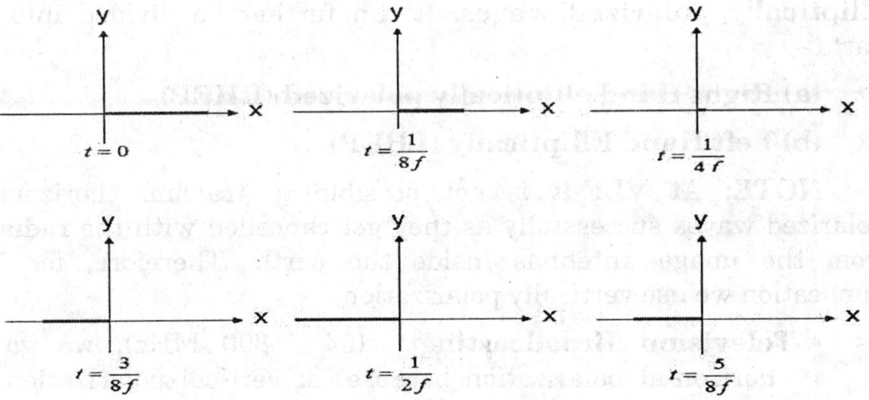

Figure 1.72 Observation of E-field at (x, y, z) = (0, 0, 0) at different times.

(At some angle `θ')

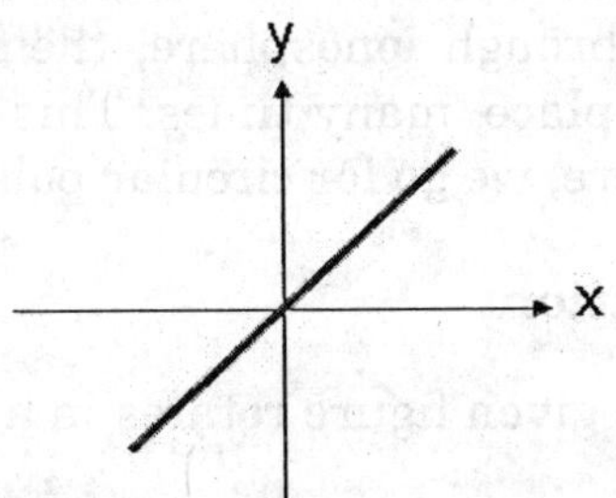

Figure 1.73 Locus of E-field amplitudes foe a linearly polarized wave at an angle

(2) Circular and Elliptical Polarization

It is useful in VHF and UHF Range applications.

If two linearly polarized waves are simultaneously produce in the same direction from the same antenna provided that these, polarization are mutually perpendicular to each other with phase difference of 90° then circularly polarized waves are produced.

They are of two types-

(a) **Right Handed Circularly Polarization:** They depend on the positive phase difference. In this type of polarization the fields rotate in counter clockwise direction.

(b) **Left Handed Circularly Polarization:** They depend on the negative phase difference. In this type of polarization the fields rotate in clockwise direction.

If their amplitudes are not equal then - these waves produces **elliptically polarized waves.** It can further be divided into two parts –

(a) Right Hand elliptically polarized (RHEP)

(b) Left Hand Elliptically (LHEP)

NOTE: At **VLF**-It is not possible to transmit horizontally polarized waves successfully as they get cancelled with the radiation from the image antennas inside the earth. Therefore, for VLF application we use vertically polarization.

- **Television Broadcasting** - (54 - 890 MHz), we go for horizontal polarization because, in vertical polarization the maximum man made noises occurs.
- **VHP/UHF/Microwave** - For minimizing the clutter echoes received from raindrops, or when signals are transmitted or received between the satellite and earth or when signals are transmitted through ionosphere, the rotation of polarization vectors takes place, many times. This on an average remains same. Therefore, we go for circular polarization.

(i) Circular Polarization

The E-field in the given figure rotates in a circle.

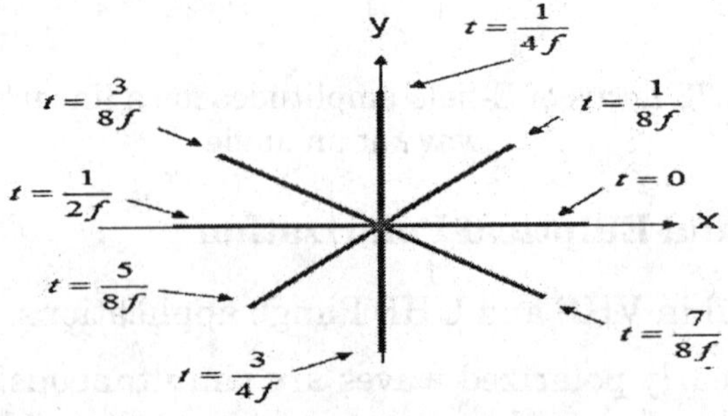

Figure 1.75 E-field strength at (x, y, z) = (0, 0, 0)

To have circular polarization, following **criteria** is to be satisfied-

(a) The E-field must have two orthogonal (perpendicular) components.

(b) The E-field's orthogonal components must have equal magnitude.

(c) The Orthogonal components must be 90 degrees out of phase.

(ii) Elliptically Polarization

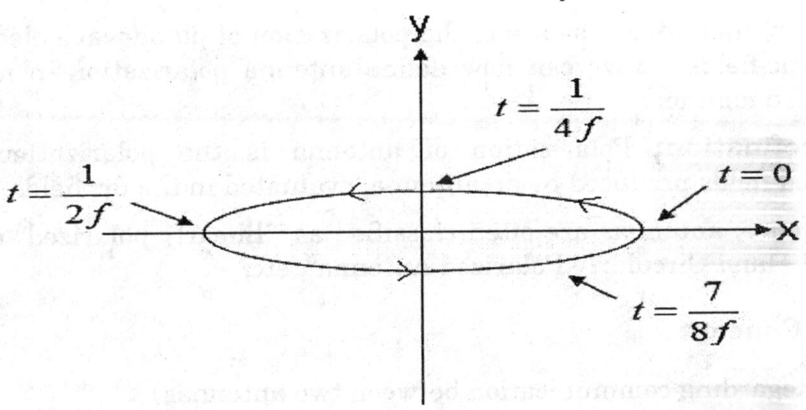

Figure 1.76 Tip of E-field for elliptical polarized wave

-RHCP- movement from x-axis towards y- axis.

-LHCP- movement from y-axis towards x- axis.

NOTE

(a) **Circular polarization and linear polarization are both special cases of elliptical polarization.**

(b) An elliptically polarization wave with an eccentricity of 1 is a circularly polarized wave.

(c) An elliptically polarized wave with an infinite (∞) eccentricity is a linearly polarized wave.

(3) Crossed Polarization

As no antenna can be designed at 100% polarization of any single type hence, two radiation patterns of an antenna are sometimes presented, as the **Co-Pol** (or desired polarization component) radiation pattern and the cross polarization pattern. It is sometimes written as **X- Pol**.

It can be said as **the polarization orthogonal to the polarization.** For instance, if the fields from an antenna are in horizontal polarization then the crossed polarization in this case is vertical polarization. Or say for the other case if the antenna is Right Handed Circularly Polarized (RHCP) then the crossed polarization in this case is Left Handed Circularly Polarized (LHCP).

It can be specified for an antenna as a power level in negative dB, indicating how many decibels below the desired polarization's power level the X-Pol level is.

1.10.19 ANTENNA POLARIZATION

Now that we are aware of the polarization of plane-wave electro-magnetic-fields, so we can now define antenna polarization in more advanced manner.

Defination: Polarization of antenna is the polarization of radiated fields produced by an antenna, evaluated in the far fields.

Hence, antennas are often classified as ``linearly polarized" or a ``Right Hand Circularly Polarized Antenna", etc.

Basic Concept

(Regarding communication between two antennas)

A horizontally polarized antenna will not communicate with a vertically polarized antenna. Due to the reciprocity theorem, antenna transmits and receives in exactly the same manner. Hence a vertically polarized antenna transmits and receives vertically polarized fields. Consequently, if a horizontally polarized antenna is trying to communicate with a vertically polarized antenna, there will be no reception.

For two linearly polarized antennas that are rotated from each other by an angle 'ϕ', the power loss due to this polarization mismatch will be described by the **Polarization Loss Factor (PLF)** is given by-

$$\text{PLF} = \cos^2 \phi$$

∴ *if both the antennas have same polarization, the angle between their radiated E-fields is zero and there is no power loss due to polarization mismatch.*

If one antenna is vertically polarized & the other is horizontally polarized, the angle is 90° & no power will be transferred.

NOTE: This explains why moving the cell phone on your head to a different angle can sometimes increase reception. Cell phone antennas are often linearly polarization, so rotating the phone can often match polarization of the phone and thus increases the reception.

Hence, circular polarization is a desirable characteristic for many antennas. Two antennas that are both circularly polarized do not suffer signal loss due to polarization mismatch.

Antennas used in GPS systems are Right Hand circularly polarized.

What if circularly polarized wave antenna is trying to receive linearly polarized wave or vice versa? What will be the resulting polarization loss factor?

The answer to the above question is that, the circular polarization is really two orthogonal linear polarized waves 90° out of phase. Hence a linearly polarized (LP antenna will simply pick up the in-phase component of circularly polarized (CP) wave. As a result, the LP antenna will have a polarization mismatch loss of 0.5 (-3 dB), no matter what the angle the LP antenna is rotated to. Therefore-

PLF (linear to circular) = 0.5 = -3 dB.

Polarization Loss Factor- is sometimes referred to as polarization efficiency, antenna mismatch factor, or antenna receiving factor.

1. Explain following: i. Antenna Polarization **(R.T.U. 2009)**
1. Explain any four terms of antenna- a) Polarization **(Raj. Univ. 2007)**
1. Define reflection factors for horizontal and vertical polarizations. **(R.T.U. 2009)**

1.10.20 ANTENNA IMPEDANCE

1. Self Impedance

Impedance at point where transmission line carrying R.F. Power from the transmitter is connected is very important, as at this point the input to the antenna is supplied and thus is called as **Antenna Input Impedance**. At this impedance the R.F. power from the transmitter is fed; ∴ it is called as **feed point impedance**. Also at this, impedance, the transmission line operates; ∴ it is called as **Driving or Terminal point impedance.**

Thus, antenna input or feed point or driving point or terminal impedance is of considerable importance because; it is generally desired to supply maximum available power from transmitter to the antenna or to extract maximum amount of received energy from the antenna.

In two terminal networks there is only the impedance offered by antenna to the transmission line.

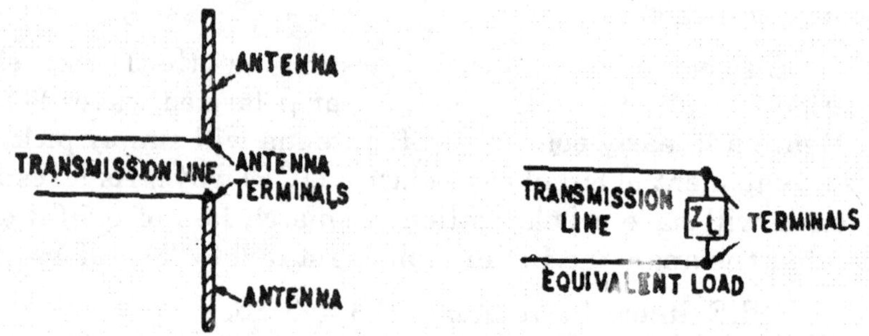

(a) Transmission line with antenna as load. *(b) Transmission line with equivalent load Z_L.*

Figure 1.77

Now if antenna is loss less (heat loss) or isolated (away from ground and other objects) then antenna terminal impedance (Z_L) is same as the self impedance (Z_{11}) is a complex quantity, the real part (R_{11}) of which is given the name of **self-resistance or radiation resistance** and the imaginary part (X_{11}) the self reactance.

However, if the antenna is in nearby, other objects like active antennas etc. then the terminal impedances can still be replaced by a two terminal network but now the terminal impedance gets modified. Besides self-impedance, the mutual impedance also comes in existence due to the currents flowing in neighboring active antennas.

∴ **"Self impedance" of an antenna is defined as its input impedance when all other antennas are completely removed that is, away from it."**

2. Mutual Impedance

Defination: The negative ratio of the voltage induced in one circuit, by a current flowing in the second circuit, to the current in the second circuit with all the circuits open circuited excepting the second circuit.

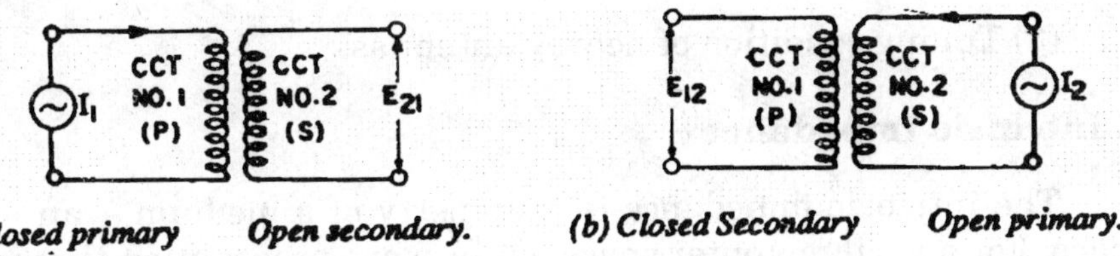

(a) Closed primary Open secondary. *(b) Closed Secondary Open primary.*

Figure 1.78

Thus by defination, mutual impedance is the negative ratio of voltage in open secondary (circuit no. b) E_{21} to the current (I_1) flowing in the closed primary (circuit no. a) or negative ratio of voltage induced in open primary (circuit no. a). E_{12} to current (I_2) flowing in closed secondary (or circuit no. b).

$$Z_{21} = -\frac{E_{21}}{I_1} \quad \text{and} \quad Z_{12} = -\frac{E_{12}}{I_2}$$

Where E_{21} = open circuit voltage across circuit no. b due to circuit I_1 in the closed circuit no. a.

E_{12} = open circuit voltage across circuit no. a due to closed current I_2 in closed circuit no. b.

Mutual impedance is not equal to the transfer impedance.

3. Transfer impedance

Ratio of voltage impressed (and not induced) in one circuit to the current produced in another circuit with all circuits closed.

If 'E' is the voltage impressed in one circuit and current produced in another closed circuit is' I', then the ratio is transfer impedance.

$$Z_T \text{ or } Z_{T_{12}} = \frac{E_1}{I_2} \quad \text{or} \quad Z_T \text{ or } Z_{T_{21}} = \frac{E_2}{I_1}$$

According to Reciprocity theorem-

$$Z_m = Z_{12} = Z_{21} \quad \text{or} \quad Z_m = \frac{E_1}{I_2} = \frac{E_2}{I_1}$$

In *antenna array* the input or driving point impedance, it depends on self as well as mutual impedance.

Mutual impedance depends on-

(1) Magnitude of induced current

(2) Phase relationships between induced and original currents.

(3) Tuning condition of nearby antennas.

4. Intrinsic Impedance

The intrinsic impedance is a property of a medium – an area of space. For a vacuum (outer space) or for wave propagation through the air around earth (often called 'free space'). It is given by-

$$Z_0 = \sqrt{\frac{\mu_0}{\varepsilon_0}} = \sqrt{\frac{4\pi 10^{-7}}{8.85.4e^{-12}}} \approx 120\,\pi \approx 377\,\Omega$$

This parameter is the ratio of the magnitude of the E-field to the magnitude of the H-field for a plane wave in a lossless medium (zero conductivity):

$$Z = \left|\frac{E}{H}\right|$$

This, relation can be derived directly from Maxwell's equations. For a general medium with permittivity and permeability given by (ε, μ) = ($\varepsilon_r\ \varepsilon_0$, $\mu_r\ \mu_0$), the intrinsic impedance is given by-

$$Z = \sqrt{\frac{\mu}{\varepsilon}} = \sqrt{\frac{\mu_0 . \mu_r}{\varepsilon_0 . \varepsilon_r}} \sqrt{\frac{\mu_r}{\varepsilon_r}} Z_0$$

For a medium with a conductivity 'σ' associated with it, the intrinsic impedance is given as-

$$Z = \sqrt{\frac{jw\mu}{jw\varepsilon + \sigma}}$$

When the conductivity is non-zero, the above intrinsic impedance is a complex number, indicating that the electric and magnetic fields are not in-phase.

The intrinsic impedance is independent of the electrical impedance of an antenna. There is no reason to have the impedance of an antenna match the intrinsic impedance of free space (no mismatch loss occurs).

1. Define the following terms regarding Antenna: (i) Antenna Impedance **(Raj. Univ. 2006, 2008)**
1. Explain any four terms of antenna- (a) Antenna impedance **(Raj. Univ. 2007)**

1.10.21 ANTENNA TEMPERATURES: (T_A)

Every object with physical temperature above absolute zero (0°K = –273°C) radiates energy.

For *loss less antenna,* antenna temperature has nothing to do with the physical temperature of the antenna proper but is related to the temperature of distant regions of space (nearer surroundings) coupled to the antenna via radiation resistance.

Defination: It is a parameter that describes how much noise an antenna produces in a given environment.

To define the environment, we'll introduce a temperature distribution - which is temperature in every direction away from the antenna in spherical coordinates.

For example: In the night, sky is roughly 4° Kelvin; the value of the temperature pattern in the direction of Earth's ground is the physical temperature of the Earth's ground. This temperature distribution is written as T (θ, ϕ). Hence an antenna's temperatures vary, depending on whether it is directional and pointed into space.

For an antenna with a radiation pattern, R (θ, ϕ), the noise temperature is mathematically defined as-

$$T_A = \frac{1}{4\pi} \int_0^{2\pi} \int_0^{\pi} R\left(\theta, \phi\right) . T\left(\theta, \phi\right) Sin\ \theta\ d\theta\ d\phi$$

Conclusion: *This states that the temperature surrounding the antenna is integrated over the entire sphere, and weighted by the antenna's radiation pattern. Hence, an isotropic antenna would have a noise temperature that is the average of all temperatures around the antenna; for a perfectly directional antenna (with a pencil beam), the antenna temperature only depend on the temperature in which the antenna is ``looking''.*

Antenna temperature is not so much inherent property of the antenna as it is a parameter that **depends on the temperature of the region's the antenna is ``looking at''.** In this sense, a receiving antenna may be regarded as a remote sensing, temperature measuring device. Both the antenna temperature (T_a) and radiation resistance (R_r) are single valued scalar quantities.

The noise power per unit band-width is available at the terminals of a resister of the resistance 'R' and temperature 'T'.

$$\rho = k.T$$

Where ρ = Power per unit bandwidth (watt/Hertz)

k = Boltzmann's constant - 1.38×10^{-23} (J/K)

T = Absolute temperature of resistor (°K)

Power per unit bandwidth 'ρ' is independent of frequency. The total power (P) is obtained by multiplying by 'B' (Bandwidth).

$$P = k.T.B$$

Where P = Total power, (Watt)

B = Bandwidth, (Hertz)

Power density per unit bandwidth or flux density (S) for the antenna having Effective aperture 'A_e', is-

$$P = S.A_e.B$$

Where, S = Power density / Bandwidth (W/m^2 Hz)

A_e = effective aperture (m^2)

B = Bandwidth in Hz

As $P = S.A_e.B$

$= k.T.B$

$$\therefore S = \frac{kT_A}{A_e} \qquad W/m^2\ Hz$$

Where, T_A = Antenna Temperature due to source in °K.

$$\therefore T_A = \frac{S\,A_e}{k} \qquad °K.$$

Dividing (received power/Bandwidth) by effective Aperture (A_e) of antenna gives flux density 'S'-

$$S = \frac{P}{Ae} = \frac{kT_A}{A_e}$$

$$\therefore T_A = \frac{S\,.\,A_e}{k}$$

Any antenna irrespective of polarization characteristic can receive only half the incident power of unpolarized wave, so the actual flux density should be twice- $S = \dfrac{2\,kT_A}{A_e}$

If celestial source extent 'Ω_s' is small compared to the antenna beam solid angle 'Ω_A'and the antenna beam is aligned with source. If the angular size of the source is small compared to Ω_A and its magnitude is known, then it is possible to determine the source temperature 'T_s', by- $T_A = \dfrac{\Omega_s}{\Omega_A} T_s$

Where, Ω_A = antenna beam solid angle (S_r)

Ω_S = Source solid angle (S_r).

T_A = Antenna noise température

T_s = Source temperature (°K)

In case of receiver itself has a certain noise temperature 'T_r', due to thermal noise in the receiver components the system noise power at the receiver terminals is given by-

$$P_s = k\ (T_A + T_r)\ B_N$$

As $T_{sys} = T_A + T_r$

$$\therefore P_s = k\ T_{sys}$$

Where, P_s = system noise power of receiver terminal

T_A = Antenna noise Temperature

T_{rs} = Receivers noise Temperature

T_{sys} = Effective system noise temperature

B_N = Bandwidth.

1. Explain any four terms of antenna-

a) Antenna temperature **(Raj. Univ. 2007)**

1.10.22 EQUIPMENT NOISE TEMPERATURE OF ANTENNA

Defination: The fictional temperature at the input of the network which would account for the noise ΔN at the output. ΔN is the additional noise introduced by the network itself. Sometimes other parameter noise figure is used.

$$F = 1 + \frac{T_e}{T_o}$$

Where T_e = Noise temperature effective (in °K)

T_o = 290°K = (273 +17) °K

f = Noise figure (dimensionless)

or $$(F - 1) = \frac{T_e}{T_o}$$

or $$T_e = T_o\ (F - 1)$$

Noise figure, in dB is given by-

$$(F)_{dB} = 10 \log_{10} F.$$

1.10.23 FRONT TO BACK RATIO:

The front to back ratio is the parameter used in describing directional radiation patterns for the antennas. If an antenna has a unique maximum direction, **the front-to-back ratio is the ratio of the gain in the maximum direction to that in the opposite direction (180 degrees from the specified maximum direction).** The parameter is usually given in dB.

$$\text{FBR} = \frac{\textit{Power Radiated in desired direction}}{\textit{Power Radiated in opposite direction}}$$

1.10.24 ANTENNA FACTOR

This parameter plays an important role in R.F. or E.M.C. antennas. It is used to describe the required electric field strength which produces 1 Volt at the terminal of an antenna, and is given by-

$$A_F = \frac{E_{incident}}{V_{Received}}$$

Antenna factor has implied impedance associated with it $\approx 50\Omega$.

Limitations of antenna factor-

- If the antenna terminals are shorted ends, then the received voltage is always zero, hence the antenna factor cannot be defined for the same.

Hence it is also given by the "open circuited", antenna factor; available voltage for an antenna with an open circuit (no receiver or matched load attached). The basic concept of the antenna factor with a terminal (port, receiver or load) impedance is given below-

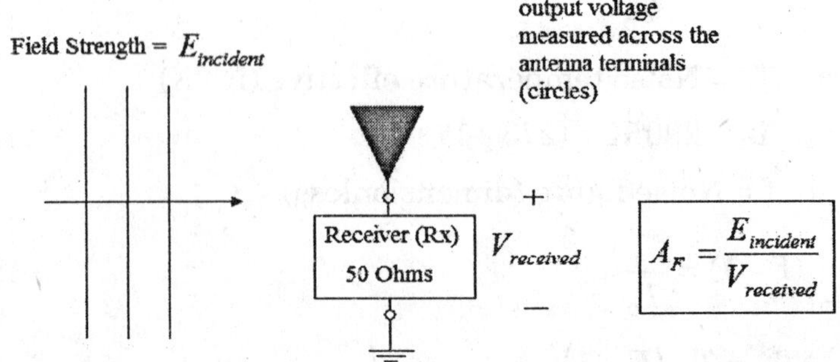

Figure 1.79 Graphical illustration of antenna factor, with Antenna terminated in a load (which represents the receiver)

In the above figure, the E-field is shown as part of a propagating wave. The antenna receives the field at a voltage, as shown on its terminals. The receiver impedance (or the load or measuring device such as a network analyzer) is shown to be connected at the antenna terminals.

Now, the **ratio of incident field strength to the output voltage is the Antenna Factor;** (only for the case where the polarization of the E-field and the antenna are matched, that is, no polarization mismatch loss).

Antenna Factor in terms of antenna gain and the operating frequency:

Let us assume that the receiver antenna receiving a propagating wave in the far field region. If an incident field is a part of the propagating wave, the power density 'S' [Watts/m²], of the wave is given by-

$$S_{inc} = E.H = \frac{|E|^2}{\eta}$$

Where η - intrinsic impedance of free space

The amount of power available from the antenna can also be determined, if effective aperture is known. Thus the power captured from the plane wave is given by-

$$P_{received} = S_{inc}A_e$$

$$= \frac{S_{inc}.\lambda^2.G}{4\pi} \qquad = \frac{S_{inc}.c^2.G}{4\pi f^2}$$

$$= \frac{E^2..\lambda^2.G}{4\pi\eta} \qquad = \frac{E^2.c^2.G}{4\pi f^2\eta} \qquad (1)$$

Where, f- frequency of plane wave

c - Speed of light

G – Gain of antenna

A_e – effective aperture of the antenna

On the receiver side the power received depends on the impedance of the receiver (Z_{load}) and the received voltage-

$$P_{received} = \frac{V^2}{Z_{load}} \qquad (2)$$

Equating the equations, (1) and (2) we have the final results-

$$\frac{V^2}{Z_{load}} = \frac{E^2.c^2.G}{4\pi f^2 \eta}$$

$$A^2{}_F = \frac{E^2}{V^2} = \frac{4\pi f^2 \eta}{Z_{load}.c^2.G} \text{ or } \frac{4\pi\eta}{Z_{load}.\lambda^2.G}$$

$$A_F = \sqrt{\frac{4\pi f^2 \eta}{Z_{load}.c^2.G}} \text{ or } \sqrt{\frac{4\pi\eta}{Z_{load}.\lambda^2.G}}$$

Now, if the load impedance is 50Ω, then-

$$A_F = \sqrt{\frac{9.73}{\lambda\sqrt{G}}}$$

NOTE

- The above relationship holds only for propagating plane waves (far field conditions), and are polarization matched to the fields. The impedance mismatch between the antenna and the receiver is included in the antenna factor; the available voltage is measured at the receiver.
- As the gain of an antenna varies with the direction, **the antenna factor also varies with the function of direction of the incoming field towards the antenna. (If no direction is specified then it is taken in the direction of maximum gain).**
- Antenna factor for near and the far fields are different in behavior, as the near fields are more reactive and complex).

Antenna factor (in terms of magnetic field)-

$$A_F = \frac{H_{incident}}{V_{Received}}$$

It is the required magnetic field to produce one volt at the terminals of the antenna.

1.10.25 AXIAL RATIO

It is the ratio of orthogonal components of an E-field. A circularly polarized field is made up of two orthogonal E- field components of equal amplitude (90^0 out of phase). As the outputs are equal in magnitude thus the axial ratio is 1 (or 0 dB).

For ellipse > 1 (> 0dB)

For linear polarization = infinite (as the orthogonal components of the fields are zero)

For circular polarization = 0

1.10.26 HIGH FREQUENCY ANTENNAS

These are the antennas which are generally operating at the frequency Range of 3 MHz to 30 MHz, that is, 100 m to 10 m of wavelength.

This wavelength, enables that the antenna system for H.F. band, can be conveniently made comparable in size with the wavelength and at the same time directional purpose can also be obtained if needed.

NOTE: When wavelength is larger as in case of LF and MF band, the high cost of structure and its large size prohibits the construction of highly directive antenna system.

Unlike LF and MF antennas, where the presence of earth affects the polar diagram, with HF antennas, it is possible to suspend shorter HF antennas at quite appreciable distance from the earth.

In HF antennas the-

(1) Ground offers considerable losses.

(2) Ionosphere offers favorable conditions for HF broadcasting.

∴ in this band ground wave coverage is not used. It means uses of vertical radiators are not practical solutions. However use of *vertical radiators sis restricted to ships and coastal areas where ground conductivity is superior* to that of normal soil conductivity.

Generally, for HF broadcasting, the simplest antenna used is half wave horizontal antenna. This antenna is suspended between two vertical **masts (pole)** of suitable height which is used for supporting the **horizontal antenna.**

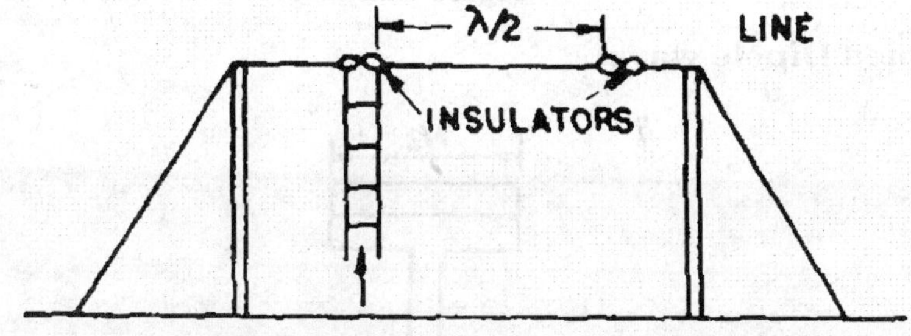

Figure 1.80 End fed half wave or Zeppelin antenna.

∴ for half wavelength antenna - the height of mast is chosen between $\lambda/8$ and $\lambda/4$.

A single half wave dipole $\lambda/2$. Either hung vertically or strung horizontally, forms the simplest antenna without much directive properties. However, **if the directive properties are required, two or more antennas can be combined to form an array**.

The units of antenna arrays are usually $\lambda/2$ antennas. Such dipole arrays may be Broadside arrays or Endfire arrays or collinear arrays etc. depending upon configuration and their feeding methods such as-

i. Centre fed

ii. End fed (Zeppelin fed)

Matching are achieved by –

i. Delta

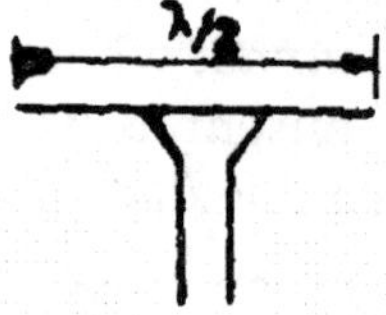

Figure 1.81

ii Wyndom,

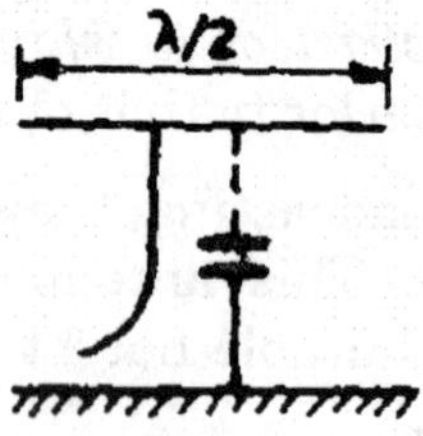

Figure 1.82

iii Folded Dipole ways

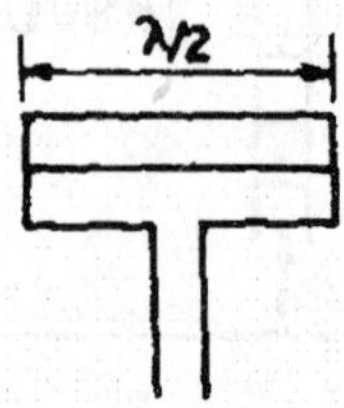

Figure 1.83

In this band for **point to point communication,** it is possible to construct antenna of large aperture which produces a beam of radio energy. The antennas which are capable of producing such beams are of two types popularly known as-

1. Resonant / Periodic Type
2. Non-Resonant/ A periodic Type/ Travelling wave type

1.11 RESONANT AND NON- RESONANT ANTENNAS

(1) Resonant Antenna

These are the antennas which correspond to a resonant transmission line that is an exact number of half wave lengths long and is open at both the ends.

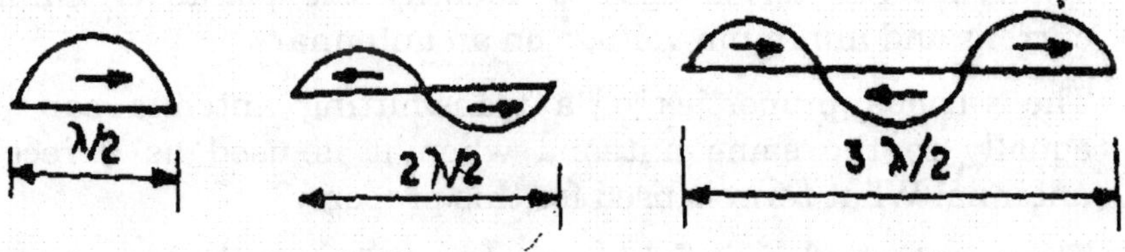

Figure 1.84 Current distribution

These are unterminated antennas and are used for fixed frequency operation.

In resonant antennas there exist forward or incident waves and backward or reflected waves that are, standing waves exist and hence the **radiation pattern of resonant antenna, are bi-directional** due to incident & reflected waves.

(2) Non-resonant Antenna

These are the antennas on which **there exists no standing wave** just like non resonant transmission line.

Thus a non resonant antenna can be said similar to the **transmission line that is excited at one end, terminated correctly at the other end, so that, no reflected waves are produced** (that is, all the incident waves are absorbed).

NOTE: As the waves travel only in one direction, ∴ it is called as **travelling wave antenna.**

It is a **wide band antenna** as it is not sharply tuned to one frequency and hence it is also called as *a-periodic antenna.*

As non-resonant antenna are terminated by characteristics impedance hence only **unidirectional radiation pattern** are produce.

SHORT QUESTIONS

ANTENNA FUNDAMENTALS

1. What are the three parts of a complete antenna system?
2. If a wave travels exactly the length of an antenna from one end to the other and back during the period of one cycle, what is the length of the antenna?
3. What are the terms used to identify the points of high current and high voltage on an antenna?
4. What are the terms used to identify the points of minimum current and minimum voltage on an antenna?
5. The various properties of a transmitting antenna can apply equally to the same antenna when it is used as a receiving antenna. What term is used for this property?
6. The direction of what field is used to designate the polarization of a wave?
7. If wave's electric lines of force rotate through 360° with every cycle of RF energy, what is the polarization of this wave?
8. What type of polarization should be used at medium and low frequencies?
9. What is an advantage of using horizontal polarization at high frequencies?
10. What type of polarization should be used if an antenna is mounted on a moving vehicle at frequencies below 50 MHz?
11. What is the radiation resistance of half wave antenna in free space?
12. A radiator source that radiates energy stronger in one direction than another is known as what type of radiation?
13. A radiating source that radiates energy equally in all the directions is known as what type of radiator?
14. A flash light is an example of what type of radiator?
15. What is primary difference between the major and minor lobes of a radiation pattern?

16. To radiate power efficiently, a long-wire antenna must have what minimum overall length?

Answers

1. Coupling device, feeder and antenna
2. One half the wavelength
3. Current and voltage loops
4. Current and voltage nodes
5. Reciprocity of antennas
6. Electric (E) Field
7. Circular polarization
8. Vertical polarization
9. Less interference is experienced by man-made noise sources.
10. Vertical polarization
11. 73 ohms
12. Anisotropic Radiator
13. Isotropic Radiator
14. Anisotropic Radiator
15. Major lobes have the greatest amount of radiation
16. One half wavelength

NUMERICALS

Q1. A thin dipole antenna is λ/15 long; if its loss resistance is 15 Ω, find the radiation resistance & the efficiency. **(Raj. Univ. 2006)**

Ans. The radiation resistance for dipole antenna is given by

$$R_r = 80\pi^2 \left(\frac{dl}{\lambda}\right)^2 \ \Omega$$

where l = Length of antenna = λ/15

λ = Operating wavelength.

Given $l = \dfrac{\lambda}{15}$ and loss resistance = R_l = 15 Ω

To find Rr and efficiency (η %)?

(i) $R_r = 80\,\pi^2 \left(\frac{dl}{\lambda}\right)^2 \,\Omega$

$$= 80\,\pi^2 \left(\frac{\lambda}{15} \cdot \frac{1}{\lambda}\right)^2$$

$$= \frac{80\,\pi^2}{15 \times 15}$$

$$Rr = 3.5\ \Omega$$

(ii) Efficiency $(\eta) = \frac{R_r}{R_r + R_l}$

$$\eta\,\% = \frac{R_r}{R_r + R_l} \times 100$$

where, R_r = radiation resistance = 3.5 Ω

R_e = loss resistance = 15 Ω

η = efficiency

$$\therefore \eta\ \% = \left(\frac{3.5}{3.5 + 15}\right) \times 100$$

$$= \frac{35}{18.5} \times 100$$

$$\eta\% = 18.92\%$$

Q2. The radial component of the radiated power density of an infinitesimal linear dipole of length l << λ at a distance r (meter) is given by ($A_o \sin^2 \theta / r^2$) Watt/m². Where, A_o is the peak value of the power density, 'θ' is usual spherical coordinate and 'λ' is wavelength. Determine the total radiated power and the maximum directivity of the antenna. **(Raj. Univ. 2005, 2004)**

Ans.

i. Total radiated power is-

$$P_r = A_o \int_0^{2\pi} \int_0^{\pi} \sin^2 \theta \sin \theta \, d\theta \, d\phi$$

$$= A_o\ 2\,\pi \int_0^{\pi} \sin^3 \theta \, d\theta$$

$$= A_o \, 2\pi \left(\frac{4}{3}\right) \qquad \because \int_0^{\pi} \sin^3 \theta \, d\theta = \frac{4}{3}$$

$$P_r = \frac{8}{3} \pi A_o$$

Maximum radiation is at $\theta = \frac{\pi}{2}$ and radiated power density is $\frac{A_o}{r^2}$

ii. Now for isotropic radiator the radiated power (P'_r) is given by-

$$P'_r = \frac{P_r}{4\pi r^2} = \frac{8/3\, \pi A_o}{4\pi r^2}$$

$$P_r' = \frac{2}{3} \frac{A_o}{r^2}$$

iii. Maximum Directivity

$$= \frac{\textit{Radiated Power density of given antenna}}{\textit{Radiated power density of isotropic radiator}}$$

$$= \frac{P_r}{P_r'} = \frac{A_o / r^2}{2/3\, A_o / r^2}$$

$$= \frac{3}{2}$$

D = 1.5

Q3. A dipole antenna is being feeded by a source having source resistance 50 Ω. If conducting loss of antenna is 5 m Watt of 20 m A current then, what is its efficiency and radiated power.

(Raj. Univ. 2005, 2004)

Ans.

Given - $R_s = 50\ \Omega$

Conducting loss = 5 mW

Current = (I) = 20 m A

To find - efficiency and Radiated Power.

(i) $I^2 R_L = 5$ mW

$$R_L = \frac{5mW}{I^2} = \frac{5 \times 10^{-3}}{20 \times 20 \times 10^{-6}}\ \Omega$$

$$= 12.5\ \Omega$$

(ii) $$\eta = \frac{R_s}{R_s + R_L} = \left(\frac{50}{50 + 12.5}\right) = \frac{50}{62.5}$$

$$\eta\% = \frac{50}{62.5} \times 100$$

$$\eta\% = 80\%$$

(iii) Power Radiated $= P_r = I^2 R_s$

$$= \left(20 \times 10^{-3}\right) \times 50$$

$$= 400 \times 10^{-6} \times 50$$

$$= 20{,}000 \times 10^{-6}$$

$$= 2 \times 10^{-2}$$

$$\therefore P_r = 20 mW$$

Q4. A 150 kH2 transmitter feed a 100 m vertical aerial against ground. If the equivalent loss resistance is 3Ω. Find-

i. Radiation resistance

ii. Radiation efficiency. **(Raj. Univ. 2002)**

Ans.

Given $f_c = 150$ kHz

$l = 100$ m

R_l = loss resistance = 3 Ω

to find - R_r and $\eta\%$?

(i) Radiation Resistance (R_r)

$$R_r = 160\,\pi^2 \left(\frac{l_{eff}}{\lambda}\right)^2$$

$$= 160\,\pi^2 \left(\frac{200}{\pi.\lambda}\right)^2 \quad as\ l_{eff} = \frac{2l}{\pi} = \frac{2 \times 100}{\pi} = \frac{200}{\pi}$$

$$\therefore R_r = \frac{160\,(40000)}{\lambda^2} = \frac{160\,(40000)}{\left(2 \times 10^3\right)^2}$$

$$\left[\text{As } \lambda = \frac{C}{f} = \frac{3 \times 10^8}{150 \times 10^3} = \frac{3 \times 10^5}{150} = 2 \times 10^3\ m\right]$$

$$R_r = \frac{16 \times 4}{40} = 1.6\ \Omega$$

$$R_r = 1.6\ \Omega$$

(ii) Radiation efficiency (η)

$$\eta = \frac{R_r}{R_r + R_L} = \frac{1.6}{1.6 + 3} = \frac{1.6}{4.6} = 0.347$$

$$\eta\% = \eta \times 100$$

$$= 0.347 \times 100$$

$$\eta\% = 34.7\%.$$

Q5. Using an example of a half wave dipole antenna shows that gain and effective aperture of an antenna are related by the following expression.

$$A_{eff} = \frac{\lambda^2}{4\pi} G_d$$ **(Raj. Univ. 2001)**

Ans. - Maximum effective area is given by-

$$A_{eff} = \frac{V_{rms}^2}{4\ R_r P} \text{ and}$$

Gain (G_d) is given by

$$G_d = \frac{4\pi P r^2}{W_r}$$

$$\therefore \frac{A_{eff}}{G_d} = \frac{V_{rms}^2 / 4 R_r P}{4\pi r^2 P / W_r} = \frac{V_{rms}^2 \cdot W_r}{\left(4\pi r^2\right)\left(4 R_r\right) P^2}$$

as $V_{rms} = E_{rms} \times l_{eff}$

where, *effective length* $= l_{eff}$

E_{rms} = *rms value of electrical field strength*

$$\because P = \frac{E_{rms}^2}{\eta} \text{ and } Wr = I_{rms}^2 \, R_r$$

$\therefore$ *Now* $\frac{A_{eff}}{G_d}$ *becomes*

$$\frac{A_{eff}}{G_d} = \frac{V_{rms}^2 \, W_r}{(4 \pi r^2)(4 R_r)(P^2)}$$

$$\frac{A_{eff}}{gd} = \frac{(E_{rms} \, l_{eff})^2 \, (I_{rms}^2 \, R_r)}{4 R_r \left(\frac{E_{rms}{}^2}{\eta}\right) 4 \pi r^2}$$

$$= \frac{E_{rms}^2 \, . \, l_{eff}{}^2 \, . \, I_{rms}^2 \, . R_r}{16 \pi r^2 \, . \, R_r . \, E_{rms}^2 \, . \, E_{rms}^2} . \eta^2$$

$$= \frac{l_{eff}{}^2 \, . \, I_{rms}^2 \, . \, \eta^2}{16 \pi r^2 \, E_{rms}^2}$$

as $$E_{rms} = \frac{120 \pi \, I_{rms} \, l_{eff}}{2.r \, \lambda}$$

where $$\eta = 120 \pi$$

$\therefore$ *Now* $\frac{A_{eff}}{G_d}$ *becomes*-

$$\frac{A_{eff}}{G_d} = \frac{l_{eff}{}^2 \, I_{rms}^2 \, \eta^2}{16 \pi r^2 \, E_{rms}{}^2}$$

$$= \frac{l_{eff}{}^2 \, I_{rms}^2 \, \eta^2}{16 \pi r^2 \left(\frac{120 \pi \, I_{rms} \, l_{eff}}{2r. \, \lambda}\right)^2}$$

$$= \frac{l_{eff}{}^2 \, . \, I_{rms}{}^2 \, . \, \eta^2 \, .4 \, r^2 \, \lambda^2}{16 \pi r^2 \, (120 \pi)^2 \, I_{rms}{}^2 \, .l_{eff}{}^2}$$

$$= \frac{120 \pi \, . \, 4 r^2 \, \lambda^2 120 \pi}{16 \pi r^2 \times 120 \pi^2 \times 120}$$

$$= \frac{\lambda^2}{4\pi}$$

$$\therefore \frac{A_{eff}}{G_d} = \frac{\lambda^2}{4\pi}$$

or $$A_{eff} = \frac{\lambda^2}{4\pi} . G_d$$

Hence proved.

Q6. Prove that radiation resistance of a current element is-

$$R_r = 80\,\pi^2 \left(\frac{dl}{\lambda}\right)^2 \Omega$$ **(Raj. Univ. 2005, 2006)**

Ans. To derive the expression for radiation resistance, it consists of the following steps-

- Obtain field components
- Obtain expression for radiated power using Poynting Vector
- Obtain average radiated power
- Obtain total power radiated from

$$P_T = \oint_s P_{av}.ds$$

- Identify expression for radiation resistance

Poynting Vector is-

$$P = E \times H \quad \text{Watts/m}^2$$

That is $P_\theta = -E_r . H_\varphi$ (1)

But E_r and H_φ are given as-

$$H_\phi = \frac{I_m \text{ dl Sin}\,\theta}{4\pi}\left[\frac{\text{Cos wt}_1}{r^2} - \frac{\text{w Sin wt}_1}{cr}\right] \quad (2)$$

$$E_r = \frac{2 I_m \text{ dl Cos}\,\theta}{4\pi\varepsilon}\left[\frac{\text{Cos wt}_1}{cr^2} + \frac{\text{Sin wt}_1}{w r^3}\right] \quad (3)$$

Where, t_1 = time delay

c = velocity of propagation

r = distance of point from element

from equation 1,2, and 3 we have-

$$P_\theta = \frac{I_m \, dl \, Sin\,\theta}{4\pi}\left[\frac{Cos\,wt_1}{r^2} - \frac{w\,Sin\,wt_1}{cr}\right] \times \frac{2 I_m \, dl \, Cos\,\theta}{4\pi\varepsilon}\left[\frac{Cos\,wt_1}{cr^2} + \frac{Sin\,wt_1}{wr^3}\right] \quad (4)$$

$$= \frac{2I^2{}_m Cos\theta\, Sin\,\theta}{16\pi^2\varepsilon}\left[\frac{Sin^2\,wt_1}{cr^4} - \frac{Cos^2\,wt_1}{r^4c} - \frac{Cos\,wt_1\,Sin\,wt_1}{wr^5} + \frac{wCos\,wt_1\,Sin\,wt_1}{cr^3}\right] \quad (5)$$

But-

$\because$ 2 Sin θ Cos θ = Sin2 θ

$$Sin\ \omega t_1\ Cos\ \ \omega t_1 = \frac{Sin\ 2\omega t_1}{2} \quad (6)$$

$Sin^2 wt_1 - Cos^2 wt_1 = Cos\ 2wt_1$

Using these identities equation (5) becomes-

$$P_\theta = \frac{I^2{}_m\ dl\ Sin2\theta}{16\pi^2\varepsilon}\left[-\frac{Cos\ 2wt_1}{r^4c} - \frac{Sin\ 2wt_1}{2wr^5} + \frac{w\ Sin\ 2wt_1}{2c^2r^3}\right] \quad (7)$$

Similarly,

$$P_r = E_\theta - H_\varphi \quad (8)$$

And E_θ is given as-

$$E_\theta = \frac{I_m\ dl\,Sin\,\theta}{4\pi\varepsilon}\left[-\frac{w\ Sin\ wt_1}{c^2\,r} + \frac{Cos\ wt_1}{cr^2} + \frac{Sin\ wt_1}{w\,r^3}\right] \quad (9)$$

From equation 3, 8, and 9, we get-

$$P_r = \frac{I^2{}_m\ dl\ Sin2\theta}{16\pi^2\varepsilon} \times$$

$$\left[\frac{Cos\,wt_1\,Sin\,wt_1}{wr^5} + \frac{Cos^2 wt_1}{r^4c} - \frac{w\,Cos\,wt_1\,Sin\,wt_1}{c^2r^3} - \frac{Sin^2 wt_1}{c\,r^4} - \frac{w\,Cos\,wt_1\,Sin\,wt_1}{c^2r^3} + \frac{w^2\,Sin^2 wt_1}{c^2r^3}\right] \quad (10)$$

Using identities of equation 6, we get-

$$P_r = \frac{I^2{}_m\ dl\ Sin^2\,\theta}{16\pi^2\varepsilon} \times \left[\frac{Sin^2 wt_1}{2c\,r^5} + \frac{Cos^2 wt_1}{r^4c} - \frac{w\ Sin2\,wt_1}{c^2r^3} + \frac{w^2\left(1 - Cosw^2 t_1\right)}{2c^3r^2}\right] \quad (11)$$

It is obvious from the equation 11, that the average value of P_r is-

$$P_r = \frac{w^2 I^2{}_m\ dl^2\ Sin^2\,\theta}{32\pi^2 r^2 \varepsilon c^2} \quad \text{or} \quad (12)$$

$$P_{r(av)} = \frac{\eta}{2}\left(\frac{\text{w } I_m \text{ dl Sin}\,\theta}{4\pi\ rc}\right)^2 \text{Watt/m}^2 \quad (13)$$

Where, $\eta_0 = \sqrt{\frac{\mu_0}{\varepsilon_0}} = 120\pi\Omega$

The total power radiated-

$$P_T = \oint P_{av}.ds \quad (14)$$

$P_{r(av)}$ is independent of φ and hence the element of area d_s on the spherical shell is given by-

$$d_s = 2\pi\, r^2 \text{ Sin } \theta \,.\, d\theta \quad (15)$$

where, θ varies between 0 to π

Now equation 14 becomes-

$$P_T = \int_0^{\pi} \frac{\eta_0}{2}\left(\frac{\text{w } I_m \text{ dl Sin}\,\theta}{4\pi\ rc}\right)^2 .2\pi r^2 .Sin\theta.d\theta$$

$$= \frac{\eta_0}{2}\frac{\text{w}^2 I^2{}_m \text{ dl}^2}{16\pi\ rc}\int_0^{\pi} Sin^3\theta.d\theta$$

But $\int_0^{\pi} Sin^3\theta.d\theta = \left(Cos\theta + \frac{1}{3}Cos^3\theta\right)_0^{\pi} = \frac{4}{3}$

So $$P_r = \frac{\eta_0 \text{w}^2 I^2{}_m \text{ dl}^2}{12\pi.c^2} \quad (16)$$

Here, I_m is the peak current.

As $I_m = \sqrt{2} I_{eff}$

$$I^2{}_m = 2 I^2{}_{eff}$$

Thus the equation 16, becomes-

$$P_T = \frac{\eta_0 \text{w}^2 I^2{}_{eff} \text{ dl}^2}{6\pi\ c^2}$$

$$P_T = 80\,\pi^2 \left(\frac{dl}{\lambda}\right)^2 I^2{}_{eff} \text{ Watts} \quad (17)$$

Hence the radiation resistance is given as- $R_r = \frac{P_T}{I^2{}_{eff}}$

$$R_r = 80\,\pi^2 \left(\frac{dl}{\lambda}\right)^2 \Omega \tag{18}$$

Hence proved.

Q7. Define the different apertures of an antenna. How is the re radiated part represented by antenna scattering aperture? (Raj. Univ. 2003)

Ans. Scattering Aperture:

The voltage induced in the antenna produces a current through both the antenna impedance Z_A and the terminal or load impedance Z_R. Power absorbed by the terminal impedance $P=I^2R_r$. The real part of impedance R_A has two parts, that is, $R_A = R_r + R_L$. Thus some of the power that is received will be dissipated as heat in the antenna as-

$$W' = I^2R_L$$

The rest of the power is dissipated in the radiation resistance is re- radiated given by-

$$W'' = I^2R_r$$

This re-radiated or scattered power is analogous to power dissipated in a generator. For maximum power transfer power dissipated in the generator will be equal to power delivered by the load. This re-radiated power is used to define a scattering aperture or scattering cross section. This aperture As is defined as ratio of the re-radiated power to the power density of the incident wave. As = scattering aperture = $\frac{W''}{P}$

Where, $W'' = I^2.\,R_r = \frac{V^2.R_r}{[R_r + R_L + R_T]^2 + [X_A + X_T]^2}$

When $R_L = 0$, $R_T = R_r$ and $X_T = -X_A$ for maximum power transfer then

$$A_s = \frac{V^2}{[4R_rP]} = A_{em} \text{ (maximum effective aperture)}$$

When load resistance is zero and $X_L = -X_A$

This zero load condition is known as a resonant short circuit (RSC) condition. Then for RSC the re-radiated power is-

$$W'' = \frac{V^2}{[R_r]}$$

$$A_S = \frac{V^2}{[R_r P]}$$

$$A_S = 4A_{em}$$

For RSC condition the scattering aperture is four times the maximum effective aperture. The ratio of scattering aperture to the effective aperture is known as scattering ratio β.

$$\beta = \frac{A_S}{A_{em}} \text{ (Dimensionless)}$$

The value of scattering ratio lies between 0 to ∞.

UNIT 2
ANTENNA TYPES

CONTENTS

ANTENNA TYPES

2.1 EFFECTS OF GROUND ON ANTENNA PERFORMANCE

Antenna at a height >5 λ is generally referred to be in free space.

Radiation patterns are highly affected by reflection from ground. (HF Range), as ground is considered as reflecting surface.

Influence of ground depends on the fact whether the antenna is actually grounded or merely situated near the ground. (i. e. Grounded and ungrounded antenna).

Ungrounded-Antennas (Ground assumed to be perfect conductors)

If a radiation source is put near a reflecting surface, the radiation received at any distance point, is the vector sum of direct and indirect (i.e. reflection from ground) radiations.

It is assumed that an *image* Antenna exists below the ground and is true mirror image of the actual antenna.

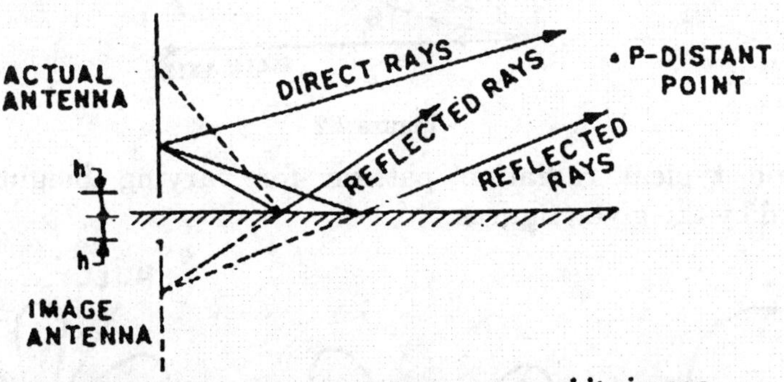

(a) Shows ungrounded antenna and its image.

Figure 2.1

The resulting radiation received at a distance point is considered to be coming from the actual antenna and image antenna.

If ground is assumed to be perfect conductor/Reflector, then the current flowing in the image have same magnitude as that of actual antenna. In ungrounded antennas the image and actual antenna forms an *array*.

Overall radiation pattern is calculated on the basis of vector sum of radiations of two nearby antennas of an array. The antenna elements of arrays are identical in length having equal currents in magnitudes and are spaced by twice the height (h) of actual antenna above ground.

The direction of current flow in *vertical* antennas is same as that of actual antenna while in case of *horizontal* antenna it is in opposite phase.

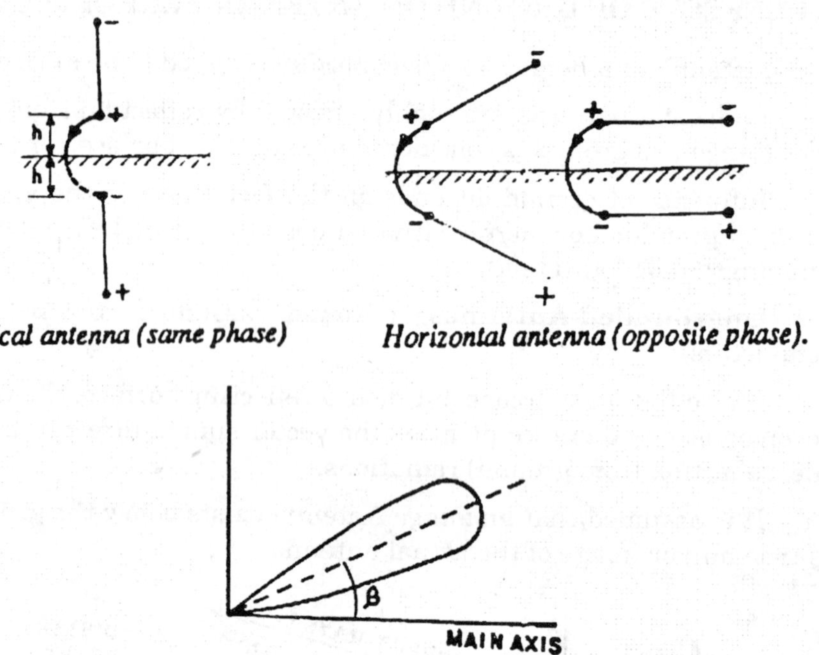

Figure 2.2

Some typical radiation pattern for varying heights can be explained by the given figures.

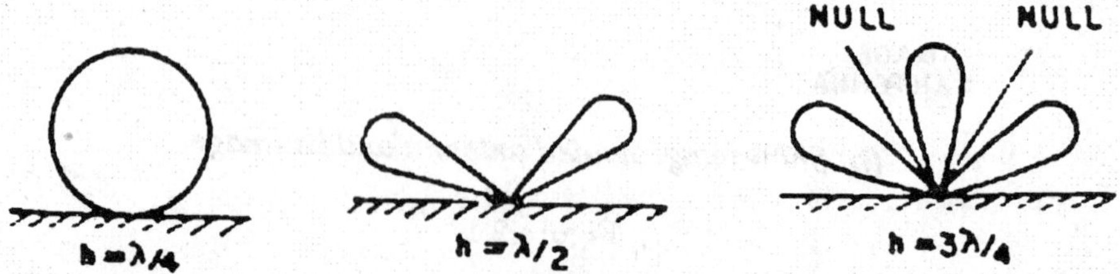

Figure 2.3

The reflection coefficient of the ground is an important factor, when ground is an imperfect-conductor (F*or a perfect conductor* - reflection coefficient is 1).

PERFECT EARTH

ACTUAL EARTH

Effects of imperfect earth on the pattern of un-grounded antennta at h = λ

Figure 2.4

Grounded Antenna

When antenna is close to the ground, then regardless of whether the antenna is grounded or not, the earth acts as a mirror and becomes part of radiating system.

The bottom of the grounded antenna is joined to the top of the image antenna and acts as an antenna of double size (2h).

For example, λ/4 (**Marconi** antenna) behaves as λ/2 **(Hertz antenna)** grounded vertical antenna while the voltage and current distribution remains same as in that of half wavelength dipole in space.

Marconi antenna has advantage over the ungrounded (Hertz antenna) to produce same radiation pattern where ground plays an important role, in producing required characteristics.

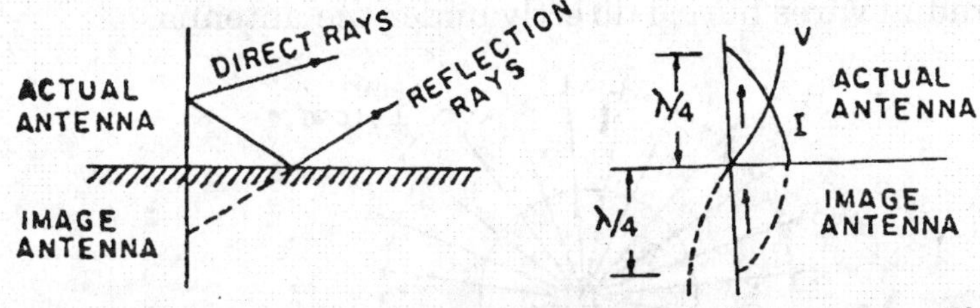

Actual and image antennas ***voltage current distribution in λ/4 antenna (Marcoini antenna).***

Figure 2.5

1. Draw the radiation pattern of the following antennas:

 (i) Resonance Antenna **(Raj. Univ. 2007)**

2.2 GROUND SYSTEM AND ANTENNA LOSSES

With our earlier discussion, it is now clear that the earth is not a perfect conductor. Therefore, a large power lost in long wave antenna occurs in ground.

Ground losses

These losses occur due to the current, charging the capacitor between the antenna and ground; leaks through the capacitance from the antenna to the earth and back through earth to the grounding point of the transmitter.

As earth is relatively a poor conductor, therefore some special provision must be made for returning these currents to grounding point on transmitter with losses.

This is why a ground system for vertical grounded antennas are required which provides a least resistance path for return current through the ground back to transmitter.

Best ground is silver sheet extending many wavelengths in all directions under the antenna. We can also use copper ground radials for this purpose. According to the nature of soil this system can be used as -

1. Radial buried ground system for conducting earth and
2. Counterpoise conductor ground system for soil of poor conductive surface, rocky earth, or at site like high buildings.

2.2.1. Radial buried ground system

It is also called as earth MAT. This grounding system includes a network of radial wires buried directly under the antenna.

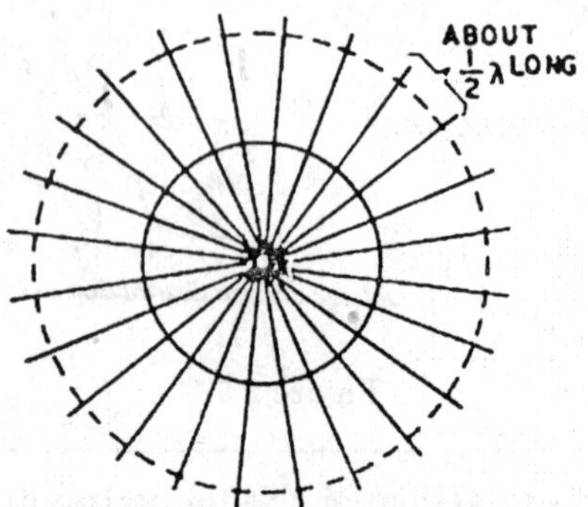

Radial buried ground system for grounded antennas.

Figure 2.6

It consists of a large number of radial wires extending from the base of tower and placed between 20 to 30 cm, below the ground by special plow which simultaneously —

(1) Digs the furrow

(2) Lays the copper wire and

(3) Cover it over.

Each radial wire is separated by 8° and has length at least $\lambda/4$.

$\therefore$ Total Radials = 120 (to save wire they are out short)

Ground system will have very small losses, if the wires are soldered or preferably welded together at the joints.

Generally a conductor is used to join all the radial wires at a distance of nearly half the radial lengths.

The far ends of radial wire grounded through a metal stake which is driven deep into subsoil, and it is better conductor than driven into the top soil if it is better conductor in case of sandy soils.

At the mast, all the 120 radials are brazed to a copper strip ring which is put around the mast base pedestal.

$$\text{Number of radials } \alpha \ \frac{1}{\text{resistance}}$$

This type of grounding system is used in low frequency antenna.

2.2.2 Counterpoise conductor ground system

- When the ground is a poor conductor i.e. Rocky area, or at the top of high buildings, then the counterpoise type of ground system is preferred which is nothing but a small edition of radial wire on earth-mat but this time above the surface of earth with insulated outer edge.

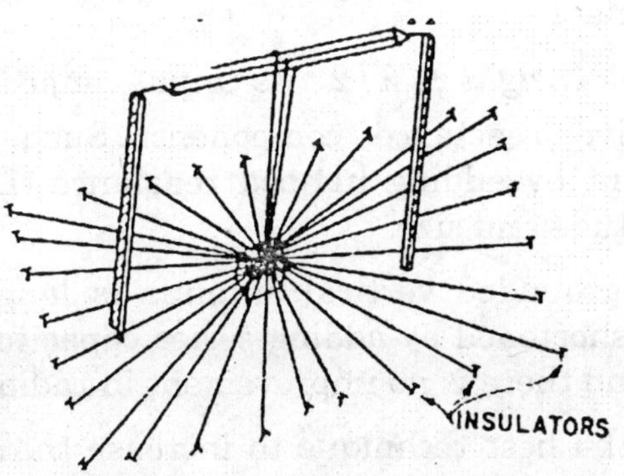

Shows Counterpoise ground system.

Figure 2.7

The supports of these wires are few, which are placed far apart above the ground. Counter piece replaces ground by conducting screen

Space between the two radials wires is not greater than the height of counterpoise above the ground.

Counterpoise wires extends from the base of antenna to the distance at least *equal to height* of vertical antenna above the ground *or* preferably twice of the antenna. *This height is kept so that, it intercept most of the electrostatic field in the vicinity of radiator.*

This radial wire mesh or screen intercepts the lines of force from antenna to earth and carry the return current on the screen wires rather than by the earth since current does not flow through the earth itself, the losses due to poor conductivity of earth are avoided.

2.3 EFFECTS OF ANTENNA HEIGHT

At *low and medium frequencies* it becomes difficult rather impracticable to use an antenna of resonant length. The heights of vertical antennas at these frequencies are very much small. Due to such a small size the radiation area of the antenna decreases hence to increase the same we go for:

Antenna Top Loading and Tuning

To understand it, first we need to know characteristic and need of different size of antennas.

1. Antenna with height$\geq \lambda/4$, this antenna is not of resonant length for given operating frequency its input impedance contains a reactive component. Hence now to make it resonant we go for top loading.
2. Grounded antenna of height of less than $\lambda/4$. This antenna has an input impedance containing a capacitive - reactance component.
3. If $\lambda/4 \leq height \leq \lambda/2$ its input impedance contains an inductive reactance component. Such system is made resonant by adding lumped reactance (L or C) of suitable magnitude and size.

Hence the grounded vertical antenna is lengthened by adding inductance and shortened by adding series capacitance (But they add less resistance and there is no improvement in radiation resistance).

Therefore, the best technique to increase the antenna radiation efficiency by increasing Radiation resistance of the system is by **TOP-LOADING.**

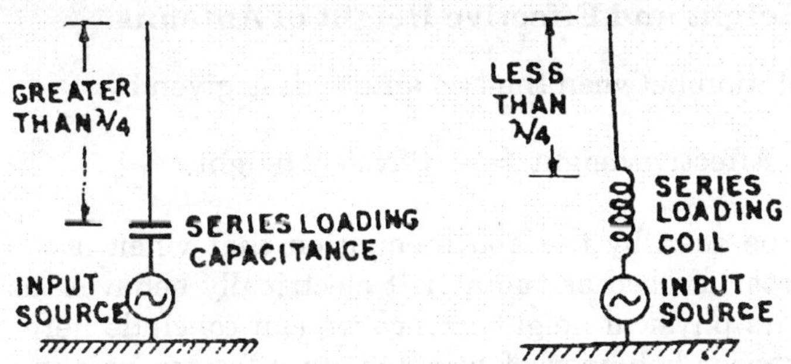

Shows use of series loading coil or capacitance to attain the resonant length λ/4.

Figure 2.8

Top loading is a method to increase the effective capacitance at the top of the antenna to obtain desired radiation pattern without replacing the antenna with the other antenna. This is done by mounting one or more horizontal conductors at the antenna top with the top loading has the form of invert L, T or top hat like structures.

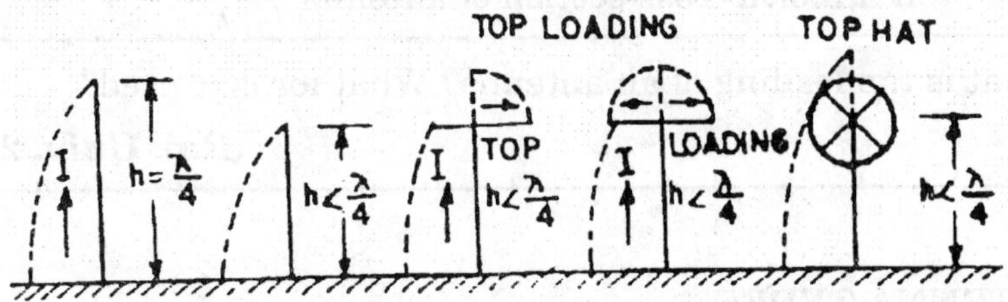

Various form of top loading with current distributions (dotted lines).

Figure 2.9

By changing top loading the current at the base of antenna can be increased and current distribution becomes more uniform.

The current in horizontal portion can be smaller than in vertical portion, which results in a vertically polarized antenna.

Line resulting current distribution along the vertical portion is similar to, the rest of the part as if extra vertical section is added. This results in increase of current in vertical/upper portion which on other hand increases both the Radiation resistance and Radiation efficiency.

Lumped reactance may also be used to maintain a resonant condition in an antenna system if the associated transmitter is operated under different frequencies. *Thus a series inductor or capacitor connected with vertical antenna may also serve as an antenna turning circuit,* which reduces the radiation efficiency.

Physical Height and Effective Height of Antenna

The relation between the two antennas is given by:-

$$\text{Effective height} = \frac{2}{\pi}\ \text{Physical height}$$

It can be seen by the above equation that when a conductor of finite diameter is used as radiator it electrically behaves, as if it was taller than its physical height. Hence we can conclude here that due to this electrical behavior of antenna its effective height is greater than its physical height by 2-8%

The difference between physical height and effective height is due to the fact-

1. That wave velocity in conductor is less than its velocity in conductor is les than its velocity in free space.
2. Effect of top-loading.
3. End effects of antenna.
4. Non uniform cross-section of antenna.

1. What is top loading of an antenna? What for is it used?

(Raj. Univ. 2004)

2.4 ANTENNA TYPES

2.4.1. Wire antennas

1. Short Dipole
2. Dipole Antenna
3. Half wave Dipole
4. Broadband Dipoles
5. Monopole
6. Folded Dipole
7. Small Loop
8. Square and Circular Loop

2.4.2. Travelling Wave Antennas

1. Yagi Uda Antenna (Frequency independent antennas)
2. Helical Antenna

3. Frequency Independent antenna LOG PERIODIC (DIPOLE ARRAY)
4. Long wire Harmonic Antenna
5. V-Antenna
6. Inverted V-Antenna
7. Rhombic Antenna

2.4.3. Aperture Antennas

1. Horn Antenna
2. Slot Antennas
3. Inverted F Antennas (IFA)
4. Slotted Waveguide Antenna

2.4.4. Reflector Antennas

1. Parabolic Reflector (Dish Antenna)
2. Corner Reflector
3. Lens Antenna

2.4.5 Microstrip Antennas

1. Rectangular Microstrip (Patch) Antenna
2. Shorting Pins: Quarter Wavelength Microstrip
3. PIFA.

2.4.1. WIRE ANTENNAS

2.4.1.1 THE SHORT DIPOLE ANTENNA

The dipole antenna can easily be said as the beginner's simple antennas. It is an open-circuited `L'-length wire which is fed at its centre, as shown in figure-

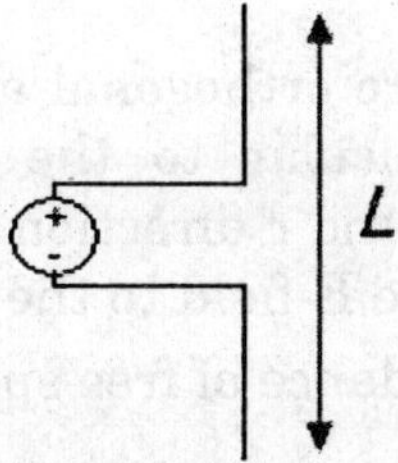

Figure 2.10 Short dipole antenna of length L.

The terminology "small" or "short" in antenna always implies, relative to a wavelength of the frequency of operation so the absolute size of the above dipole does not matter. Typically, a dipole is small dipole if its length is less than a tenth of a wavelength-

$$L < \frac{\lambda}{10}$$

If the antenna is oriented along the z-axis with the center of the dipole at z = 0, then the current distribution on a thin, short dipole is given by-

$$I(z) = I_o\left(1 - \frac{2|z|}{L}\right)$$

The current distribution is plotted in the given Figure, I (z) is the amplitude of the current distribution, it is oscillating at the frequency `f'

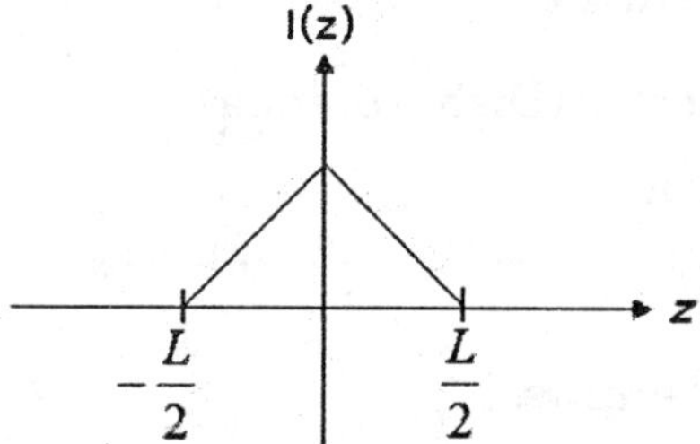

Figure 2.11 Current distributions along a short dipole

The fields radiated from this antenna in the far field, only Eθ and Hϕ fields are non zero, rest all the components are zero which are given by-

(i) $E_r = H_r = E_\phi = H_\theta = 0$

(ii) $E_\theta = \dfrac{j\eta k I_o L\, e^{-jkr}}{8\pi r}$

(iii) $H_\phi = \dfrac{E\theta}{\eta}$

1. Eθ and Hϕ fields are orthogonal and in-phase. Further, the fields are perpendicular to the direction of propagation which is always in the r direction (away from the antenna). Also, the ratio of the E-field to the H-field is given by η (the characteristic impedance of free space).

$$\eta = \frac{E\,\theta}{H\,\phi}$$

It indicates that in the far-field region the fields are propagating like a plane wave.

2. These fields die off as `r' increases, which indicates the power falls given by-

$$P(r) \; \alpha \; \frac{1}{r^2}$$

3. (a) The fields are proportional to `L' which indicates, that a longer dipole will radiate more power (This is true as long as increasing the length, does not cause, the short dipole assumption to become invalid).

   (b) The fields are proportional to current Io, which says more current means more power.

4. The exponential term: e^{-jkr} describes the phase-variation of the wave versus distance. The fields oscillate in time at the frequency `f' in addition to the above spatial variation.

In the last, the spatial variation of the fields as a function of direction from the antenna is given by sin θ.

For a vertical antenna oriented along the Z-axis, the radiation will be maximized in the x-y plane. Therefore, theoretically, there is no radiation along the Z-axis far from the antenna.

Directivity of short dipole antenna

The directivity of the center fed short dipole antenna depends only on the `$\sin\theta$ ' component of the fields. (Which is nearly 1.5 (1.76 dB), infact it is very low to realize any antenna). This antenna has an **Omni directional Radiation** pattern as the fields are only a function of the polar angle, again they have no azimuthal variation and therefore, the half power Beam width is 90°.

Polarization of short dipole antenna

This antenna has a linear polarization. For the x-y plane, this antenna is described as **vertically polarized,** because the E-field is vertically oriented, along the Z-axis.

Input impedance of short dipole antenna

It depends on the radius `a' of the dipole. The impedance 'Z' is made up of three components representing the stored energy in the fields they are-

1. R_{rad} - Radiation Resistance

2. R_{loss} - Loss Resistance and

3. X - Reactive or imaginary component

Hence the impedance is given by -

$$Z = R_{rad} + R_{loss} + jX$$

The radiation resistance (R_{rad}) is given by:-

$$R_{rad} = 20\pi^2 \left(\frac{L}{\lambda}\right)^2$$

The resistance representing loss due to the finite - conductivity of the antenna is:-

$$R_{loss} = \frac{L}{6\pi a}\sqrt{\frac{\pi f \mu}{2\sigma}}$$

where $\sigma \rightarrow$ high conductivity of dipole

$f \rightarrow$ Frequency due to skin effect

The imaginary part of the impedance i.e. reactance component of the dipole impedance is given by:-

$$X = \frac{-120\lambda}{\pi L}\left[\ln\left(\frac{L}{2a}\right) - 1\right]$$

Example 1. Assume that the radius is 0.001 λ and the length of dipole is 0.05 λ. Suppose further that this antenna is to operate at f = 3 MHz, and that the metal is copper having its conductivity (σ) as 59, 600, 000 S/m. Then find the impedance.

Solution: given a = 0.001 λ

L = 0.05 λ

f = 3 MHz

σ = 59,600,000 s/m

To find Z= impedance?

(i) $R_{rad} = 20\pi^2 \left(\frac{L}{\pi}\right)^2 = 20\pi^2 \left(\frac{0.05\lambda}{\lambda}\right)^2$

$= 0.49\ \Omega$

(ii) $R_{loss} = \frac{L}{6\pi a}\sqrt{\frac{\pi f \mu}{2\sigma}}$

$$= \frac{0.05\lambda}{6\pi(0.001\lambda)}\sqrt{\frac{\pi \times 3 \times 10^6 \times \mu}{2 \times 59{,}600{,}000}}$$

$$= \frac{1}{12}\sqrt{\frac{15\mu}{596\pi}}$$

$= 4.83\ m\ \Omega$ (*negligible in comparison to radiation resistance*)

(iii) $$X = \frac{-120\lambda}{\pi L}\left[\ln\left(\frac{L}{2\sigma}\right) - 1\right]$$

$$= \frac{-120\lambda}{\pi(0.05\lambda)}\left[\ln\left(\frac{0.05\lambda}{2(59{,}600{,}000)}\right) - 1\right]$$

$$= \frac{-2 \times 10^3}{\pi}\left[\ln\left(\frac{\lambda}{4(59600000)}\right) - 1\right]$$

$= 1695\ \Omega$

∴ Now $Z = R_{rad} + R_{loss} + jX$

$= 0.49 + (\text{negligible}) + j\,1695$

$= (0.49 + j\,1695)\ \Omega$

∴ We now see that, for this antenna it is very difficult to have proper impedance matching. Even if the reactance is properly cancelled out, hence only a small amount of power will be delivered from 50 Ω sources to a 0.49 Ω load.

Conclusion (1): For short dipoles that are smaller fractions of a wavelength, the radiation resistance becomes smaller than the loss resistance, and consequently this antenna becomes inefficient at higher frequencies.

Conclusion (2): As the input impedance varies widely with the frequency because of the reactance component of the input impedance. Hence these antennas are typically used in narrow band applications and therefore, the bandwidths for these antennas are very difficult to define.

2.4.1.2. DIPOLE ANTENNA

Dipole antennas are very thin radius antennas, where the dipoles are similar to the short dipole antennas with the only difference that, it is not required for the dipoles to be small compared to wavelength (frequency at which the antenna is operating) as in short dipole antennas.

For a dipole antenna of length L, which is oriented along Z-axis, centered at Z = 0, the current flows in the Z-direction, with the amplitude which is mathematically calculated as -

$$I(z) = \begin{cases} I_o \sin\left[k\left(\frac{L}{2} - z\right)\right], & 0 \le z \le \frac{L}{2} \\ I_o \sin\left[k\left(\frac{L}{2} + z\right)\right], & -\frac{L}{2} \le z \le 0 \end{cases}$$

This current oscillates in time, sinusoidal at the frequency `f'. The current distribution for a quarter wavelengths is

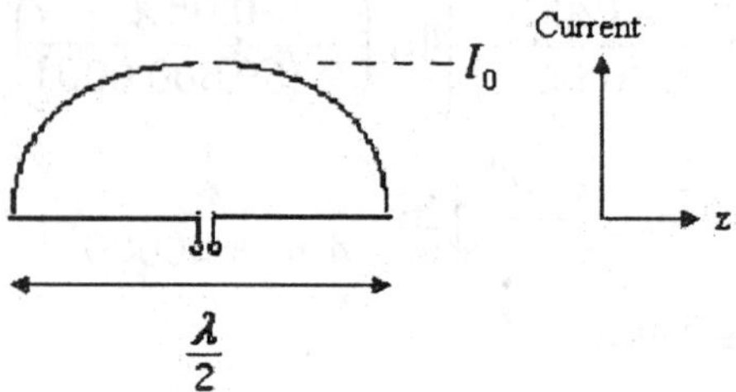

Figure 2.12(a) Quarter Dipole Antenna

And current distribution for full wavelength is

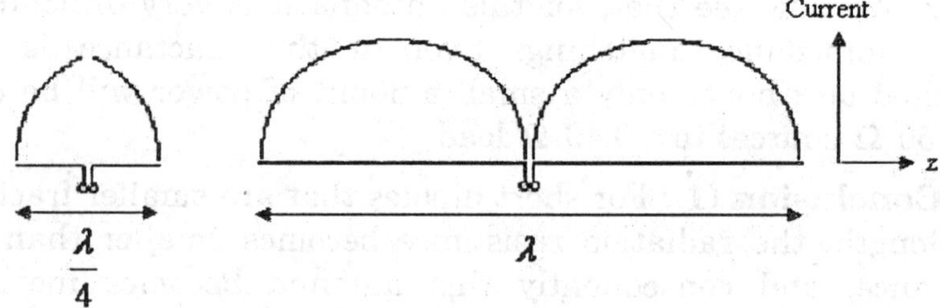

Fig. 2.12(b) Dipole Antenna

Input Impedance of Dipole Antenna

Let us consider the *input impedance of a dipole* as a function of its length, plotted in the figure (3) below, where, the input impedance is specified as -

$$Z = R + jX$$

where $R \rightarrow$ *resistance*

$X \rightarrow$ *reactance.*

(**NOTE:** for very small dipoles, the input impedance is capacitive (as it can store energy), which means the impedance is leaded by the negative reactance value (and relatively small real impedance).

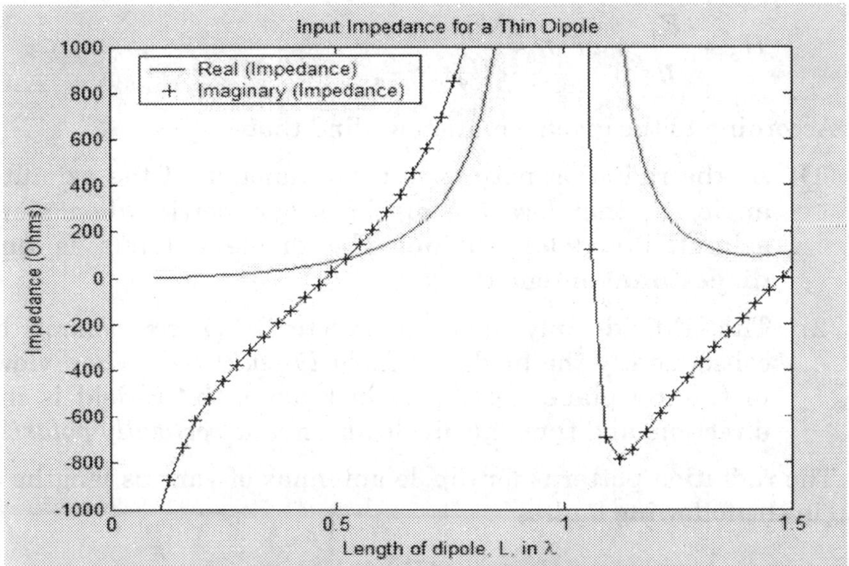

Fig.2.13 Input impedance as a function of the length (L) of a dipole antenna

Whereas the dipole gets larger, the input resistance increases, along with the reactance

It can be seen by the above given figure; that, at slighting less than 0.5 λ antenna has zero imaginary component to the impedance (reactance X = 0), and thus the antennas is said to be resonant.

If the dipole antenna's length becomes close to one wavelength, the input impedance becomes infinite (As per high frequency transmission line theory); which can be explained as under:

Consider a single wavelength dipole shown in Figure (A.2.2.), where its current distribution is shown. Since the current at the terminal is zero, the input impedance (V/I) is surely infinite. Consequently, infinite impedance occurs whenever, the dipole is an integer multiple of a wavelength.

Radiation pattern of dipole antenna

The far fields from the dipole antenna of length L are given by

$$E_r = H_r = E_\phi = H_\theta = 0$$

$$E_\theta = \frac{j\eta I_0 e^{-jkr}}{2\pi r}\left[\frac{\cos\left(\frac{kL}{2}\cos\theta\right) - \cos\left(\frac{kL}{2}\right)}{\sin\theta}\right]$$

$$H_\phi = \frac{E_\theta}{\eta} \quad or \quad \eta = \frac{E_\theta}{H_\phi}$$

According to the given equation we find that-

(1) As the radiation pattern is not a function of the azimuthal angle `ϕ', therefore the dipole is symmetric when viewed azimuthally. (**Conclusion:** The dipole antenna is **omni directional** antenna).

(2) The E-field only has one vector (E_ϕ) component and consequently the fields are *linearly polarized.* When viewed in the x-y plane (for dipole in Z-axis), the E-field is in y-direction, and thus the dipole antenna is *vertically polarized.*

The radiation patterns for dipole antennas of various lengths are given in the following figure-

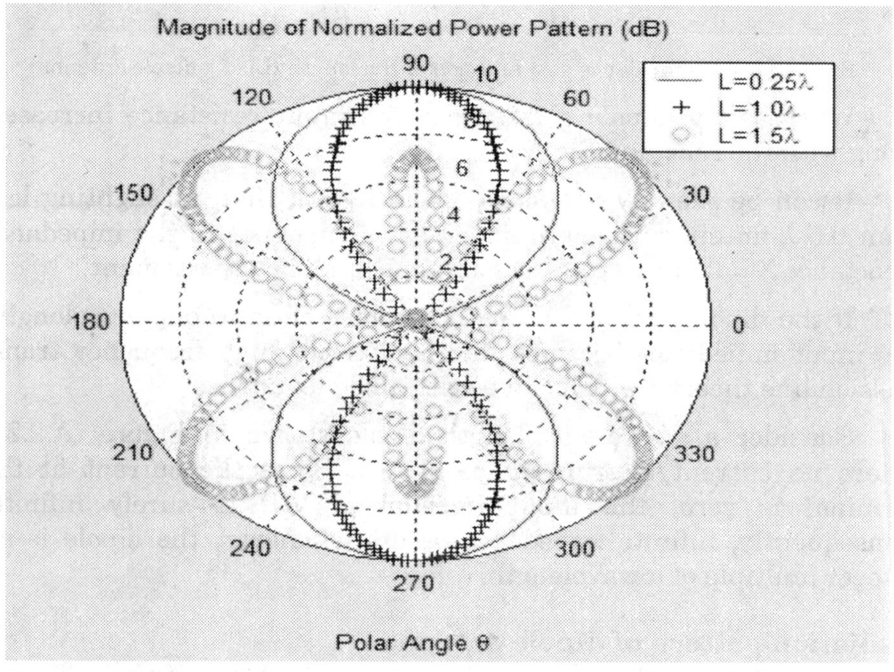

Figure 2.14 Normalized radiation patterns for dipoles of specified length.

According to the given Figure, the 1.5 wavelength dipole pattern is plotted, where the pattern is maximum at +45° and –45°.

∴ It can be well said, that **the full wave dipole is more directional than the quarter wave dipole.**

(NOTE: *In antenna theory a larger antenna in general (not always) has a greater directivity).*

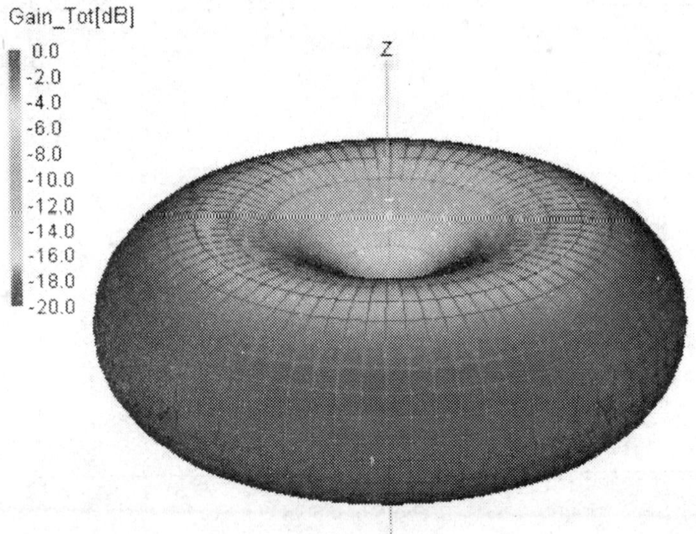

Figure 2.15 Normalized 3d radiation pattern for the 1-wavelength dipole.

The 3-D pattern for full wave dipole is given in figure. This pattern is similar to the pattern for quarter and half wave dipole.

The 3-D radiation pattern for the 1.5 wave dipole is somewhat different from full wave dipole, and is shown in the next figure-

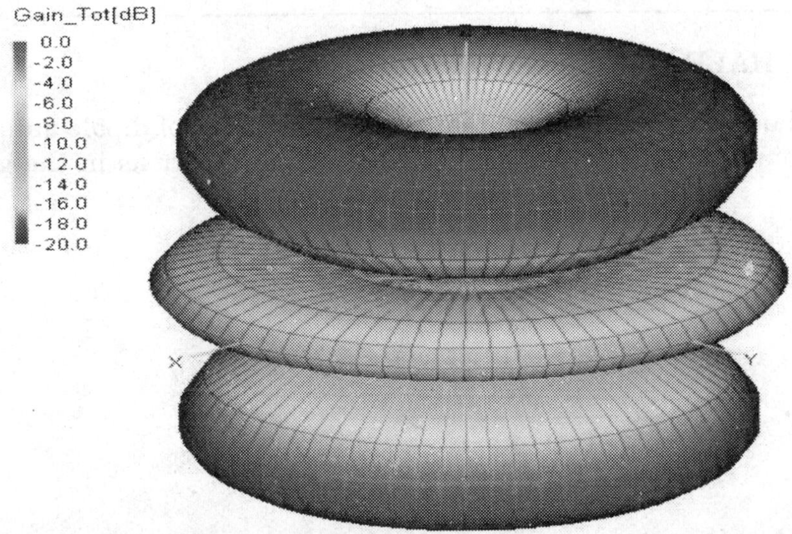

Figure 2.16 Normalized 3-d radiation pattern for the 1.5 wavelength dipole.

Directivity of dipole antenna

The peak directivity variation of a dipole antenna is given in the figure below –

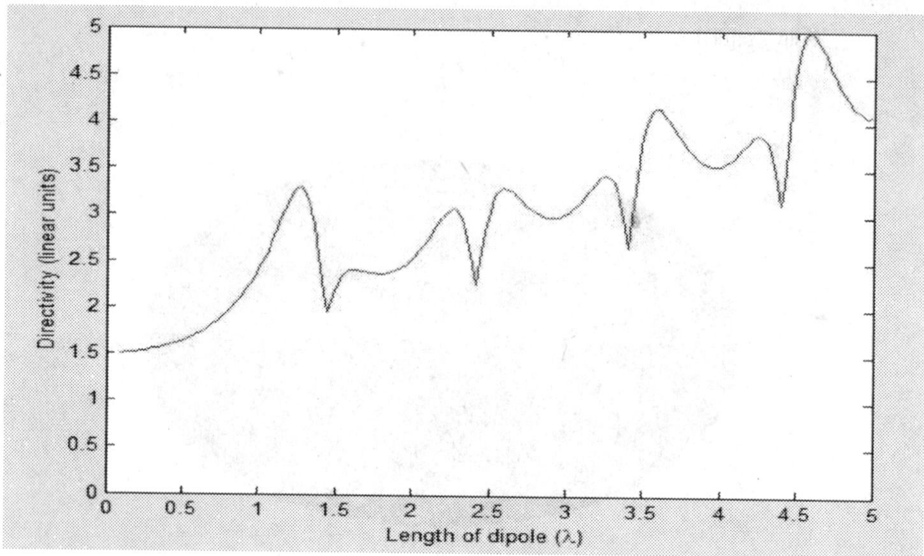

Figure 2.17 Dipole directivity as a function of dipole length

This figure indicates that at L = 1.25 λ, the directivity increases, with length. However, for longer length the directivity has an upward trend but is no longer monotonic.

1. Draw the radiation pattern of the following antennas:
   (i) Dipole (Half wave) **(Raj. Univ. 2008)**

2.4.1.3. HALF-WAVE DIPOLE ANTENNA

The Half wave dipole antenna is a special case of dipole antenna. It is simply a half wavelength wire fed at the center as in the given figure-

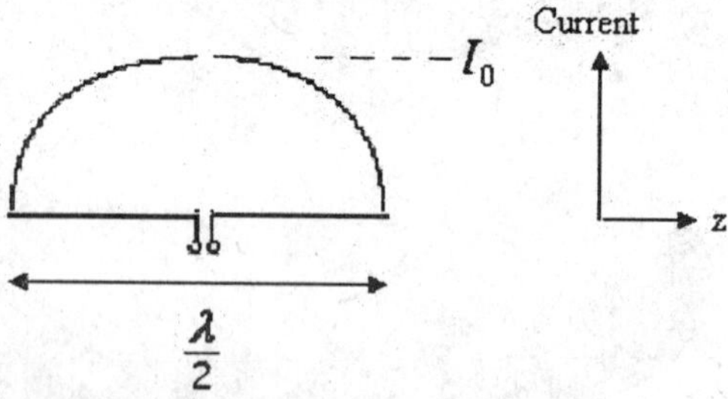

Figure 2.18 Current on a half-wave dipole.

The calculated standard input impedance is given by

$$Z_{in} = 73 + j\,42.5\ \Omega$$

The fields of the half wave dipole are given by-

$$E_\theta = \frac{j\eta I_0 e^{-jkr} \cos\left(\frac{\pi \cos\theta}{2}\right)}{2\pi r \sin\theta}$$

$$H_\phi = \frac{E_\theta}{\eta}$$

The directivity of a half wave dipole antenna is 1.64 (2.15 dB) and the Half Power beam width is 78°.

(**NOTE:** By reducing the length of antenna slightly, the antenna can become resonant).

Say if the dipole's length is reduced to 0.48 λ, the input impedance of the antenna becomes $Z_{in} = 70\ \Omega$, with no reactive component, whereas, the radiation pattern remains the same.

Above length is valid if the dipole is very thin. But in reality, the dipoles are made with thick material, which increases Bandwidth of antenna; reducing the resonance length slightly depending on thickness but close to 0.47 λ.

2.4.1.4. BROADBAND DIPOLE ATENNA

Thumb Rule in antenna design

(**NOTE:** By increasing the occupying volume of an antenna, the same can be used for the broadband applications). ∴ A dipole antenna can be used in more broadband applications only by increasing the radius \`A' of the dipole antenna.

Example: On dipole of length 1.5 m, the dipole is half wavelength long t 100 MHz we consider 3 cases

(1) $A = 0.001\text{ m} = \frac{1}{3000}$ of wavelength at 100 MHz

(2) $A = 0.015\text{ m} = \frac{1}{100}$ of wavelength at 100 MHz

(3) $A = 0.05\text{ m} = \frac{1}{30}$ of wavelength at 100 MHz

The resulting S11 for each of these 3 cases is plotted versus frequency (assuming matched load = $50\,\Omega$)

By looking at the figure we can conclude that-

(1) Thicker the dipole is made, larger the Bandwidth becomes.

(2) Thicker the dipole gets, lower the resonant frequency becomes.

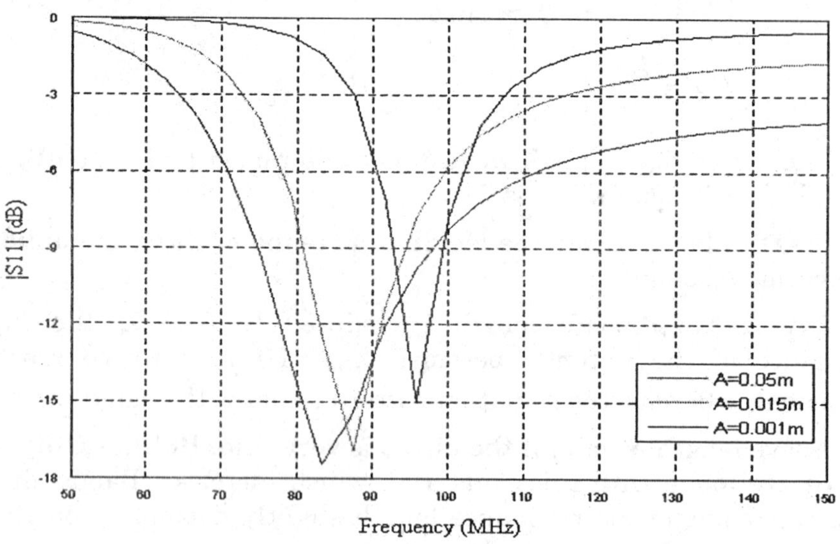

Figure 2.19 Magnitude of S11 for Dipoles of Varying Radii.

-Other way, if an antenna is designed to resonate at 150 MHz, the resonant length decreases as dipole gets thicker.

2.4.1.5. MONOPOLE ANTENNA

A monopole antenna is exactly half of a dipole antenna. It is always mounted on some sort of ground plane surface. The special case of a monopole antenna of length `L' is mounted above an infinite ground plane, which is Perfect Electric Conductor (PEC), is given in the figure -

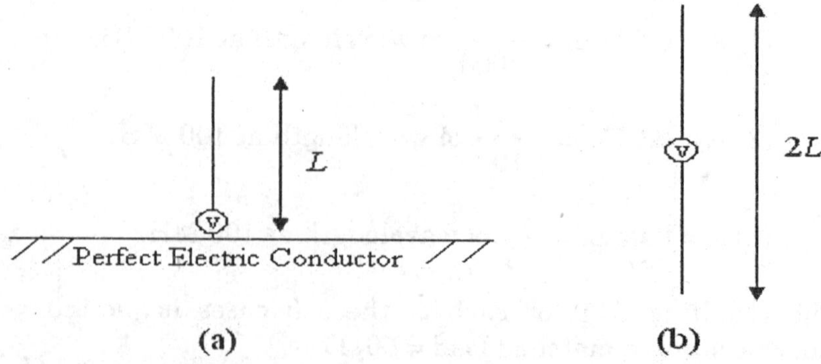

Figure 2.20 Monopole above a PEC (a), equivalent source in free space (b)

To find the fields above the ground plane is completed by, using image theory, by using the equivalent source (antenna) in free space as shown in figure (b). Here in figure (b) the antenna is simple a dipole with twice the length of monopole antenna in figure (a). The fields above the ground plane in figure (a) are same as that of dipole antenna. But the fields below the ground surface are null.

As it is already known that the radiation pattern of monopoles above ground are known from dipole result; with only change, that the impedance of monopole is one half of that of full dipole antenna.

For the quarter-wave monopole ($L = \lambda/4$), the impedance is half of half wavelength dipole,

$$\therefore Z_{in} = 36.5 + j21.25\Omega.$$

It can be understood as, only half the voltage is required to drive a monopole t same current as a dipole (A dipole having + V/2 and –V/2 applied to its ends, whereas a monopole needs to apply +V/2 between monopole and the ground, to drive the same current); the impedance is halved. As-

$$Z_{in} = V/I$$

Directivity of a monopole antenna

Directivity of a monopole antenna is directly related to that of a dipole antenna. Say if D_1 (dB) = Directivity of dipole of length 2L then the directivity of monopole (D2) antenna of length `L' is-

$$D_2 = D1 + 3 \text{ (dB)}$$

NOTE : Directivity of monopole is twice the directivity of dipole antenna of twice the length

Reason: As no radiations occur below the ground, hence monopole antenna is effectively twice as ``directive".

Monopoles are half of the size of dipoles, therefore are preferred where smaller antennas are to be used.

For e.g. antennas on older cell phones, with infinite ground, are approximated by a small metal plate below the antenna.

Effects of a finite Ground Plane

(1) The *impedance* of monopole antenna is minimally affected by ground plane of at least a few wavelengths in size.

(2) *Radiation Patterns* is strongly affected by finite sized ground. Due to which pattern radiation in ``skewed" direction is away from the horizontal plane.

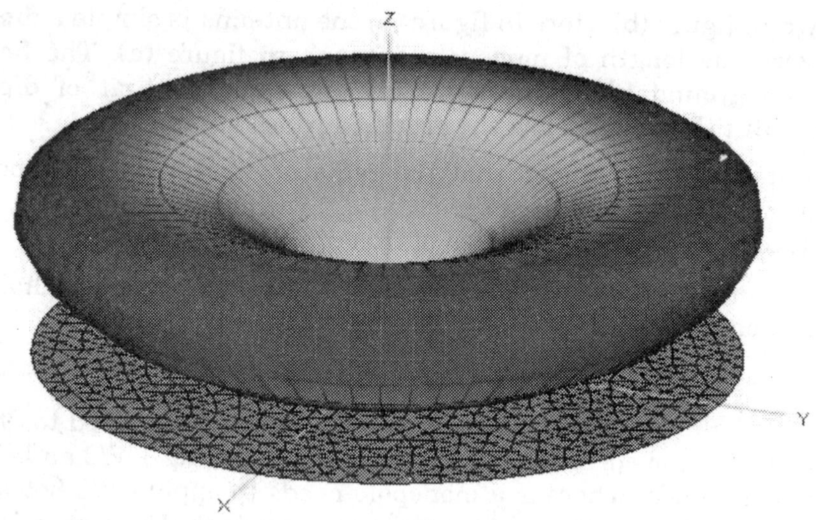

Figure 2.21 Radiation pattern for a quarter wave monopole antenna into direction on ground plane with diameter = 3 wavelengths.

(**NOTE:** Even after mounting the antenna over a finite sized ground plane, the radiation pattern is still omnidirectional). But the direction has changed from x-y plane to angle elevated from that plane.

Conclusion: The larger the ground plane becomes, the lower is the directivity that is as the ground plane approaches infinite size the radiation pattern increases in x-y plane.

2.4.1.6. FOLDED DIPOLE ANTENNA

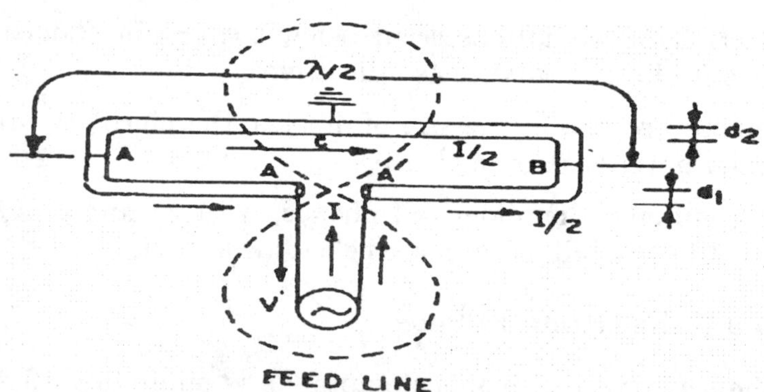

AB = Minimum current points
C = Maximum current point or Minimum voltage point $d_1 = d_2$.
Two wire folded dipole with
(i) Current distribution, (ii) Radiation pattern.

Figure 2.22

Construction

There are 2 half wave dipoles - one continuous and the other split at the centre-have been folded at the end and joined together in parallel at the ends.

The split dipole is fed at the centre by a balanced transmission line. Thus, the two dipole antennas have the same voltages at their ends.

They are essentially two dipoles in parallel as far as radiation fields are concerned. **The radiation pattern of a folded dipole and a conventional half wave dipole is same but the input impedance of the folded dipole is higher.** It differs from the conventional dipole mainly in two respects

1. Directivity
2. Broadness in Bandwidth.

Directivity of folded dipole is **bi-directional** but become of the distribution of currents in the parts of folded dipole the input impedance becomes higher.

Input Impedance of Folded Dipole Antenna

If the radii of the 2 conductors are equal, then equal currents flow in both the conductors, in same direction i.e. *currents are* equal in magnitude and phase in the two dipoles. Since the total power developed in folded dipole is equal to that developed in the conventional dipole; the input (or terminal) impedance of folded dipole is greater than that of the conventional dipole.

It can be proved that the input impedance at terminal of folded dipole antenna is equal to square of number of conductors, comprising the antenna times the impedance at the terminals of a conventional dipole.

A folded dipole antenna has a **built in impedance transforming properties,** which makes it easy to match a transmission line that feeds the antenna.

Alternatively, instead of changing the no. of dipole, it is also possible to change the input impedance by keeping the radii of two dipoles unequal which on other hand changes the currents are flowing. By doing so larger current flows in thicker dipole and thus it becomes possible to attain any input impedance that may be desired with the dipole of unequal radii transformation ratio of 1.5 to 25 can be achieved and this ratio can still be boost up by increasing the no. of dipoles.

The folded dipole does not accept power at any even harmonics (i.e. 2nd, 4th etc.) of fundamental frequency so it appears as a continuation of transmission line.

Equation of Input Impedance

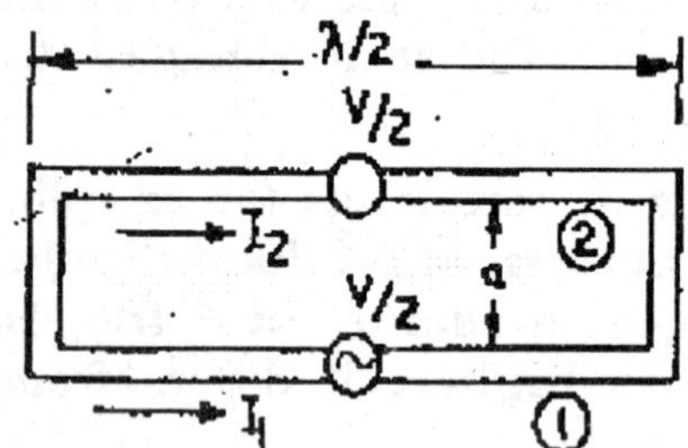

Equivalent diagram of two wire folded $\frac{1}{2}$ wave dipole

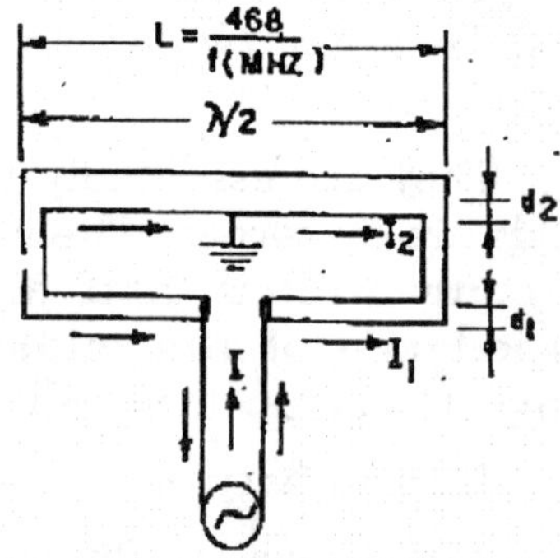

Folded dipole with unequal radii, $I_1 < I_2$ and $I = I_1 + I_2$.

Figure 2.23 (a, b)

V, is the emf applied at antenna terminals AA'

Let this V is been divided equally in each dipole. Hence Voltage in each dipole is V/2 as shown in the above figure, and by nodal analysis-

$$\frac{V}{2} = I_1 Z_{11} + I_2 Z_{12}$$

Where, I_1, I_2 – Currents flowing at terminals of dipole no. 1 and 2

Z_{11}, Z_{12} – Self impedance of dipole no. 1 and mutual impedance between dipole 1 and 2 respectively.

But as $I_1 = I_2$

$$\therefore \frac{V}{2} = I_1 \left(Z_{11} + Z_{12}\right)$$

The two dipoles in system are close to each other, thus the spacing between the two dipoles is of the order λ/100-

And as Z_{11}, Z_{12} are nearly equal to-

$$\therefore \frac{V}{2} = I_1 \left(2Z_{11}\right)$$

or $$\frac{V}{I_1} = 2\left(2Z_{11}\right)$$

$$Z = \frac{V}{I_1} = 4Z_{11} = 4 \times 73$$

where $Z_{11} = 73\,\Omega$ for dipole

$$\therefore Z = 292\,\Omega$$

Because the characteristic impedance of twin lead transmission lines are roughly 300 Ω, thus this dipole is often used when connecting to the above mentioned line, for optimal power transfer.

Similarly, folded dipole of 3 wires, as in the given figure-

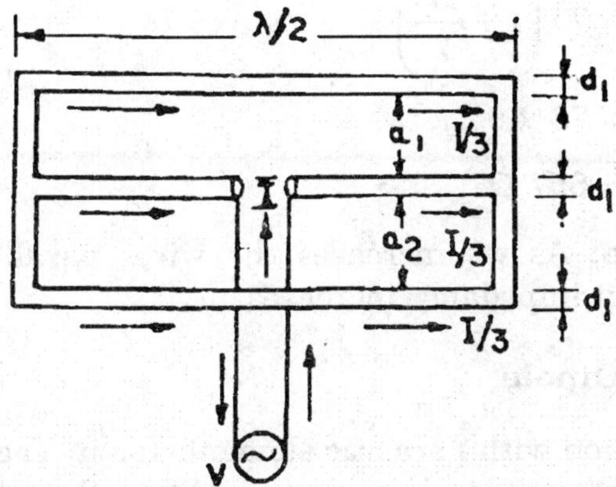

Three wire folded dipole or Tripole with current distribution. The currents in the transmission line is equal and opposite and hence have cancelling effect.

Figure 2.24

It can be proved that termination impedance is-

$$\frac{V}{3} = I_1\,(3\,Z_{11})$$

$$\therefore\; Z = \frac{V}{I_1} = 3^2\,Z_{11} = 9 \times 73$$

$$Z = 657\,\Omega$$

∴ Now generalized equation can be given by-

$$Z = n^2 \times 73$$

n = number of half wave dipoles

Since the impedance transformation is possible by making unequal radii of two dipoles, hence formula (generalized) input impedance is-

$$Z = Z_{11}\left(1+\frac{r_2}{r_1}\right)^2$$

$$Z = 73\left(1+\frac{r_2}{r_1}\right)^2$$

Where r_2 and r_1 = radii of elements

As $r_2 = 2\,r_1$

$$\therefore\ Z = 73\left(1+\frac{2r_1}{r_1}\right)^2$$

$$= 73 \times 9$$

$$= 657\ \Omega.$$

Conclusion: As we increases the wires parallel to the folded antenna the input impedance increases.

Uses of folded Dipole

In Conjunction with parasitic elements (as in Yagi-Uda antenna), the folded dipole antenna is used (as driven element) for the wide band operation such as television.

Advantages

(i) High input impedance

(ii) Wide Band frequency

(iii) Acts as built in reactance compensation network.

1. Draw the radiation pattern of the following antennas: (i) Folded Dipole Antenna **(Raj. Univ. 2007)**
2. Discuss the following Antenna: (i) Folded Dipole Antenna **(Raj. Univ. 2006)**

2.4.1.7. SMALL LOOP ANTENNAS

The small loop antennas are closed loop shaped antennas as shown in the figure-

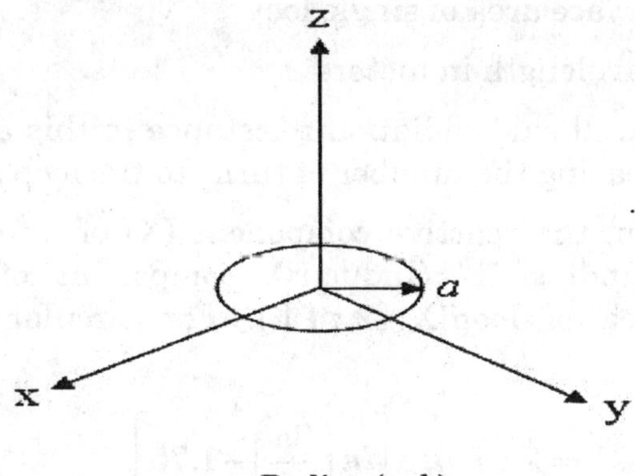

a- Radius (a«λ)

Figure 2.25 Small loop antennas, lying in x-y plane

Small loop antennas are also called as **"DUAL of the dipole antenna"** because the small dipole had a magnetic current flowing (as opposed to electric current as in a dipole), the fields resemble that of small loop.

Differences between small loop and Dipole

As the loop is electrically small, the current in the loop is nearly constant along it, ∴ I = Io,

The fields from small circular loops are given by-

$$E_\phi = \frac{\eta K^2 a^2 I_o e^{-jkr} \sin\theta}{4r}$$

$$H_\theta = \frac{E_\phi}{\eta}$$

Variation of the pattern with the direction is given by "sin θ", so that the radiation pattern of a small loop antenna and short dipole has same pattern. However, the fields of small dipole have the E- fields and H-fields switched relative to short of that dipole; E-field is horizontally polarized in x-y plane.

The impedance of short dipole is capacitive (imaginary part having negative sign) while the small loop antenna have an inductive impedance (having imaginary part with positive sign).

Radiation Resistance of loop antenna-

$$R_r = \left(\frac{177\,NS}{\lambda}\right)^2$$

where N = turns of small loop antenna

S = surface area of single loop

λ = wavelength in meters

∴ It can be well said radiation resistance in this antenna can be increased by increasing the number of turns to the loop antenna.

In small loop, the reactive component (X) of impedance can be determined by finding `L' (Inductive component of Loop) which depends on shape of loop $(X = 2\pi f L)$. For circular loop reactive component is

$$X = 2\pi f.\mu a\left[ln\left(\frac{8a}{\rho}\right) - 1.75\right]$$

where a = radius of loop

ρ = wire radius.

Application

These antennas have low radiation resistance and high reactance, thus their impedance is difficult to match to a transmitter. As a result, these antennas are generally used as a receiver antenna, where impedance mismatch loss is of least importance, as in-

- Pagers
- Field strength probes used in wireless measurements.

2.4.1.8. LOOP-ANTENNAS

Loop antennas are radiating coil of any convenient cross-section of one or more turns carrying RF current. They can be of any shape with ferrite or air core.

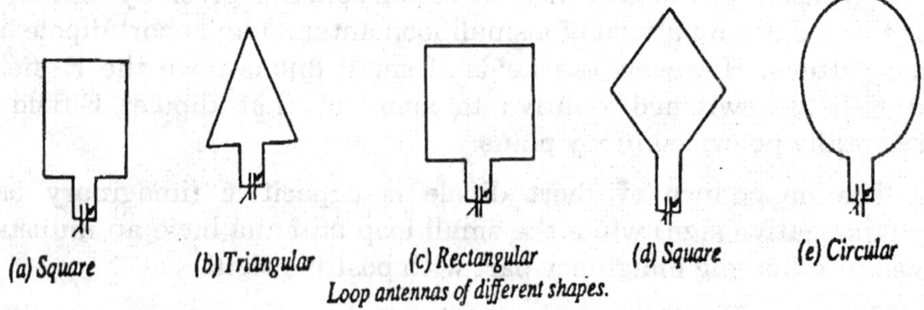

Loop antennas of different shapes.

Figure 2.26

Applications

These antennas are used in-

1. Radio receivers
2. Aircraft receivers
3. Direction finders
4. UHF -Transmitters.

Frame

It is a loop with more than one turn. Dimension of frame is small in comparison to the wavelength. Due to this, the current is of same magnitude and phase throughout the loop.

Case (1)

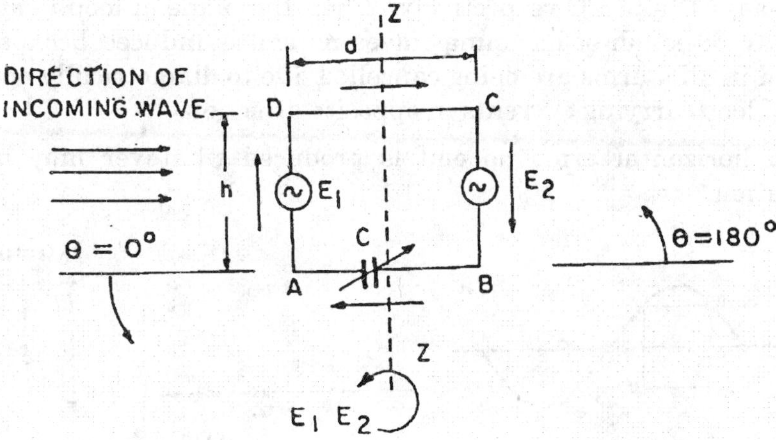

Loop antenna with rotational axis.

Figure 2.27

-Let us consider a rectangular loop antenna with single loop having the rotation along ZZ' axis in the vertical plane.

Out of four arms of rectangular loop, the horizontal arms (AB and DC) acts as horizontal antennas while the two vertical arms (AD and BC) acts as vertical antennas.

If the plane of the loop is set at right angles to the direction from which vertically polarized waves are coming, then the same voltages will be induced in each of the vertical sides of the antennas (AD and BC). But these voltages in the two arms, send current around the loop in the opposite directions, hence they gets cancelled out as both the currents are same in magnitude and opposite in phase. This happens because during normal position of loop, the planes of antenna with respect to the incoming waves both the sides (all 4 arms) are at equidistant from the transmitter.

(a) Perspective view *(b) Plan view. O, BC = O, AD i.e. distance of AD and BC* *(c) Vectors vector difference between E_1 and E_2 is zero.*

Figure 2.28

E1 and E2 represent, the induced voltages in the vertical antennas, AD and BC respectively. When the plane of loop is at right angles to direction of incoming waves no emf is induced because emf induced in all 4 arms are being cancelled due to diametrically opposite sides of loop carrying currents in opposite direction.

In horizontal arms no emf is produced whatever may be the orientation.

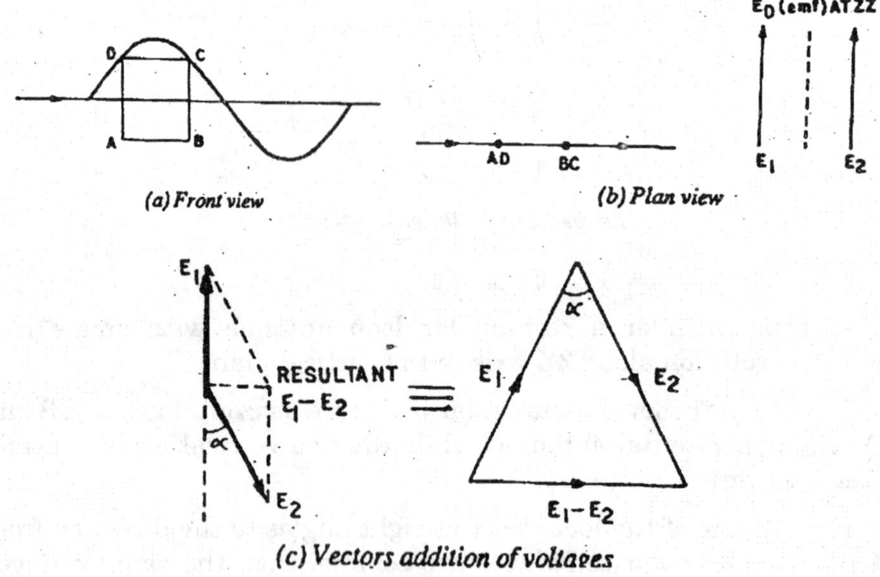

(a) Front view *(b) Plan view*

(c) Vectors addition of voltaees

Figure 2.29

Case (2)

When the loop is rotated by 90° so that plane of the loop is along the direction of incoming radio waves. The voltage induced in each vertical sides of loop are not now cancelled because of the distance involved between two vertical sides equal to width of loop i.e. CD.

The radio waves take some time to reach from one side to another and hence a little phase exists between the induced voltages E1 and E2. At any instant, there is some emf around the loop equal to the difference between emf induced in vertical sides. The r. m. s. value of emf induced in two vertical sides will be same in magnitude but they will be out of phase by amount says α.

So the resultant induced emf across axis (ZZ') = $E_1 \sim E_2$

In any other position of plane loop except the right-angle position, the vertical sides of antenna will be at different distance with respect to the transmitter and hence vectors E1 and E2 are not in phase. The resultant emf, equal to vector difference of E1 and E2 will be produced around the loop.

Loop emf is maximum when the plane of loop is along the direction of incoming radio waves from the transmitter.

Radiation Pattern

Radiation Pattern of loop is independent of exact shape of the loop, if loop is small compared with 'λ'. It is similar to radiation pattern of (Hertzian dipole antenna) with the only difference of the Electric Vector `E' and magnetic field `H' at interchanged position. This is why **small loop is generally referred to as magnetic dipole.** The loop is surrounded by a magnetic field everywhere at right angle to the loop.

Direction finding by Rotation Loop

As compared to a simple vertical antenna, loop antenna has directional properties.

We use loop antenna as compared to frame, to determine the direction of unknown transmitter, which is capable of rotation about vertical axis. **When loop with Receiver is turned 360°, there are two maxima at $\theta = 0°$ and 180° and two minima at $\theta = 90°$ and 270°** ($\therefore \cos\theta = \pm 1$ at 0 and π, and $\cos\theta = 0$ at $\pi/2$ and $3\pi/2$).

Thus when coil is rotated, maximum signal is heard in receiver when loop antenna coil is set along the direction of transmitter and no signal is heard when loop is at right angle to the direction of incoming waves.

This is sufficient to ascertain the direction of the transmitter. It is observed that while rotating the loop antenna change in signal strength is more pronounced in neighborhood of zero signals rather than the maximum signal strength. Thus in practice an accurate bearing is obtained from zero signal position.

When the loop is rotated to either null or maximum signal position, there is still uncertainty whether signal is arriving from left or right (front or back) direction of loop. This uncertainty exists because there are two position of loop, for zero signals and is known as **180° ambiguity. This means loop antenna when used as direction finder is unable to distinguish between bearing of a distant transmitter and its reciprocal bearing.**

NOTE: When the transmitting station is tuned in and the loop is rotated for minimum position then the loop will be in right angles to the transmitting antenna and accuracy of bearing depends on sharpness of zero obtained.

We have only, sensed a line along which the unknown transmitting station bears (i.e. lies) and the actual direction of transmitter that which of may be determined by "**Sensing**", which determines that which of the two directions is correct one i.e. in left or in right direction for the presence of the transmitter.

Determination of sense

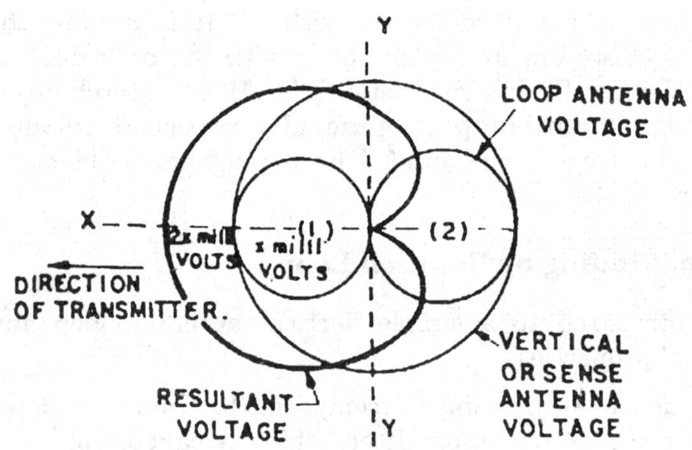

Horizontal radiation pattern of sense and loop antenna.

Figure 2.30

The 180° ambiguity in sense of bearing can be resolved **by employing a open wire or vertical antenna in conjunction with loop antenna. ∴ Vertical antenna is called as "sense antenna".** Apparatus used for finding sense is called as **"sense finder".**

Sense finding is achieved by combining the output of loop antenna with a voltage of equal magnitude, which is derived from sense antenna.

∴ When the output of loop antenna is to be combined with the voltage of vertical antenna then the loop antenna voltage must be passed through 90° phase shifter circuit before the combination so that two output either combines in phase 0° or 180° out of phase.

Now where sense antenna voltage is merged into the loop antenna voltage, an unbalance is created in resultant voltage. The unbalance of voltage makes one of the directions of maximum signal louder than the other. Since the radiation pattern of a loop antenna is figure of eight and that of open wire antenna or sense antenna is a circle, as it receives well in all the directions. Hence, if the two outputs are combined property resultant pattern of heart shape or cardioids are formed in which there is only one minimum and only one maximum.

Receiver connections are generally such that the transmitter bearing is indicated by the position of the receiver, which gives maximum signal i.e. transmitter lies in the x-direction (as in figure).

The emf's of loop and vertical antennas are additive or subtractive, depending upon whether these voltages are either 0° in phase or 180° out of phase respectively. The actual effects, needed to be additive or subtractive are not emfs themselves but potential differences they produce across the RF amplifier of receiver. The previous figure shows graphical variation of resultant input potential difference of RF amplifier due to both emf's as loop antenna is rotated. The potential difference due to loop is figure of eight and due to sense antenna is circle.

The amplitudes of two emf's are adjusted to equality i.e. radius of vertical antenna circle is equal to the diameter of each circles of figure of 8. ∴ Addition takes place towards left and cancellation towards right and variation of resultant potential difference with the direction of top is HEART SHAPE. The radii vector of resultant figure is almost doubled.

First a null position is obtained without the sense antenna is brought into circuit with the help of switch, for determination of sense and loop is rotated until the direction of loudest signal is obtained. A pointer on the loop then shows the correct direction of unknown transmitting station.

The objective of sense finder design is to change in phase by 90° of one effect, relative the other. Because the voltage induce in by wave in sense antenna is 90°. Out of phase with respect to the loop resultant induced voltage e_θ. Thus the steps involved in design are-

(i) To produce a change in phase 90° in the effect of sense antenna emf, relative to that of loop antenna emf and

(ii) Adjustment of effects to equality in magnitude.

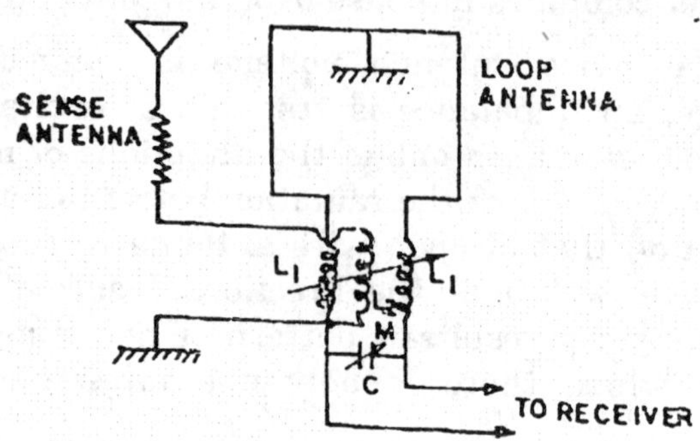

Figure 2.31

Errors in loop direction finding

Loop antennas are subjected to certain errors. Most of these errors are minimized by proper calibration and construction. The prominent errors are-

(1) Antenna or vertical effects.

(2) Night effect or Airplane error or polarization error.

1. Antenna effects

Loop direction finding entirely depends on the fact, that when loop is perpendicular to the direction of propagation, no signal is heard. But anything which causes signal to be heard in this position is harmful for accurate direction finding. One such effect is vertical effect or Antenna effect arises due to unequal action of two vertical sides of loop antenna. The loop antenna gives false bearings if it is not electrically balanced with respect to the ground.

In other words, two zero signal positions will not exactly be spaced 180° apart with respect to each other. The minima may be indistinct as.

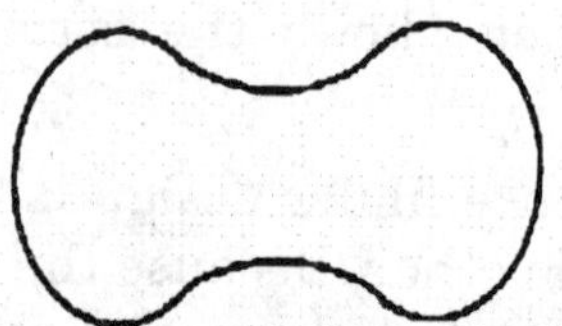

Indistinct minima due to vertical effect.

Figure 2.32

This will introduce errors in the bearing reading. Various methods of minimizing vertical effects are-

1. Screened loop
2. Earthed centre point loop
3. Push pull receiver input circuit
4. Transformer coupling between loop and Receiver.

2. Night Errors

A loop antenna will give correct bearing only when the passing Electro Magnetic waves are vertically polarized i.e. Electric vector is in vertical plane and magnetic vector is in horizontal plane with respect to the earth and there is no horizontally polarized down coming waves present. Although horizontally polarized down coming waves are normally present, induce emfs in the horizontal arms of the loop antenna which do not cancel out even when the plane of the loop is at right angles to transmitter direction. This affect null position and gives incorrect bearing an indistinct minima or both at night. Causing night effect or error, as the frequencies generally used in direction finding work is below 2000 KHz. At these frequencies, the strength of down coming sky wave, reflected from ionosphere is much pronounced in night than during the day.

As the frequency and distance increases, error also increases because, and then the ratio of sky wave to ground wave field strengths is higher. A loop is more frequently used at lower and broadcast frequencies and troubles are introduced by horizontally polarized down coming waves, so it is also called as **polarization error**. The night or polarization errors limit the utility of loop antenna as direction finder. The transmission distances over which it can be used as direction finder become limited. Error also makes this transmission distance dependent on the time of day and night and frequency as well. Even night effect dominates to such an extent that it becomes difficult to use loop antenna at high frequency. Thus, loop antenna due to night effect, is limited only for a very small distance.

The night effect is also produced if the signal is received from an aircraft radiating with a trailing antenna. Under such condition loop antenna is subjected to horizontally polarize down coming waves. These horizontally polarized down coming wave again produced polarization error in the bearing reading and in this case, it is called, **Airplane error or Airplane effect.**

The night effect is minimized by using Adcock system of direction finder, due to constructional feature. The horizontal arms in Adcock system is shielded so that no induced emf in it is due to horizontally polarized down coming waves.

Disadvantages of loop direction finders

A Loop direction finder suffers from the following disadvantages:

(i) It is only suitable for low and medium frequency and not for higher frequencies except for a very small distance.

(ii) Transmission efficiency is very poor and hence cannot be used for transmission purpose unless the loop designs are modified for UHF uses.

(iii) Spurious induced voltages are produced if nearby loop, wires and conductors are present.

(iv) It is subjected to Antenna effect and Night effect which of course are minimized by balancing and using Adcock Antenna. However, bearing accuracy upto 1° or 2° is obtained if horizontally polarized down coming waves are absent.

1. How is the loop antenna useful for direction finding? **(Raj. Univ. 2003)**
2. Write short note on circular loop antenna. **(Raj. Univ. 2001)**
3. Describe in which frequency range and where the following antennas are normally used (in not more than 10 lines for each): (i) Loop antenna **(Raj. Univ. 2004)**

2.4.2. TRAVELLING WAVE ANTENNA

Travelling Antennas

Resonant or tuned antennas (in HF band) have narrow band width. High frequency tuned antennas involve considerable ringing and expense and hence they are suitable only for fixed frequency operation on an allotted frequency.

Radio communication, with reflections from ionospheres needs to operate on widely spaced frequencies. It becomes necessary to shift rapidly from one frequency to another for a day and night working. Such needs are being met by these travelling wave radiators.

When antennas are to be operated on a number of allotted frequencies, travelling wave antennas are employed.

Travelling wave Antenna: - are those antennas in which there are no standing waves and waves travel in only one direction. This is obtained simply by terminating one end of antennas by a non-inductive resistance of value equal to Z_0 such antennas are called as - *A periodic or non resonant antennas*. These antennas have increased bandwidth.

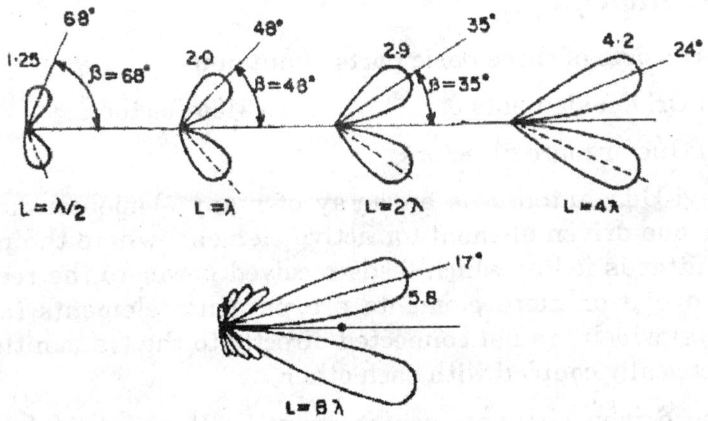

Radiation patterns for travelling wave antenna of different lengths.

Figure 2.33

2.4.2.1. YAGI - UDA ANTENNAS

These antennas are one of the oldest and most common antenna design known to all. It is very simple to construct. These antennas operate in the HF to UHF bands (about 3MHz to 3GHz). The Bandwidth of Yagi Uda antennas is very small, of the few percent of center frequency.

It has high gain greater than 10 dB.

Figure 2.34 Roof Top Antennas for Receiving T.V. Signals

History

Yagi Antenna was invented in Japan, with results first published in Japanese in 1926, by the scientist Shintaro-Uda. His work was presented for the first time in English by H.Yagi, who went to America and gave first English speech on this antenna, which led to wide use of this antenna. Hence, even though the antenna was invented by S. Uda, then also the antenna is generally called as Yagi antenna.

Construction

It consists of three basic parts of antenna-

(1) Driven elements (2) Reflector

(3) One or more directors

Yagi-Uda antenna is an array of many elements; out of which there is one driven element (or active element) where the power from Transmitter is fed or which feeds received power to the receiver. The rest and one or more elements are parasitic elements (i.e. passive elements) which are not connected directly to the transmitter line but are electrically coupled with each other.

The driven elements are resonant, half wave dipole usually of metallic rod at frequency of operation. The parasitic elements of continuous metallic rods are arranged parallel to driven element and at same line of sight level. They are arranged collinearly and close together as in figure, given with one reflector and one director.

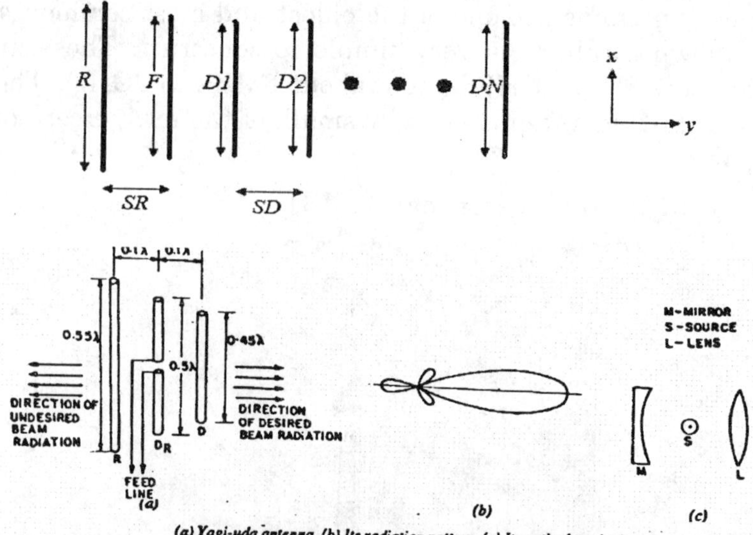

(a) Yagi-uda antenna, (b) Its radiation pattern, (c) Its optical equivalent
R = Reflector (Parasitic element) ; D_R = Driven element ; D = Driector (Parasitic element).

Figure 2.35 Geometry of Yagi-Uda Antenna.

F- Driven element (Active Feed)

D1, D2,..., Dn- Director (Parasitic)

R- Reflector (Parasitic)

SR- Distance between R and F

SD- Distance between the two directors

The parasitic elements receive their excitation from the voltages induced in them by current flow in driven element. The phase and currents flowing due to induced voltage depend on the spacing between elements and upon reactance of elements (i.e. length). The reactance may be varied by dimensioning the length of parasitic element.

The spacing between driven and parasitic elements that are usually used, in practice are of order of $\lambda/10$, i.e., 0.10 λ to 0.15 λ. The parasitic element in front of **driven** element is known as **director** and its number may be more than one, whereas element in back of its known as director. Generally, both directors and reflectors are used in same antenna. **The reflector is 5% more and director is 5% less than the driven element which is $\lambda/2$ at resonant frequency.** In practice, for three-element array of Yagi-antenna the following formula gives length which work satisfactorily-

$$\text{Reflector length} = \frac{500}{f(MHz)} \textit{ feet}$$

$$\text{Driven element length} = \frac{475}{f(MHz)} \textit{ feet}$$

$$\text{Director length} = \frac{450}{f\ (MHz)} \textit{ feet}$$

Dipole impedance $\Rightarrow$ *between* 23 Ω *to* 73 Ω

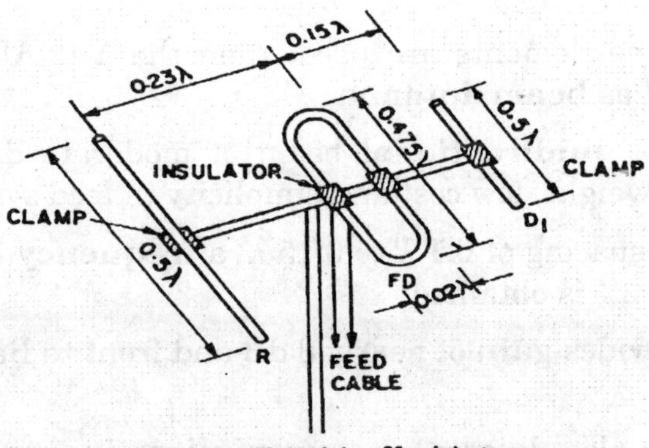

A typical Television Yagi Antenna.

Figure 2.36

Working

The spacing between elements and lengths of parasitic elements determine the phase of currents. Parasitic antenna in vicinity of radiating antenna is used either to reflector to direct the radiated energy so that a compact directional antenna system could be obtained.

A parasitic element of equal or greater length than $\lambda/2$ will be inductive while elements of length less than $\lambda/2$ will be capacitive. Hence phases of currents in former case (i.e. length > $\lambda/2$) will ***lag*** induce voltage whereas in latter case (i.e., length < $\lambda/2$) will ***lead*** induced voltage. Properly spaced dipoles shorter than $\lambda/2$ acts as director and add the fields of driven element in direction away from driven element. If more than one director is employed, then each director will excite next. On the other hand an element of length equal or greater than $\lambda/2$ acts as reflector and sums up fields of driven element in direction from reflector towards driven element, if properly spaced. **Additional gain is achieved by using additional directors in beam direction.** The distance between two elements may range from 0.1 λ to 0.3 λ, close spacing of elements are used in parasitic arrays to get good excitation. **Greater distance between driven and director elements, the greater the capacitive reactance needed to provide correct phasing of parasitic current.** $\therefore$ the length of rod is tapered off to achieve the capacitive reactance.

The **driven** element radiates in front to rear direction (i.e. from reflector to directors) If distance between driven and director elements is decreased, then it will load driven element, irrespective of its length. Thus input impedance at input terminals of driven element reduces. This is why a folded dipole is invariably used as driven element so that reduction input in impedance is compressed i.e. raised.

General characteristics

(1) If three-elements are used, then the Yagi-Uda antenna is called as **beam antenna.**

(2) It has **unidirectional** beam of moderate directivity with light weight, low cost and simplicity in feed system design.

(3) With spacing of 0.1 λ to 0.15 λ, a **frequency bandwidth** of order 2% is obtained.

(4) It provides **gain** of nearly 8 dB and front to Back Ratio of 20 dB.

(5) It is also known as ***super directive or super gain antenna*** due to its high gain and beam width per unit area of an array. An antenna or array which provides directive

gain, appreciable greater than that obtainable from uniform distribution is known as **super directive antenna**.

(6) If greater directivity is desired, further elements may be used.

(7) It essentially is a fixed frequency device i.e. frequency sensitive and bandwidth of about 3% can be obtained. This bandwidth is sufficient for T.V. Reception.

(8) Generally we use only one reflector, but by adding more reflector improve the performance, slightly, as it changes the front to back Ratio of antenna.

(9) Reflector are kept little longer than resonant (directors), due to the fact that-

(a) Larger the element is, better reflection can take place.

(b) If the reflector is longer than its resonant (directors) length, the impedance of reflector will be inductive. Hence the current on the reflector **lags** the voltage induced on the reflector. The director, will be shorter than the resonant making them capacitive, so that current **leads** the voltage; which cause a phase distribution to occur across the elements, simulating the phase progression of wave across the array. Thus arrays are said to be traveling wave antenna.

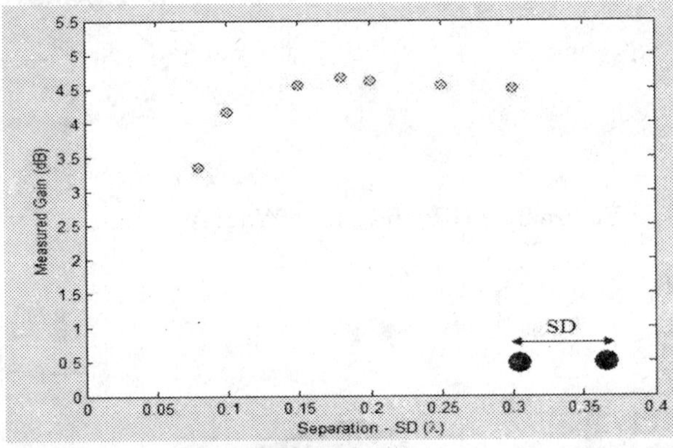

Figure 2.37 Gain versus separation for 2-element Yagi antenna.

(10) Now it is interesting to know here that, Yagi antenna with-

1 - Reflector

1 - Feed element and

0 – Directors can very simply be called as, Half wave dipole antenna.

(11) Gain versus separation for Yagi Antenna

The above graph shows that the gain increases by 2.5 dB, if the separation SD is changed between 0.15 λ to 0.3 λ.

Adding on additional director always increases the gain; however the gain in directivity decreases as the number of elements gets larger. For instance, if there are eight directors, and one more director is added, the increase in gain will be less than 0.5 dB.

(12) Radiation Pattern of Yagi-Uda Antenna

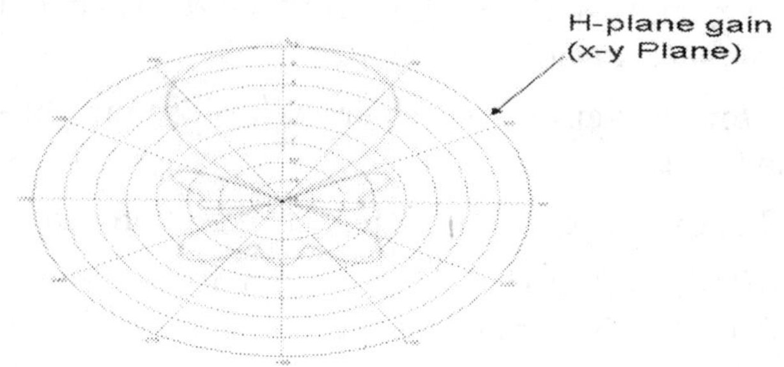

Figure 2.38 E-plane gain of Yagi antenna.

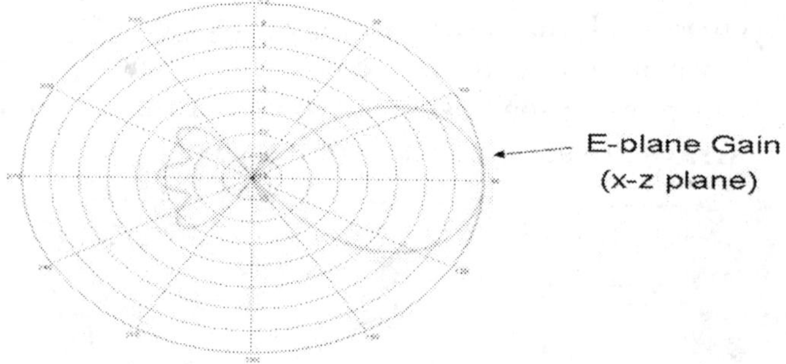

Figure 2.39 H-Plane gain of Yagi Antenna.

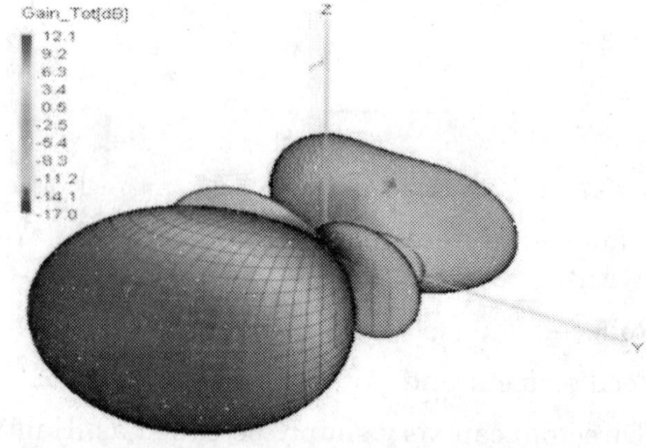

Figure 2.40 3-D Radiation Pattern of Yagi Antenna

(13) Gain can be increased (and pattern could be made more directional) by adding more directors or optimizing space or sometimes (not generally) by adding another reflector.

(14) Front to Back Ratio is 19 dB.

1. Draw the radiation pattern of the following antennas: (i) Yagi-Uda (Five elements) Antenna **(Raj. Univ. 2007, 2008)**
2. Describe in which frequency range and where the following antennas are normally used (in not more than 10 lines for each): (i) Yagi antenna **(Raj. Univ. 2004)**

2.4.2.2. HELICAL ANTENNAS

The most popular helical antenna (Helix), is a travelling wave antenna in the shape of corkscrew which produces long distance radiations along the axis of helix. Helixes have a different shape as seen in the picture –

Figure 2.41

They are also referred to as axial-mode helical antennas.

Benefits:

(1) Wide Bandwidth.

(2) Easily constructed.

(3) Real input impedance.

(4) Has circularly polarized fields.

(5) Helical antenna is another basic type of radiator and perhaps it is simplest antenna to provide circularly polarized waves.

(6) Helical antenna is used in Extra Terrestrial communication in which satellite relays etc. ere involved.

NOTE: In the given picture of the Helical antenna, if along the helix curve, one curves the finger of hand it will be only right hand fingers to curl against the helix with thumbs up in the sky (So the waves emitted from the antenna is Right Hand Circularly Polarized) so the given picture is Right Handed helical antenna.

Now if we wound the helix antenna the other way, it will emit out as Left Handed Circularly Polarized wave and thus it will be Left Handed helical antenna.

Travelling wave antenna means that the current travels along the antenna and the phase waves are in phase.

Construction

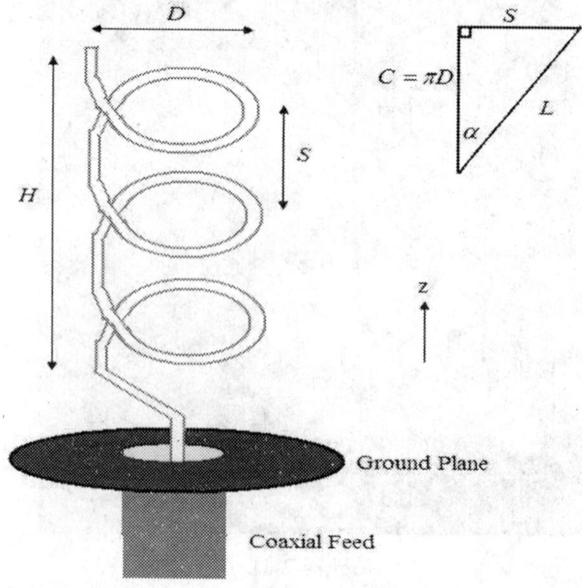

Figure 2.42

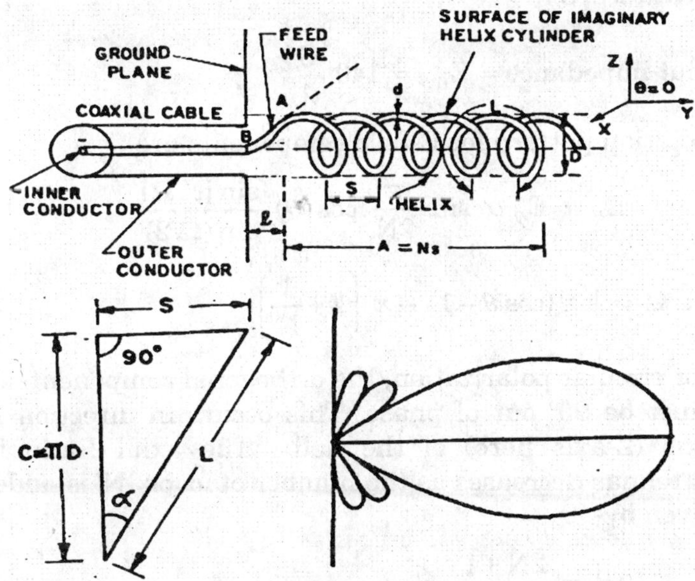

Figure 2.43

It consists of a helix of thick copper wire or turning wound in the shape of screw thread and used as an antenna in conjunction with a flat metal plate called as ground plate.

D = Diameter of a turn on the helix

C = Circumference of helix (πD)

α = Pitch angel $\alpha = \tan^{-1}\left(\frac{S}{\pi D}\right) = \tan^{-1}\left(\frac{S}{C}\right)$

d = dia. of helix conductor

A = Axial length = NS

N = No. of turns

S = Length of one turn

l = Spacing of helix from ground plane

$$L = \sqrt{S^2 + C^2} \qquad L = \sqrt{S^2 + (\pi D)^2}$$

Pitch angle (α) = angle between line tangent to helix wire and plane normal to axis.

$$\tan\alpha = \frac{S}{C} = \frac{S}{\pi D}$$

$$\therefore \alpha = \tan^{-1}\left(\frac{S}{\pi D}\right) = 12^0 - 14^0$$

Basic Parameters

- Input impedance = $Z_{in} = 140 \frac{C}{\lambda}$
- Radiation pattern for E-field components are-

$$E_\theta \,\alpha\, E_\phi \,\alpha\, \sin\frac{\pi}{2N}\cos(\theta)\frac{\sin(N\,\Omega)}{\sin(\Omega/2)}$$

where $\Omega = k\,S(\cos\theta - 1) - \pi\left(2+\frac{1}{N}\right)$

- For circular polarization, the orthogonal components of E-field must be 90° out of phase. This occurs in direction near the axis (Z-axis here) of the helix. The axial Ratio for helix antennas decreases as the number of loops `N' is added and is given by-

$$AR = \frac{2N+1}{2N}$$

- Gain of helix

$$G = \frac{6.2\,c^2\,N\,S}{\lambda^3} = \frac{6.2\,c^2\,N\,S\,f^3}{c^3} \quad \text{As } \lambda = \frac{c}{f}$$

- Half power beam width of Helical antenna

$$HPBW = \frac{65\,\lambda}{C\sqrt{\frac{NS}{\lambda}}}$$

A helical antenna may radiate in many modes but prominent modes of radiation are two i.e.

1. Normal or perpendicular mode of radiation.

2. Axial or endfire or beam mode of radiation.

1. Normal mode of radiation

- In this mode of operation the radiation field is maximum in broadways.
- Radiations are **circularly polarized** waves.
- Basic equations-

$$\text{Axial Ratio} = \frac{2S\lambda}{\pi^2\,D^2}$$

$$S = \frac{c^2}{2\lambda}$$

$$\alpha = \tan^{-1}\left(\frac{c}{2\lambda}\right)$$

- Mode of operation is narrow in Bandwidth.
- In this we have small radiation efficiency.
- It is used only in limited operations and is rarely used.

2. Axial or beam mode of radiation

This mode of operation is generally used. In this the radiation field is maximized in end fire direction i.e. along the helix axis. It has either circularly or nearly circularly polarized wave.

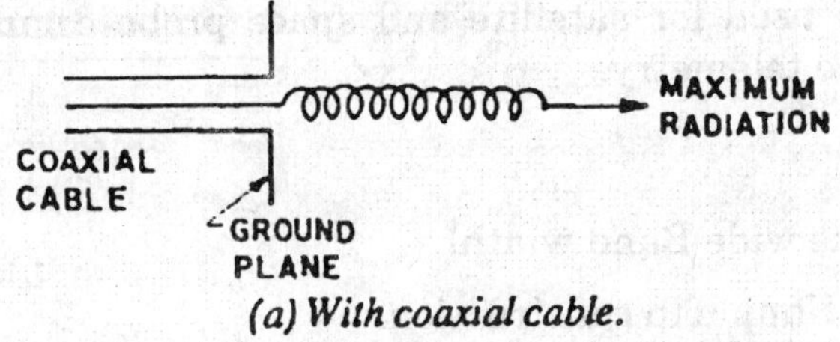

(a) With coaxial cable.

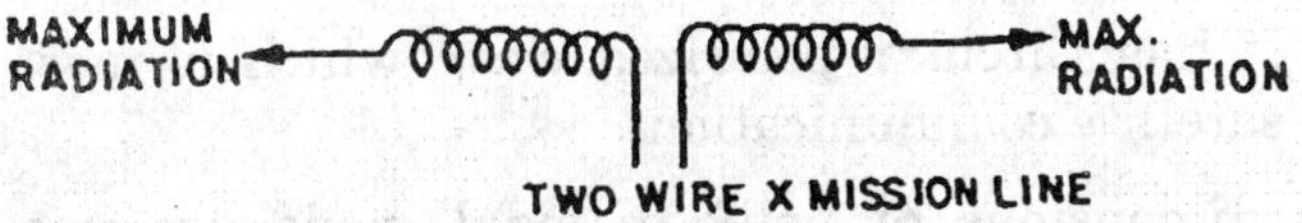

(b) Two wire transmission line.

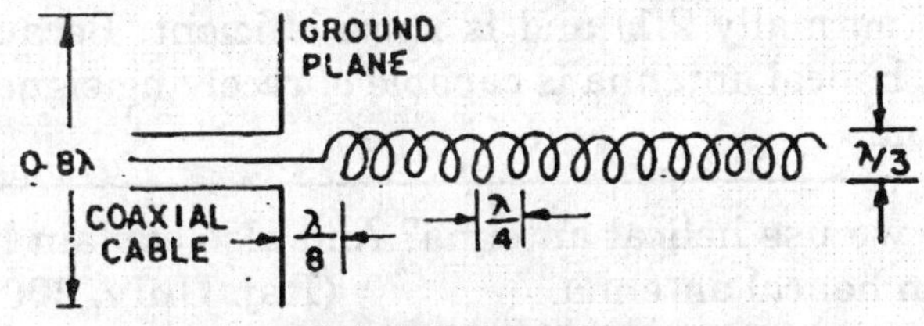

(c) Typical dimensions.

Arrangement for generating axial mode.

Figure 2.44

Arrangement for generating Axial mode

$$R = \frac{140c}{\lambda}\Omega$$

$$(\text{HPBW})_\theta = \frac{52}{c}\sqrt{\frac{\lambda^3}{NS}}\ \deg ree$$

$$(\text{BWFN})_\theta = \frac{115}{c}\sqrt{\frac{\lambda^3}{NS}}\ \deg ree$$

$$\text{D} = \frac{15NSc^2}{\lambda^3}\ \Omega \text{ and}$$

$$\text{AR} = \frac{2\text{N}+1}{2\text{N}}$$

Applications

1. This antenna is commonly used to transmit or receive the VHF signals through the ionospheric layer of the sky.
2. It is used for satellite and space probe communication i.e. space telemetry.

Advantages

1. It has wide Band width.
2. It is Simple in construction.
3. It has highest directivity.
4. It has Circular polarized wave which is most suitable for satellite communication.

(The dimensions of helix in axial mode are not critical and thereby resulting in greater Bandwidth. This end fire axial mode is most practical as it can achieve circular polarization over wider Bandwidth, (normally 2:1) and is more efficient. Because of circular polarization, helical antenna is capable of receiving signals of any kind of polarization.

1. Why do we use helical antenna? And also explain helical pitch angle in helical antenna. **(Raj. Univ. 2007)**
2. What is helical antenna? What are its different modes? What is type of polarization? Where is it used? **(Raj. Univ. 2002, 2003)**
3. Explain two major modes of operation of a helical antenna. Which are the two modes preferred and why? List out its major advantages over conventional line antennas. **(Raj. Univ. 2001)**

4. Describe in which frequency range and where the following antennas are normally used (in not more than 10 lines for each): (i) Helical Antenna **(Raj. Univ. 2004)**

2.4.2.3. LOG PERIODIC (DIPOLE ARRAY)

Frequency Independent antennas-

Condition

It was pointed out by V.H. Rumsey that the performance of a loss less antenna is independent of frequency, if their dimension in terms of wavelength remains constant.

Such a result could be achieved if antenna could be specified in terms of angles. This can be fulfilled by any antenna, whose equation in spherical co-ordinates is of the form-

$$r = e^{a(\phi + \psi_o)} f(\phi)$$

$$f(\phi) = function\ of\ \theta$$

in case of planar antenna the above equation is reduced to-

$$r = e^{a(\phi + \theta_o)}$$

This is equation of an equiangular or logarithmic spiral, where, `a' is rate of expansion.

Where θ_o = is orientation.

According to V.H. Rumsey, an antenna which is defined only in terms of angles should have characteristics of frequency independently. Frequency independent (F.I.) antenna may be defined as **the antenna for which the impedance, pattern and the directivity, remains constant as a function of frequency.** *Thus in order that an antenna to be frequency independent, the antenna should expand or contract in proportion to the wavelength or if antenna structure is not mechanically, adjustable, the size of active or radiating region should be proportional to wavelength.*

Duttamel reasoned that it is possible to force radiation from the rear of Angle structure, with properly located discontinuities; and log periodic antennas are one of those antennas. Log- periodic antennas are broadband antennas where bandwidth of 10:1 can be achieved easily and even 100:1 is feasible if theoretical design closely approximated. The broadband characteristic of Log periodic antenna includes both, the impedance and the pattern. *Radiation pattern may* *be **bi-directional** or **uni-directional** of low to moderate direction gain. Much higher directive gain can be achieved by using them as elements of an array.*

1. Write short note on frequency independent antenna. (Raj. Univ. 2003)
2. Discuss the following Antenna: (i) Frequency Independent Antenna (Raj. Univ. 2006)

Log periodic antenna structure is so chosen that electrical properties must repeat periodically with logarithm of frequency. Frequency independence can be obtained when the variation of properties is over one period, and hence all the periods, is small (change in size of structures). The structure size changes with each repetition by a constant scale factor so that the structure expands or contract.

The log periodic principle can be understood by the **"Log Periodic Dipole Array" (LPDA). All the dimensions increase in proportion to distance from the origin.** It has a number of dipoles of different lengths and spacing, and is fed by a balanced two-wire transmission line which is transposed between each adjacent pairs of dipoles. It is fed at narrow end and the maximum beam radiation. The dipole length increases along the antenna such that included angle 'α' is constant. The lengths (L) and spacing(R or S) are graduated in such a way that certain dimensions of adjacent elements bear a constant ratio to each other like-

Scale factor or design ratio (τ) <<1

$$\frac{R_1}{R_2} = \frac{R_2}{R_3} = \frac{R_3}{R_4} = - \text{----} = \frac{R_n}{R_{n+1}} = \tau$$

$$\frac{L_1}{L_2} = \frac{L_2}{L_3} = \frac{L_3}{L_4} = \text{------} = \frac{L_n}{L_{n+1}} = \tau$$

$$\therefore \frac{Rn}{R_{n+1}} = \frac{L_n}{L_{n+1}} = \tau \qquad (1)$$

Where

τ = design Ratio = Scale Factor = Periodicity Factor.

Again, $\frac{S_{n+1}}{S_n} = \frac{L_{n+1}}{L_n} = \frac{1}{\tau} = k$

$$\frac{S_{n+1}}{S_n} = \frac{L_{n+1}}{L_n} = k; \quad k > 1 \qquad (2)$$

These two conditions cause the ends of dipoles to lie along two straight lines, that meet at an angle `α' at one end and converges at the other end of the structure.

Typical value of $\alpha = 30^\circ$ and $\tau = 0.7$.

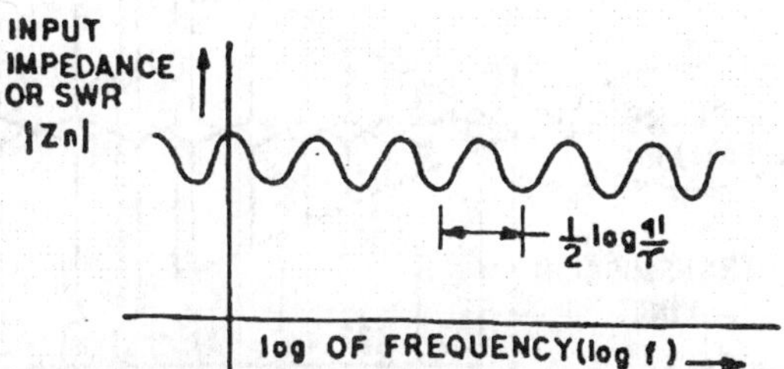

Plot of Input Impedance | Zin | versus logarithm of frequency.

Figure 2.45

A graph is plotted between input impedance $|Z_{in}|$ (and SWR on the feed line) and frequency a repetitive variation is observed. If this plot is made against the logarithm of frequency, instead of frequency itself, then this variation will be periodic. In other words input impedance $|Z_{in}|$ will go through identical cycles of variations; where each cycle is exactly like the preceding one; but this does not implies that the variation is necessarily a sinusoidal wave. **It is this behavior of log periodic antenna that has given the name "Log periodic" which implies that the impedance is logarithmically periodic function of frequency. Not only this, all electrical properties under-go similar periodic variations, particularly radiation pattern, directive gain, side lobe level, beam width and beam direction.**

Let the lower and upper frequencies be, f_1 and f_2 respectively. These frequencies can be correlated by the formula-

$$\log \frac{f_2}{f_1} = \log \frac{1}{\tau}$$

or $$\frac{f_2}{f_1} = \frac{1}{\tau}$$

or $$f_1 = \tau f_2$$

$$f_2 > f_1$$

Therefore, it can be concluded, that, **when the log periodic antenna is operated at a given frequency it is observed that all**

structure does not radiated but only a certain portion radiates known as "active region".

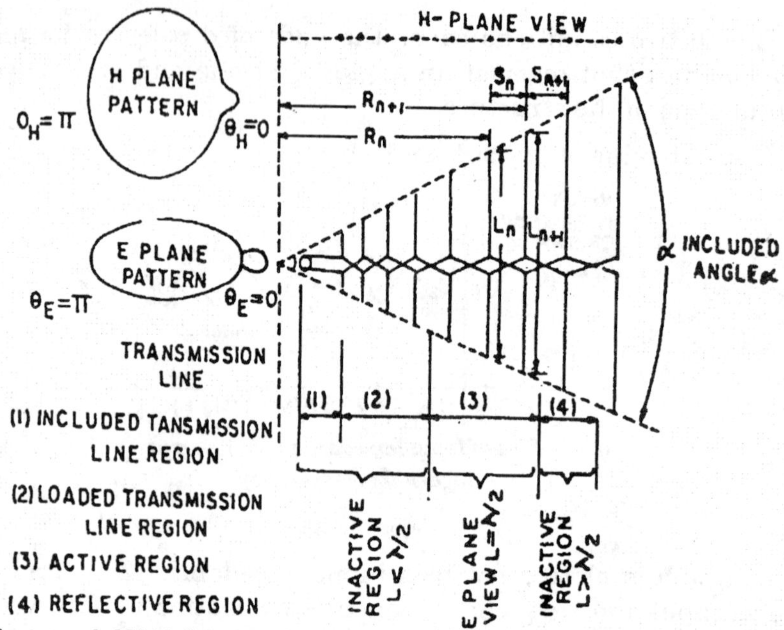

(b) Radiation pattern of a LPDA in E-plane and H-plane. *(a) A log periodic dipole array with main region of operation.*

Figure 2.46

The Log- Periodic antenna can be divided into four basic regions, which are-

1. Transmission line region
2. Loaded Transmission line region
3. Active Region
4. Reflective Region

Cut off frequencies

They are those frequencies at which the longest and the shortest dipoles are nearly equal to that of resonant length i.e. $\lambda/2$.

Active region of antenna is towards the (apex) shorter element, for the highest frequencies at the middle for intermediate frequencies and near the longest elements for lowest frequencies. In other words phase centre from longest end to shortest end as frequencies change from minimum to maximum. Maximum to minimum ratio of frequencies determines the bandwidth.

Radiation pattern is-

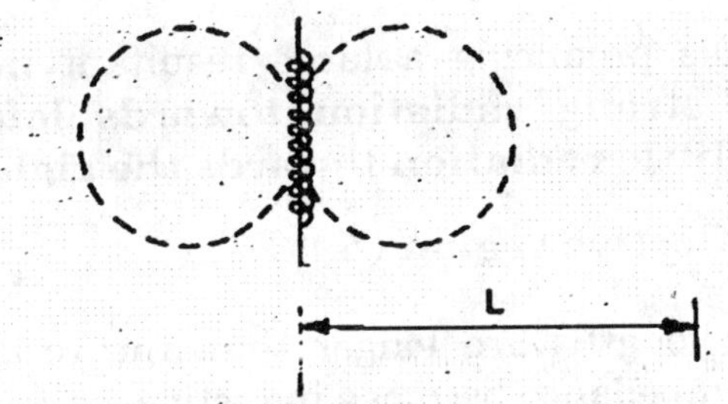

Normal mode of radiation (Circular polarization).

Figure 2.47

for structures having two active regions and fed at centre;

And

O , radiating for structures having only one active region.

The analysis of a log-periodic dipole array can be done by considering three regions of antenna-

(i) **Transmission - line region** (Inactive region L < λ/2)

(ii) **Active Region** (L Ω λ/2)

(iii) **Reflective region** (Inactive region (L > λ/2).

(i) Inactive transmission line Region

At middle of operating range, the antenna elements are short with resonant length i.e. L ≤ λ/2, therefore, the elements present a relatively high capacitance impedance. The element current is small and leads the base voltage supplied by the transmission line by 90° (approx.). The element spacing in wavelength is also small. By transposition of transmission line introduces 180° phase shift between adjacent dipoles. **Hence currents in elements of these regions are small and hence small radiations are in backward direction (left).**

(ii) Active Region (L = λ/2)

In this region the dipole lengths are approximately (L = λ/2) and therefore, the impedance offered by dipoles of this region are resistive. Hence, the element's current are large and in phase with base voltage. The current just below resonance is slightly, ***leading*** and above resonance slightly ***lagging***. The spacing between the two elements is now sufficiently large, causing the phase in a particular element to lead approx. by 90°.

eg. – By the time field radiated from element l_{n+1} reaches l_n, the phase of l_n advances by 90⁰ and its fields add to the field of l_{n+1}

elements, in phase producing a large resultant field towards left. Hence **there is strong radiation towards left in backward direction and a little radiation towards the right.**

(iii) Inactive Reflective region ($L > \lambda/2$)

The element lengths are longer than the resonant length (i.e. $L \geq \lambda/2$) hence impedance becomes inductive, causing the current in elements to lag base voltage. The base voltage supplied by transmission line is now very much small as almost all energy transmitted down line has been attracted and radiated by the active region. This region presents large reactive impedance to the line and thus, **any small amount of incident wave front of active region is reflected back towards backward direction.**

General characteristics

(1) Log Periodic antenna or array is excited from shorter length side or high frequency side for one active region, antenna and at the centre for two active region. They are fed by a balanced two wire transmission line.

(2) For **unidirectional** log Periodic antenna, ***structure fires in backward direction*** (i.e. towards shorter element) and forward radiations are very small or zero (i.e. towards right).

For **bidirectional** log periodic antenna, the maximum radiation is in broadside direction (i.e. normal to the surface of antenna).

(3) Transmission line inactive region must have proper characteristics impedance with negligible radiation; as the input impedance depends chiefly on the characteristic impedance of transmission line that feeds the antenna.

(4) In active region, the currents magnitude and phasing should be proper so that strong radiations occur along background direction and zero or negligible radiation along forward direction (in case of unidirectional) and broadside for bidirectional pattern. Typical values for spacing is $\lambda/4$ and phase is 90° for unidirectional zero phase for bidirectional.

(5) In inactive reflective region, there should be rapid decay of current within this range for a successful Frequency Independent Antenna, i.e., here the structure should be truncated, effectively.

(6) Out of all log periodic antennas, all are not frequency independent; which are broadband antennas.

(7) Gain of antenna lies between 7.5 dB to 12 dB.

Applications

1. Used in **HF Communication** (where multiband steerable and fixed antennas are generally used).
2. This antenna is also used in **T.V. reception,** where only one log Periodic design, suffices for all the channels (even upto UHF band).

Advantage

1. No power is wasted in terminating resistance.
2. It is best suited for **all round monitoring** in which case a single Log Periodic antenna will cover all the higher frequencies bounds, if the cost in the installation is no problem.

1. Why we say log-periodic antenna and also explain log-periodic antenna. **(Raj. Univ. 2008)**
2. Write short note on log periodic antenna. **(Raj. Univ. 2000)**
3. Describe in which frequency range and where the following antennas are normally used (in not more than 10 lines for each): (i) Log-periodic antenna **(Raj. Univ. 2004)**
4. What means frequency independent antenna and also define Log-Periodic antenna. **(R.T.U. 2009)**

2.4.2.4. LONG WIRE / HARMONIC ANTENNAS

Condition for resonant antenna = Length of antenna = $\frac{n\lambda}{2}$

Definition

As antenna is more than half wave long it is called as **long wire antenna.** They are single wire $(4\lambda/2 \; or \; more \; \lambda/2)$ antenna with length $n\,\lambda/2$. As *n*, increases the directivity also increases.

Characteristics

1. The directional characteristics split into various lobes at different angles, from wire axis, as against ``doughnut shape" of a single $\lambda/2$ antenna.

2. It radiates **horizontally polarized** wave at low angles from about 17° to 24° relative to earth surface.
3. In long wire antenna, the currents in adjacent half wave section must be out of phase. This condition can be satisfied if long wire antenna if fed at either end or at any current loop. A long wire is made up of half wave length long at lowest frequency of operation and fed at the end.
4. The long wires are of two types –

   (i) Resonant (unterminated)

   (ii) Non-Resonant (terminated)

(i) Resonant long wire antenna

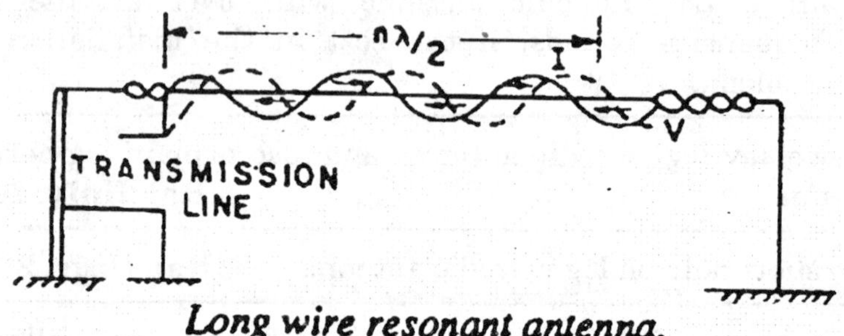

Long wire resonant antenna.

Figure 2.48

They are unterminated antennas.

(a) Standing wave exists along the length of this antenna.

(b) Pattern is bidirectional (corresponding to incident waves and reflected waves.

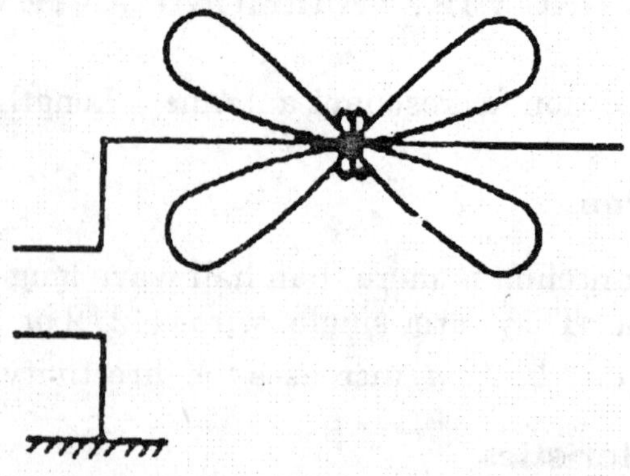

Bi-directional due to resonant antenna.

Figure 2.49

(ii) Non-Resonant long wire antenna

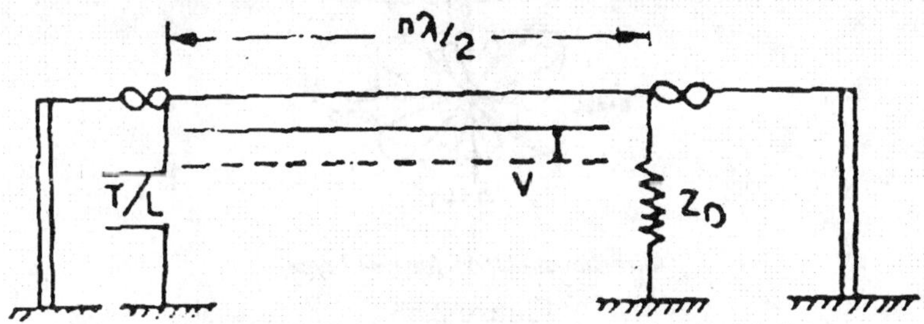

(b) Long wire non-resonant antenna.

Figure 2.50

(a) These antennas are terminated by characteristics impedance.

(b) All the incident waves are absorbed in terminating impedance and these are no reflected back.

(c) The pattern is only due to incident waves i.e. unidirectional only and (uniform current and voltage exist along the axis of wire).

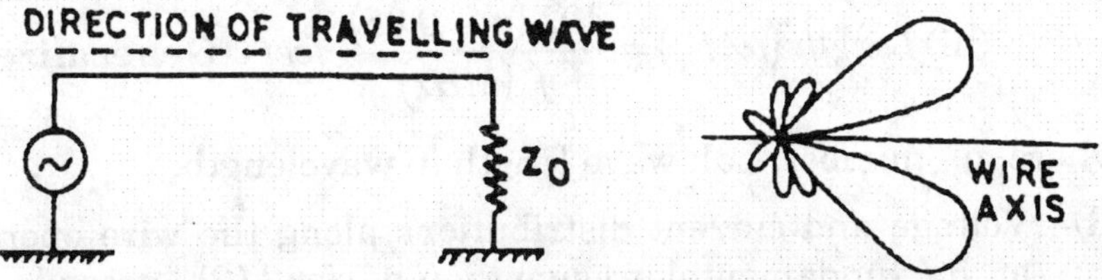

(a) A travelling wave radiator. *(b) Radiation pattern.*
Travelling wave radiator and its radiation pattern.

Figure 2.51

The angle of radiation with reference to wire axis depends on number of wavelength (i.e. even or odd). Directional Pattern for-

n = 3

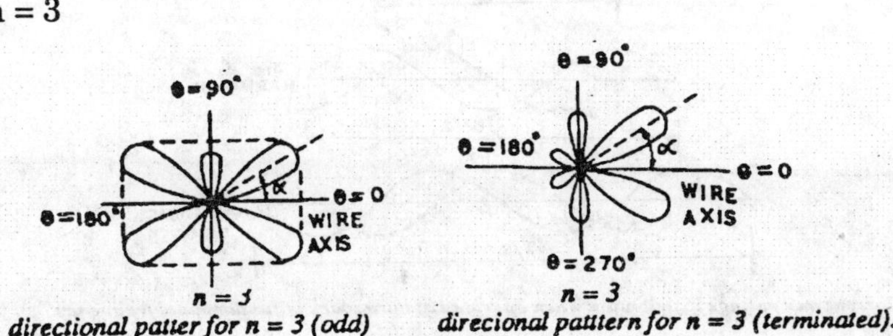

directional patter for n = 3 (odd) *direcional pattern for n = 3 (terminated).*

Figure 2.52

n = 4

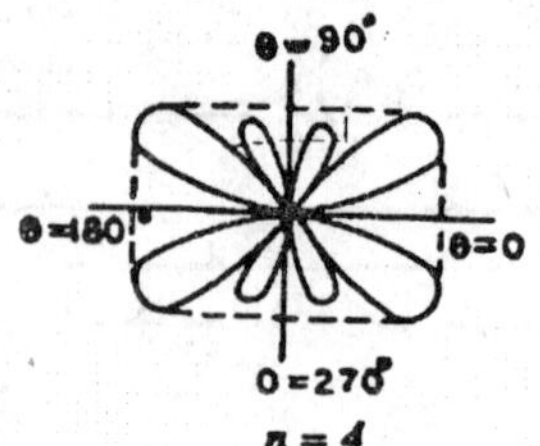

directional pattern for n=4 (even)

Figure 2.53

(1) Max. Radiation from long wire antenna of 8 λ long with respect to wire and minor lobes is 17.5°.

(2) Physical length of long wire antenna can be extended from physical length of l/2 antenna

(i) $Length\ \lambda/2 = \dfrac{492 \times (1-0.05)}{f\ (MHz)}\ feet$

(ii) $Length\ \dfrac{2\lambda}{2} = \dfrac{492\ (2-0.05)}{f\ (MHz)}\ feet\ (2\ half\ \lambda/2)$

(iii) $Length\ (n\lambda/2) = \dfrac{492\ (n-0.05)}{f\ (MHz)}\ feet$ **(Generalized)**

where, n - number, half wave length in wavelength.

(3) Voltage and current distributions along the wire operating at its fundamental frequency (i.e., for $\lambda/2$), second, third and fourth harmonic are-

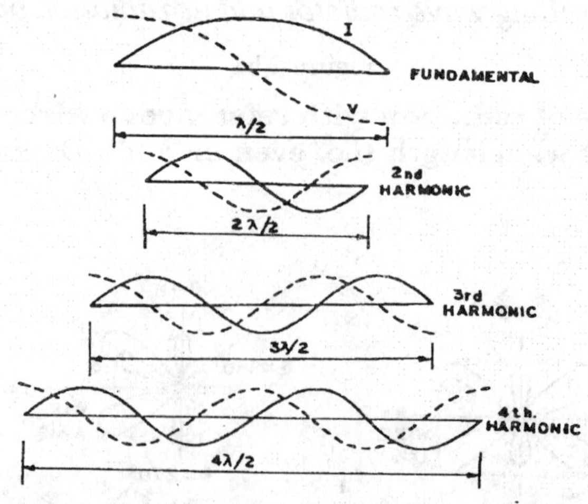

Current and voltage distribution when operated at harmonics of fundamental frequency.

Figure 2.54

Current and voltage distribution when operated at Harmonics of fundamental frequency

(1) The Radiation Resistance of a resonant long wire antenna, with 'n' half wavelength long, in free space is-

$$R_r = 73 + 69 \log_{10} n$$

where, space(n) - number of half wavelength

(2) The angle of maximum radiation i.e. angle between maximum lobe and wire axis is-

$$Cos\ \theta_{max} = \frac{n-1}{n}$$

2.4.2.5. THE V-ANTENNA

V- Antennas are the extension of long wire antennas.

Construction

(i) Two long wire antennas are arranged in the form of horizontal V, fed at the apex.

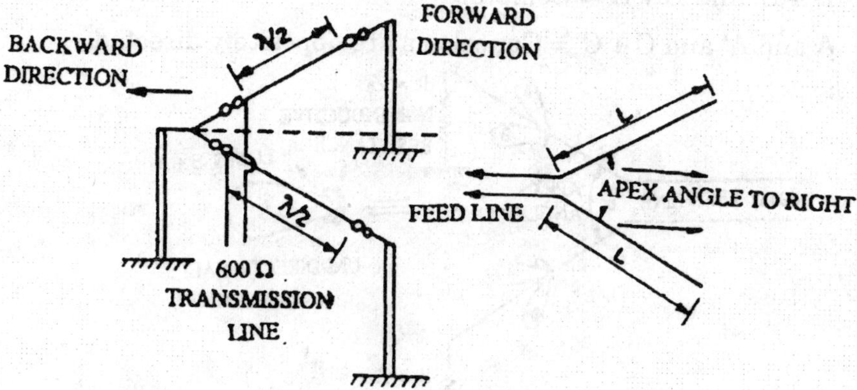

Figure 2.55 High frequency tuned or resonant V-antenna

(ii) Apex angle varies according to length of leg. It varies between 36° to 72° for a V-antenna structure of 8λ to 2 λ long. If V-antenna is to be operated over a wide range, the apex angle is made the average between optimum for the highest and lowest frequencies in terms of no. of λ/2 in each leg.

(iii) If the angle between the two sides of the V, is equal to twice the angle that the cone of maximum radiation of each wire makes with the axis of that wire, then the two cones will add up in the direction of the line bisecting the apex angle of V, and there produce a maximum lobe of radiation.

(iv) The two wires are fed 180° out of phase with each other which provides required gain and directivity. With increase in the length of legs of antenna, the directivity and gain also increases. It can be achieved by cancelling oppositely directed corresponding radiation lobes in each leg and by adding similarly directed corresponding lobes in each leg. The resultant is a bi-directional pattern which is sharper, than the same length single long wire.

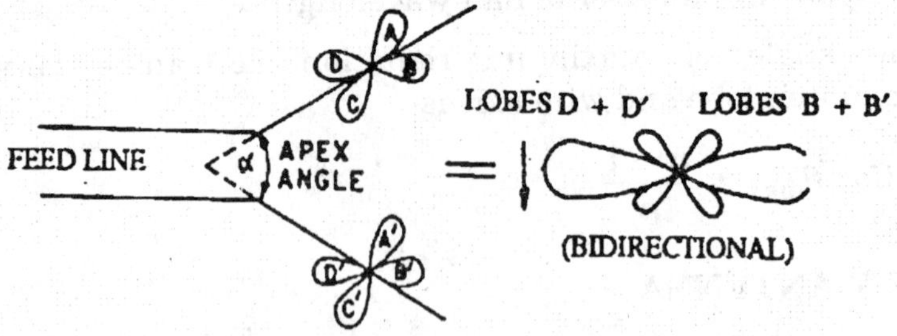

Directional pattern of resonant Antenna in which lobes D and D' and B and B' combine while A cancels C' and C cancels A' as they are oppositely directed.

Figure 2.56

Bi-Directional Pattern of resonant Antenna

D + D' and B + B' = Combine

A and A' and C + C' = Cancels as are oppositely directed

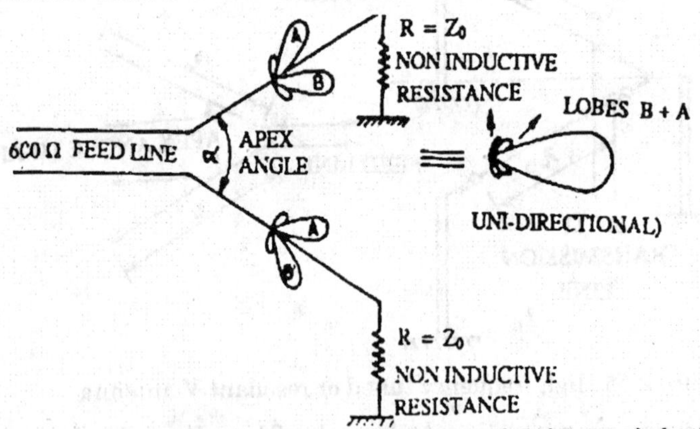

Unidirectional pattern obtained by terminating the two legs by a noninductive characteristic Impedance. Lobes B and A' combine and A and B' cancelled out due to opposite direction.

Figure 2.57

The gain achieved with the V-antenna is nearly twice in comparison to the single long wire antenna, which has a length equal to that of legs of V-antenna.

Radiation from 2 legs Combine to from a well defined beam-

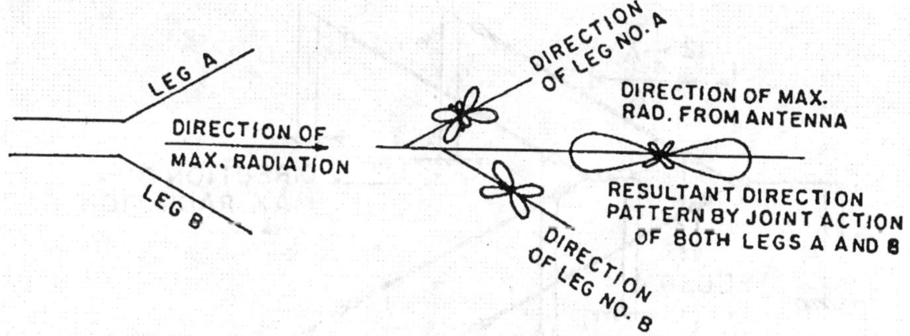

Radiations from two legs combine to form a well defined beam.

Figure 2.58

(v) One way to convert the bidirectional to unidirectional pattern is termination with a non-inductive resistance of valve equal to Z_0 of leg.

(vi) The other way to increase the directivity is to use an array of V, the element of which is V-antenna itself; or in other words the directivity in vertical plane may be increased by skating two or more V antennas one above the other.

Commercial design of V-antennas

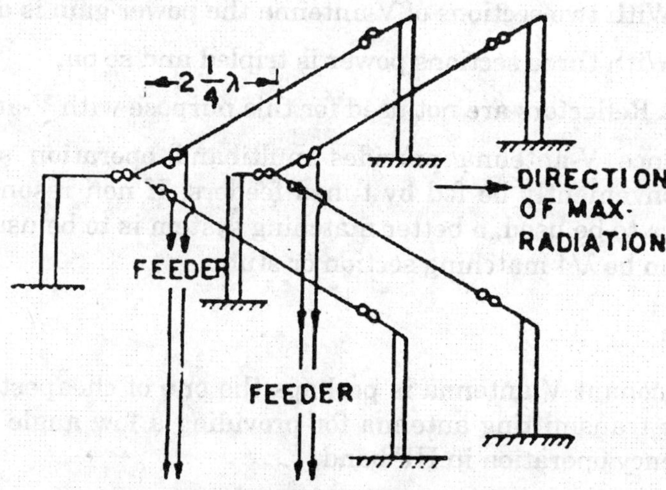

Commercial design of V antennas.

Figure 2.59

(vii) Unidirectional can be made by using another V-antenna placed at an *odd multiple of λ/4* in back of the first and exciting. The two with the phase difference of 90° so that endfire action takes place. Endfire action doubles the power gain obtained.

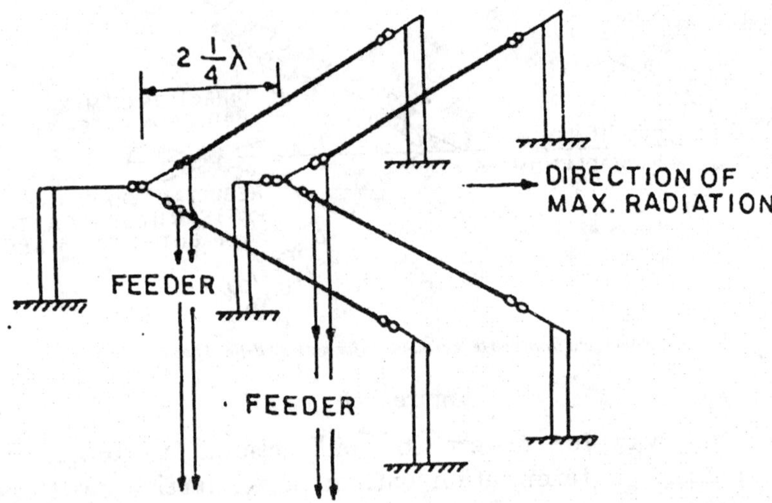

Figure 2.60 Endfire case, using another at an odd multiple λ/4

(viii) The directional pattern is narrowed in a horizontal plane by arranging two or more V-antennas in broadside. In this case power gain is approximately equal to the number of V-antenna sections, used.

For e.g.-

- With two sections of V-antenna the power gain is doubled.
- With three sections power is tripled and so on.

NOTE: Reflectors are not used for this purpose with V-antenna.

(ix) Since V-antenna provides multiband operation so it can conveniently be fed by tuned feeders. If non resonant lines are to be used, a better matching system is to be used, which can be λ/4 matching section or stub.

Advantage

The resonant V antenna is perhaps the one of cheapest forms of receiving or transmitting antenna for providing a low angle beam for fixed frequency operation in HF band.

Disadvantage

Strong minor lobes are observed in V-antenna radiation patterns.

> 1. Draw the radiation pattern of the following antennas:
>
> (i) V-Antenna **(Raj. Univ. 2007, 2008)**

2.4.2.6. INVERTED V-ANTENNA

Inverted V-antenna used in high frequency band, is one of travelling wave antenna. A travelling wave antenna for low frequency reception is *wave or Beverage antenna.*

Inverted V-antenna requires three mast, i.e. non-conducting low heighted pillars. Direction of maximum radiation is towards the terminated end.

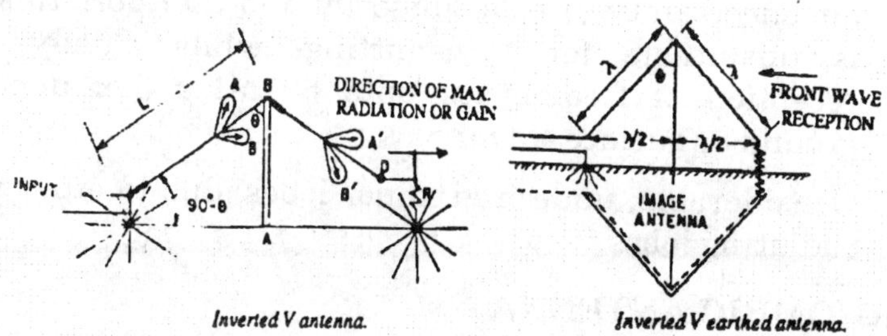

Figure 2.61

The input from transmitter is fed at point C through a transmission line. The other end is connected to a number of radial earth wires. The next end of the antenna wire D is terminated with a resistor which itself is connected to radial earth wire.

High frequency length of this antenna = $8\lambda/2$

Low frequency length of this antenna = $4\lambda/2$

Very Low frequency length of this antenna = $2\lambda/2$

$\Rightarrow$ *HF* ; $4\lambda/2$ *LF* ; $2\,\lambda/2$ *VLF Range.*

Angle θ - **tilt angle**

$\theta = 36°$ to $71°$ generally

For a given number of L/λ wavelengths in leg CB, is a compromise between two factors which are-

(1) Angle of major lobe corresponding to L/λ, and

(2) Tilt angle for which the fields from BC and BD will combine to give maximum gain.

The radiation pattern of *A B' cancels* and *A' B combines* as they are of opposite and in same phase respectively.

It has **vertically polarized** wave.

If the antenna is constructed over normal or poor conductivity ground, then the polar diagram is modified, and the resultant gain along the ground is reduced.

In practice maximum gain occurs in forward direction at half *Brewster Angle* or Little lower. **Brewster angle** is critical angle of incidence for vertically polarized waves after which a change of π occurs in the reflected wave.

Disadvantage

(i) Vertically V-antenna has a lobe in vertical plane which lies between Brewster angle and horizontal which generates a bit uncertainty in polar diagram and therefore this antenna is unsuitable for transmitting, while chiefly used for reception of frequencies upto 60 MHz (i.e. upper HF of ground or surface wave)

(ii) Considerable, undesired minor lobes, due to other portion of radiating lobes.

2.4.2.7 RHOMBIC ANTENNA

This antenna is based on travelling wave radiator's principle. The inverted V and its image behave as rhombic antenna whose maximum gain is along the ground.

Two wire conductors are pulled apart at one point so that the rhombus or diamond shape is formed.

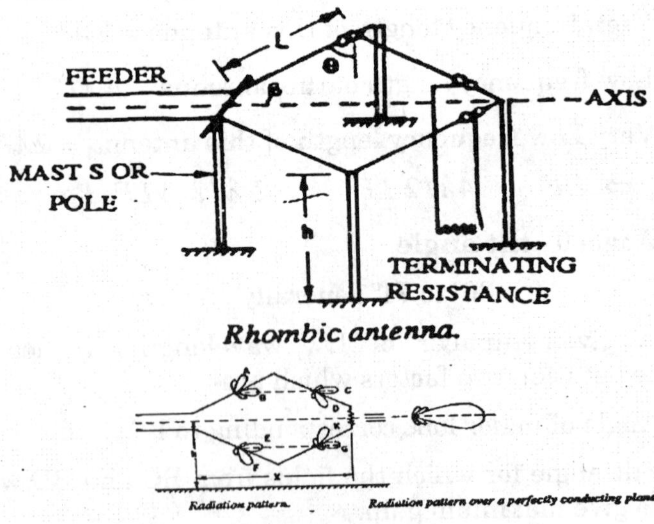

Figure 2.62

If two wires (i.e., two sides of rhombus) are pulled apart to such an extent that four lobes BCEG are combined, then additional gain is achieved.

To avoid reflections the antenna should be properly terminated.

Tilt angle = 90 – θ (angle of major lobe)

With length 4 λ, the angle of major lobe = 17° and with length 8 λ, the angle of major lobe = 24°

∴ Rhombic antenna operates at *Wide band of frequency* and hence it is-

(1) Used in all radio communication facility, where enough space necessary, for its large structure is easily available. For this purpose it is designed, as two *inverted V*-antenna in series with two V-antennas connected end to end forming an obtuse angle.

(2) It is ideally suitable for HF Transmission and reception especially in commercial *point to point communication*

Input is fed through a balanced line and terminating non-inductive resistor is adjusted so that travelling waves are set up in the four legs of rhombic antenna.

Maximum gain is along main axis direction through feed point to termination in space.

It is **horizontally polarized** antenna. However, presence of earth brings an elevation a bit, in the upward direction.

Four legs of rhombic antenna, when terminated by a resistance of value equal to 'Z_o' of system, give a non-resonant current distribution. Then two sides out of four are considered as two wire transmission line.

It is assumed that current flowing through each leg is not being attenuated but in practice current fed point and other end of leg are different i.e. there is some drop across leg. Therefore to counteract this current drop construction is made where each leg is made from three wires to keep current throughout the antenna uniform as impedance B and D decreases.

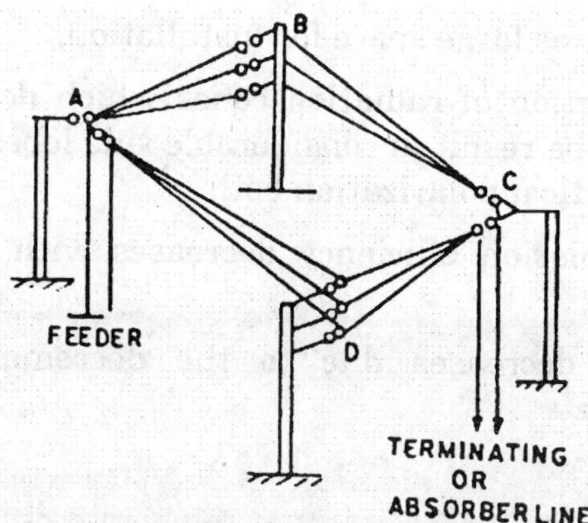

Figure 2.63

Unidirectional pattern is converted to bidirectional pattern. By simply removing the terminating resistance power loss in terminating resistor are 35 or 50%.

For Rhombic Antenna with 2λ length the directivity is 20 and power gain is 50; whereas, the same antenna with 4λ length, the directivity is 90 and power gain is 60.

Hence, we conclude that as the length increases, the directivity and power gain of rhombic antenna also increases.

Advantage

1. Array of Rhombic antenna (**MUSA**) (Multiple Unit Steerable Antenna) is used for long distance, short wave reception of horizontal polarized down coming waves.
2. Input impedance and radiation pattern does not change over considerable frequency Range.
3. It is highly directional broadband antenna.
4. It is used in Radio communications.
5. Simple and cheap to install.
6. Input impedance of rhombic antenna is twice the input impedance of single side radiator.
7. Low vertical angle of radiation makes the antenna suitable for long distance, F-layer propagation.
8. It is short wave or low heighted antennas.
9. Untuned and useful as wideband antenna, which can rapidly switch the working from one frequency to the other.

Disadvantage

1. It requires large space for installation.
2. The portion of radiation cones which do not combine with main lobe result in considerable side lobes having horizontal and vertical polarization both.
3. Transmission efficiency decreases with increase in minor lobes.
4. Power decreases due to the decrement in terminating resistor.

1. Describe the principle of operation of the Rhombic antenna. Explain how various parameters of the antenna affect in radiation pattern. Compare it with a log-periodic antenna. **(Raj. Univ. 2004)**
2. Write short note on use of Rhombic antenna. **(Raj. Univ. 2003)**

2.4.3 APERTURE ANTENNAS

2.4.3.1 HORN ANTENNA

It is considered as flared out or opened out waveguide (***Waveguide:*** A waveguide is simply a hollow, metal cavity. Waveguides are used to guide electromagnetic energy from one place to another. They can also be said as guides, which are capable of radiating radiation into open space, provided the same is excited at one end and opened at the other end).

Radiation is much greater through waveguide than the two wire transmission line.

NOTE: Fields inside the waveguide propagates in same manner as in free space, but with the difference that in waveguide the waves are constrained by the walls of waveguide from being spherically spreading. On reaching at the waveguide mouth, these propagating fields continue to propagate in same general direction but now also start spreading laterally according to **Huygens's Principle,** and the wavefront eventually results spherical. Near the mouth of the waveguide, there is **"near field"** region in which wavefront is more complicated.

However, this may be treated as transition region where the change over from the guided propagation to free space propagation occurs. As the waveguide impedance and free space impedance are not equal, hence to avoid standing wave ratio, flaring of walls of waveguide is done besides matching of impedance, also provide concentrated radiation pattern i.e. greater directivity and narrower beam width. It is the flared structure that is given the name electromagnetic horn radiator.

In waveguide small portion of the incident wave is radiated and the large portion is reflected back by the open circuit. The open circuit is a discontinuity which matches the waveguide to space very poorly besides this diffraction around the edges provides poor radiation and non-directive radiation pattern. Hence in order to avoid these difficulties, the mouth of waveguide is opened out which assume the shape of electromagnetic horn, just like an opened out transmission line which gives a dipole. If the waveguide is terminated by any type of horn, the abrupt discontinuity existed is replaced by **gradual transformation**, where by all the energy incident in forward direction in waveguide is now radiated, provided impedance matching is proper. This improves directivity and reduces diffraction.

Flaring

1. If flaring is done in **only one direction**, then **sectorial horn** is produced. Flaring in the direction of electric vector and magnetic vector, the sectorial E-plane horn and sectorial H-plane horn are obtained respectively.
2. If flaring is done along **both the walls** (E and H) of rectangular waveguide, then pyramidal horn in obtained.
3. By flaring the walls of circular waveguide a conical horn is formed.

Different kinds of Horn Antennas

1. E-plane Horn antenna

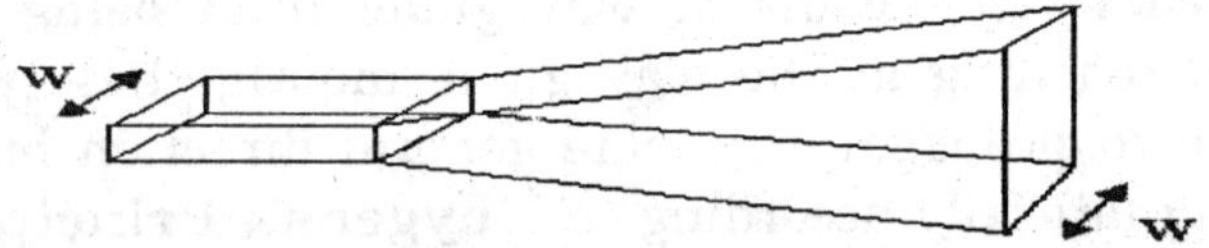

Figure 2.64 Diagram of E-plane Horn antenna

- This horn is flared in E-plane.
- It has constant width for the waveguide and horn that is `w'.

2. H-plane Horn

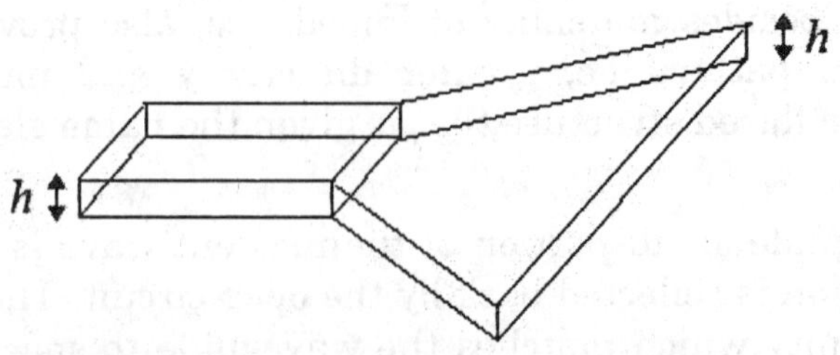

Figure 2.65 Diagram of H-plane Horn antenna

- This horn is flared in H-plane.
- It has constant height for the waveguide and horn that is `h'.

3. Pyramidal Horn antenna

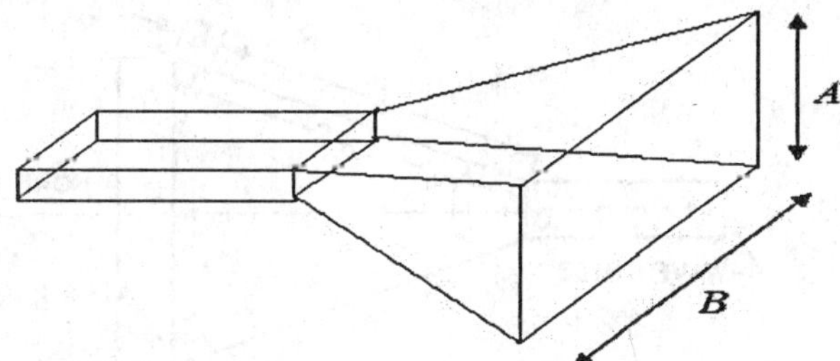

Figure 2.66 Diagram of Pyramidal Horn antenna

- It is the most popular type of horn antenna.
- It is flared in both the planes (i.e. E and H planes).
- It has height A and width B at the end of horn.

4. Conical Horn Antenna

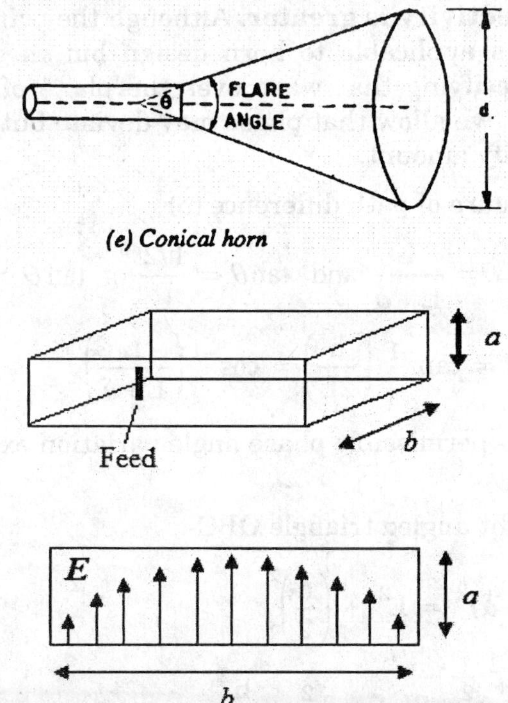

Figure 2.67 Diagram of Electric vector inside the waveguide

Horns are typically fed by a section of a waveguide. The wave-guide itself are often fed with short dipole antenna. The waveguide here in the figure, is a rectangular waveguide of height `a' and width `b', with b > a. The E-field distribution for the dominant mode is shown, in the lower part of the figure.

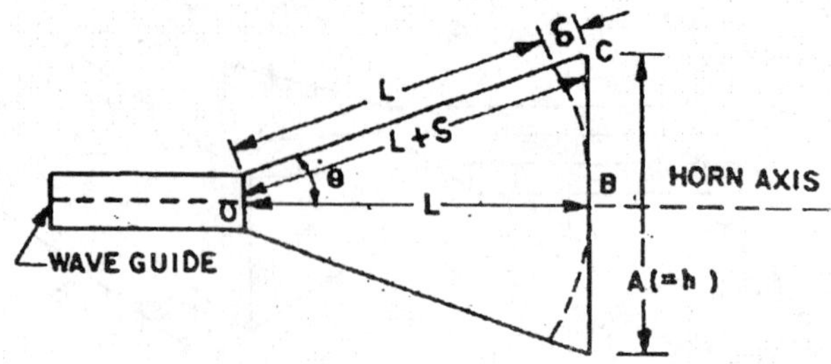

Path difference δ

L = Axial length ; A = Aperture ; θ = Flare anlge (half of).

Figure 2.68

Function

Electromagnetic horn is used to produce a uniform phase front with a larger aperture in comparison to waveguide and thus the directivity is greater. Although the principle of equality of path length is applicable to horn design but in different sense i.e. instead of specifying that wave over the plane of horn mouth is in phase exactly, we allow that phase may deviate but by an amount less than the specify amount.

By the figure of path difference (δ)

$$\text{Cos}\,\theta = \frac{L}{L+\delta} \text{ and } \tan\theta = \frac{h/2}{L} \text{ or } \tan\theta = \frac{h}{2L}$$

$$\therefore \theta = \tan^{-1}\left(\frac{h}{2L}\right) = \cos^{-1}\left(\frac{L}{L+\delta}\right)$$

Where δ - permissible phase angle variation expressed as fraction of 360°.

From right angled triangle OBC-

$$(L+\delta)^2 = L^2 + \left(\frac{h}{2}\right)^2$$

$$L^2 + \delta^2 + 2L\delta = L^2 + \frac{h^2}{4}$$

Again, If δ is small than, $\delta^2 << 1$

$$\therefore 2\,L\delta = \frac{h^2}{4}$$

or $$L = \frac{h^2}{8\delta}$$

If flare angle (2θ) is *very large* then, the wavefront on the mouth of horn will be curved rather than plane. This will result in non uniform phase distribution over the aperture, resulting increased beamwidth and decreased directivity, and vice-versa occurs if the flare angle is ***very small*** aperture area for a specified length L. The **directivity is proportional to the aperture size** for a given aperture distribution. Thus there is optimum aperture angle θ:

Values of, δ for plane is 0.25, for conical is 0.32 and H-plane horn is 0.40.

Directivity of parabolic antenna is more than horn antenna.

There are no resonant elements involved in horn antenna and hence they can be operated over a broad band of frequencies.

(1) Half power Beamwidth (optimum flare horns)

$$\theta_E = \frac{56\lambda}{h} \text{ degrees E-direction}$$

$$\theta_H = \frac{67\lambda}{w} \text{ degrees H-direction}$$

(2) Directivity

$$D = \frac{(7.5)\,h.w}{\lambda^2} = \frac{(7.5)\,A}{\lambda^2}$$

Where, A = h x w

= area of horn mouth opening (aperture), and

(3) Power Gain

$$G_p = \frac{(4.5)\,h.w}{\lambda^2} = \frac{(4.5)\,A}{\lambda^2}$$

Horn Antenna Field Equations

The E-field distribution across the aperture of the horn is what is responsible for the radiation.

The radiation pattern of a horn will depend on `B' and `A' (Horn opening) and `R' (length of the horn) along with `b' and `a' (dimension of the waveguide). `R' also **affects the flare angles of the horn.** These parameters are optimized to find the performance of the antenna, which are shown in the figure-

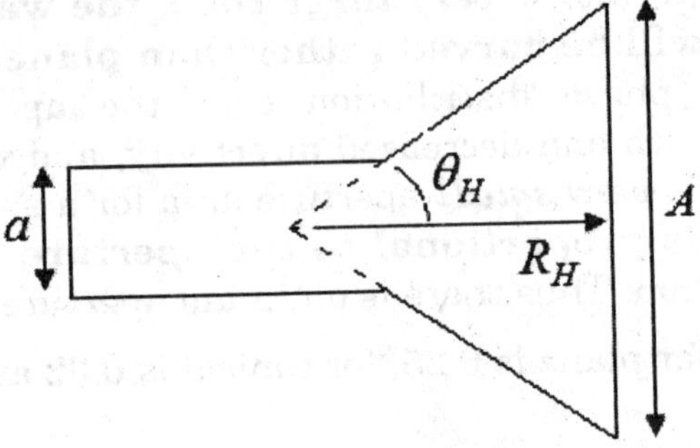

Figure 2.69 Cross section of waveguide cut in the H-plane.

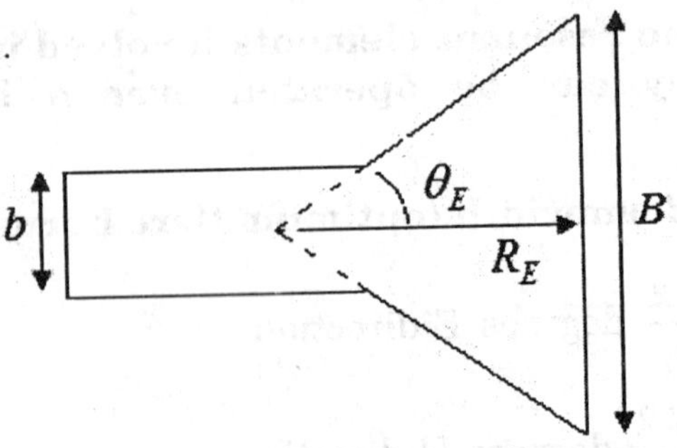

Figure 2.70 Cross section of waveguide cut in the E-plane.

Where θ_E and θ_H - flare angles.

Flare angles depends on the height, width and the length of the horn antenna. Given the coordinate system in the next figure (which is centered at the opening of horn), the radiation is maximum in +z - direction.

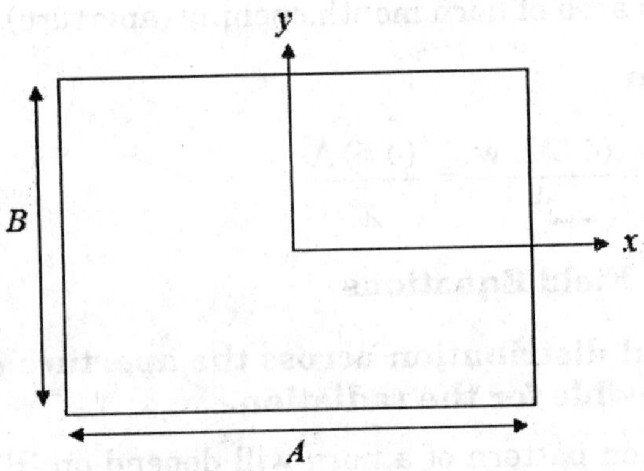

Figure 2.71

E-field distribution across the opening of horn can be given by-

$$E_A = \hat{y} E_o \cos\left(\frac{\pi x}{A}\right) e^{-j\frac{k}{2}\left(\frac{x^2}{R_H}+\frac{y^2}{R_E}\right)}$$

E-field in the far field is linearly polarized, with magnitude-

$$|E| = \frac{k}{4\pi r}(1+\cos\theta)\int_{-B/2}^{B/2}\int_{-A/2}^{A/2} E_A(x, y)\, e^{jk(x\sin\theta\cos\phi + y\sin\theta\sin\phi)}\, dx\, dy$$

These equations states that the far-fields, of the horn antenna are Fourier Transform of fields at the opening of the horn.

Radiation Pattern of Horn Antenna

Let us find the radiation pattern with the help of horn antenna which is large, and one which work above 2GHz; as we know that the horn antenna made for higher frequencies are smaller. The horn antenna, with a waveguide feed is given in the figure-

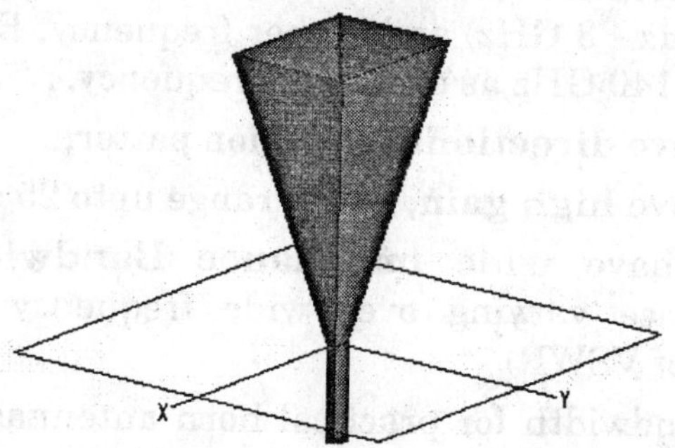

Figure 2.72 Horn antenna at high `f'

The radiation patter at 2 GHz is given in the figure-

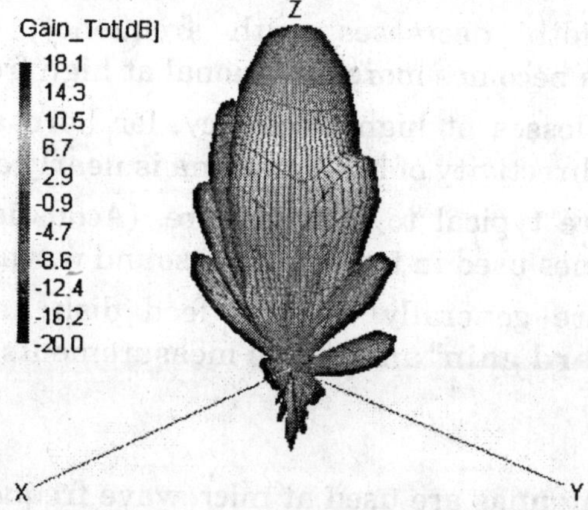

Figure 2.73 Horn radiation pattern at 2 GHz

- Gain of the antenna from the given figure = 18.1 dB in +z - direction.
- Half power Beamwidth is 15° in x-z-plane (H-plane) and 11° in y-z-plane (E-plane).

The horn geometry affects the gain of the antenna. (Geometry = length and aperture size of horn). As horns are to design to operate over a wide frequency band, they are designed to have optimal gain at the lowest frequency in the band. At higher frequencies, the geometry is no longer optimal, so E-field across the aperture is not optimal. However, the horn's aperture becomes electrically larger at higher frequencies (the aperture is more wavelengths long as the frequencies increases or the wavelength decreases). Consequently, the loss of an optimal aperture field is offset by an electrically larger horn, and the gain actually increases as the frequency increases.

General Characteristics

1. Horn antennas are very popular for the application at **UHF** (300 MHz - 3 GHz) and higher frequency. Even, they can be used at 140 GHz as operating frequency.
2. They have **directional** radiation pattern.
3. They have **high gain**, which range upto 25 dB.
4. Horns have **wide impedance Bandwidth** with input impedance varying over wide frequency range (i.e. low values of VSWR)
5. The Bandwidth for practical horn antennas can be order of 20: 1.
6. The gain increases as frequency of operation is increased, therefore gain of horn antenna is very high.
7. Beamwidth decreases with frequency; therefore horn antenna becomes more directional at high frequencies.
8. As the losses at high frequency, for horn antenna, is very less ∴ directivity of horn antenna is nearly equal to its gain.
9. They are typical to manufacture. (Acoustic horns are also sometimes used in transmitting sound waves).
10. They are generally used to feed dish antenna, or as a **"standard gain"** antenna in measurements.

Uses

- Horn antennas are used at microwave frequencies with moderate power gain (as for high power horn dimension becomes large) so we require parabolic reflectors.

1. Write short note on Horn Antennas. **(Raj. Univ. 2004)**
2. What do you understand by E-plane and H-plane sectorial horn antenna? Explain with the help of neat diagram. **(Raj. Univ. 2000)**
3. Discuss the following Antenna: (i) Slot Antenna **(Raj. Univ. 2006)**
4. Describe in which frequency range and where the following antennas are normally used (in not more than 10 lines for each): (i) Horn antenna **(Raj. Univ. 2004)**

2.4.3.2.a. SLOT ANTENNAS

They have, simply an opening cut in a sheet of conductor which is energized in some appropriate manner, such as coaxial cable of waveguide.

A simple type of slot antenna is half wavelength long with narrow width and excited via 50 Ω coaxial cable, normally connected about 0.05 λ from one end of the slot to achieve reasonable matching condition. A horizontal slot so energized produces vertical polarization in the direction normal to slot and vertical slot produces horizontal polarization. **Radiation occurs from both the sides of conductive sheet but if the slot is "boxed", with internal dimension of depth d= λ/4, the radiation is in outward direction from the opening of box.**

To gain an intuition about slot antennas, first we'll learn ***Babinets principle*** (put into antenna terms by H.G. Booker in 1946).

(A) This principle relates the radiated fields and impedance of an aperture or slot antenna to that of the field of its dual antenna. The dual of a slot antenna would be if the conductive material and air were interchanged that is, the slot antenna became a metal slab in space. An example of dual antenna is shown in figure-

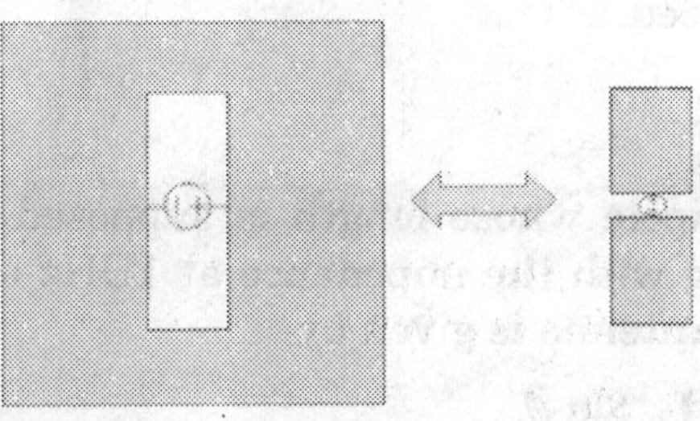

Figure 2.74

A voltage source is applied across the short end of the slot. This induces an E-field distribution within the slot, and currents that travel around the slot perimeter, both contributed to radiation. **The dual antenna is similar to a dipole antenna.** The voltage source is applied at the centre of the dipole, so that the voltage source is rotated.

Babinet's principle relates these two antennas. The first result states that the impedance of slot (Z_s) is related to the impedance of its dual antenna (Z_c) by the relation-

$$Z_c Z_s = \frac{\eta^2}{4}$$

where η = intrinsic impedance of free space.

(B) Major result of Babinet's/Bookers principle is that the fields of the dual antenna are almost the same as the slot antenna (inter changed field components are called "duals"). The fields of slot antenna (S) are related to the fields of its complement (C) by-

$$E_{\theta S} = H_{\theta C}$$

$$E_{\varphi S} = H_{\phi C}$$

$$H_{\theta S} = \frac{-E_{\theta C}}{\eta^2}$$

$$H_{\phi C} = \frac{-E_{\phi C}}{\eta^2}$$

From these equations, we can say that if we know the fields from one antenna, we know the fields of the other antennas. Since, it is easy to know the fields from the dipole antenna, thus the fields and impedance from slot antenna can also be derived, if Babinet's principle is well understood.

NOTE: The polarization of dual antennas is reversed, that is as the dipole antenna is vertically polarized, and the slot antenna will be horizontally polarized.

Duality Example

Consider a dipole whose length is supposed to be 14.4 cm and width is 2cm along with the impedance at 1GHz as (65 + j15) Ω. The fields from dipole antenna is given by-

$$E_{\theta d} = E_o \sin\theta$$

Example: What are the fields from a slot at 1 GHz with the same dimensions as the above dipole?

Solution: Using Babinet's principles-

(i) $\text{Impedance}(Z_s) = \dfrac{\eta^2}{4Z_c} = \dfrac{(120\,\pi)^2}{4(65+j15)} = 519 - j120\ \ \Omega$

Here,

- Impedance of slot is greater than dipole impedance.
- Impedance of dipole is inductive (positive imaginary part) while that of slot it is capacitive (negative imaginary part)

(ii) The E-fields for the slot can be easily found by-

$$E_\phi = \frac{E_o \sin\theta}{\eta}$$

Conclusion: *E-fields only contain a phi (azimuth) component. The antenna is therefore horizontally polarized.*

To enhance gain and directive properties of basic slot antenna, it is common to have arrays of slots in a manner similar to arrays of dipoles. Some VHF transmitters employ cylindrical arrays of slots to produce **Omni-directional radiation** in the horizontal plane with horizontal polarization.

- In slot antenna, energy is radiated when a high frequency field exists across a narrow slot in a conducting plane.

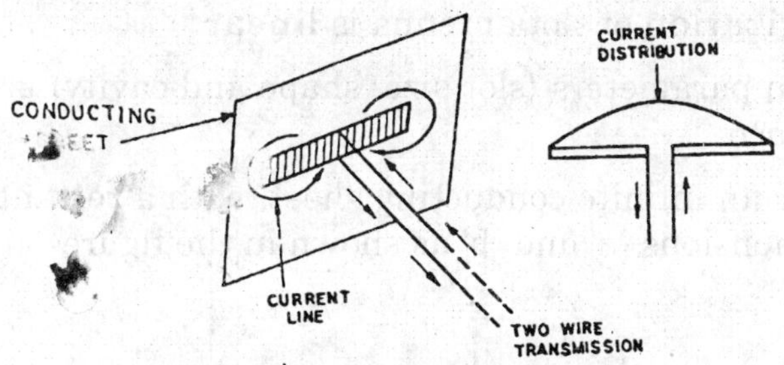

(a) Slot antenna, (b) Complementary wire antenna.

Figure 2.74(2)

- The fields are excited by a two wire transmission line.
- The *electric field* across the slot is maximum at the centre and tapers off towards the edges, while sometime currents flow in conducting plane in general manner indicated when slot is exactly half wavelength long, the electric field distribution is sinusoidal and the impedance offered by slot to two wire transmission line is 56 Ω.

Radiation produced by a narrow slot in conducting plane has exactly the same directional pattern as the radiation from a thin, flat, wire type-antenna, having a shape corresponding to slot and a current distribution that is same as distribution of electric field across the slot.

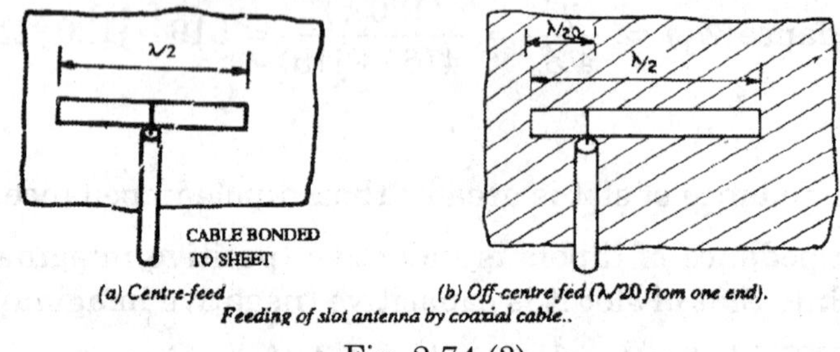

Fig. 2.74 (3)

The only difference in slot and its complementary antenna are -

(1) Polarization are different i.e. electric fields associated with slot antenna are identical to magnetic field of other antenna.

(2) Radiation from back side of conducting plane has the opposite polarity from that of complementary antenna, because of the way in which the fields are directed.

General Characteristics

1. Slot antennas are used typically at frequencies between 300 MHz and 24 GHz.

2. It has radiation pattern which is nearly **Omni directional**.

3. **Polarization** of slot antenna is **linear**.

4. Design parameters (slot size, shape and cavity) is used to tune the performance.

Consider an infinite conducting sheet, with a rectangular slot cut out of the dimensions `a' and `b' as shown in the figure-

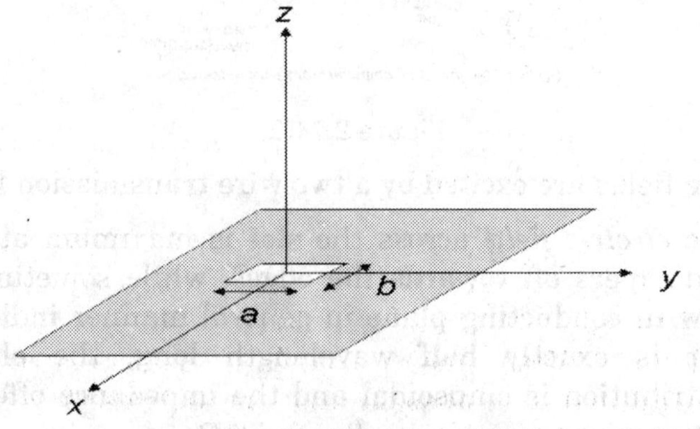

Figure 2.75 Rectangular Slot antenna with dimension a and b.

If we can excite some reasonable fields in the slot (often called the aperture), we have an antenna.

1. Draw two different types of slot antenna. What is the average size of Microstrip antenna? Where are such antennas being used? **(Raj. Univ. 2003)**
2. Describe in which frequency range and where the following antennas are normally used (in not more than 10 lines for each): (i) Slot antenna **(Raj. Univ. 2004)**

2.4.3.2. b. CAVITY-BACKED SLOT ANTENNAS

A practical slot antenna is the cavity backed slot antenna. Let us first place, an idea of cavity Backed slot antenna. The basic cavity backed slot antenna is shown in figure-

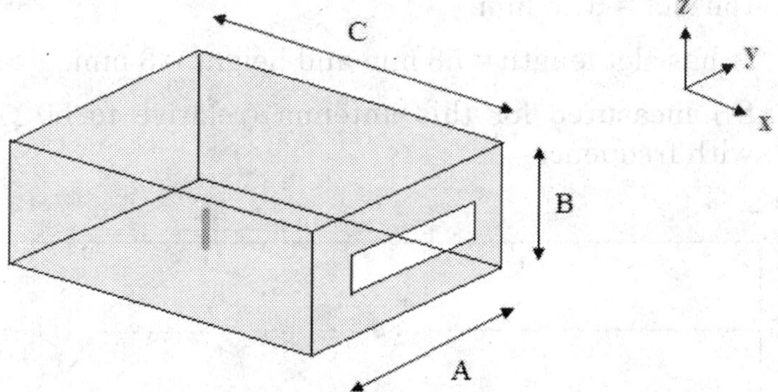

Figure 2.76 Cavity backed slot antenna

In the hollow metallic (whose walls are electrically conducted), rectangular cube with size A* B * C, a cut is made at any side; which is called as a Slot.

The cavity is typically excited by a probe antenna in the interior of the cavity, which typically modeled as a **monopole antenna** (in the centre of the rectangular block)

Conclusion

(1) The volume of the *cavity* is generally filled with air but it can be filled with the **dielectric; as material; it reduces the resonant length of the antenna** (means minimizes the size of antenna).

(2) Again the **Bandwidth is directly proportional to the *Cavity* size** i.e., if the cavity is more the antenna works at higher Bandwidth.

(3) **With the dielectric cavity medium the Bandwidth and efficiency of this antenna decreases.**

Analysis of the antenna

1. Let us take, the slot with-
   - Cavity (C) = 36 mm
   - Length (A) = 87 mm
   - Height (B) = 36 mm
   - Ratio = 3.5 C : 9A : 3.5 B
2. The height of feeding monopole antenna is say = 29.5 mm, which means that the monopole antenna is a quarter wave long at 2.55 GHz.
3. Monopole antenna is placed at the centre of y-axis, behind the slot = 61.5 mm.
4. It has slot length = 58 mm and height = 3 mm.
5. S_{11} measured for this antenna (relative to 50 Ω source) with frequency-

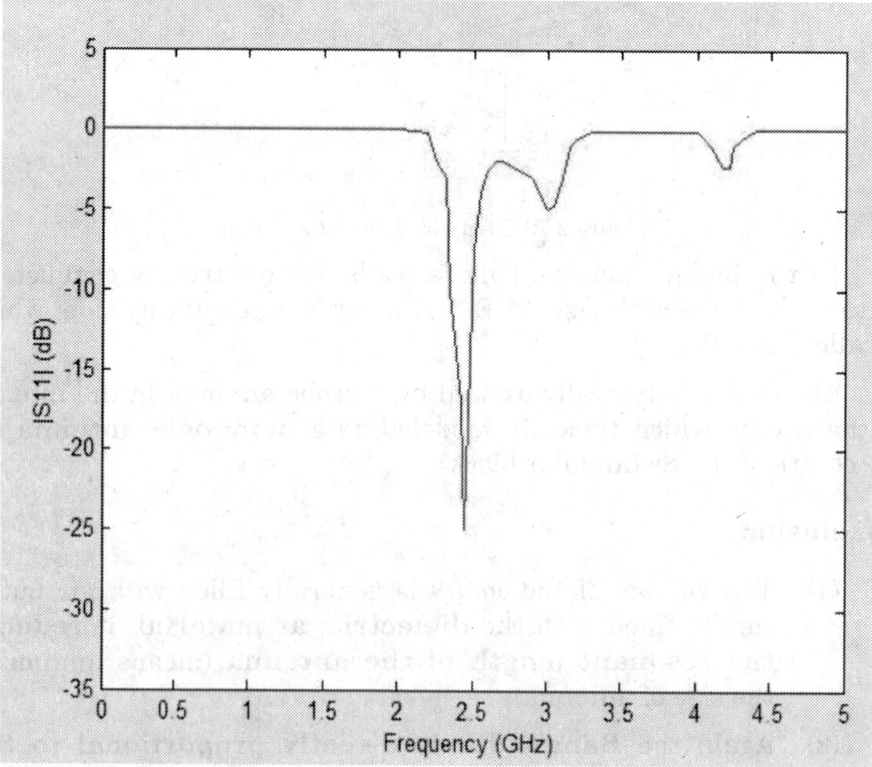

Figure 2.77 as a function of Frequency for Cavity backed Slot

Analyzing this graph

From the plotted graph, we conclude that the-

(1) First resonance of antenna is nearly 2.45 GHz

(2) At this frequency, slot antenna is nearly 0.474 λ long. (This is nearly equal to resonant dipole antenna)

(3) S11 drops below - 20 dB, at this frequency, most of the power is radiated away.

(4) Bandwidth < - 6dB as-

$$f_1 = 2.35 \text{ GHz}$$

$$f_2 = 2.55 \text{ GHz}$$

$$\therefore \text{Bandwidth} = f_2 - f_1$$

$$= (2.55 - 2.35) \text{ GHz}$$

$$= 0.20 \text{ GHz}$$

Hence, Fractional Bandwidth = 8%

(5) Other drops are at = 3 GHz and 4.18 GHz, for which the slot lengths are 0.58 λ and 0.18 λ respectively.

NOTE: With increase in frequency the slot length increases.

Radiation pattern

(1) H-plane (xy-plane)

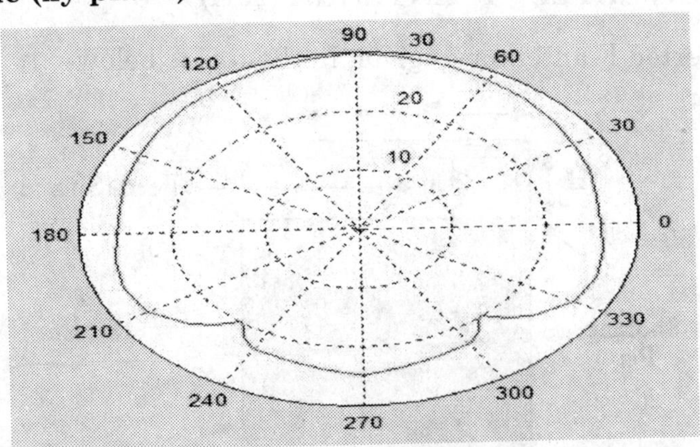

Figure 2.78 H-plane (xy plane). Angle measured off x-axis towards y-axis

H-plane

Radiation pattern is nearly same as that of the radiation pattern of dipole antenna in forward H-plane. Its beamwidth is 60°. In the rear of H-plane, the back lobe is about 6 dB down from the peak of the main beam.

2. E-plane (xz-plane)

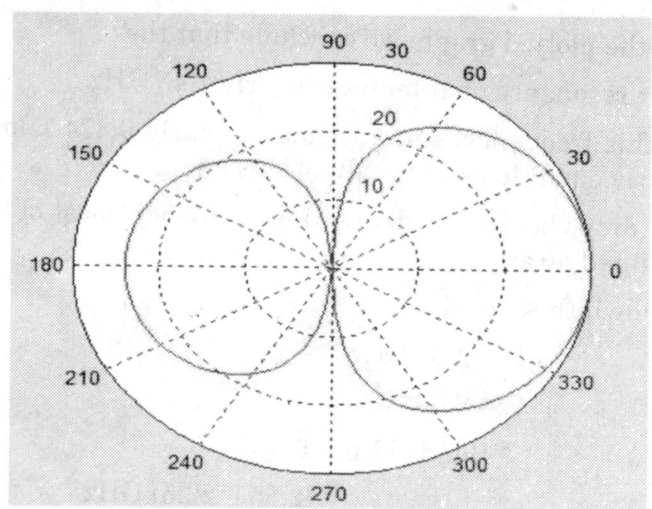

Figure 2.79 E-plane (xz plane). Angle measured off z-axis (to x-axis)

E-plane

It has a broad pattern with beamwidth as 120° and gain of 2-3 dB.

As Beam width is very large ∴ they can be used in antenna arrays

2.4.3.3. INVERTED - F- ANTENNA (IFA)

Inverted F-antenna, is given in the figure below-

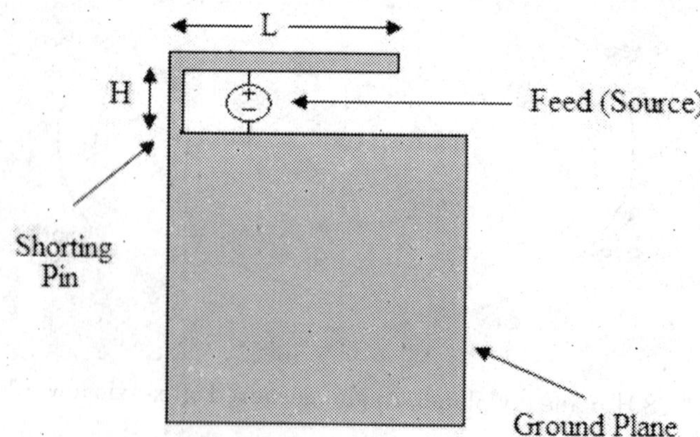

Figure 2.80 Geometry of Inverted F-Antenna (IFA)

The top of the antenna is like the wire antenna, but after the analysis we can conclude that it is the aperture antenna.

Construction

This antenna is basically divided into three parts-

(1) Upper Arm

(2) Ground Plane

(3) Shorting Pin

a) The feed is connected between the ground plane and the upper Arm of the IFA.

b) The length `L' of the dipper arm of the IFA is nearly equal to the quarter of wavelength.

c) The feed is closer to the shorting pin, than to the open end of the upper arm.

d) Polarization of IFA is **vertical polarized antenna**.

e) Radiation pattern is nearly a **donut shape** with its axis in vertical direction.

f) The width of ground plane should be at least as wide as the upper arm, to avoid radiation losses.

g) The height of ground plane should be at least $\lambda/4$. If it is smaller than the Bandwidth of antenna the efficiency of IFA will decrease.

h) The height of IFA, is small fraction of wavelength

i) The Radiation properties and impedance are independent of the height of IFA.

Analysis

Let us focus our discussion to a common question, whether how and why this antenna structure radiates energy.

To understand this, let us start with the slot antenna with the geometry as given in the figure-

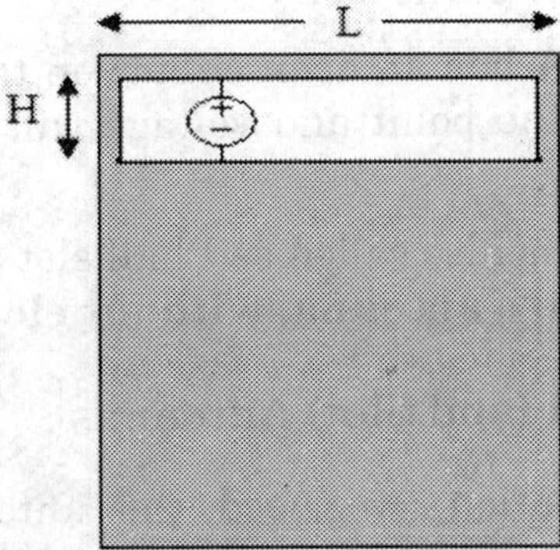

Figure 2.81 Geometry of Slot Antenna to be analyzed

Working: The slot in the given antenna should be half wavelength long for proper radiation. The antenna is excited by the voltage source, where it can be said that at the ends of slot i.e. across the aperture, voltage is zero as shown in the given figure, because of the shorting post on each side.

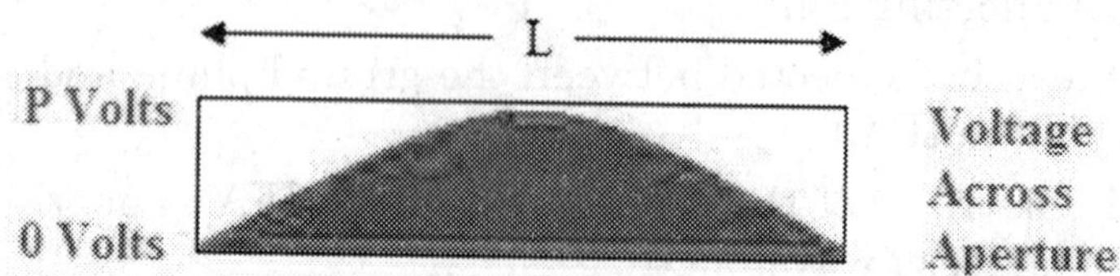

Figure 2.82 Voltage distributions along half wave Slot

Where P-Peak Voltage (Volts)

So according to the voltage, the given slot has the property of the transmission line which is short circuited, and hence the voltage and current are 90° out of phase. As a result, the current is zero at the centre of slot and maximum at the edge.

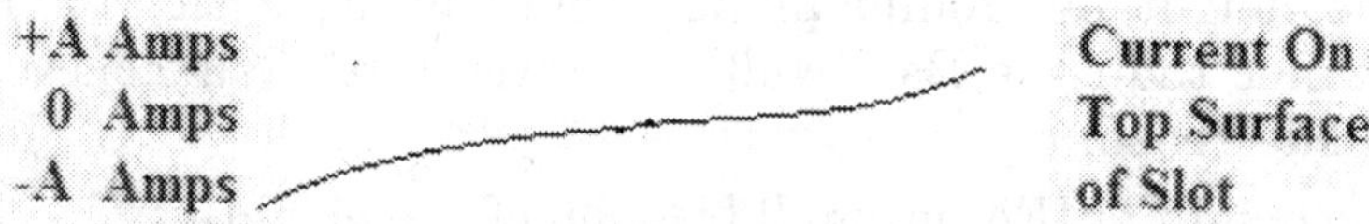

Figure 2.83 Current Distribution along a Half-wave Slot

As the voltage is in phase, across the entire aperture so the slot antenna radiates. In these radiations E-field is vertical so is the polarization. E-field lines are up everywhere along the slot.

So, now if we want to convert slot antenna to IFA, we can break the slot in half and remove its right side. The current at the centre of slot is zero, therefore the slot antenna can be thought of having an open circuit at the centre of the slot.

Since, now the IFA has an open circuit on the right side of feed, the current is zero at the point and voltage will be maximize in the slot antenna.

So sometimes IFA is also called as ``half slot antenna". **Only IFA antenna is the aperture antenna, without close ends.**

Circuit Model for IFA (and Slot) Antennas

For effective radiation, we need the antenna to be a good radiating structure (the currents or electric fields add up in phase) and we need to be able to get the energy down the transmission line and onto the antenna. This means we need the impedance of the antenna to be roughly 50 Ohms. To accomplish this, it is desirable of have the reactive component of the impedance (imaginary part) to be zero. For the IFA or slot antenna, note that the feed sees a shorted transmission line a small fraction of a wavelength from the antenna. A shorted transmission line that is a small fraction of a wavelength creates an inductive reactive component.

Similarly, the open circuit on the IFA creates a capacitance to the right of the feed. The feed location is chosen to ``balance out" the capacitance (to the right of the feed) and the inductance (to the left of the feed as shown in the previous figure.). An equivalent circuit model of the IFA is shown in Figure -

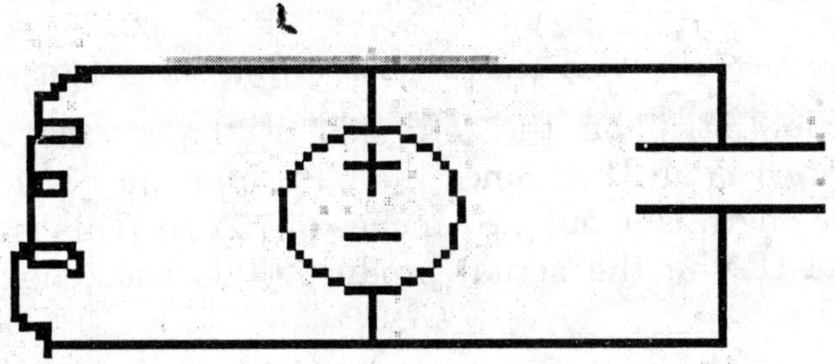

Figure 2.84 Circuit Model for IFA antenna.

NOTE: that for a slot antenna, the short circuit to the right of the feed still creates a capacitance, because the length of the slot to the right of the feed is larger than a quarter-wavelength.

In regards to the real part of the impedance for an IFA or slot antenna, note that *if the slot antenna is fed at the center of the slot (where the voltage is maximum and the current is zero (as Z = V/I)), the impedance becomes practically infinite, so that the antenna would not radiate.*

By moving the feed away from the center of a slot, this also allows the current to decrease from zero, which allows the impedance to drop to a more desirable value. The same is true on the IFA antenna. Hence, **the feed location is a critical factor in designing an IFA or slot.**

IFA in the Real World

IFAs are commonly used in mobile phones, due to their small size (quarter-wavelength). These antennas are visible once the back cover is removed, shown in Figure-

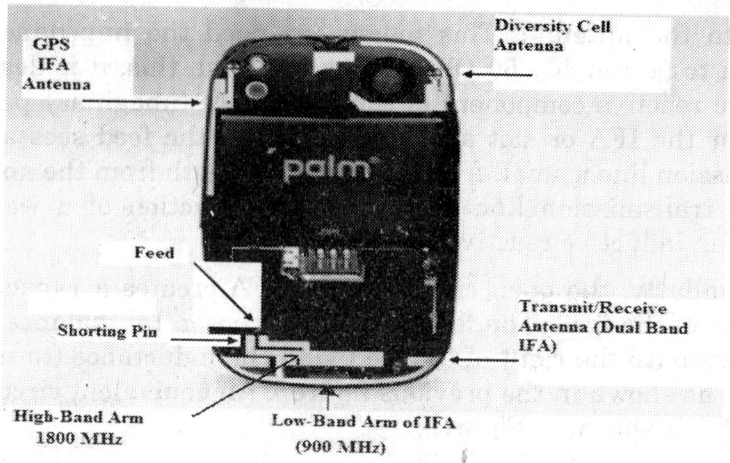

Figure 2.85 Palm Pre Antennas viewable by Removing Back Cover

The yellow strip on the left side of the mobile is the **GPS antenna**, which is an IFA. Since the GPS frequency is 1.575 GHz, a quarter wavelength is about 1.87 inches (4.75 cm). This is roughly the length of the IFA in the actual product (it is shortened for proper tuning).

Two other antennas are visible in Figure-

1) In the upper right side, there is the **diversity cell antenna**, and; at the bottom of the device is a **dual band IFA**. This antenna is the transmit/receive cell antenna, which are designed to operate at 900 MHz and 1800 MHz bands. To do this, the antenna engineers make an IFA for the high band (1800 MHz), which is the shorter arm shown in the Figure. Using the same feed and shorting pin, the designers branched off another arm for the low band (i.e. the longer arm is wrap around itself). Because the designers have space limit (which is a big challenge in mobile phone antenna development), they wrapped the IFA around itself on the edges. By doing this, they to obtain the required length for a 900 MHz IFA.
2) However, **by wrapping the antenna around itself, the bandwidth and radiation efficiency decrease.**

Advantages

1) Small size.
2) Easy construction.

2.4.3.4. SLOTTED WAVE GUIDE ANTENNAS

Slotted waveguide antenna arrays are generally **used in** –

1. Navigation.
2. RADAR
3. Other high frequency systems.

Properties of antenna

(1) Simple to fabricate

(2) Low loss antenna

(3) High efficiency

(4) Radiate **linear polarization** with **low cross polarization**

(5) They are used in aircraft applications because they can be made fixed to the surface on which they are mounted.

(6) The slots are typically thin nearly less than 0.1 of wavelength and 0.5 wavelengths long, at the center frequency of operation. The slots on the waveguide are assumed to have narrow-width. Increasing the width increases the bandwidth Fractional Bandwidth for thin slots = 3–5% and for wide slot = 75%

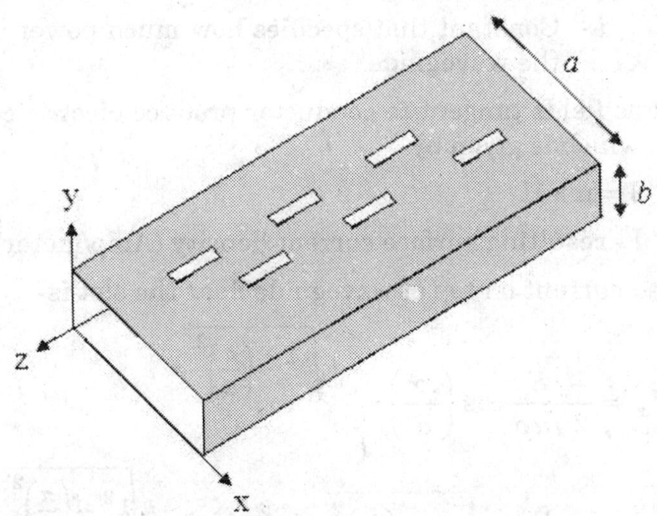

Figure 2.86 Basic geometry of a slotted waveguide antenna

As in cavity backed slot antenna, each slot can be independently fed with voltage source across the slot, but this is not practically possible. Hence, generally for this purpose a waveguide is used as the transmission line to feed the elements.

Radiation depends on the-

1. Position of the slot

2. Shape of the slot
3. Orientation of the slot
4. Shape of the waveguide, and
5. Operational frequency

The fields that exist within the waveguide, are given by-

$$H_x = \frac{-E_o}{2\pi f \mu}\sqrt{k^2 - \left(\frac{\pi}{a}\right)^2}\ \sin\left(\frac{\pi x}{a}\right) e^{-j\sqrt{k^2 - \left(\frac{\pi}{a}\right)^2}\, z}$$

$$E_y = E_o \sin\left(\frac{\pi x}{a}\right) e^{-j\sqrt{k^2 - \left(\frac{\pi}{a}\right)^2}\, z}$$

$$H_x = \frac{jE_o}{2af\mu} \cos\left(\frac{\pi x}{a}\right) e^{-j\sqrt{k^2 - \left(\frac{\pi}{a}\right)^2}\, z}$$

where f – frequency.

k - Wave number.

a- Constant that specifies how much power is added to the waveguide.

Magnetic fields tangent to conductor produce electric currents on the surface, which is given by-

J = n x H

where J - resulting surface current density (Amp/meter)

Induced current on top of waveguide near the slot is-

$$J_x = \frac{-jE_o}{2 f\mu a} \cos\left(\frac{\pi x}{a}\right) e^{-j\sqrt{k^2 - \left(\frac{\pi}{a}\right)^2}\, z}$$

$$J_z = \frac{-E_o}{2\pi f \mu}\sqrt{k^2 - \left(\frac{\pi}{a}\right)^2}\ \sin\left(\frac{\pi x}{a}\right) e^{-j\sqrt{k^2 - \left(\frac{\pi}{a}\right)^2}\, z}$$

Radiation occurs when the currents "around" the slots in order to continue on their desired direction. As an example, consider a narrow slot in the center of the waveguide, as in the given figure-

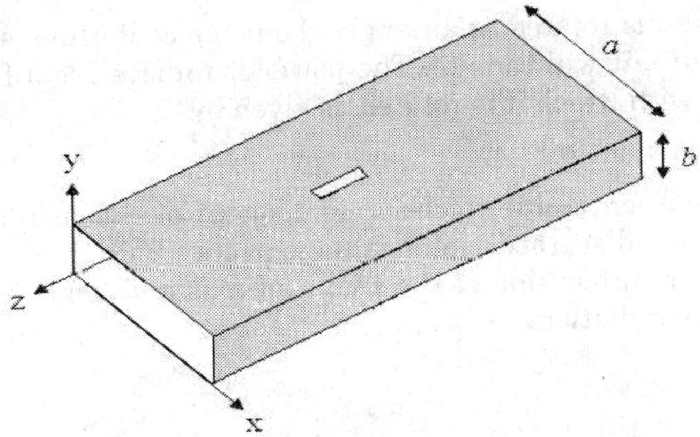

Figure 2.87 Waveguide with a thin slot centered about its width

The Z-component of the current will not be disturbed, because the slot is thin and the Z-current would not need to travel around the slot. Hence, the x-component of the current will be responsible for the radiation. However, at this location (x = a/2), the component of the current density is zero - i.e. no current and thus followed by no radiation.

Hence, **slots cannot be placed in the center of the waveguide,** in the above figure.

If the slots are displaced from the centerline, then the current will not be zero, and will need to travel around the slot. Hence radiation occurs. The distance from the edge determines the magnitude of the current. Therefore, the power that the slots radiate can be altered by moving slots close or far from the edge. Hence, phased array can be designed with varying excitation to each element.

For the slot given in the below figure, it disturbs the Z-component of the current density and thus the slot starts its radiation. **If the slot is displaced away from the center line, the amount of power that antenna radiates can be adjusted.**

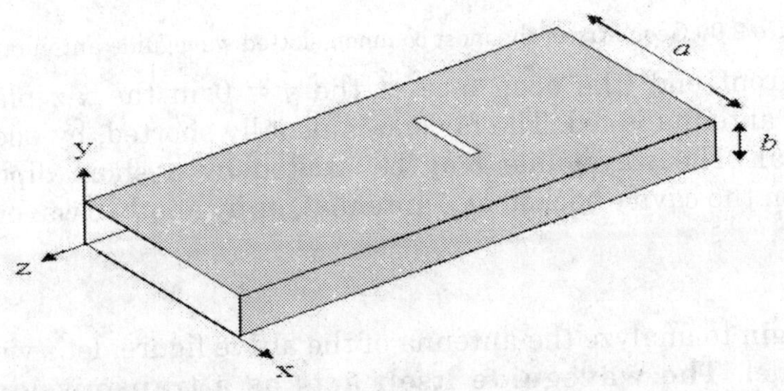

Figure 2.88 Horizontal slots in a waveguide

If the slot is rotated at an angle about the centerline, as given in the next figure, it will radiate. The power it radiates, is a function of the angle `f' with which it is rotated, is given by-

$$P_r = \sin^2\phi$$

The radiation is due to the, Z-component of the current. The x-component is disturbed, and the current will have opposite magnitudes on either side of the center of axis and hence it tends to cancel out the radiation.

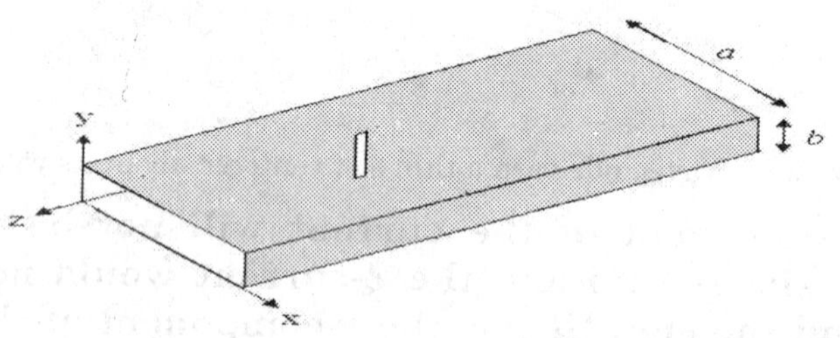

Figure 2.89 Rotated slot antenna in a waveguide

Common Slotted Waveguide Antennas

The most common slotted waveguide resembles that is shown in Figure.

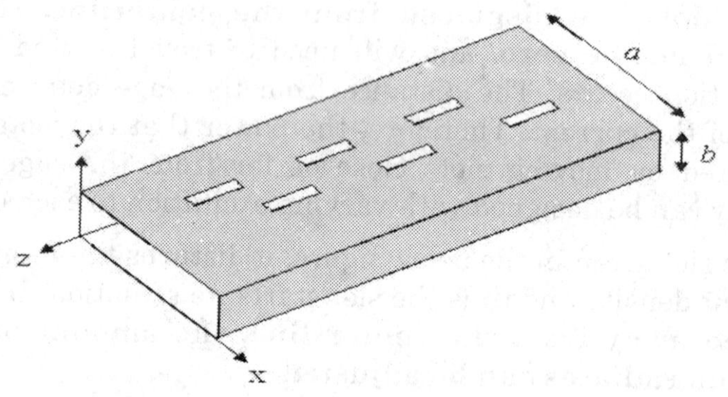

Figure 2.90 Geometry of the most common slotted waveguide antenna

The front end (the open face at the y = 0 in the x-z plane) is where the antenna is fed. The far end is usually shorted, by enclosing if, in metal. The waveguide may be excited by a short dipole (as observed on the cavity backed slot antenna), or by another waveguide.

Analyze

To begin to analyze the antenna of the above figure, let's view the circuit model. **The waveguide itself acts as a transmission line** and the slots in the waveguide can be viewed as parallel (shunt)

admittances. The end of the waveguide is short circuited, so a rough circuit model of the above Figure is-

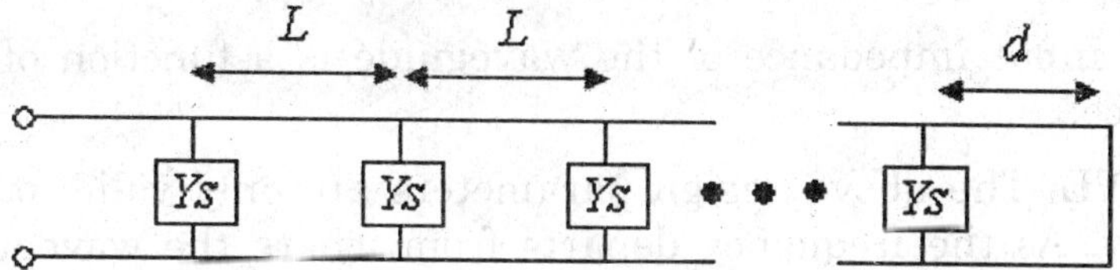

Figure 2.91 Circuit model of slotted waveguide antenna

The last slot is the distance'd' from the end (which is short-circuited, as seen in the Figure), and the slot elements are spaced at distance L from each other.

Before we discuss according to the sizes, which will be given in terms of the guide-wavelength, i.e. is the wavelength within the waveguide. The guide **wavelength (λ_g) is a function of the width of the waveguide (a) and the free space wavelength (λ).** For the dominant TE_{01} mode, the guide wavelength is given by-

$$\lambda_g = \frac{\lambda}{\sqrt{4a^1 - \lambda^2}}$$

The distance between the last slot and the end'd' is often chosen to be a quarter-wavelength. Transmission line theory states that the **impedance of a short circuit a quarter-wavelength down a transmission line is an open circuit.** Hence, above Figure then reduces to-

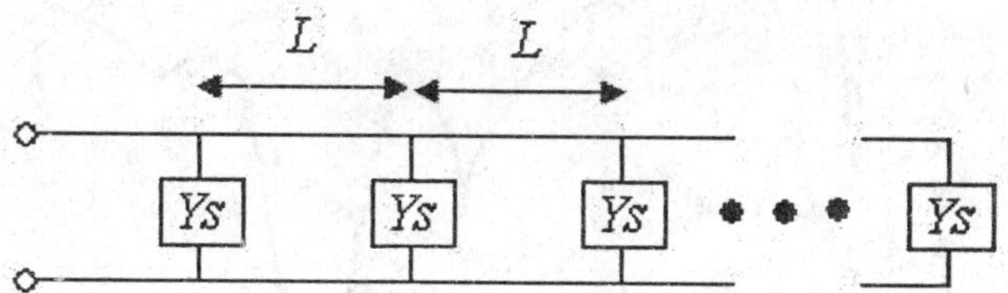

Figure 2.92 Circuit model of slotted waveguide using quarter-wavelength transformation

If the parameter 'L' is chosen to be a half-wavelength, then the input impedance of 'Z' Ohms viewed a half wavelength away is 'Z' Ohms. 'L' is designed to be about a half-wavelength for this reason. If the waveguide slot antenna is designed in this manner, then all of the slots can be viewed as parallel combination. Hence, the input admittance and input impedance for an 'N' element slotted array can be quickly calculated, by-

$$Y_{in} = NY_S$$

$$Z_{in} = \frac{I}{NY_S}$$

The input impedance of the waveguide is a function of the slot impedance.

NOTE: The above design parameters are only valid at a single frequency. As the frequency departs from where the waveguide was designed to work, there will be degradation in the performance of the antenna. To give an idea of the frequency characteristics, of a slotted waveguide, a sample measurement of S11 as a function of frequency is to be shown. The waveguide is designed to operate at 10 GHz. It is fed by a coaxial feed at the bottom as shown in Figure given below-

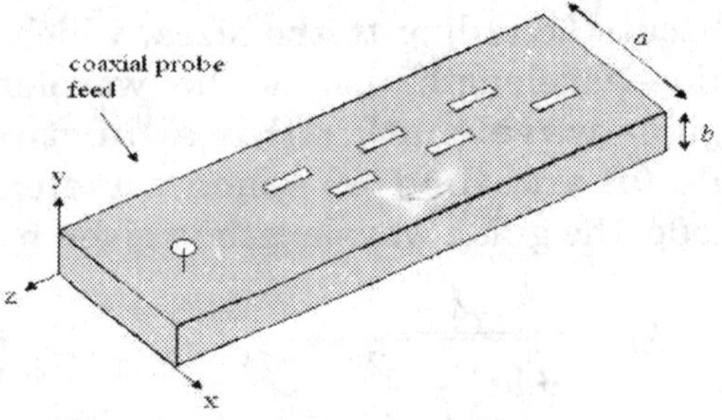

Figure 2.93 Slotted waveguide antenna fed by a coaxial feed

The resulting S-parameter graph is shown in the following figure-

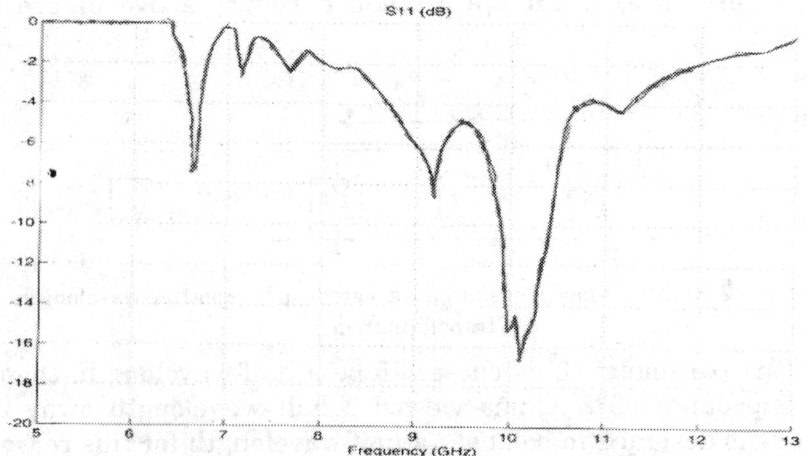

Figure 2.94 Resulting S-parameter graph

NOTE: The antenna has a very large drop in S11 around 10 GHz. This indicates that most of the power is radiated away at this frequency. The bandwidth of the antenna (if defined as where S11 is less than -6dB) extends from about 9.7 GHz to 10.5 GHz, giving a Fractional Bandwidth of 8%.

NOTE: There is also a resonance at about 6.7 and 9.2 GHz. below 6.5 GHz, the waveguide is below the cutoff frequency and virtually no energy is radiated. The S-parameter graph shown above gives a good idea of what the bandwidth and frequency characteristics of a slotted waveguide will resemble.

The 3 D radiation pattern for the slotted waveguide with gain= 17 dB (approximately) is shown in the following figure-

Figure 2.95 3-D radiation patterns for slotted waveguide

x-z plane (or h-plane) the beamwidth is very narrow (2-5°)

y-z plane (or e-plane) the beamwidth is larger.

Slotted waveguide Antennas (As special array antennas)

To obtain more pencil-type beam radiation pattern, we are revising the slotted antennas. In the previous section, on slotted waveguides, it was shown that for a single waveguide strip, the radiation pattern has a wide and beamwidth in E-plane and smaller beamwidth in H-plane.

It can be explained by the fact that, **longer antenna produces narrower beam,** and as in the previous antennas, the physical dimensions along the **E-plane is shorter than along H-plane** (long and thin).

Now, let us arrange slotted waveguide in parallel, as shown in Figure-

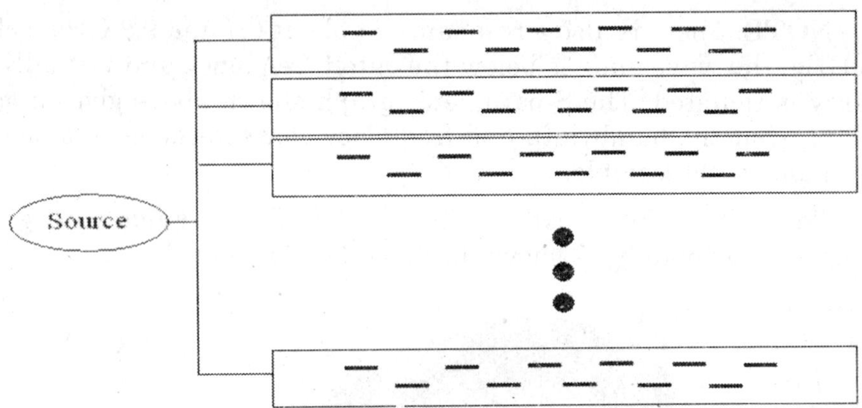

Figure 2.96 Array of slotted waveguides fed by a single source.

By putting waveguides on top of other as shown in the above figure, the E-plane beamwidth can be greatly reduced; other then this by adding a phase delay to each waveguide, the array of waveguides can be steered in E-plane. The phase delay can be summed up by varying line lengths (different frequencies produces different phase delays, allowing scanning simply by changing the frequency).

Antenna arrays made up of hundreds or even thousands elements are often slotted waveguides, which are typically narrowband deviation away from the design; frequency changes, the impedance of the individual slots. Therefore, impedances of many slots add up producing highly reactive impedance, associated with the waveguide away from the resonant frequency.

Figure 2.97 Antenna Array Applications

Top of the airplane, contains large slotted waveguide array, for scanning in horizontal (azimuth) plane, around the airplane.

Power Handling Capabilities of slotted antenna

Slotted waveguides, are used as they **are capable of transmitting high power levels.** Maximum power is a function of altitude & is further divided in CW (continuous wave) and microsecond pulse.

CW gives maximum power handling capability of a slotted waveguide when array continuously transmits in Mega Watts (MW)

Microsecond pulse gives the maximum power when the waveguide radiates a brief pulse and then off.

$$\text{Power handling capabilities } \alpha \frac{1}{\text{altitude}} \alpha \frac{1}{\text{frequency}}$$

Table 1 Power versus altitude for a typical X-band (10 GHz) waveguide

Altitude [ft]	**Max. Power (CW) [MW]**	**Max. Power (1 μS Pulse) [MW]**
0	0.95	1.05
10,000	0.55	0.62
20,000	0.25	0.30
30,000	0.15	0.20
40,000	0.06	0.10
50,000	0.03	0.05

NOTE: Power decreases rapidly with altitude.

This is important to keep in mind in designing RADARS for aircraft.

2.4.4. REFLECTOR ANTENNAS

VHF, UHF, SHF, Antennas

VHF, UHF and SHF antennas are those antennas which are operating between the frequency range of 30 to 300 MHz, 300 t0 3000 MHz and 3000 t0 30,000 MHz respectively the corresponding wavelengths of these antennas are 10 to 1 meter, 1 to 0.1 meter (or 100 cm to 10 cm) and 100 m to 1 cm. Antennas operating above the frequency 2000 MHz are often called as microwave Antennas.

Elevated wires and tubing rods of aluminum or Copper is used in VHF and UHF bands and besides these arrays VHF and UHF antennas are also used to increase the directivity. As the frequency has increased, so the problems and hence new methods become practicable. In general frequency ranges from 30-3000 MHz i.e. VHF and UHF will be considered together and the SHF afterwards (as in these bands wavelengths reaches in cm. range).

∴ Design of SHF antennas changes altogether.

MICROWAVE ANTENNAS

As, the transmitting and receiving antennas, tend to be directive i.e. high gain and narrow beamwidth in both horizontal and vertical planes.

As the frequency increases, the wavelength decreases and thus becomes easier to construct an antenna system, that are large in terms of wavelengths, and which ∴ can be made to have greater directivity. At microwave frequencies the physical size, of high gain antenna becomes small enough to make practical, the use of suitably shaped metallic reflectors to produce the desired directivity.

Some of the microwave antennas are-

(i) Parabolic reflector (Paraboidal or microwave dish)

(ii) Lens antenna

(iii) Horn antenna

(iv) Corner Reflector

2.4.4.1. PARABOLIC REFLECTOR (Satellite Dish)

Basic

The most commonly used reflector antennas are of parabolic shape generally called as **satellite dish** or **parabolic reflector** antenna.

Figure 2.98 the ``big dish" of Stanford University.

Figure 2.99 random ``direct TV dish" on a roof.

Parabola

It is defined as the locus of a point, which moves in such way that its distance from the fixed point (called **focus**) and its distance from a straight line (called **directrix**) is constant.

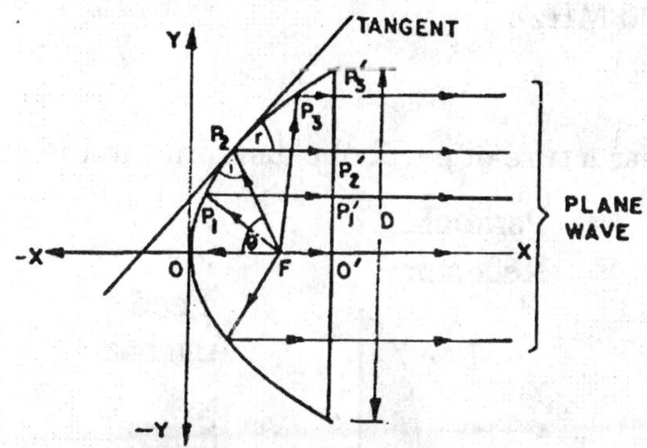

Geometry of Parabolic reflector.

Figure 2.100

Where OF = f = Focal length

k = Constant = depends on the shape of Parabola curve

F = focus

O = Vertex

OO' = Axis of parabola

D = Open mouth of the parabola = Aperture

Now, by the definition of a parabola-

$$FP_1 + PP'_1 = FP_2 + P_2P_2' = FP_3 + P_3 P_3' = Constant\ (k)$$

Hence, the equation of parabola-

$$Y^2 = 4fe$$

(F/D) = The ratio of focal length to Aperture size which is nearly 0.25 to 0.50.

Characteristic of Parabolic Reflector

(1) It has highest gain, ranging from 30-40 dB.

(2) It has low cross polarization.

(3) Reasonable Bandwidth is especial feature of this antenna.

(4) Fractional Bandwidth is 5%, on commercially available models of the antenna.

(5) As we know that the band increases as the dish size increases. Therefore, these giant antennas have large band-width, which vary from 150 MHz to 1.5 GHz; while the smaller dish antennas, operates nearly between 2 and 28 GHz, whereas the large dishes can operate in VHF region (30-300 MHz).

Construction

The basic structure of parabolic dish antenna is-

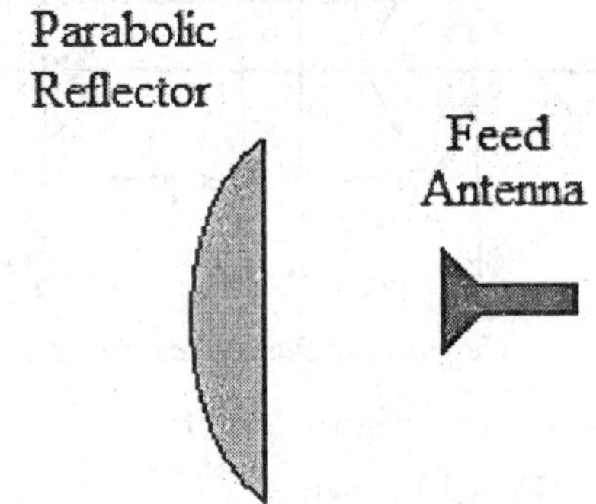

Figure 2.101 Components of a dish antenna

(1) It consists of feed antenna (i.e. horn antenna with a circular aperture) pointed towards the parabolic reflector, unlike alter antennas, (dipole antenna of half wavelength long at the frequency of operation), the reflecting dish must be much larger than a wavelength in size.

(2) The dish of at least several wavelengths in diameter, (where the diameter can be of the order of 100 wavelengths) for very high gain dishes (>50 dB gain).

The distance between the feed antenna and the reflector should be of several wavelengths long. This is similar to the corner reflector antennas where the antenna is roughly a half wavelength from the reflector.

Parabolic Dish Geometry

In this section we can get the answer to the question like –

- Why parabola is of desired shape for reflector?
- Why paraboloid makes a great reflection?

To begin with the explanation for the above questions, let us look into the equation of parabola with focal length (F), in (x, z) plane given as-

$$x^2 = 4F\,(F - Z),$$

as

$$[x] \le \frac{D}{2}$$

The plot obtained by this equation is given below-

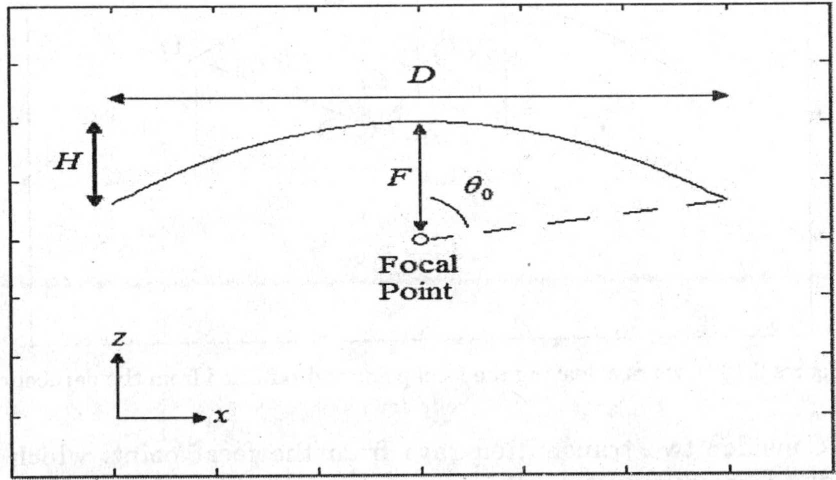

Figure 2.102 Illustration of parabola with defining parameters

Now as we know that, the parabola can be completely described by **two main parameters** –

- The diameter (D)
- The focal length (F)

Again here we can define **two auxiliary parameters**-

- Vertical height of the reflector (H)
- The maximum angle between the focal point and the edge of dish (θ_0).

All of the parameters can be correlated by the following equations-

$$\frac{F}{D} = \frac{1}{4\tan\,(\theta_o\;/2)}$$

$$F = \frac{D^2}{16\;H}$$

To further analyze the reflectors, we have to use approximations from optical geometry. Let us assume that the reflector is larger than the wavelength. Hence, with this assumption we can next analyze the structure via straight line rays from the focal point, with each ray acting as a plane wave.

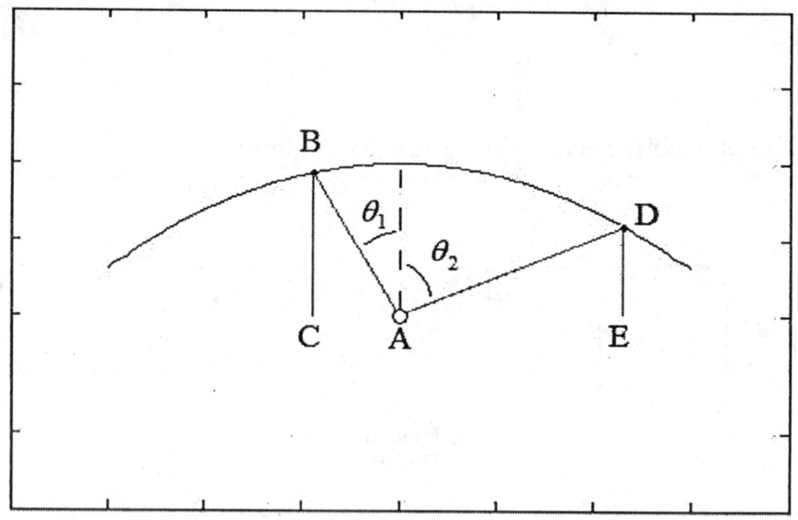

Figure 2.103 Two rays leaving the focal point and reflected from the parabolic reflector

Consider two transmitted rays from the focal point, which are arriving from different angles as given in the above figure. Here the reflector is assumed to be perfectly conducting, thus the rays are completely reflected back.

To further analyze, let us closely observe two rays-

(1) Both the rays end up travelling in the downward direction (which can be determined because the incident and reflected angles, relative to the normal of the surface must be equal) such rays are called as **collimated rays.**

(2) The path lengths ADE and ABC are equal; which can be proved by different geometrical theorems.

The above facts can be proved for any set of angles chosen. Hence, we conclude that-

(1) All rays emanating from the focal point (the source or feed antenna) will be reflected in same direction.

(2) The distance each ray travels from the focal point to the reflector and then to the focal plane is constant.

As a result of these observations, we can say that, *the distribution of the field on the focal plane will be in phase and travelling in the same direction. This gives rise to the parabolic dish antennas* ***highly directional radiation*** *pattern. This is why the shape of the dish is parabolic.*

Beam Formation by Parabolic Reflector

Focusing or beam formation action of parabolic reflector can be understood by considering a source of radiation at the focus.

Let a ray starts from the focus (F) at an angle (θ), with respect to parabolic axis (OO'). The curve strikes at a point P_2, on the curve of the parabola.

Let a tangent be drawn at point P_2 on the curve. According to the law of reflection, the angle of incidence (angle i) and angle reflection (angle r) will be equal. This results that the reflected ray is parallel to the parabolic axis, regardless of the particular value of 'θ' that may be considered.

The entire wave originating from the focus is reflected parallel to the parabolic axis. This implies that the entire wave thus, reaching at the aperture plane is in phase. This shows that a constant phase wave-front a surface is created in the aperture plane. ∴ Rays are parallel to parabolic axis, because rays are always perpendicular to wavefront.

NOTE: As all the waves are in phase, so a very strong and concentrated beam of radiation is there along the parabolic axis.

The entire wave emanating from the source at focus and reflected by parabola are travelling the same distance, in same time, to reach the directrix and hence they are in phase. The principle of equality of path length is maintained between all rays of two wavelengths. There is a path length difference between two rays, and hence cancellation action takes place.

(Parabola converts a spherical wavefront coming from the focus into plane wavefront. The part of radiation from focus, which does not, strikes the parabolic curve as spherical wave appears as minor lobes. This is waste of power; which is minimized by partially shielding the source).

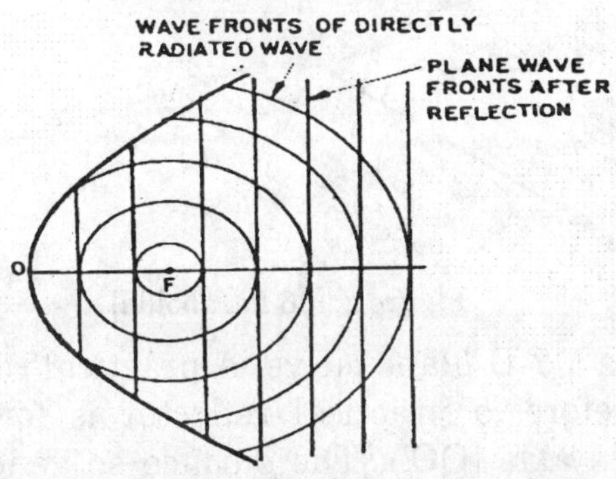

Figure 2.104 Production of plane wavefront by reflector with omni – directional, source at the focus

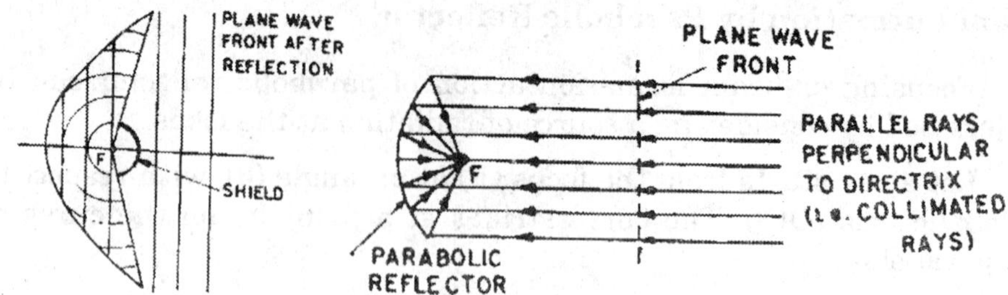

Figure 2.105 Focusing by a parabolic reflector (Receiving case)

Further if beam of parallel ray is incident on a parabolic surface, they will be focused at a point i.e. **Focus**. This is in effect due to principle of reciprocity theorem already discussed which says that - ``properties of an antenna are independent whether it is for transmission or reception." The parabolic reflector is directional, for the reception case. Rays coming perpendicular to directrix will be focused at the focus and not other, due to path length difference. These parallel rays are known as **Collimated Rays**.

Paraboidal reflector or Microwave dish

Now if we revolve the parabola about Z-axis, we obtain a paraboloid as given in the figure-

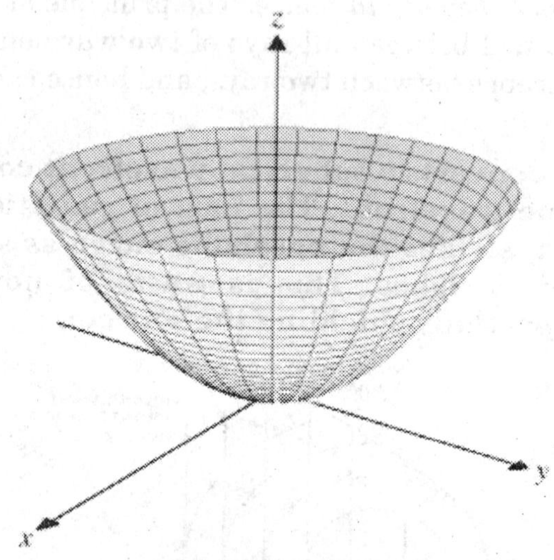

Figure 2.106 Paraboidal

A parabola is a 2-D plane curve. A practical reflector is 3-D curve surface and therefore, a practical reflector is formed by rotating a parabola about its axis (OO'). The surface so generated is known as **"Paraboloid"** which is often called as **``microwave dish"** or **``Parabolic reflector"** also.

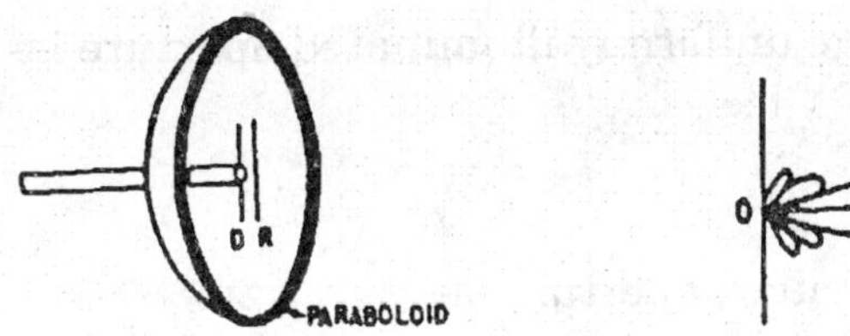

(a) Full paraboloidal reflector with dipole source at the Focus *(b) Radiation pattern of a Paraboloid of aperture* $D = 10\lambda$.

Figure 2.107

Paraboloid produces a parallel beam of circular cross-section, because the mouth of paraboloid is circular. If a third Cartesians coordinate `z' has its axis perpendicular to both x-axis and y-axis, then the equation" of paraboloid is-

$$y^2 + z^2 = 4fx$$

The intersection of any plane perpendicular to x-axis with paraboloid surface is a **circle**. Beam produced is like the one produced in the search light.

The **Radiation pattern** of an antenna employing the paraboloid reflector has a very sharp major lobe accompanied by a number of minor lobes, which of course, are smaller in size, in the direction of paraboloid axis. The 3-D pattern is a figure obtained by revolving about OO' and actual shape is like a thick cigar.

If the feed or primary antenna is isotropic, then paraboloid will produce a beam of radiation. Assuming, circular aperture is large,

Beamwidth between first null is given by-

$$BWFN = \frac{140\lambda}{D} \text{ (degree)}$$

where λ = free space wavelength, (m)

D = Diameter of aperture mouth diameter, (m)

The beamwidth between first nulls for a large uniformly illuminated rectangular aperture is given by-

$$BWFN = \frac{115\,\lambda}{L} \text{ (degree)}$$

Where, L = Length of Aperture, (λ)

Also width between half power points for a large circular aperture is given by-

$$HPBW = \frac{58\,\lambda}{D} \text{ (degree)}$$

Directivity D, of a large uniformly illuminated aperture is-

$$D = \frac{4\,\eta\,A}{\lambda^2}$$

Directivity D, for a circular, aperture is-

$$D = \frac{4\pi}{\lambda^2}\left(\frac{\pi D^2}{4}\right) = \pi^2\left(\frac{D}{\lambda}\right)^2$$

$$D = 9.87\left(\frac{D}{\lambda}\right)^2$$

Where D = Diameter of the aperture, (λ)

Design

The parabolic antenna it is important to note that-

(1) By increasing the value of the diameter (D), the gain of the antenna increases

(2) The focal length (F) is a free parameter.

Hence, range of the ratio F/D is generally between 0.3 and 1.0.

Let u now study the gain Calculations for a parabolic reflector antenna.

In practice, **the primary (or feed) antenna is not an isotropic and thus does not radiate, uniformly which introduces distortion.** Surface of Paraboloid is gradual tapering towards the edge, which result in less capturing area which is smaller than the actual area, that is-

$$A_o = kA$$

Where Ao = Capture area

A = Actual area of mouth

k = Const. depends on type of antenna used for feed

= 0.65 (approx.) for dipole antenna

The antenna radiation depends on the fields across the aperture of the parabolic reflector. The maximum possible gain of the antenna depends on the physical area of the aperture and is given by

$$G_{max} = \frac{4\pi}{\lambda^2}A = \frac{(\pi D)^2}{\lambda^2}$$

The actual gain is in terms of the effective aperture, which is related to the physical area by the efficiency term (ε). `ε' is of the order of 0.6 to 0.7 for a well designed dish antenna-

$$G = \varepsilon \frac{4\pi}{\lambda^2} A = \varepsilon \frac{(\pi D)^2}{\lambda^2}$$

The power gain of the circular Aperture is given by-

$$G_p = \frac{4\pi A_o}{\lambda^2} = \frac{4\pi kA}{\lambda^2}$$

$$\therefore A = \frac{\pi o^2}{4}$$

$$= \frac{4\pi k}{\lambda^2}\left(\frac{\pi D^2}{4}\right)$$

$$= 0.65 \frac{\pi D}{\lambda}$$

$$= 0.65\,(3.14)^2 \left(\frac{D}{\lambda}\right)^2$$

$$\therefore G_p = 6.389 \left(\frac{D}{\lambda}\right)^2$$

or $$G_p \approx 6\left(\frac{D}{\lambda}\right)^2$$

Where, D/λ = Aperture Ratio.

Effective Radiated Power (ERP) of an antenna is multiplication of an input power fed to antenna and its power gain.

For effective use, reflector must have an open circular mouth aperture which is possible at low frequency, but the size of mouth becomes too large, heavy and bulky and hence purposely avoided for T.V. broadcast band etc.

The important practical implication of this property is that reflector can focus parallel rays on to the focal point or conversely it can produce a parallel beam from radiation originating from the focal point. *It is seen that some of desired rays are not captured by reflector and these constitute ``**split over**" which increases noise which is particularly troublesome in satellite ground stations.*

Some radiations from primary radiations occur in forward direction in addition to desired parallel beam. These are known as backlobe radiations, which are undesirable as they can interfere destructively with reflected beam, and hence the practical radiators are designed to minimize f\D (Aperture Number).

$$A_e = \eta A$$

Where, Ae = Effective area of primary antenna.

n = Aperture efficiency or illumination efficiency.

A = Physical area of reflector.

$$A = \frac{\pi D^2}{4} = \pi r^2$$

Where, D = Diameter

r = Radius

The **efficiency** is important in understanding the tradeoffs involved in the design of a parabolic reflector. The efficiency can be written as the product of a series of terms-

$$\varepsilon = \varepsilon_r \varepsilon_{AT} \varepsilon_S \varepsilon_O$$

Let us study each of these terms one by one-

(1) Radiation efficiency (ε_r)

\`ε_r' is the usual efficiency which deals with the ohmic losses. As the horn antennas are most commonly used as the feed antennas (having very little loss), thus the parabolic reflector is typically made up of metal with very high conductivity. Therefore the efficiency here is generally kept close to one (which can be considered as negligible value).

(2) Aperture Taper Efficiency (ε_{AT})

\`ε_{AT}' measures as to how a uniform field is across the antenna's aperture, if the E-field is uniform in the phase and the amplitude across the aperture (the far field is roughly the Fourier Transform of the aperture fields), the antenna will have the maximum gain. However, the aperture fields tend to diminish away from the main axis of the reflector, which leads to the lower gain.

This efficiency can be improved by increasing the F/D ratio, which on the other hand lowers the cross polarization of the radiated fields. But its adverse effect says that by increasing F/D ratio, it reduces the spillover efficiency.

(3) Spillover Efficiency (ε_s)

`$\varepsilon_s$' measures the amount of radiations from the feed antenna, reflected by the reflector. Due to the finite size of the reflector, some of the radiation from the feed antenna travels away from the main axis at an angle greater than `θ_o', and hence they are not reflected back. This problem can be improved upon, by moving the feed closer to the reflector, or by increasing the size of the reflector.

(4) Other efficiencies (ε_o)

The parameter `ε_o' is a major parameter out of all the other parameters mentioned here. It degrades the gain of the antenna and consists of the effects such as-

(i) **Surface error:** With the slight variation in the shape of the reflector, the performance of antenna degrades (especially at higher frequencies having small wavelength) and also the rays are scattered by small surface anomalies.

(ii) **Cross Polarization**: The loss of gain due to cross polarized (undesirable) radiation.

(iii) **Aperture Blockage:** Some of the radiations are blocked by the feed antenna, which would have been transmitted by the reflector. (This can be considered as loss).

(iv) **Non-Ideal feed Phase center:** The parabolic dish has a desirable properties relative to a single focal point. As the feed antenna cannot be a point source, therefore there will be some loss due to a non-perfect phase center for a horn antenna.

Efficiency

The efficiency is the function of the feed point of an antenna (in terms of F and D) and the feed antenna's radiation pattern. Their relationship is given by-

$$R(\theta) = \begin{cases} \cos^n \theta, & 0 \le \theta \le 90^\circ \\ 0, & 90^\circ \le \theta \le 180^\circ \end{cases}$$

This result was given by S. Silver in 1949. He calculated the aperture efficiency for a class of radiation pattern. For some value of `n' in the above function, the feed (horn) antenna, will not have a pattern exactly like what should be the output from the equation above, but its approximation. Using the above equation the aperture efficiency of a parabolic reflector can be calculated and can be plotted as in the given figure below for varying values of 'θ_o' and F/D ratio.

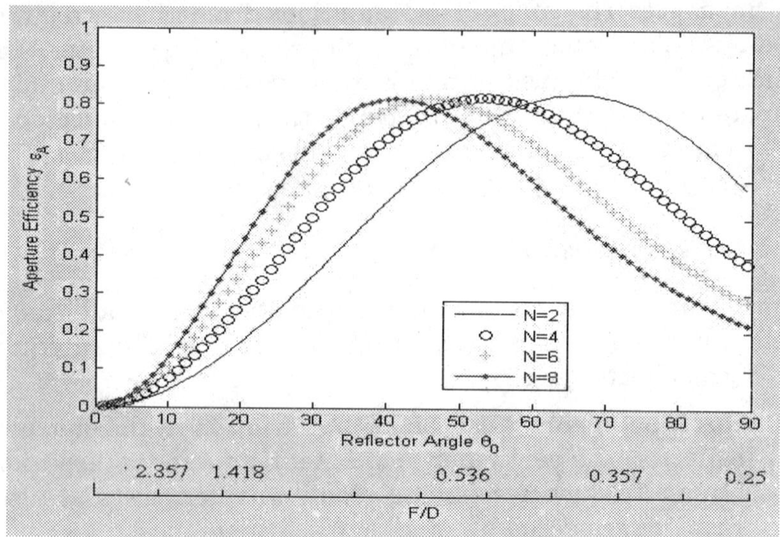

Figure 2.108 Aperture Efficiency of a Parabolic Reflector as a function of F/D or the angle θ_o, for varying feed antenna radiation patterns.

According to the given figure-

(1) D is made as large as possible so as to increase the physical aperture.

(2) From the above graph we can find, the F/D ratio of the antenna, which maximizes the aperture efficiency.

F/D ratio and angle θ_o relationship equation (has already been discussed).

Other Types of Parabolic Reflectors

Except the full Parabola, there are other three prominent parabolic reflectors, which are-

(a) Truncated Paraboloid

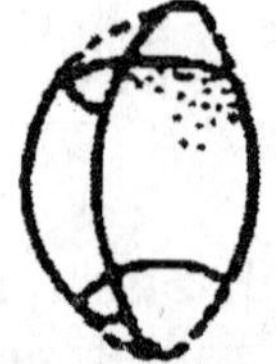

Cut Paraboloid Reflector or Tuncated Paraboloid

Figure 2.109

It is such that, its appearance is not circular when viewed from a point on parabolic axis.

(b) Cylinder Paraboloid

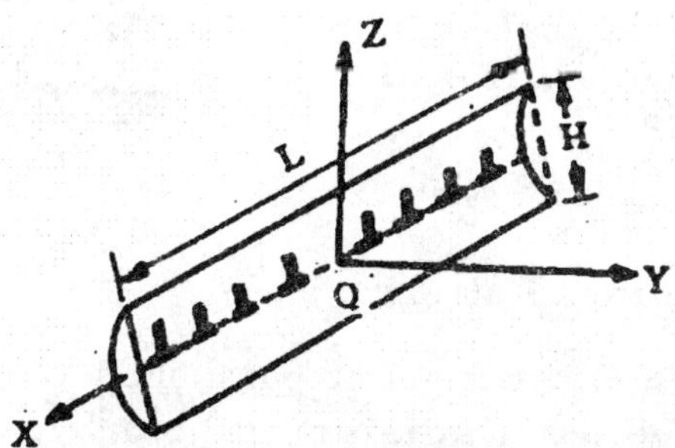

Parabolic Clyinder Reflector with line Source of eight dipoles

Figure 2.110

It is formed by moving the parabola in sideways that is, a plane sheet curved to parabolic shape in 1-D only forms a cylinder. This has a "**focal line**" instead of a focal point and a **"vertex line"** instead of vertex. If source is put along the focal line then this curvature is illuminated uniformly. **Focus beam is in vertical plane.** Inside the paraboloid are eight in phase dipoles. Beam width is less than full Paraboloid, and the wide Beam is in E-plane (xy) and a narrow Beam in H-plane (yz).'

(c) Pill Box

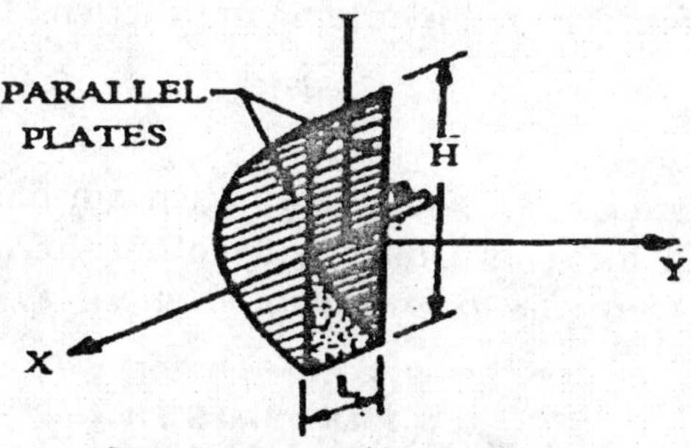

Pill box or cheese antenna

Figure 2.111

It is the shortest parabola, enclosed by parallel plates. It is fed with probe, excited by a coaxial line. It produces a beam which is wide in E-plane and narrow in H-plane.

The Radiation of pill type is just as the **fan type** shape and has an advantage of it **small size.**

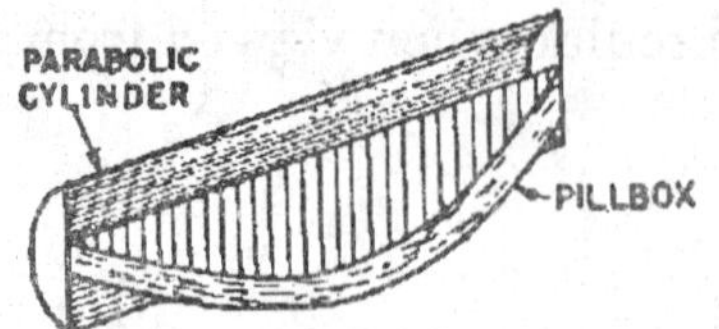

Feeding of parabolic cylinder with Pillbox cheese antenna as line source.

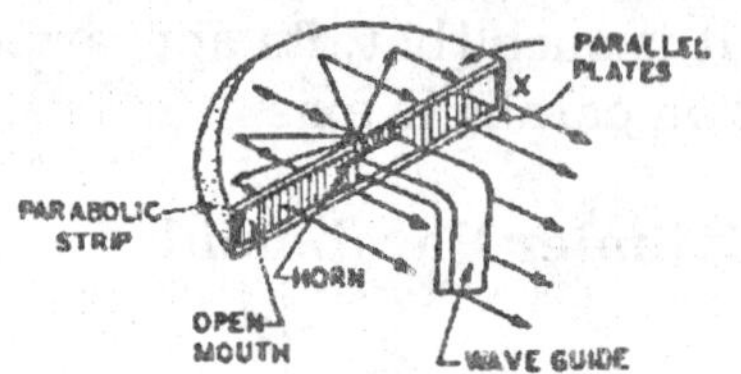

Pillbox cheese antenna producing line source of radiation through open mouth.

Figure 2.112

Primary and Secondary Pattern

The antenna placed at focus of paraboloid is known as "**feed Radiator**" or "**Primary Radiator**" or simply a **feed** and its radiation pattern is called as *Primary pattern.*

Parabolic reflector is known as "**Secondary radiator**" and its radiation pattern of entire system is called as *Secondary pattern.*

Feed systems

There are two types of basic components, of primary radiation at the focus, that are-

Source = primary radiation/feed radiator/simply feed.

Reflector = Secondary radiator.

An ideal "feed" would be that radiator which radiates towards reflector in such a way that it illuminates the entire surface of reflector and no (or zero) energy is radiated in any other directions. Such an ideal radiator is not available in practice.

(1) Dipole Feed

Dipole antenna is also not very much suitable for the feed but occasionally used for Yagi-Uda or any other Array. *Feeding with a dipole involves, changing from an unbalanced system to a balanced system.*

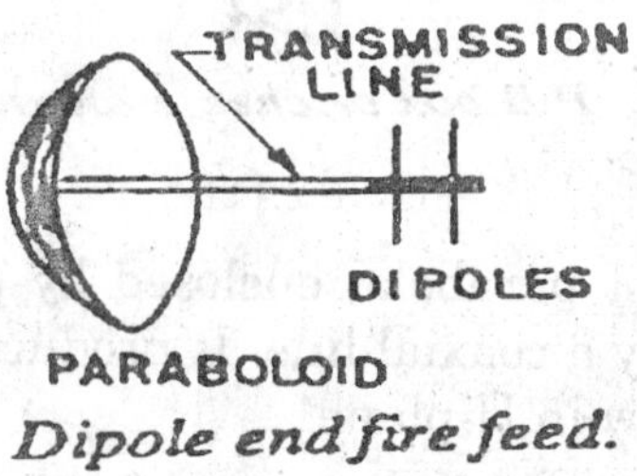

Dipole end fire feed.

Figure 2.113

(2) Horn feed

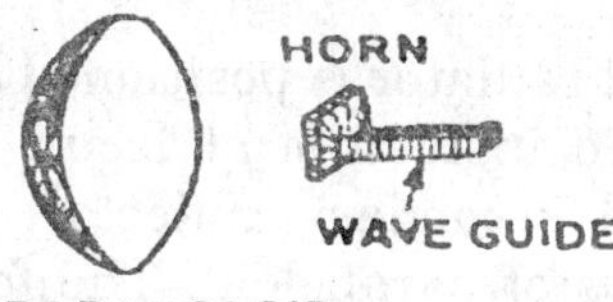

(b) Horn feed.

Figure 2.114

Most common feed for paraboloid reflector antenna is a "waveguide horn". Horn feed is a waveguide feed. Horn antenna points the Paraboloid and radiation pattern of **horn antenna** is mild and in the same direction. Thus direct radiation from horn antenna is minimum. If the circular polarization is required then, **conical horn or helix antenna** can be used as feed at the focus of paraboloid. **For getting maximized beam pattern along the parabolic axis, feed is placed at the focus.** But if feed is moved laterally from focus, that is, perpendicular to axis, then beam deteriorates that is, limited beam motion can be obtained. On the other hand **if the feed is moved along axis, then pattern is broadened.** Thus important ***position of feed is focus*** and for the small reflector of short focal length the position of feed is shown in the figure-

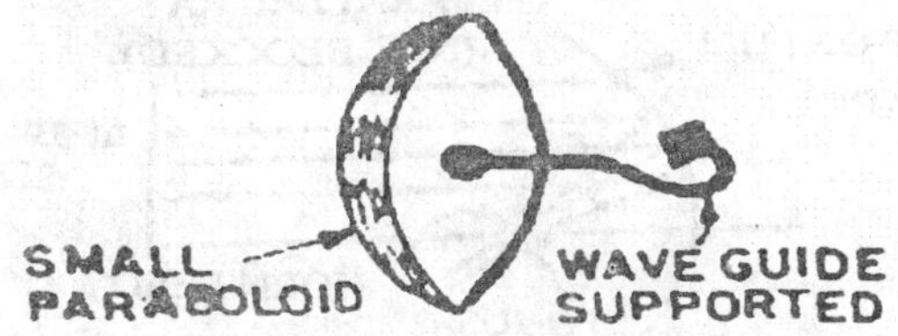

Feed support for a small paraboloid reflector.

Figure 2.115

(3) Cassegrain feed

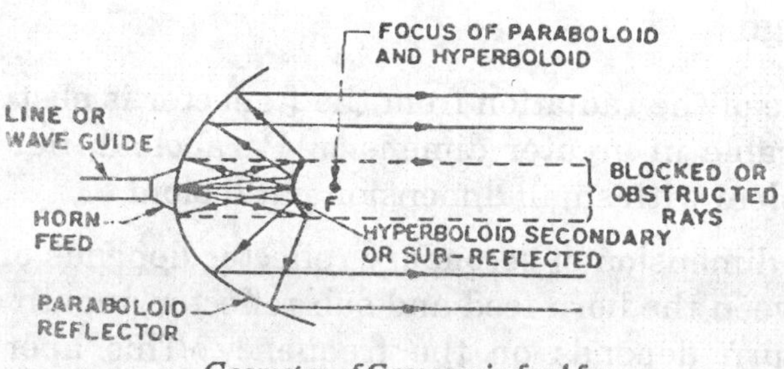

Geometry of Cassegrain feed for a paraboloid reflector.

Figure 2.116

It is named after the name of an astronomer from the 18th Century.

In Cassegrain feed radiator is positioned around an opening near, the vertex of paraboloid instead of at focus. Cassegrain feed system employs a hyperboloid secondary reflector, whose one of the focii coincides with the focus of paraboloid. The feed radiator is aimed at the *secondary hyperboloid reflector* or **sub reflector**. The radiations emitted from the feed radiator, are reflected from Cassegrain-secondary reflector, which illuminates the main paraboloid reflector as if they had originated from the focus. Then paraboloid reflector colliminates the rays (render parallel) as usual.

Sometimes, it becomes important, to **minimize the length** of transmission line or waveguide connecting feed radiator with receiver or transmitter. This **is needed specially to avoid** losses. Although there could be a solution of this problem by placing RF amplifier stages of Receiver near the focus which minimizes losses at the reception. But this is not practicable for transmitter, as RF amplifier of transmitter is bulky, heavy and enough power, so it is not possible to place at feed point. Hence the practical solution in such a cases is Cassegrain feed when the transmission line or waveguide length between feed the transmitter and the receiver, is required to be short.

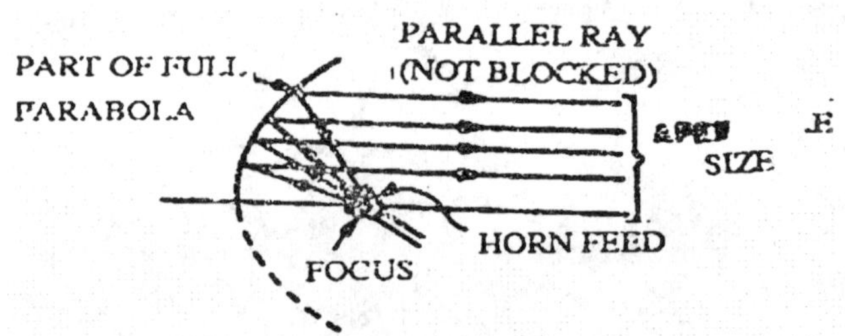

Off set Paraboloid reflector showing no blocking of rays.

Figure 2.117

Disadvantage

(1) Some of the radiation from the Reflector is obstructed. This is tolerable in greater dimension Paraboloid; but this becomes problem with small dimension paraboloid.

(2) The dimension of secondary reflector depends on the distance between the horn feed and sub-reflector, mouth of horn which in turn depends on the frequency. This aperture blocking deflect can be avoided by using an **offset reflector** which is applicable to focal point.

(3) The other method is to use a polarization twisting scheme, in which hyperboloid reflector is made up of a wire grating (i.e. transparent) instead of solid.

Advantages

(1) Reduction in spill over and minor lobe radiations.

(2) Ability to get an equivalent focal length much greater than the physical length.

(3) Ability to place a feed in convenient location.

(4) Capability for scanning or broadening of beam by moving one of reflecting surfaces.

Other feed is Georgina feed.

Radiation Pattern of Parabolic Antenna

3-D radiation pattern for the parabolic dish reflector with diameter D = 11λ, F/D ratio = 0.5. In this antenna the feed system is a circular horn antenna.

Figure 2.118 Side View

Figure 2.119 Front View

3-D Radiation patterns of parabolic dish Reflector

From the diagram (3-D pattern), we see that-

1. Physical aperture $=11\ \pi^2\ =1194 = 30.7\ dB$.
2. Actual gain = 29.3 dB = 851.
3. Overall efficiency = 77%.
4. Radiation Pattern is highly directional.
5. Half Power Beamwidth is nearly 5°.
6. Front to Back Ratio = 33 dB.

1. Why do we use parabolic reflector in microwave ranges? Also explain feed system in parabolic reflector antenna. **(Raj. Univ. 2007)**
2. Draw the radiation pattern of the following antennas: (i) Parabolic Reflector **(Raj. Univ. 2008)**
3. Calculate the approximate gain and beam width of a paraboloidal reflector antenna at operating frequency 4 GHz, diameter 20 meters and illumination efficiency 55% **(R.T.U. 2009)**
4. Calculate angular aperture for a paraboloid reflector antenna for which aperture number is- i. .25 ii. .50 iii. .60 Given that diameter of reflector mouth is 10m. Calculate the position of focal point with reference to reflector mouth in each case. **(R.T.U. 2009)**
5. What is microwave range and explain Parabolic Reflector antenna with feeding system. **(R.T.U. 2009)**
6. Draw the different feeding mechanism in parabolic reflector. **(R.T.U. M. tech. 2010)**

2.4.4.2. THE CORNER REFLECTOR ANTENNA

To increase the directivity of an antenna, we use reflectors. For an example, if we take a wire antenna, lets say a half wave dipole antenna, with a conductive sheet behind it to direct radiation in the forward direction. To further increase the directivity, a corner reflector may be used, as shown in figure.

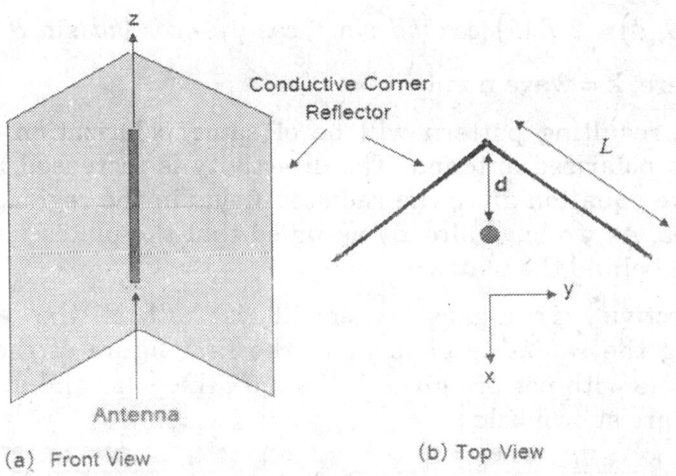

Figure 2.120 Geometry of Corner Reflector

The *angle between the plates is generally 90°*. Now let us use image theory to understand the radiation pattern of this antenna, and then calculating the result by array theory. We start the analysis by the assumption that the, *reflecting plates extends to infinite.* Equivalent source distribution, valid for the region in front of the plates which is given by-

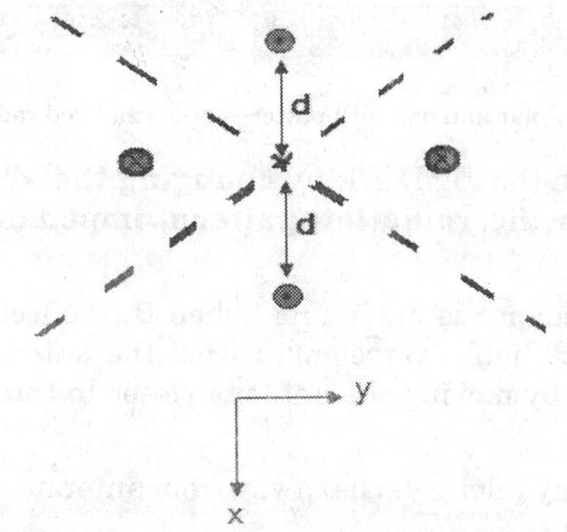

Figure 2.121 Equivalent sources in free space

Here O Antennas are in phase with the actual antenna.

⊗ Antenna which are 180° out of phase.

For the antenna having an omni-directional pattern equals to f (θ). The radiation pattern (R) of the "equivalent set of radiators" is given by -

$$R(\theta, \phi) = 2 f(\theta) [\cos(kd \sin\theta \cos\phi) - \cos(kd \sin\theta \sin\phi)]$$

Where, k = wave number

The resulting pattern will be of same polarization, as that of vertically polarized antenna. The directivity is increased by 9-12 dB. The above equation gives the radiated fields in the region, in front of the plates. As we have already assumed that the plates were infinite, the fields behind the plates are zero.

Directivity is highest when 'd' is half of the wavelength. Assuming the radiating element in the first figure of the topic is a short dipole with pattern given by "sin θ", the polar and azimuth field patterns are shown below -

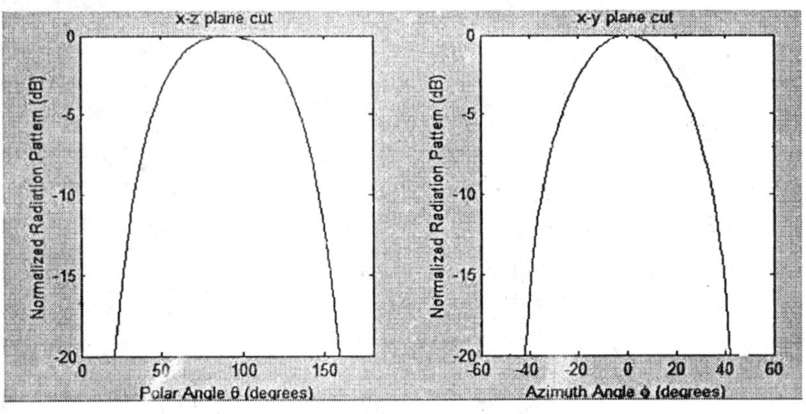

Figure 2.122 Polar and azimuth patterns of normalized radiation pattern

As shown in the Fig.D.2.1; **by changing the 'd' distance, there is a change in the radiation pattern, impedance and gain of antenna.**

Input impedance is maximum, when the reflector is spaced at a distance equal to half a wavelength from the antenna, which can be further reduced by moving the antenna closer to the reflector. (Where **L = 2d**)

A tracing ray (along y-axis) away from antenna, is reflected if **the length (L) is at least** $\sqrt{2}\, d$.

NOTE: The height of the reflecting plates should be greater than the radiating element.

2.4.4.3. LENS ANTENNA

Lens antennas as compared to Reflectors have same applications, but are used at higher frequency (Starting from 1000 MHz) but its greatest use is at 3000 MHz. As *at lower frequency Lens*

antenna becomes bulky and heavy. They act, just as glass lenses are used in optics.

Principles

The collimating action, of a simple optical lens can be illustrated by the given figure-

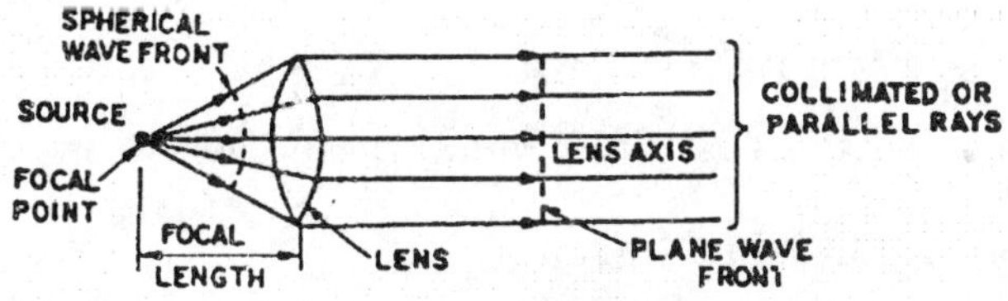

Operation of simple converging lens.

Figure 2.123

Assuming a source at focal point, at a distance of focal length, along the lens axis, it is seen that Collimated or parallel rays are obtained (plane wavefront) on right hand side of lens. From optical point of view, a divergent beam is collimated because refraction takes place as a result of which, ray at the centre are refracted less than at the edges. **An optical lens** (usually of glass) **operates by virtue of having a refractive index more than unity.**

Similarly a lens made up of dielectric material, operates for radio frequencies e.g. an electromagnetic source at focus on left hand side will produce collimated i.e. parallel rays on right hand side. Converge is also true i.e. incoming parallel rays on right hand side will converge to a point on the focus at left hand side. The two cases correspond to the use of lens antenna for transmitting and receiving. **This also `illustrates the principle of reciprocity theorem.**

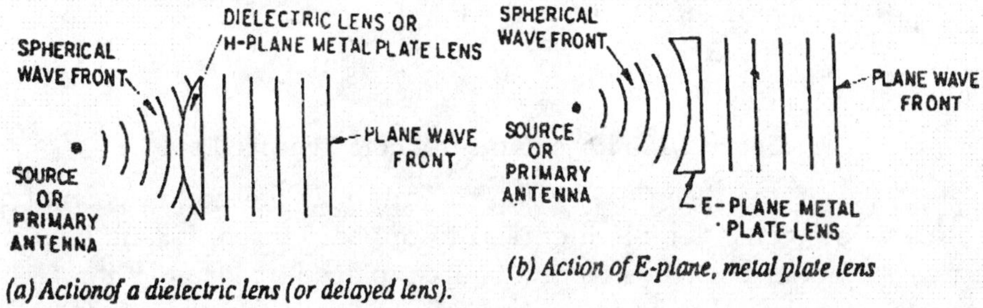

(a) Actionof a dielectric lens (or delayed lens).

(b) Action of E-plane, metal plate lens

Figure 2.124

Further, it is possible to construct a medium having refractive index less than unity, at radio frequencies so that focusing properties can be achieved. Thus whatever may be the refractive index of lens material (whether high or low), the purpose is to straighten out the wavefront. At microwave frequencies, the microwave lens is used in conjunction with horn antenna for correction into plane waveform or to focus parallel rays at one point. For lens antenna also the geometric optics principle - "**the principle of Equality of path length**" - is applicable.

Types of Lens Antenna

(a) E-plane Metal Plate lens.

(b) Dielectric lens/ H-Plane metal plate /Delay lens.

(a) E-plane Metal Plate lens

These are the lenses in which the traveling wavefront are **speed** up by the lens media.

(b) Dielectric lens/ H-Plane metal plate /Delay lens

Lenses in which the travelling wave fronts are **retarded or delayed** by lens medium. There are two types of dielectric-

- Non-Metallic Dielectric (Polystyrene / Lucite).
- Metallic or Artificial Dielectric.

(j) Non Metallic Dielectric lens Antenna

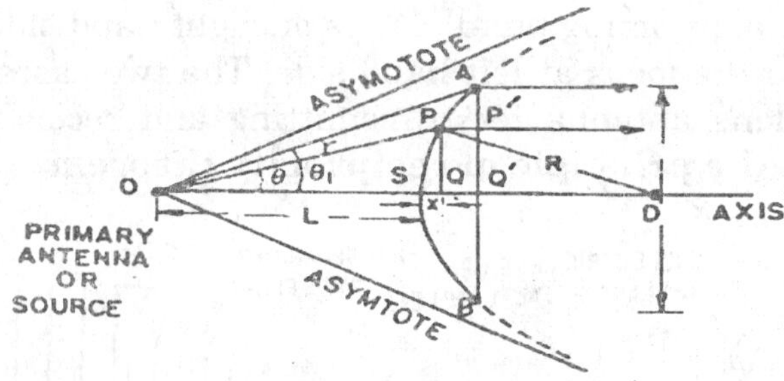

Figure 2.125 Rays Paths in Plano Concave Lens

This may be designed by the ray analysis method of geometrical optics as dielectric lens is identical to optical lenses. If this lens is to convert a spherical wave from a source, placed at focus `F' to a plane wavefront, then all the rays' path from `O' to the plane surface (AB) of lens, should have equal electrical length. A Plano-concave lens solves this purpose and primary antenna is supposed to be isotropic. **According to principle of equality of electrical (or optical) path lengths all paths from source to plane surface of lens should be equal, to so that field over entire plane surface is in phase.**

Hence wave emerging from source, will have a constant phase across Aperture (D), if lens is so shaped that time taken by ray to travel from `O' to aperture plane (AB) is same for all possible ray paths.

Let velocity of wave in air and in lens medium be 'c' and 'v', respectively, then for equal time condition-

$$OP + PP' = OS + SQ' \qquad \text{Since } PP' = QQ'$$

$$= OS + SQ + QQ'$$

or $\quad OP = OS + SQ \qquad \text{Since } QS = x$

or $\quad \frac{r}{c} = \frac{L}{c} + \frac{x}{v}$

or $\quad r = L + \left(\frac{c}{v}\right)^{x} \qquad \therefore \mu = \frac{c}{v} \quad But\ x = OQ - OS = r\cos\theta - L$

$$r = L + \mu x$$

$$r = L + \mu(r\cos\theta - L)$$

or $\quad r(1 - \mu\cos\theta) = L(1-\mu)$

or $\quad r = \frac{L(\mu - 1)}{(\mu\cos\theta - 1)}$

and $\quad \mu = \frac{c}{v} = \frac{Velocity\ of\ air}{Velocity\ in\ lens\ medium}\ ; where\ \mu > 1$

Equation (1) represents the outline of lens in polar co-ordinate and hence gives shape of lens. This is equation " of a hyperbola whose focal length is (L) and radius of curvature (R)-

$$R = L(\mu - 1)\ provided\ \theta\ is\ small$$

The asymptote of hyperbola is at angle θ with respect to the axis and can be obtained by tending to ∞ (*i.e.*, $r \to \infty$)

$$(\mu\cos\theta - 1) = \frac{L(\mu - 1)}{\lambda} = \frac{L(\mu - 1)}{\infty} = \frac{Finite}{\infty} = 0$$

$$\therefore \cos\theta = \frac{1}{\mu}$$

The wave emanating from such lens antennas produces the radiation pattern which is characteristic of lens aperture and of **uniformity of illumination** just like a parabolic reflector. The radiation of uniformly illuminated circular lens antenna of `D' diameter is like that of paraboloid and its gain is given by-

$$r = \frac{L(\mu - 1)}{(\mu \cos \theta - 1)}.$$

However, it is difficult to get illumination for a lens antenna. **A long focal length lens provides more uniform illumination rather than a short focal length lens.**

NOTE: For a frequency less than 10,000 MHz, lens antennas have excessive thickness; which is undesirable. This problem however is curved by designing the less antenna of zoned or stepped dielectric. It can be shown that thickness `t' of such a stepped lens is given by-

$$t = \frac{\lambda}{\mu - 1}$$

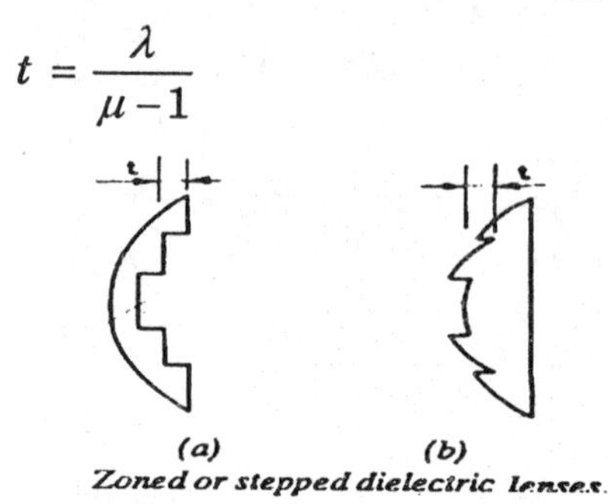

Figure 2.126

Primary feed of Lens Antenna

The primary sources for feeding the lens antenna whether dielectric or metal plate should satisfy similar requirement as in case of Parabolic Reflectors. **Lens antenna is usually fed with horn antenna.** *The focal length of lens is approx. equal to the aperture dimension, is better choice and thus* **lens feed is more directive than for a reflector.** To avoid stray radiations from the horn, the sides of horn is continued upto lens. If focal length and aperture dimension are equal, the flare angle of horn corresponds to 53°.

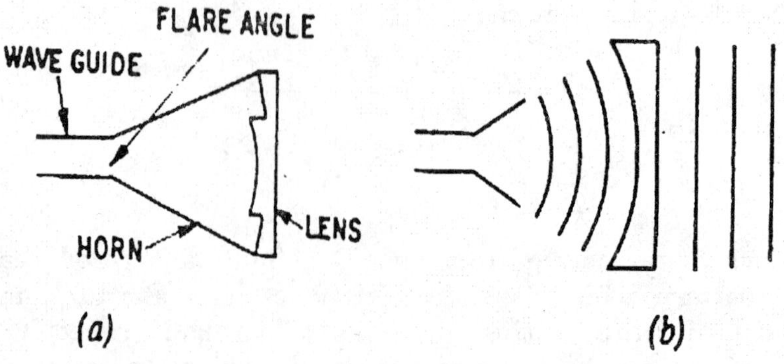

Figure 2.127 Feeding of Lens antenna with horn feed.

Uses

(1) Dielectric lens which is not stepped a wideband antenna. Typical bandwidth for non stepped and stepped lens antenna is 12% and 5%.

(2) Reflector and Lens antenna are used above 1000 MHz. (Lens for it usually > 3000 MHz and not below it.)

Merits

(1) Feed and feed support does not obstruct the aperture, as rays are transmitted away from the feed.

(2) It has **greater design tolerance**, i.e. larger amount of wrapping and twisting is possible in lens antenna; as wave enters from one side and emerges out from the other, maintaining the electrical path length.

(3) It can be used to feed, off the optical axis and hence useful in application where beam is required to be moved angularly with respect to the axis.

Demerits

(1) Lenses are usually, bulky and heavy which generates complications in the design of antenna.

(2) Costlier for same gain and Beamwidth, in comparison with the reflectors.

(ii) E - plane Metal plate Lens Antenna

A metal plate lens makes use of waveguide theory which states that the guide wavelength (λ_g) is related to free space wavelength (λ via)-

$$\frac{1}{\lambda_g^2} = \left(\frac{1}{\lambda}\right)^2 - \left(\frac{1}{2a}\right)^2 \quad \text{For } H_{10} \text{ mode in rectangular waveguide}$$

Where λ_g = guide wavelength

λ = free space wavelength

a = wider internal dimension of rectangular waveguide or spacing of plates

The phase velocity of waves in the guide is given by-

$$v = \frac{c\lambda_g}{\lambda}$$

This velocity is always > c.

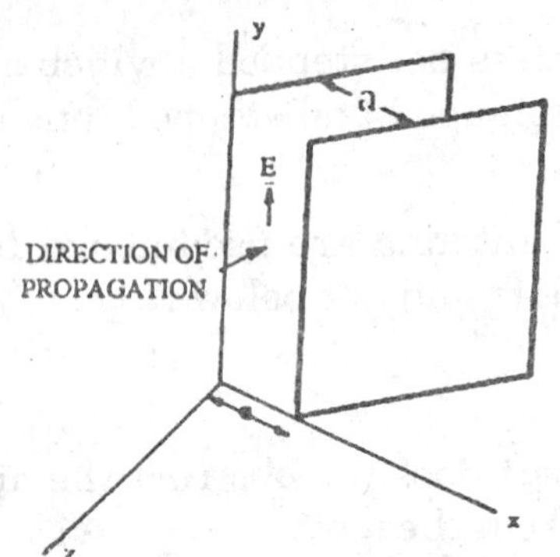

Propagated wave between two plates of metal plate lens

Figure 2.128

Let us consider a wave propagated between two infinite parallel plates spaced at a distance "a" apart; having its electric vector parallel to plates. This may be regarded as a part of rectangular wave guide carrying H_{10} wave with its dimension "b" infinitely large. A structure consisting of many such parallel plates with spacing "a" can be regarded as uniform medium with an equivalent or effective index of refraction 'μ' given by the ratio of velocities as -

$$\mu = \frac{c}{v} = \frac{c}{c}\frac{\lambda}{\lambda_g}$$

$$\mu = \frac{\lambda}{\lambda_g} = \frac{c}{v}$$

as we know that-

$$\left(\frac{\lambda}{\lambda_g}\right)^2 = \left(\frac{\lambda}{\lambda}\right)^2 - \left(\frac{\lambda}{2a}\right)^2$$

$$\therefore \frac{\lambda}{\lambda_g} = \sqrt{1 - \left(\frac{\lambda}{2a}\right)^2}$$

$$\mu = \frac{c}{v} = \frac{\lambda}{\lambda_g} = \sqrt{1 - \left(\frac{\lambda}{2a}\right)^2}$$

or $$\mu = \sqrt{1 - \left(\frac{\lambda}{2a}\right)^2}$$

It is apparent that the refractive index is always less than unity Therefore, the value of (a) must not be less than its critical value i.e.-

$$\sqrt{1-\left(\frac{\lambda}{2a}\right)^2} = 0\ ;\ 1 = \frac{\lambda}{2a}$$

or $$a = \frac{\lambda}{2}$$

1. Explain the basic principle of operation of a dielectric lens antenna showing how it converts a spherical wave front into plane wave front? **(Raj. Univ. 2004)**
2. Write short note on: Lens Antenna **(R.T.U. M. tech. 2010)**

2.4.5. MICROSTRIP ANTENNAS

2.4.5.1 MICROSTRIP OR PATCH ANTENNAS

Rectangular Microstrip Antenna

Due to-

(1) Ease of fabrication

(2) Low cost

(3) Low profile

(4) Small size

(5) Less Weight

(6) High Performance

They can be printed directly onto circuit board. They are most commonly used in mobile phones. It is also due to their ease of installation and aerodynamics profile that these antennas are used in the field of communications.

Applications

- In Space crafts
- In Aircrafts

Construction

According to the microstrip antenna given in the figure-

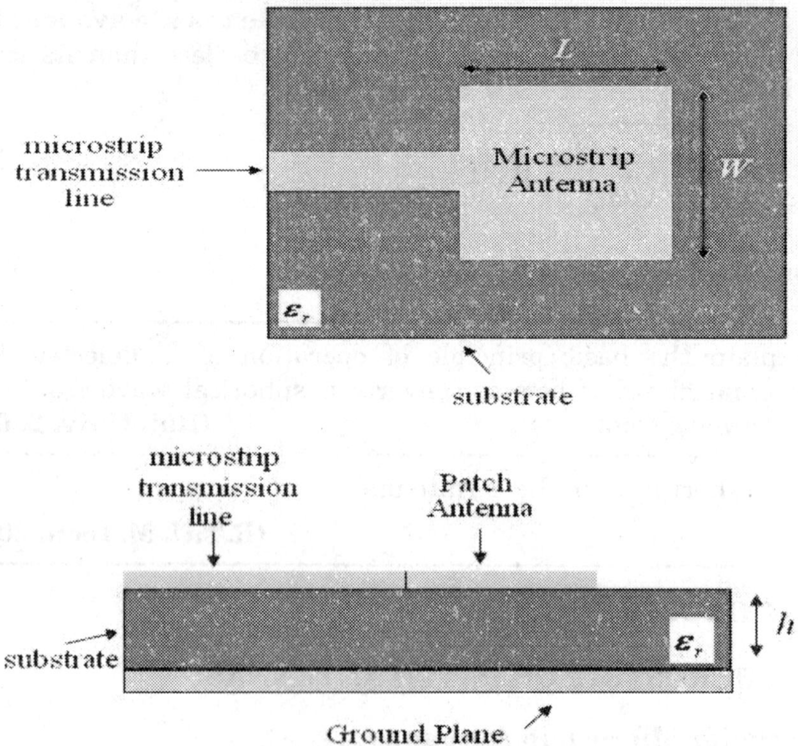

Figure 2.129 Geometry of Micro strip (Patch) Antenna

Rectangular patch on dielectric- coated, ground plane antennas, are most commonly used in the field of communications-

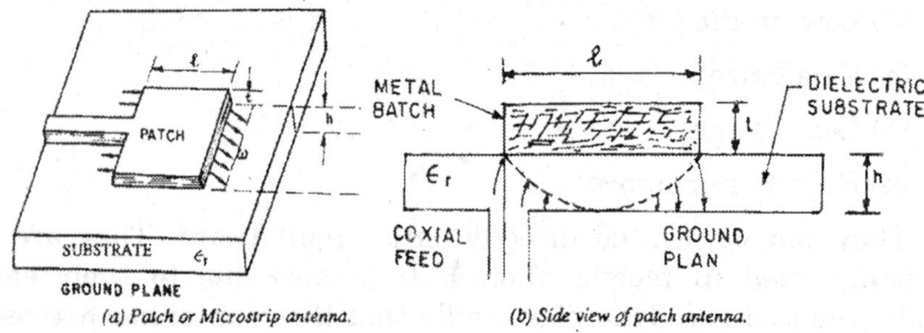

(a) Patch or Microstrip antenna. (b) Side view of patch antenna.

Figure 2.130

This antenna is fed by transmission line. The patch, microstrip and ground plane are made up of-

(1) High Conductivity

(2) Patch has length `L' and Width `W'

(3) Patch is mounted on substrate of thickness `h' and permittivity `ε_r'

(4) Ground plane with height `h' (which is smaller than the operating wavelength)

(5) Frequency of operation of patch antenna, is determined by length `L' i.e. -

$$f_c = \frac{C}{2L\sqrt{\varepsilon_r}} = \frac{1}{2L\sqrt{\varepsilon_o\ \varepsilon_r\ \mu_o}}$$

Where, f_c = center frequency

It means, patch antenna should have length equal to one half of wavelength within dielectric medium of substrate.

(6) Microstrip antenna consists of a **very thin metallic strip (patch)** (t << λ) placed on a small fraction of λ (h<<λ) above a ground plane.

(7) The strip and ground plane are separated by a dielectric sheet called as **Substrate.**

(8) The Radiating patch may be, square, circular, elliptical rectangular or **any shape**. Out of which square, Rectangular or circular are mostly preferred because of ease of analysis and fabrication.

(9) **Feed line** is also a conducting strip normally of smaller width. Coaxial feeds where the inner conductor of coaxial line is attached to the radiating patch are widely used.

(10) **Linear and circular polarization** can be achieved with microstrip antennas and arrays of strip elements, with single or multiple feed may be used for **greater directivity.**

(11) The width of `W' of the antenna controls the input impedance is nearly 300 ohm. By increasing the width of antenna the input impedance can be reduced. The width of the antenna also controls the radiation pattern. Where the normalized pattern is given by-

$$E_\theta = \frac{\sin\left(\frac{kW\sin\theta\sin\phi}{2}\right)}{\frac{kW\sin\theta\sin\phi}{2}}\cos\left(\frac{kL}{2}\sin\theta\cos\phi\right)\cos\phi$$

$$E_\omega = \frac{\sin\left(\frac{kW\sin\theta\sin\phi}{2}\right)}{\frac{kW\sin\theta\sin\phi}{2}}\cos\left(\frac{kL}{2}\sin\theta\cos\phi\right)\cos\theta\sin\phi$$

$$k = \frac{2\pi}{\lambda} = free\ \ space\ wave\ number.$$

Working

As the thickness of microstrip is normally very small, the waves generated within the dielectric substrate, undergoes reflections to some extent when they arrive at edge of strip, resulting radiation of only small fraction of incident energy. Therefore, antenna is considered to be very inefficient and it behaves as more like a cavity rather than a radiator.

Patch antenna acts as a resonant λ/2 parallel plate microstrip transmission line with characteristic impedance equal to reciprocal of number `n' of *parallel field cell transmission Lines.* Each field transmission line has a characteristic impedance Z_o equal to the intrinsic impedance of medium, that is-

$$Z_o = n_i = \sqrt{\frac{\mu}{e}} = \sqrt{\frac{\mu_o}{\varepsilon_o}} \sqrt{\frac{\mu_r}{\varepsilon_r}}$$

As $\mu = \mu_o \mu_r$ *and* $\varepsilon = \varepsilon_o \varepsilon_r$

$$\therefore Z_o = 120\,\pi \sqrt{\frac{\mu_r}{\varepsilon_r}}$$

It is obvious from patch (left to right) that the cross section has 10 field cells transmission lines, hence for $\varepsilon_r = 2$, the characteristic impedance of patch antenna is given by-

$$Z_o = \frac{Z_o}{n\sqrt{\varepsilon_r}} = \frac{376.7}{10\sqrt{2}} = 26.63\ \Omega$$

This can be written as-

1. $$Z_c = \frac{Z_o t}{n\sqrt{\varepsilon_r}}$$

$$as \quad n = \frac{l}{t}$$

2. **Beam Area =** $\Omega_A = \frac{1}{2} = \Pi$

3. Radiation pattern = Broad.

4. $D = Directivity = \frac{4\,\Pi}{\Omega_A} = \frac{4\Pi}{T_1} = 4$

$$D_{dB} = 10\log_{10} 4 = 10 \times 0.6021 \quad D = 6.021 \text{ dB}$$

5. Radiation conductance (G)-

$$G = \frac{1}{90}\left(\frac{\omega}{\lambda}\right)^2 \quad \textit{for } w << \lambda$$

$$G = \frac{1}{120}\left(\frac{\omega}{\lambda}\right)^2 \quad \textit{for } w >> \lambda$$

Disadvantage

1. They are inefficient, as they work for a, very narrow frequency bandwidth. This Frequency Range is for low profile application above 100 MHz only.
2. The limitation of microstrip antenna is, its narrow frequency bandwidth which is controlled by characteristics of parallel plate transmission line.

This *bandwidth can be increased by-*

(a) Increasing thickness `h' of parallel plate transmission line.

(b) Using **high dielectric constant,** `ε_r' substrate, so that physical dimensions of parallel plate lines are decreased.

(c) Increasing **inductance** of microstrip by cutting holes or slot in it.

(d) Adding **reactive component** to reduce VSWR.

(e) In order to increase directivity of antenna, **multiple radiators** are used in cascade to form an array.

Rectangular Microstrip Antenna

Consider a square patch antenna fed at the end as before. Assume the substrate is air (or Styrofoam, with permittivity equal to 1) and that,

L = W = 1.5 meters, So that the patch is to resonate at 100 MHz. The height `h' is taken to be 3 cm.

NOTE: The **microstrips are usually made for higher frequencies,** so that they are much smaller in practice. When matched to a 200 Ohm load, the magnitude of S11 is shown in Figure 2.131.

Some noteworthy observations are noticeable from the Figure Fig 2.131

First, the bandwidth of the patch antenna is very small. Rectangular patch antennas are narrowband antenna with bandwidth typically 3%.

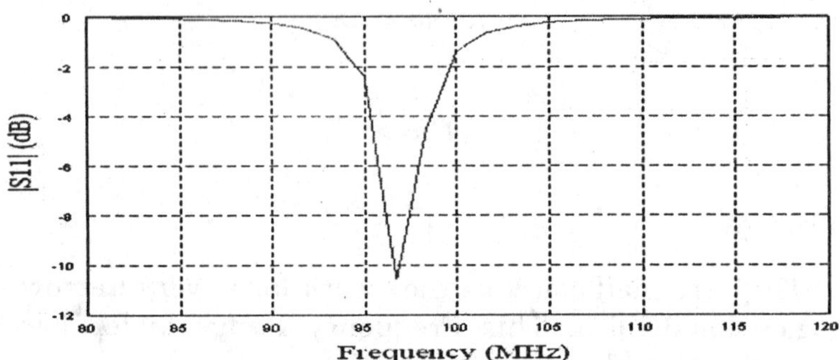

Figure 2.131 Magnitude of S11 versus Frequency

Secondly, the antenna is designed to operate at 100 MHz, but it is resonant at approximately 96 MHz. This shift is due to fringing fields around the antenna, which makes the patch seems longer. Hence, when designing a patch it is typically trimmed by 2-4% to achieve resonance at the desired frequency.

The magnitude of fields is given by-

$$f(\theta, \phi) = \sqrt{E_\theta^2 + E_\phi^2}$$

The fields are plotted in the next figure, is for W = L = λ/2.

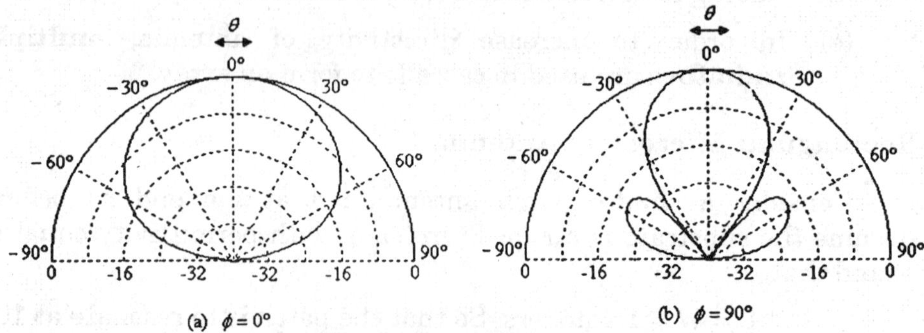

Figure 2.132 Normalized Radiation Pattern for Microstrip (Patch) Antenna.

From the figure- The directivity of patch antenna is nearly 5 - 7 dB, and the fields are *linearly polarized.*

Effect of E-field in design of Patch Antenna

The fringing fields around the antenna can help us to explain why the microstrip antenna radiates. The side view of patch antenna with E-field, shown in figure. As the current at the end of the patch is zero (open circuit end), the current is maximum at the center of half wave patch and zero at the beginning of the patch. This low current value at the feed explains why the impedance is high when fed at the end.

This antenna behaves as an open circuited transmission line having the voltage reflection coefficient equal to one. Here the voltage and current are out of phase. Also the voltage at the end of patch the voltage is maximum (+V volts). At the half wavelength the voltage is minimum (-V volts).

The field in the patch is given in the figure-

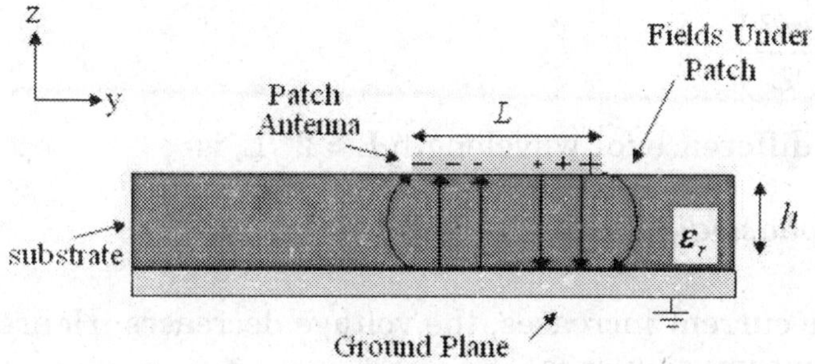

Figure 2.133. Side view of patch antenna with E-fields shown underneath.

These fields are responsible for the radiation. It is very important to know here that these fields near the surface of patch are both in +y direction; which results in summation of fields in phase and produce the radiation of microstrip antenna.

As the `ε_r' decreases, the fringing fields bends more, thus they extend away from the patch. Therefore, using a smaller permittivity for the substrate yields better radiation.

But when microstrip transmission line, where no power is to be radiated, a high valve of `ε_r' is desired, so the fields are confined, thus resulting in less radiation.

Feeding Methods

(A) Inset Feed

Earlier, the patch antenna was fed at the end, which yielded a high input impedance. Thus there was requirement of modification in the feed. As the current is low at the ends of half wave patch and increases in magnitude towards the center, the input impedance (Z= V/I) could be reduced if the patch was fed closer to the center. One method of doing this is by using an inset feed i.e., apply feed at a distance 'R' from the end as given in the figure-

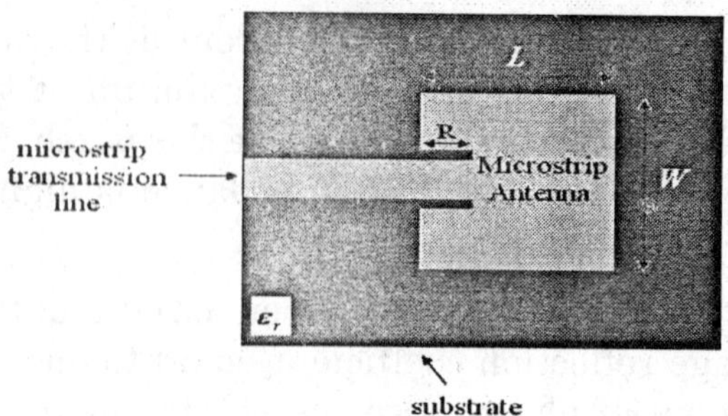

Figure 2.134 Patch Antennas with an Inset Feed.

A sinusoidal distribution of the current, moving from the end at the distance `R' is given by-

$$\text{Cos}\left(\frac{\pi R}{L}\right)$$

The phase difference for wavelength L = 2* L, is-

$$phase\ difference = \frac{2\pi R}{L} = \frac{2\pi R}{2L} = \frac{\pi R}{L}$$

As the current increases, the voltage decreases. Hence, using Z = V/I, the input impedance is -

$$Z_{in}(R) = \cos^2\left(\frac{\pi R}{L}\right) . Z_{in}(0).$$

where Z_{in} (0) - input impedance (if the patch is fed at the end).

Hence, by feeding the patch antenna, in this style, decreases the input impedance.

Example: $if\ R = \frac{L}{4}\ then\ -\cos\left(\frac{\pi R}{L}\right) = \cos\left(\frac{\pi L}{L4}\right) = \cos\left(\frac{\pi}{4}\right)$

$$\therefore Z_{in} = \cos^2\left(\frac{\pi}{4}\right) = \left(\frac{1}{\sqrt{2}}\right)^2 = \frac{1}{2}$$

Now with λ/8 length inset feed decreases the input impedance further by 50%.

Hence, this method can be used to tune the input impedance to the desired value.

(B) Fed with a Quarter Wavelength Transmission Line

By using a quarter wavelength transmission line of characteristic impedance `$Z_1$', the microstrip antenna can be matched to a transmission line of characteristic impedance `Z_o' as given in the figure-

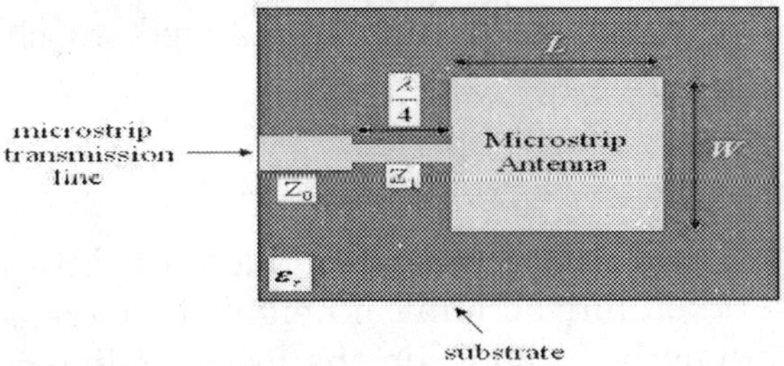

Figure 2.135 Patch antennas with a quarter wavelength matching section.

This feed, is specially designed to match the input impedance (Z_{in}) to the transmission line (Z_o). If the impedance of the antenna is (Z_A), then the input impedance viewed from the beginning of the quarter wavelength line becomes-

$$Z_{in} = Z_o = \frac{Z_1^2}{Z_A}$$

Z_{in} is altered by selection of Z_1, so that $Z_{in} = Z_o$ & the antenna impedance is matched. Where, the Z_1 can be altered by changing width of quarter wavelength strip.

Wider the strip is, lower is the characteristic impedance (Z_o). (For the selection of line)

(C) Coaxial Cable (probe) feed

In this type of feed the Microstrip antenna, are fed from underneath via a probe as shown in figure-

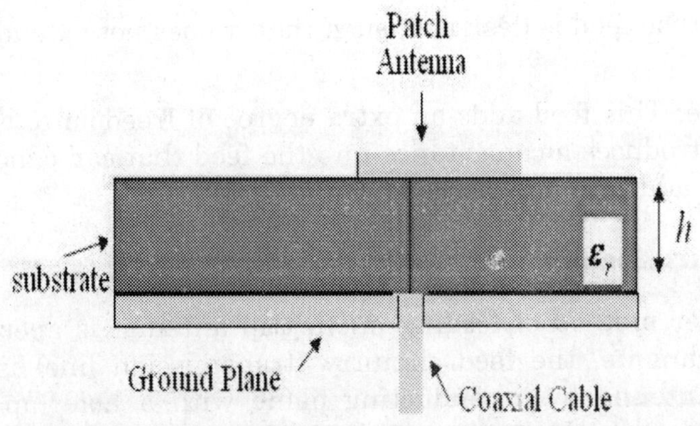

Figure 2.136 Coaxial cable feed of patch antenna.

The outer conductor of the coaxial cable is to the ground plane, and the centre conductor is extended up to the patch antenna.

The position of the feed can be changed, to change the input impedance.

Disadvantage

The coaxial feed introduces an inductance into the feed, which may need to be taken into account, if height `h', increases nearly equal fraction of wavelength. In addition the probe radiates, which leads to radiations in undesirable directions.

(D) Coupled/Indirect Feed

The probe fed is attend in such a manner that feed do not directly touch the antenna, as in the given figure-

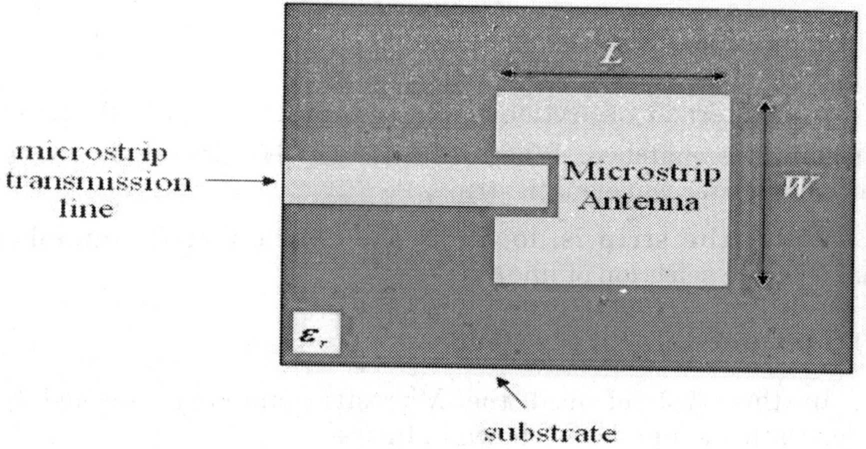

Figure 2.137

The probe feed is designed, such that it does not extend up to the antenna.

Advantage: This feed adds an extra degree of freedom to the design. The gap introduces a capacitance into the feed that can cancel out the inductance added by the probe feed.

(E) Aperture feeds

Another method of feeding microstrip antenna is aperture feed. In this technique, the feed circuitry (transmission line) is shielded from the antenna by a conducting plane with a hole (aperture) to transmit energy to the antenna as in the figure-

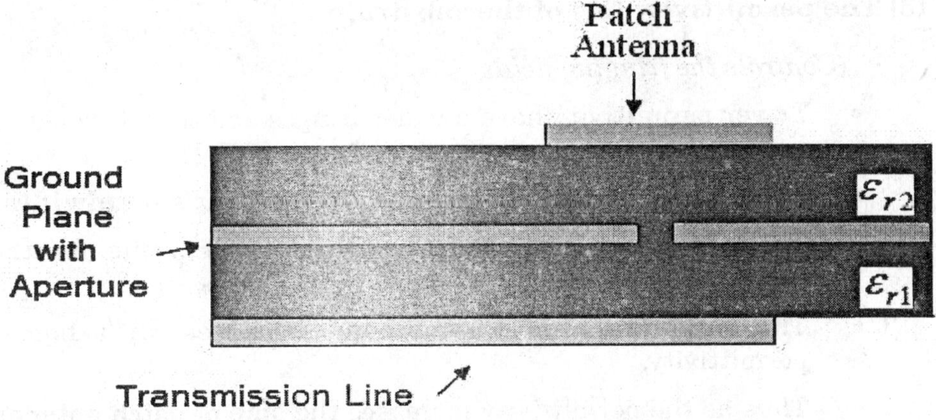

Figure 2.138 Aperture Coupled Feed

The upper substrate is made with a lower permittivity to produce loosely bound fringing fields, yielding better radiation. The lower substrate can be independently made with a high value of permittivity for tightly coupled fields that produces radiations as desired.

Disadvantage: Difficulty in fabrication.

MICROSTRIP ANTENNA (Design Parameters and tradeoffs)

All the parameters in a rectangular patch antenna designs are-

- L – Length
- W – Width
- h – Height
- E_r - Permittivity.

In this section, we brief a general idea of how the parameters affect the performance in order to understand the design process.

(1) The length of the patch (L) *(Controls the resonant frequency)*

The length of the longest patch on microstrip controls the lowest frequency of operation which can be explained by the equation-

$$f_c = \frac{c}{2L\sqrt{\varepsilon_r}} = \frac{1}{2L\sqrt{\varepsilon_0\ \varepsilon_r\ \mu_0}}$$

(2) The width (W)

(Controls the input impedance and the radiation pattern)

Wider the patch becomes, the lower the input impedance is.

(3) The permittivity (E_r) of the substrate

(Controls the fringing fields)

- Lower permittivity have a wider fringes and therefore better **radiation**.
- Decreasing permittivity increases the antenna's **bandwidth.**
- The **efficiency** is increased with a lower value for the permittivity.
- The **impedance** of the antenna increases with higher permittivity.
- Thus as the permittivity increases the **size** of patch antenna reduces. Therefore, it is used in cell phones where in little space antenna of half wavelength is to be designed; where L, frequency and permittivity can be given as-

$$L = \frac{1}{2 f_c \sqrt{E_o E_r \mu_0}}$$

Hence by the formula we can say that, by increasing the permittivity by factor of 4, the length required decreases by a factor of 2. Thus it results in antenna miniaturization.

(4) The height (h) of the substrate

(Controls the bandwidth)

Increasing the height increases the bandwidth. As we know that an, antenna occupying more space in a spherical volume have wider bandwidth.

- Increasing the height increases the efficiency of antenna.
- Increasing the height induces the surface waves that travel within the substrate; which is undesired radiation and may couple to other components.

Relation of Bandwidth with other parameters can be given by-

$$B \alpha \frac{E_r - W}{Er^2 L} . h$$

2.4.5.2. SHORTING PINS USED IN PATCH ANTENNAS

To improve the performance of patch antenna, the designers commonly uses, the method of introducing the shorting pins, from the patch to the ground plane at various locations.

(1) In Quarter-Wavelength Patch

The figure given here shows a quarter wavelength patch shorted at the far end.

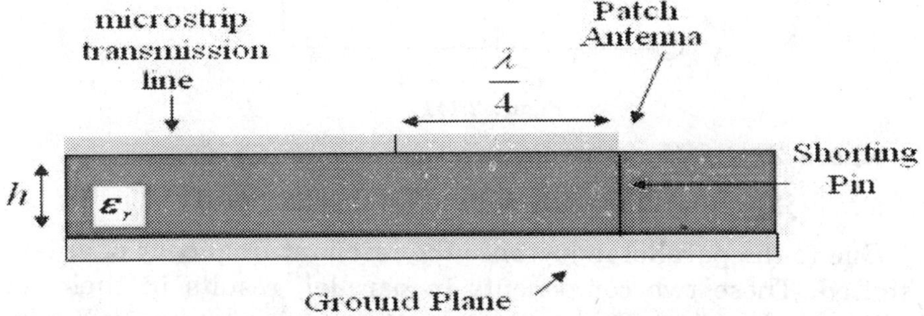

Figure 2.139 Quarter Wavelength patch with shorting pin at the end

Because the patch is shorted at the end, the current at the end of the patch is no longer forced to be zero. As a result, this antenna actually has the same current-voltage distribution as a half-wave patch antenna. However, the fringing fields which are responsible for radiation are shorted on the far end, so only the fields nearest the transmission line radiates. Thus, the gain is reduced maintaining the same basic properties as a half-wavelength patch, simultaneously the size is reduced to 50%.

(2) Shorting pin at the feed to a patch

A shorting pin can be used at the feed to the patch antenna, as shown in figure-

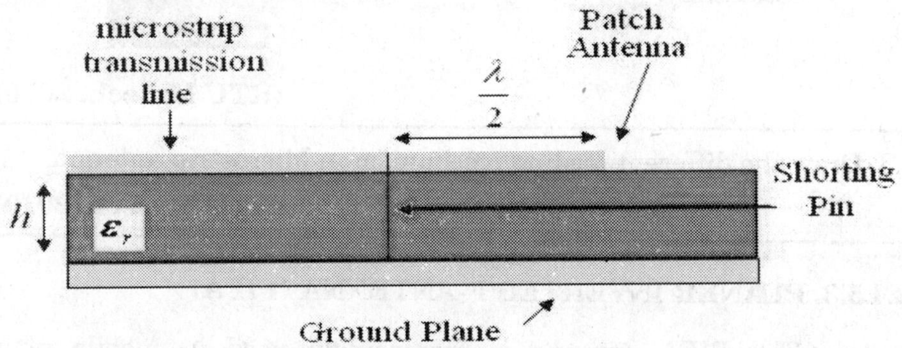

Figure 2.140

It is wrong to think here that the shorting pin nullify any power delivered to the antenna. However, because patches are high frequency devices (>1GHz), these pins actually introduce a parallel inductance to the antenna impedance, which can be, shown by the equivalent circuit of the given patch antenna as in the figure below-

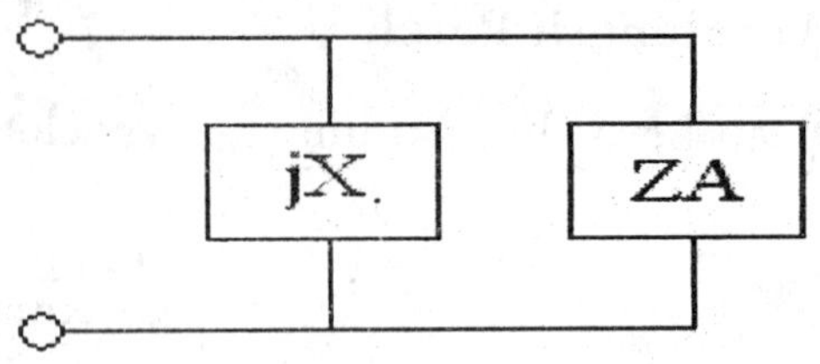

Figure 2.141

where Z_A = antenna impedance

jX = Reactance (introduced by shorting pin)

Due to the parallel inductance the resonant frequency of antenna is shifted. These two components in parallel, results in their total admittances (y = 1/Z). Hence, the admittance of patch has 1/jX added to it. In this manner, the resonant frequency can be altered.

Again it is very interesting to know here that, the shorting pin can become capacitive if instead of extending all the way to ground plane, it is left floating a small amount above. This introduces another design parameter to optimize performance.

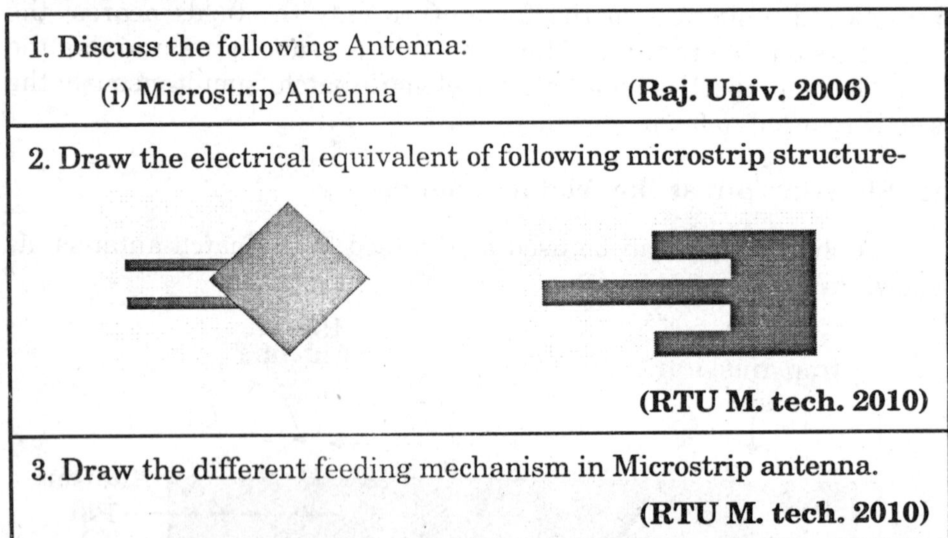

1. Discuss the following Antenna:

   (i) Microstrip Antenna **(Raj. Univ. 2006)**

2. Draw the electrical equivalent of following microstrip structure-

   (RTU M. tech. 2010)

3. Draw the different feeding mechanism in Microstrip antenna.

   (RTU M. tech. 2010)

2.4.5.3. PLANER INVERTED F-ANTENNA (PIFA)

- The PIFA antenna is increasingly used in mobile phone market. This antenna resembles an inverted-F. It has advantage of
  - o Low profile.
  - o Omni-directional pattern.

- Due to shorting pins at the end, the PIFA is resonant at a quarter wavelengths.
- In this antenna the feed is placed between the open and shorted end, and the position controls the input impedance.

The structure of PIFA is given below –

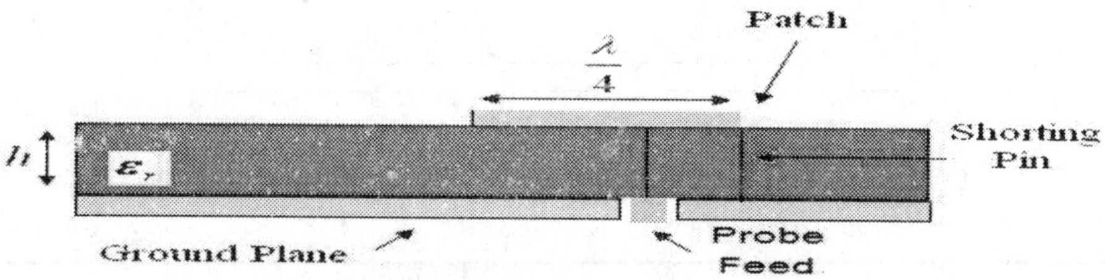

Figure 2.142 PIFA – Antenna

Microstrip antenna - transmit fields

In this section, we consider a short pulse which starts from the end of a microstrip, and travels towards the patch antenna. Some of the pulse radiates away, and some of the power reflects back down the microstrip line. This type of simulation gives a little bit better idea of what is going on with a patch antenna, specifically when short pulses (short waveforms, or brief applied voltages) are incident upon a microstrip antenna).

Specifically, consider a patch antenna that is mounted on a ground plane, with a dielectric with *permittivity* equal to 2.2. The thickness of this dielectric is 0.795 mm (millimeters). The patch antenna will be 1.25 centimeters wide and 1.56 centimeters. The microstrips antenna will feed the patch offset from the center, as shown in Figure-

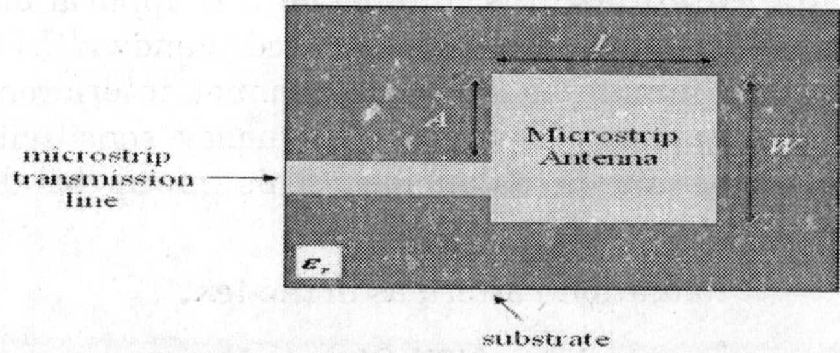

Figure 2.143 Offset feed for Patch Antenna.

The transient pulse will be of the form given by exp (-(t-T0)/T) ^2, where T_0 is the time delay and will be 45 pS (picoseconds, 10^-12), and T is a parameter that controls the rate of rise and fall, which is 15 pS. This function is plotted in Figure-

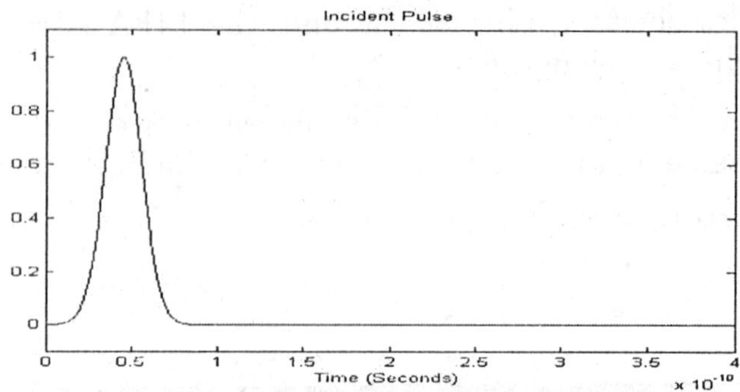

Figure 2.144 Incident (transient) pulse fed to a Patch Antenna.

1. Write short note on Microstrip antenna. **(Raj. Univ. 2002, 2005)**

UNIVERSITY QUESTIONS

THEORITICAL

1. Which type of antenna is suitable for following applications and why? Also draw the radiation pattern for each of these antennas:

(i) To receive a single T.V. Channel

(ii) To receive more number of TV Channels

(iii) To provide circular polarization

(iv) For direction finding in HF/VHF Range **(Raj. Univ. 2005)**

Ans.

(i) To receive a single T.V. Channel

Yagi – Uda Antennas: It is suitable for this application because it has narrow bandwidth and this limited bandwidth is major advantage since it minimizes adjacent channel interference. It is essential a fixed frequency device i.e. frequency sensitivity and a bandwidth of about 3% is obtainable. This much bandwidth is sufficient for T.V. reception.

Radiation Pattern as in the text

(ii) To receive more number of TV Channels

Log-Periodic Antennas: It provides good gain and wide bandwidth because it is really sum of many separate dipoles each tuned to a slightly different frequency. So to receive more number of T.V. channels is easy.

Radiation Pattern as in the text

(iii) To provide circular polarization

Helical Antennas: It is indispensible for space communication due to its wide bandwidth, simplicity, higher directivity and circular polarization. Thus it is used in transmitting telemetry data from moon to earth.

Since the dimension of the helix in axial mode are not critical and therefore result in greater bandwidth. Thus end fire axial mode is most practical as it can achieve circular polarization over a wider bandwidth and is more efficient. The helical antenna is capable of receiving signals of arbitrary polarizations due to circular polarization.

Radiation Pattern as in the text

(iv) For direction finding in HF/VHF Range

Loop Antennas: It has the radiation pattern maximum in a direction perpendicular to the centre axis through the loop and very low in the broadside of the loop. Du e to this reason the loop antenna is often used in direction finding. The loop is rotated until a minimum signal is observed and therefore the transmitter is either on one side or on the other side of the loop.

Radiation Pattern as in the text

2. Some short question:

i. Which antenna is used for radio receiver?

ii. Which antenna has highest directivity?

iii. Which antenna has high power gain?

iv. Difference between λ/2 dipole and folded dipole.

(R.T.U. 2009)

NUMERICAL

Q1. Calculate the approximate gain and beamwidth of a Paraboloid reflector antenna at operating frequency 4 GHz, diameter 20 meters and illumination efficiency 55%. **[Raj. Univ. 2008]**

Ans. Given, D = 20 m or radius = r = 10 m

$f_c = 4$ GHz

k = 55% = 0.55

$$\therefore i)\ \lambda = \frac{c}{f} = \frac{300}{4\times10^3} = 0.075 = 75\times10^{-3}\ m. \quad \text{and}$$

(ii) $\text{Gain (G)} = \frac{4\pi}{\lambda^2}\cdot(kA)$

$$= \frac{4\pi}{(0.075)^2}\times 0.55\times \pi r^2$$

$$= \frac{4\ \pi^2\times 0.55\times(10)^2}{(0.075)^2}$$

$$= \frac{4\ (3.14)^2\times 0.55\times 100}{(0.075)^2} = 96{,}404.98$$

(iii) Now G (dB) = 10 $\log_{10}$ 96,404.98

= 10 × 4.98 =49.8

= 49.8 dB

(iv) $\text{Beam width} = \frac{140\lambda}{D} = \frac{140\times 0.075}{20}$

= 0.525 = 5.25°

Q 2. A Parabolic antenna having a circular mouth is to have a power gain of 1000 at λ = 10 cm. Estimate the diameter of mouth and half - power beam width of antenna. **(Raj. Univ. 2007)**

Ans. Given, λ = 10 cm

$G_p = 1000$

To find out - Diameter (D) and HPBW =?

(i) $$G_p = 6\left(\frac{D}{\lambda}\right)^2$$

or $$\left(\frac{D}{\lambda}\right)^2 = \frac{G_p}{6}$$

$$\frac{D}{\lambda} = \sqrt{\frac{G_p}{6}}$$

(ii) $$HPBW = \frac{1}{2}(BWFN) = \frac{1}{2}\times\frac{140\lambda}{D}$$

$$= \frac{70\,\lambda}{D}$$

or $$\frac{D}{\lambda} = \sqrt{\frac{G_p}{6}}$$

$$\therefore HPBW = 70\sqrt{\frac{6}{G_p}} = 70\sqrt{\frac{6}{1000}}$$

$$= 5.426^{\circ}$$ Ans.

Q.3. Calculate the beamwidth between half power points and between nulls and the gain in dB of a 2 m paraboloid reflector used at 6 GHz. **(Raj. Univ. 2005)**

Ans. given - D = 2 m

f_c = 6 GHz

To find - gain in dB, HPBW.

(i) $$\text{HPBW} = \frac{70\lambda}{D} = \frac{70 \times 0.05}{2} = 1.75$$

(ii) $$G_p = 6.4\left(\frac{D}{\lambda}\right)^2$$

$$= 6.4\left(\frac{2}{0.05}\right)^2 = \frac{25.6}{25 \times 10^{-4}}$$

$$= 10240$$

$$G_p\,(dB) = 10\,\log_{10}\,10240$$

$$= 10 \times 4.25$$

$$= 42.5\ dB$$

Q.4. What should be the minimum distance required for far field measurement for a 2.4 m parabolic dish antenna operating at 4 GHz? **(Raj. Univ. 2005)**

Ans. Given d = 2.4 m

f_c = 4 GHz

To find out - minimum distance=?

$$R = \frac{2d^2}{\lambda} = \frac{2\,(2.4)^2}{\lambda}$$

as $$\lambda = \frac{c}{f} = \frac{3\times10^8}{4\times10^9} = \frac{3}{40}$$

$$R = \frac{2\,(2.4)^2 \times 40}{3} = 153.6m$$

Q.5. Two space crafts A and B are separated by 3000 km each has a Paraboidal reflector antenna of 0.9 m diameter operating at a frequency of 2 GHz with an aperture efficiency of 65% (i.e. k = 0.65). Calculate the path loss and antenna gain in dB. What should be the transmitted power in dBm from space craft A, if the receiver on space craft B requires 90 dBm? **(Raj. Univ. 2004)**

Ans. Given distance = d_{km} = 3000 km

$$f_{MHz} = 2\text{ GHz} = 2 \times 10^3\ MHz$$

Diameter = 0.9 m = D

k = 0.65

W_R = 90 dBm

To find - Path loss, Gain (dB), Transmitted power (W_T)=?

(i) Path loss = 32.45 + 20 $\log_{10} f_{MHz}$ + 20 $\log_{10} d_{km}$

= 32.45 + 69.54 + 66.01

= 168 dB.

(ii) Antenna gain $= \frac{4\pi}{\lambda^2} \cdot \frac{\pi D^2}{4} \times 0.65$

$$= \left(\frac{\pi D}{\lambda}\right)^2 \times 0.65$$

where $$\lambda = \frac{c}{f} = \frac{3 \times 10^8}{2 \times 10^9}$$

= 0.15 m

and D = 0.9 m

$$= \left(\frac{\pi \times 0.9}{0.15}\right)^2 = 354.9456$$

$$G(dB) = 10 \log_{10} 354.9456$$

= 23.635 dB.

Transmitted power (W_T) = Received power (W_r) – Gain of transmitting antenna (G_T) – Gain of receiving antenna (G_R) + Path Loss (L_S)

$$W_t(dB) = W_r\,(dB) - G_T\,(dB) - G_R\,(dB) + L_s\,(dB)$$

$$= -90 - 23.635 - 23.635 + 168$$

$$= 30.73 \text{ dBm.}$$

Q.6. Parabolic reflectors used for both the receiving and transmitting aerials of a microwave circuit are replaced by other having a diameter of 1.5 times that of the original. Calculate the overall improvement in dB. **(Raj. Univ. 2002)**

Ans. Gain of parabolic reflectors $\Rightarrow \dfrac{4\,\pi D}{\lambda^2} = \dfrac{4\,\pi\,kA}{\lambda^2}$

where, $\quad A = \dfrac{\pi D^2}{4}$

$$G = \frac{4\,\pi k}{\lambda^2}\left(\frac{\pi D^2}{4}\right)$$

$$= 0.65\left(\frac{\pi D}{\lambda}\right)^2$$

$$= 0.65 \times (3.14)^2 \left(\frac{D}{\lambda}\right)^2$$

$$= 6.3\left(\frac{D}{\lambda}\right)^2$$

$$G = 6.3\left(\frac{D}{\lambda}\right)^2$$

Say now original gain G1 = $6.3\left(\dfrac{D_1}{\lambda}\right)^2$, and new gain for D_2 = 1.5 D_1 is say G_2 given by-

$$G_2 = 6.3\left(\frac{D_2}{\lambda}\right)^2$$

$$G_2 = 6.3\left(\frac{1.5\,D_1}{\lambda}\right)^2$$

$$= 6.3\ (1.5)^2 \left(\frac{D_1}{\lambda}\right)^2$$

$$G_2 = 14.175 \left(\frac{D_1}{\lambda}\right)^2$$

Or $\quad G_2 = 2.25\ G1$

SHORT QUESTIONS

TYPES OF ANTENNA

1. What are the two basic classifications of antenna?
2. What three factors determine the type, size and shape of an antenna?
3. What terms are often used to describe basic half wave antennas?
4. If a basic half wave antenna is mounted vertically. What type of radiation pattern will be produced?
5. In which plane will the half wave antenna be operating if it is mounted horizontally?
6. Because the radiation pattern of a dipole is similar to that of a doubt, what will happen to the pattern if the length of the doublet is increased?
7. What is the simplest method of feeding power to the half-wave antenna?
8. What is radiation pattern of a quarter wave antenna?
9. What is the difference in the amount of impedance between a three wire dipole and a simple centre fed dipole?
10. Which has a wider frequency range, a simple dipole or a folded dipole?
11. What two factors determine the directivity pattern of the parasitic array?
12. What two main advantages of parasitic array can be obtained by combining a reflector and a director with the driven element?
13. The parasitic array can be rotated to receive or transmit in different directions. What is the name given to such an antenna?
14. What are the disadvantages of the parasitic array?
15. What is the advantage of adding parasitic elements to Yagi array?
16. The Yagi antenna is an example of what type of array?
17. What is another name for the Beverage antenna?

18. What is the polarity of the currents that feed the V-antenna?
19. What is the main disadvantage of the rhombic antenna?
20. What are the primary reasons for the development of the turnstile antenna?

Answers

1. Half wave (Hertz)
   Quarter waver (Marconi)
2. Frequency of operation of the transmitter, amount of power to be radiated and general direction of the receiving set.
3. Dipole, Doublet and Hertz
4. Non directional
5. Vertical plane
6. The pattern would flatten
7. To connect one end through a capacitor to the final output stage of the transmitter.
8. A circular radiation pattern in the horizontal plane, or same as a half wave.
9. Nine times the feed point impedance
10. Folded dipole.
11. Length of the parasitic element (tuning) and spacing between the parasitic and driven elements.
12. Increased gain and directivity
13. Rotary array
14. Their adjustment is critical and they do not operate over wide frequency range.
15. Increased gain
16. Multi-element parasitic array
17. Wave antenna
18. Opposite
19. It requires a large antenna site
20. For omnidirectional VHF communications.

UNIT 3
ANTENNA ARRAYS AND MEASUREMENTS

CONTENTS

ANTENNA ARRAYS AND MEASUREMENTS

3.1 ANTENNA ARRAYS

It is a system of similar antennas oriented similarly to get greater directivity in a desired direction.

A radiating system consisting of several spaced and properly phased radiators.

It is the method of combining radiations from a group or array of similar antennas in which phenomena of wave-interference are involved.

The total field (not power) produced by an antenna array system at a great distance from it, is the vector sum of the fields produced by the individual antennas of array system.

Relative phases of individual field component depend on the relative distance of individual elements of the antenna array and in turn depend on the direction.

An antenna array is a set of `N' spatially separated antennas. The number of antennas in an array can be as small as two, or as large as several thousand elements.

Advantages

An antenna array (often called as ``**phased array**") is a set of two or more antennas. The signals from the antennas are combined or processed in order to achieve improved performance over that of a single antenna working in the same environment. **The antenna arrays are used-**

(1) To **increase** the overall **gain.**

(2) To provide **diversity reception.**

(3) To **cancel out interference/noise** from a particular set of directions.

(4) ``Steer'' the array so that it is most **sensitive in a desired direction** and least in the undesired direction.

(5) To **determine the direction** of arrival of the incoming signals.

(6) To **maximize** the signal to interference/Noise ratio (SIR/SNR).

Drawbacks

(1) Increased cost

(2) Increased size

(3) More complex.

The following Fig. 1 show some examples of the antenna arrays-

Figure 3.1 Cell-tower array
(Two receiving antennas and one transmitting antennas)

3.2 VARIOUS FORMS OF ANTENNA ARRAYS:

3.2.1 LINEAR ARRAY

If individual antennas of array are equally spaced along straight line then it is called as linear array of antennas. Individual antenna of an array is termed as **elements.**

Uniform linear array- It is a kind of linear array. In which equally spaced antennas are fed with the current in equal magnitude with uniform progressive phase shift along line.

Array of antennas can be configured as-

i. Straight line

ii. Circles

iii. Triangles

iv. Rectangle

Elements generally used are half wave dipole antenna.

NOTE: Variation is electrical length from λ/2 is within 5% and hence the radiating properties of the element are not affected.

3.2.2 BROADSIDE ARRAY

It is the array in which the numbers of identical parallel antennas are set up along a line, drawn perpendicular to their respective axis.

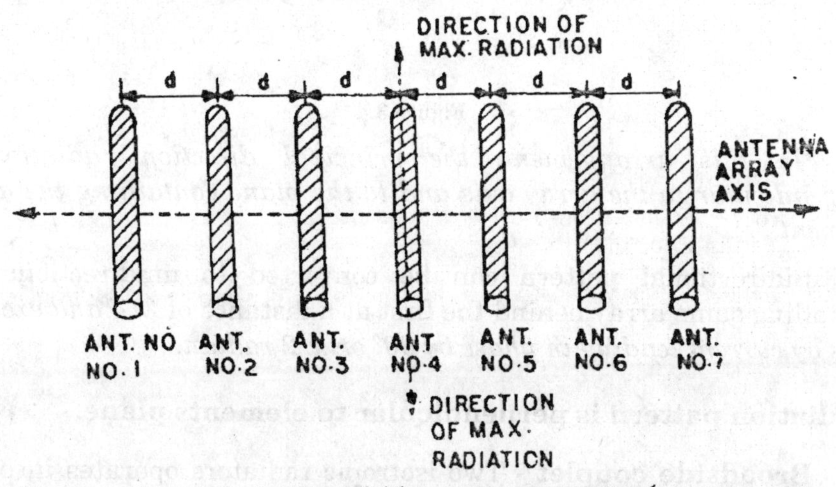

(a) Broadside array arrangement.

Figure 3.2

In broad side array, individual antennas (elements) are equally spaced along a line and each element is fed with **current of equal magnitude** all in same phase.

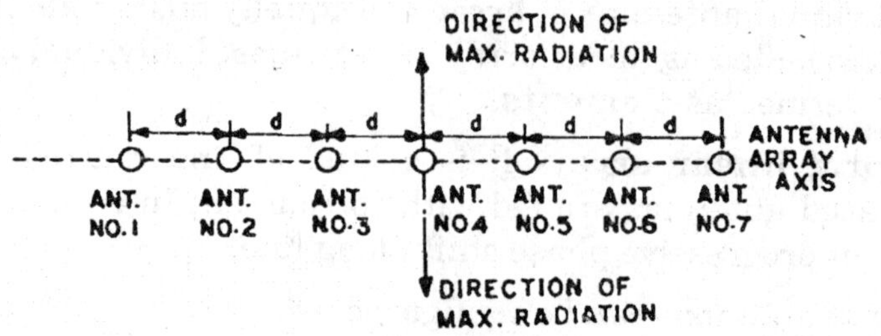

(b) Front view of the array.

Figure 3.3

By doing so, this arrangement fires in broadsides directions (i.e. perpendicular to the line of array axis) where there are maximum radiations and relatively a little radiations in other directions and hence **radiation pattern of broadside array is bidirectional,** (which radiates equally in either direction of maximum radiations).

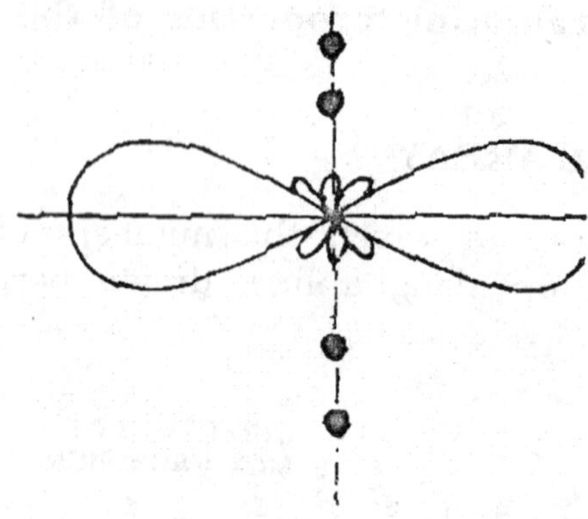

Figure 3.4

"In this arrangement, the principal direction radiation is perpendicular to the array axis and to the plane containing the array element."

Bidirectional pattern can be converted to unidirectional by installing same array behind the first at a distance of *$\lambda/4$ and exciting this by current leading in phase by 90° or $\pi/2$ radian.*

Radiation pattern is perpendicular to elements plane.

Broadside couplet - Two isotropic radiators operates in phase thereby they reinforce each other most strongly in plane right angle to the line joining them.

> **1.** Write short notes on :
> (i) Broadside Array

3.2.3 ENDFIRE ARRAY

These arrays are similar to the broadside array except that individual elements are **fed out of phase (usually 180°) and equal in magnitude.** In this individual elements are excited by current having progressive phase difference between adjacent elements (in cycle) thus becomes equal to the spacing (in λ) between elements. It has a **unidirectional pattern.**

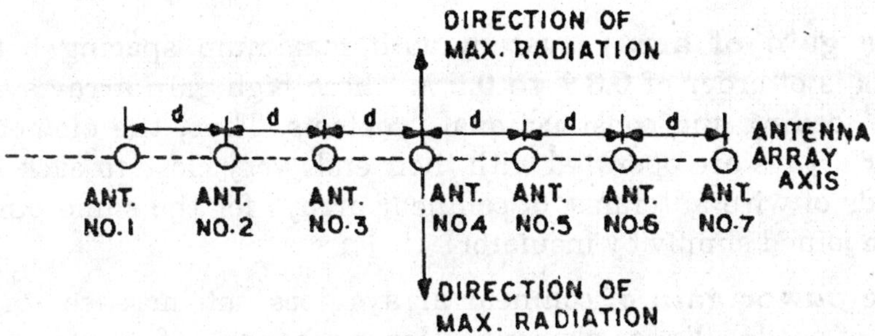

(b) Front view of the array.

Figure 3.5

In this arrangement the principal direction of radiation coincides with the direction of array axis.

NOTE: (It may be bidirectional, when two elements are fed in 180° out of phase).

Endfire couplet – This couplet is formed, when two equal radiators are operated in phase quadrature at a distance of λ/4 apart.

> **1.** Write short notes on :-
> (ii) Endfire Array

3.2.4 COLINEAR ARRAYS

They are arranged **coaxially**, that is, antennas are mounted end to end in a single line or in other words one antenna is stacked over the other.

In these arrays all the elements are fed with same phase current. They are similar to the Broadside antennas, where the direction of maximum radiation is perpendicular to the line of antenna. It has an **Omni directional** radiation pattern.

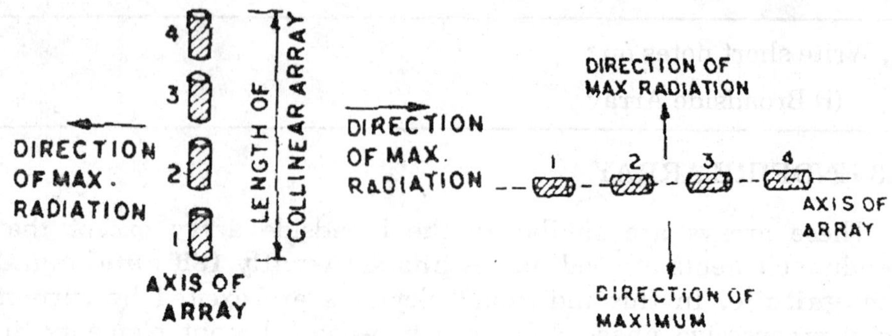

Figure 3.6

NOTE: These arrays are also called as broadcast or omni-directional array.

The **gain of array system** with maximum spacing between elements is of order of **0.3 λ to 0.5 λ.** These high gain array systems face the feeding and constructional problems. Thus, the elements in collinear arrays are operated with their ends very close to each other. (The ends of wire antennas, of collinear arrays for the same purpose, can even joined simply by insulator).

The **power gain** of collinear arrays does not increase with the number of used collinear elements. Hence more than four elements in this array are generally not used as more gain can also be obtained by some otherwise device; but the two elements collinear arrays are mostly used as it helps in multiband operation. *Two elements collinear array is also known as two half waves in phase.*

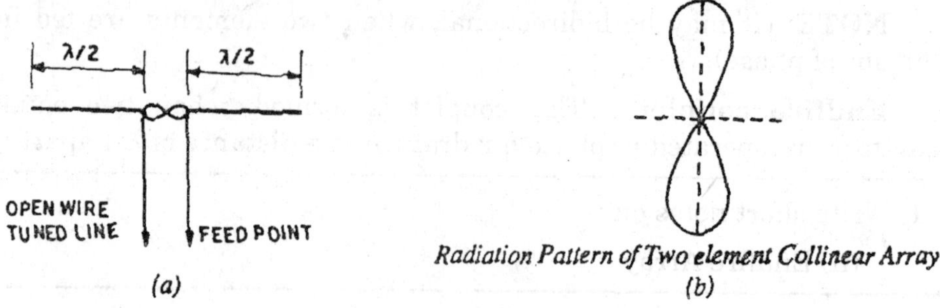

Figure 3.7

With increase in length of array the directivity of the system increases.

1. Explain collinear and combinational arrays.

3.2.5 PARASITIC ARRAY

These antennas are also called because in these antennas the feeding style is similar to parasitic feeding; where the direct feed is provided to only one element (active element) of the array while others are energized indirectly by the radiation produced by the active antenna. There are two types of basic elements-

i. **Driven element**- These elements are supplied power directly from source (i.e. Transmitter) usually through the transmission line.

ii. **Parasitic element**- These are the elements where the feeding is indirect, that is they derives power by radiation from nearby driven element, whereas the parabolic elements obtain power solely through electromagnetic coupling with a driven element because of its proximity to that driven element.

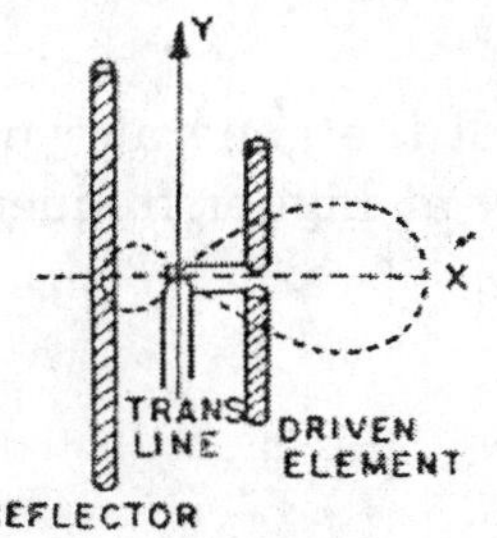

(a) Parasitic element lengthened.

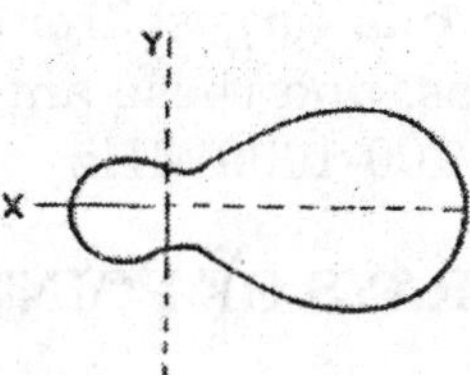

(b) Radiation pattern (Cross-Section).

Figure 3.8

Multi element arrays: or ``Parasitic Arrays" may have driven elements more than one. The most common example of parasitic array is YAGI-UDA antenna.

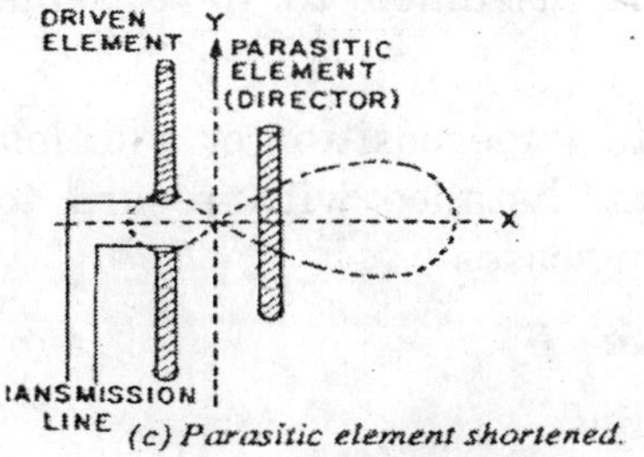

(c) Parasitic element shortened.

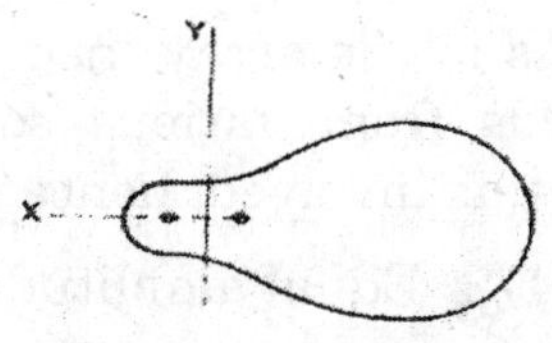

(d) Cross-Sectional radiation pattern.

Figure 3.9

NOTE

i. **Amplitude and phase of current induced**, in a parasitic elements depends on its tuning and interfacing between parasitic and driven element to which it is coupled.

ii. Variation in the **distance** between the driven element and parasitic element changes the relative phases.

iii. A distance of $\lambda/4$ between elements and phase difference of $\pi/2$ radian (90°) provides the unidirectional pattern.

It is **very important** to note here that-

A parasitic element lengthened by 5% with respect to the driven element acts as REFLECTOR and shortened by 5% acts as DIRECTOR

Where, **Reflector,** makes the radiation maximum in perpendicular direction from parasitic element towards driven element and

Director helps in making maximum radiation in perpendicular direction from the driven element.

Features

It has largest front to Back Ratio over all the general types of antennas, and these antennas works efficiently at higher frequencies that is, 100-1000 MHz

3.3 ARRAYS OF POINT SOURCES

Assumption

1. Here, we consider a hypothetical antenna which acts as a point source or a volume less radiator.
2. It is assumed that the two point sources are separated by distance'd', and
3. The two point sources here are assumed to have same polarization.

As in the array theory of antennas, the superposition or addition of fields from various sources at a great distance with regard to phases, is involved, hence we go for following cases-

(1) Equal amplitude and equal phase

(2) Equal amplitude and opposite phase

(3) Unequal amplitude and opposite phase.

CASE (1) Array of two point sources with equal phase and equal amplitude.

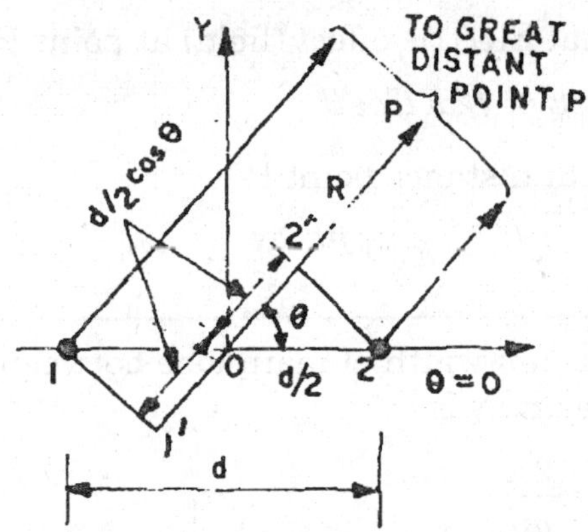

(a) Two isotropic point sources situated symmetrically w.r.t. origin O same amplitude and phase.

Figure 3.10

To calculate fields at point 'P', at distant (say R) from the origin `O' which is taken as the reference point for phase calculation. Waves from source 1 reaches the point P with some delay than the wave from source 2, it is because of the path difference (1' 2') involved between the two waves. Thus the fields due to source 1 *lags* while that due to source 2 *leads*. Therefore,

Path difference = (1'2') (in meters)

$$= \left(\frac{d}{2} Cos\,\theta + \frac{d}{2} Cos\,\theta\right) metres$$

$$= (d\ Cos\ \theta)\ (metres)$$

$$= \left(\frac{d}{\lambda} Cos\ \theta\right) wavelength.$$

Now as we know that the-

(ψ) Phase angle = 2 π (Path difference)

$$\therefore \psi = 2\pi\left(\frac{d}{\lambda} Cos\ \theta\right) = \frac{2\pi d}{\lambda} Cos\ \theta, \quad \text{(In radians)}$$

Now if $\frac{2\pi}{\lambda} = \beta$

$$\therefore\ \psi = \beta.d.\ Cos\,\theta \quad (1)$$

Also let E_1 = Electric field at distance point P due to source 1

E_2 = E-field at distance point P due to source 2

E = Total Electric effect (field) at point P

$$\psi = \beta . d . \operatorname{Cos} \theta$$

∴ Total far field at distance point P-

$$E = E_1 \, e^{-j\psi/2} + E_2 \, e^{j\psi/2} \qquad (2)$$

---

1. Prove that total beam width of main lobe between first nulls for a long broadside array is :

$$\theta_{BW} = \frac{2\lambda}{nd}$$

where, λ = wavelength

n = no. of isotropic point sources

d = distance between isotropic sources. **(Raj. Univ. 2006)**

---

2. Explain what is meant by collinear Endfire and Broadside disposition of antenna arrays? What do you mean by directivity and gain? **(R.T.U. 2009)**

CASE (1) Point source of equal amplitude and phase

According to the given situation in equation (2), the $E_o = E_1 = E_2$

Therefore, now the equation becomes-

$$E = E_o (e^{-j\psi/2} + e^{j\psi/2})$$

On multiplying the numerator and the denominator-

$$E = 2E_o \left(\frac{e^{-j\psi/2} + e^{j\psi/2}}{2} \right)$$

$$E = 2\, E_o \cos \psi/2$$

$$E = 2\, E_o \cos \left(\frac{\beta}{2} d \cos \theta \right)$$

Now $$E = 2\, E_o \left[\cos \left(\frac{\beta\, d \cos \theta}{2} \right) \right]$$

The amplitude of the above equation is = 2 E_o, which is equal to the maximum value that is one.

$$\therefore\ E_o = \frac{1}{2}$$

Thus, $$E = Cos\left(\frac{\beta d \cos\theta}{2}\right)$$

Now, by substituting the value of-

$\beta = \frac{2\pi}{\lambda}$ *and* $d = \frac{\lambda}{2}$, in the above equation we have-

$$E = Cos\left(\frac{2\pi}{\lambda}.\frac{\lambda}{2}.\frac{1}{2}\ Cos\ \theta\right)$$

$$E = Cos\left(\frac{\pi}{2}\ Cos\ \theta\right)$$

Hence, now by the help of the above equation we can draw the field pattern.

(1) Maxima direction

E is maximum when-

$$Cos\left(\frac{\pi}{2}\ Cos\ \theta\right) = \pm 1 \qquad \text{(Condition of maxima)}$$

$$\therefore \frac{\pi}{2}\ Cos\ \theta_{max} = \pm n\pi$$

Where $n = 0, 1, 2....$

$$\frac{\pi}{2}\ Cos\ \theta_{max} = 0$$

$$Cos\ \theta_{max} = 0$$

Now, $\theta_{max} = 90°$ *and* $270°$

(2) Minima direction

E is minimum when-

$$Cos\left(\frac{\pi}{2}\ Cos\theta\right) = 0 \qquad \text{(Condition of minima)}$$

$$\therefore \frac{\pi}{2}\ Cos\ \theta_{min} = \pm(2n+1)\frac{\pi}{2} \qquad \text{where, n = 0, 1, 2 ...}$$

$$\frac{\pi}{2} Cos\, \theta_{min} = \frac{\pi}{2}$$

$$Cos\, \theta_{min} = \pm 1$$

$$\theta_{min} = 0^o \text{ and } 180^o$$

(3) Half power point direction

$$Cos\left(\pi/2\ Cos\theta\right) = \pm 1/\sqrt{2}$$

$$\frac{\pi}{2} Cos\, \theta_{HPPD} = \pm\left(2n+1\right)\frac{\pi}{4} \quad \text{where, n = 0, 1, 2...}$$

$$\frac{\pi}{2} Cos\, \theta_{HPPD} = \pm\, \pi/4$$

$$Cos\ \theta_{HPPD} = \pm\ \frac{1}{2}$$

$$\theta_{HPPD} = 60^o,\ 120^o$$

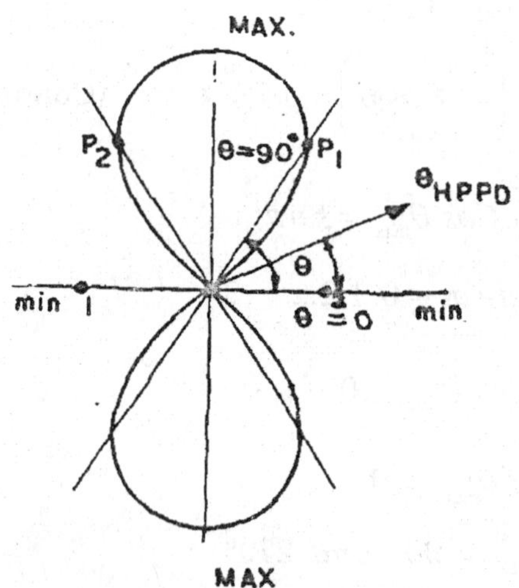

(b). Field pattern of Fig. 7.9 (a) i.e. same amplitude and phase with d = λ/2.

Figure 3.11

NOTE (1): Thus the above case can be considered as that of a broadside array with doughnut shape radiation pattern.

NOTE (2): If the reference point is shifted to source 1 instead of the origin, then-

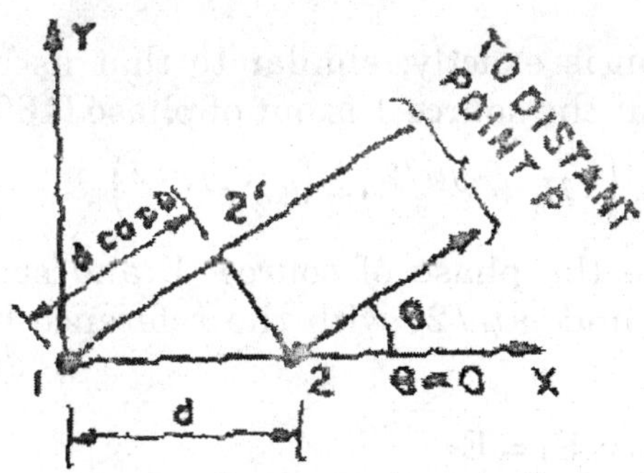

Figure 3.12

$E = E_1 . e^{j\psi} + E_2 . e^{j\psi}$

Now, as the source1 is at the origin, $e^{j\psi}$ is at the zero position, therefore- it becomes e^{j0}

$$E = E_1 . e^{j0} + E_2 . e^{j\psi}$$

$$E = E_1 + E_2 . e^{j\psi}$$

As both the current amplitudes are same, thus-

$$E_0 = E_1 = E_2$$

Then, $\quad E = E_0 \left(1 + e^{j\psi}\right)$

By multiplying and dividing the equation by 2 we have-

$$E = 2E_0 \left(1 + e^{j\psi}\right)$$

$$E = 2E0 \left(\frac{1 + e^{j\psi}}{2}\right)$$

Taking $e^{jw/2}$ common-

$$E = 2E0\, e^{j\psi/2} \left(\frac{e^{-j\psi/2} + e^{j\psi/2}}{2}\right)$$

$$E = 2E0\, e^{j\psi/2} \operatorname{Cos} \psi/2$$

Rest analysis remains the same.

1. Derive an expression of normalized electric field for two isotropic point sources of some amplitude and phase, if source `1' is placed at origin. **(Raj. Univ. 2006)**

CASE (2) Point sources with equal amplitude & opposite phase

This situation is exactly; similar to that as above one with the only difference that the source 1 is out of phase (180°) to source 2-

$$\therefore E = \left(-E_1\, e^{-j\psi/2}\right) + \left(E_2\, e^{j\psi/2}\right)$$

Now, because the phase of source '1' and source '2' at distance point P in $-\psi/2$ and $+\psi/2$, with the reference being at midway of the two sources.

Again $\quad E_o = E_1 = E_2$

$$\therefore E = E_o\, 2j\left(\frac{e^{j\psi/2} - e^{-j\psi/2}}{2j}\right)$$

$$E = E_o\, 2j \;\; Sin\, \psi/2$$

$$E = E_o\, 2\, j\, Sin\left(\frac{\beta d}{2}\, Cos\, \theta\right)$$

Again, the difference between the previous case value and this value is that-

(1) The above equation is a sine function instead of the cosine function.

(2) `j' here represents the opposite phase of 90°.

Therefore, $$E = 2\, j\; E_o\; Sin\left(\frac{2\pi}{\lambda}.\frac{\lambda}{2}.\frac{1}{2} Cos\, \theta\right)$$

$$E = 2\, j\; E_o\; Sin\left(\frac{\pi}{2}\, Cos\, \theta\right)$$

The amplitude of the above given equation is 2j E_o which is kept equal to the maximum value=1

$$2E_o\, j = 1$$

Substituting this value in the above equation-

$$E_{norm} = \text{Sin}\left(\frac{\pi}{2}\, Cos\, \theta\right)$$

(1) Maximum direction

$$Sin\left(\frac{\pi}{2} Cos\, \theta\right) = \pm 1 \qquad \text{(Condition of maxima)}$$

$$\therefore \frac{\pi}{2} \text{Cos } \theta_{max} = \pm(2n+1)\frac{\pi}{2} \qquad \text{where, } n = 0, 1, 2....$$

$$\text{Cos } \theta_{max} = \pm 1 \qquad \textit{for } n = 0$$

$$\therefore \theta_{max} = 0^{\circ} \textit{ and } 180^{\circ}$$

(2) Minima direction

$$\text{Sin}\left(\frac{\pi}{2} \text{Cos}\theta\right) = 0 \qquad \text{(Condition of minima)}$$

$$\frac{\pi}{2} \text{Cos } \theta_{min} = \pm n \pi \qquad \text{Where, } n = 0, 1, 2...$$

$$\text{Cos } \theta_{min} = 0$$

$$\theta_{min} = 90^{\circ} \textit{ and } -90^{\circ}$$

(3) Half power point direction

$$\text{Sin}\left(\frac{\pi}{2} \text{Cos } \theta\right) = \pm \frac{1}{\sqrt{2}}$$

$$\frac{\pi}{2} \text{Cos } \theta_{HPPD} = \pm(2n+1)\frac{\pi}{4}$$

$$\frac{\pi}{2} \text{Cos}\theta_{HPPD} = \pm \frac{\pi}{4} \qquad at, n = 0$$

$$\text{Cos } \theta_{HPPD} = \pm \frac{1}{2}$$

$$\theta_{HPPD} = 60^{\circ}, \pm 120^{\circ}$$

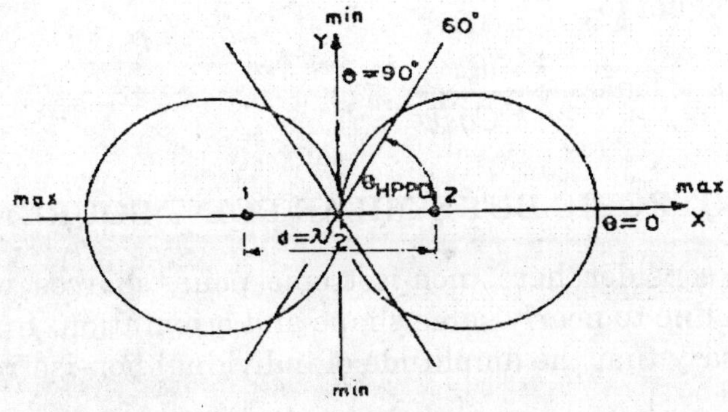

Figure 3.13

Case (3) Unequal amplitude and any phase

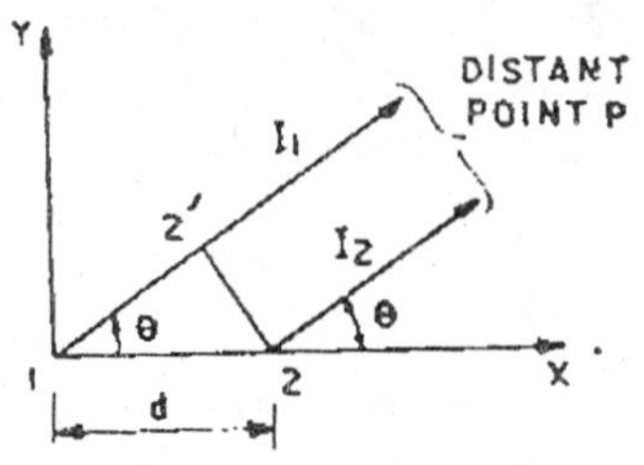

(a). Two point sources with unequal amplitude and any phase difference δ.

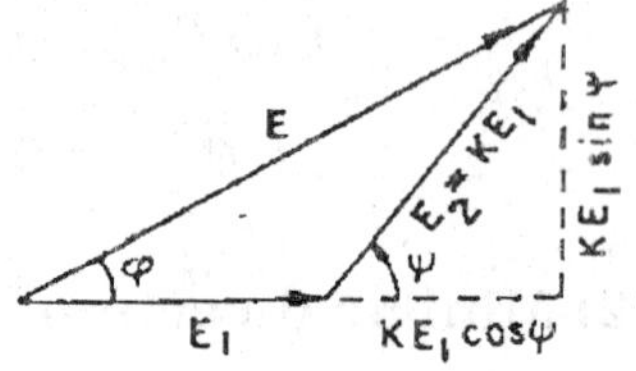

(b). Vector diagram

Figure 3.14

Here, $E_1 > E_2$ and $\psi = 2\frac{\pi}{\lambda} d \cos\theta + \alpha$

Where, α = phase by which current I_2 leads I_1 of any value.

Now according to the equation-

$$E = E_1 e^{jo} + E_2 e^{j\psi}$$

$$E = E_1 \left(e^{jo} + \frac{E_2}{E_1} e^{j\psi} \right)$$

$$E = E_1 \left(1 + k\, e^{j\psi} \right) \quad \text{Where } k = \frac{E_2}{E_1}$$

As $E_1 > E_2$, *thus* $k < 1$ *or* $0 \le k \le 1$

$$\therefore E = \left| E_1 \left\{ 1 + k \, Cos\, \psi + jk \, Sin\, \psi \right\} \right|$$

$$E = E_1 \sqrt{(1 + k \, Cos\, \psi)^2 + (k \, Sin\, \psi)^2} \angle \phi$$

where θ = Phase angle at point P

This is equal to-

$$\theta = \tan^{-1} \frac{k \, Sin \psi}{1 + k \, Cos \psi}$$

3.4 NON-ISOTROPIC BUT SIMILAR POINT SOURCES

Let us consider here, non-isotropic point sources with similar field pattern due to nearly same shape and orientation. In this case, it is not necessary that the amplitude of individual non-isotropic sources is equal.

Let's consider here two short dipoles, which are super imposed over the two isotropic point sources and are separated by the distance 'd', the field pattern of this arrangement is given by-

$$E = 2E_o \cos \psi/2 \quad \text{(Equation for two point sources)}$$

$$E_o - E_1 \sin 0 \quad \text{(Equation for a single point source)}$$

$$\psi = Bd \cos \theta + \alpha$$

Substitute the second and the value of ψ equation in first, we have-

$$E = 2\ E_1\ Sin\ \theta\ Cos\ \psi/2$$

$$\therefore E_{norm} = (Sin\theta)\times(Cos\ \psi/2)$$

$$E_{norm} = (\text{Pattern of Individual isotropic source}) \times (\text{Pattern of Array of two isotropic point source})$$

This leads to the **principle of multiplication of pattern,** as the multiplication of individual point sources and pattern of array of isotropic point sources **gives the field pattern of non-isotropic but similar point sources.**

3.5 MULTIPLICATION OF PATTERN

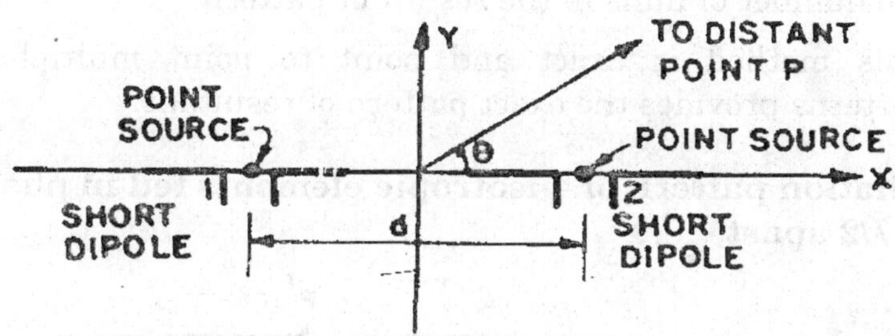

Figure 3.15

*"The **total field pattern of an array** of non-isotropic but similar sources is the **multiplication of the individual source patterns and the pattern of an array of isotropic point sources** each located at the phase centre of individual source and having the relative amplitude and phase.*

*Whereas the **total phase pattern is the addition of the phase pattern of the individual sources and that of the array of isotropic point sources."***

$$E = \{E_i\ (\theta,\phi)\times E_a(\theta,\phi)\}\times\{pi\ \ (\theta,\phi) + pa\ \ (\theta,\phi)\}$$

Where, E_i = field pattern of individual source

E_a = field pattern of array.

p_i = phase pattern of individual source

p_a = Phase pattern of array.

θ = Polar angle

ϕ = Azimuth angles.

$$E = 2\, E_o \; Cos\; \psi / 2$$

$$E = 2\, E_1 \; Sin\theta \; Cos\psi / 2$$

$$E = E(\theta)\, Cos\; \psi / 2$$

Key features

1. Multiplication of patterns provides a speedy method for sketching the pattern complicated arrays.
2. The width of principle lobe (that is, width between nulls) and corresponding width of array pattern are same.
3. Secondary lobes are determined from number of nulls in resultant pattern.
4. The sum of nulls of individual pattern and the array pattern are the number of nulls in the resultant pattern.
5. This method is exact and point to point multiplication of patterns provides the exact pattern of resultant.

(1) Radiation pattern of 4-isotropic elements fed in phase, placed λ/2 apart

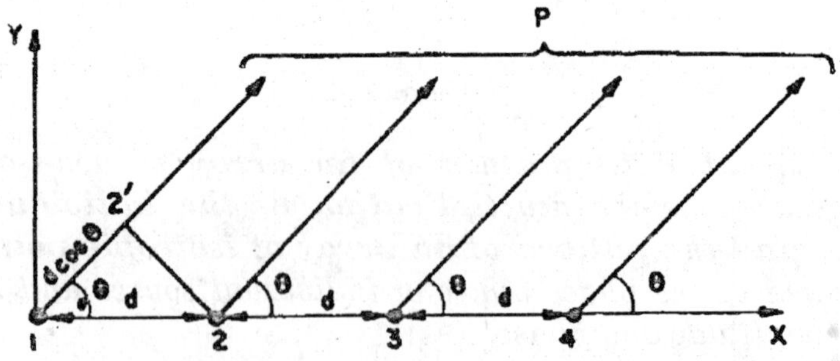

Figure 3.16

Let us consider 4-elements of isotropic (or non-directive) radiators, which are linear arrays in which elements are placed at a distance of λ/2 and are fed in phase.

To get Radiation pattern of array in general is to add the field of individual 4-elements, at a distance point 'P' vectorially-

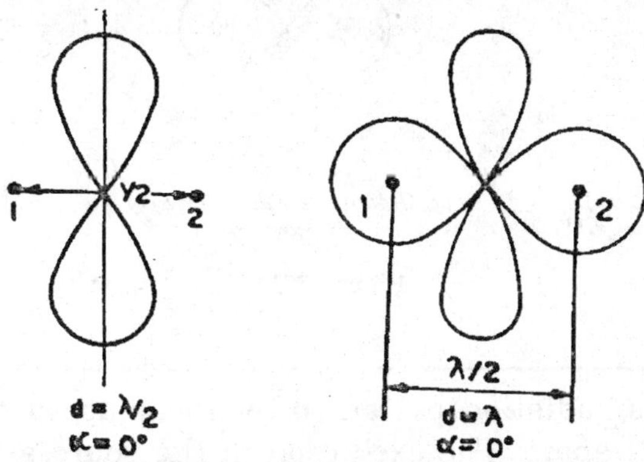

Figure 3.17

But, instead of that an alternative method, using the principle of multiplicity of pattern is used.

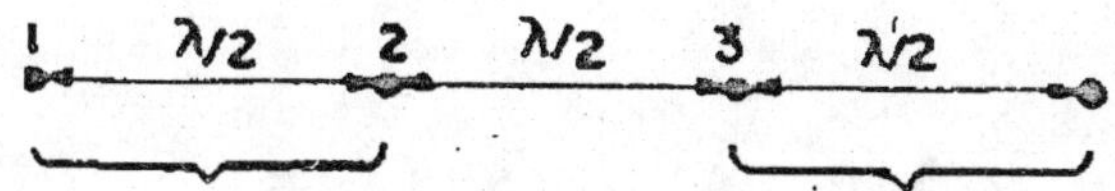

(a) 4-isotropic elements spaced λ/2.

Figure 3.18

For this the elements (1) and (2) are considered as one unit, and are to be placed between the midway of elements. Similarly, the elements (3) and (4) are assumed to be as another unit which is to be placed between the two elements.

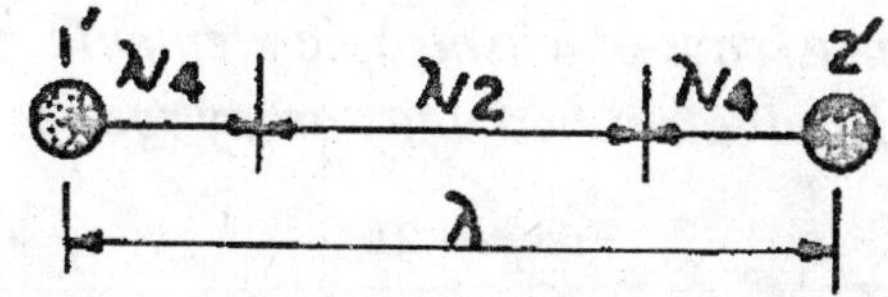

(b) 2-units array, where one unit
(O = 0 ← λ/2 → 0) spaced λ.

Figure 3.19

These two units have the same radiation pattern as previous (Figure of 8) and radiation pattern of 2 isotropic antennas.

Therefore, the combined radiation pattern of 4-isotropic elements is

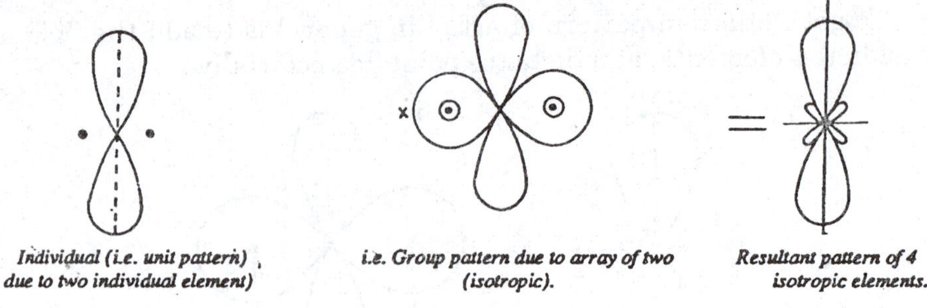

Figure 3.20

> 1. Discuss the radiation pattern of a linear array of 4-isotropic sources λ/2 apart. The excitation in the sources are in phase and have amplitude ratios 1 : 3 : 3 : 1. **(Raj. Univ. 2004)**

(2) Radiation pattern of 8-isotropic elements fed in phase, placed λ/2 apart

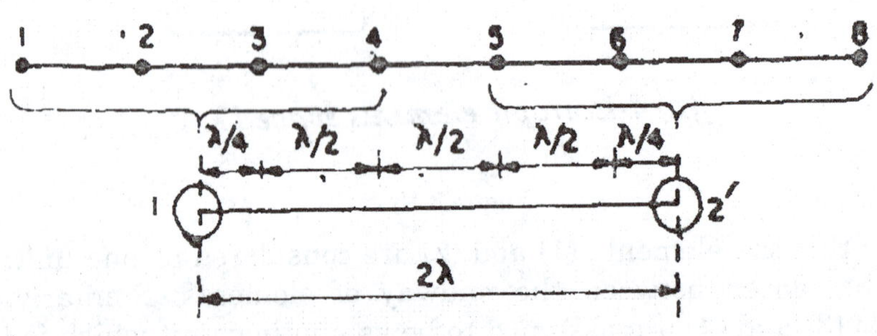

(a) Linear Arrays of 8-isotropic elements spaced λ/2.
(b) Equivalent two units array spaced 2 λ.

Figure 3.21

These are linear array of 8-isotropic elements which are spaced at λ/2 distance. The equivalent two unit's arrays are spaced 2λ away from each other.

Now let us, assume 4 isotropic elements to be one unit and then to find the radiation pattern of two such units spaced a distance 2λ apart by using the multiplication of pattern.

Therefore, the Radiation pattern of isotropic element is given by-

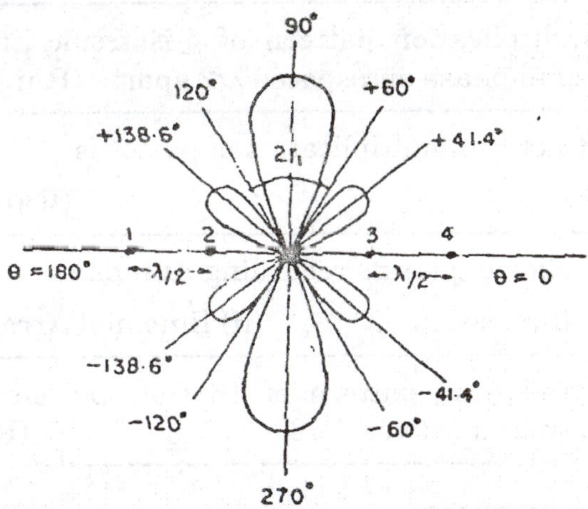

Figure 3.22

And the radiation pattern of 2-isotropic antenna spaced 2λ apart fed in phase is given by-

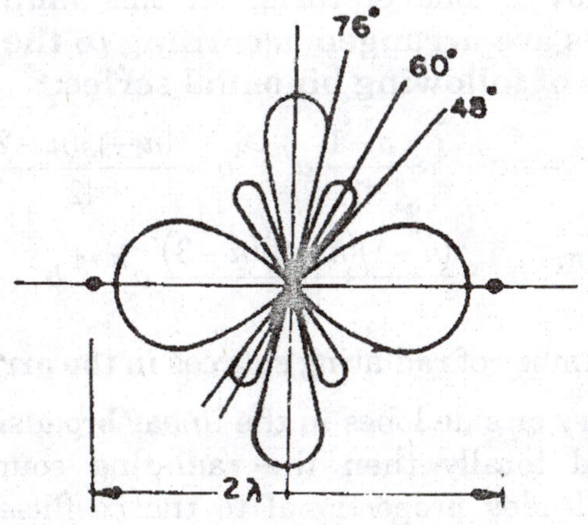

Figure 3.23

Thus, the radiation pattern for the whole arrangement of 8-isotropic antenna array is plotted to be-

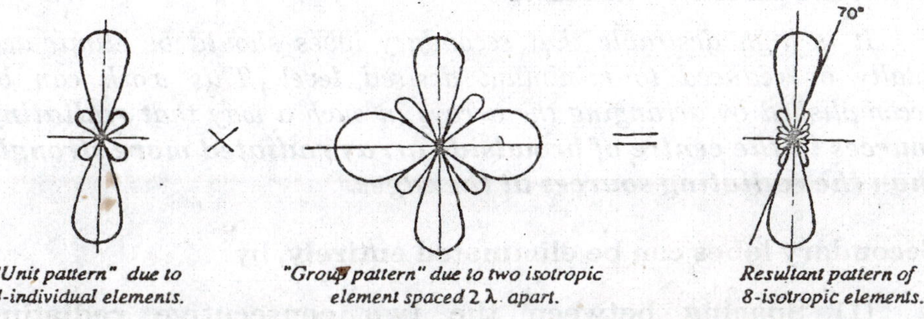

"Unit pattern" due to 4-individual elements. *"Group pattern" due to two isotropic element spaced 2 λ apart.* *Resultant pattern of 8-isotropic elements.*

Figure 3.24

1. Explain multiplication pattern of 4-isotropic and 8-isotropic element fed in phase and spaced λ/2 apart. **(Raj. Univ. 2008).**
2. Write short note on multiplication of patterns. **(Raj. Univ. 2005)**
3. Define the following terms regarding Antenna : (i) Pattern multiplication (ii) Binomial Arrays
4. Draw the radiation pattern of 16 isotropic elements fed in phase and spaced between λ/2. **(R.T.U. 2009)**

3.6 BINOMIAL ARRAYS

Till now we studied linear arrays of n-isotropic sources of equal amplitude but **arrays of non-uniform amplitudes** are also possible and binomial array is one of them. In this **amplitudes of the radiating sources are arranged according to the coefficients of successive terms of following binomial series:**

$$(a+b)^{n-1} = na^{n-1} + \frac{n-1}{!1} a^{n-2}b + \frac{(n-1)(n-2)}{!2}$$

$$a^{n-3}b^2 + \frac{(n-1)(n-2)(n-3)}{!3} a^{n-4} b^3 + \ldots$$

where, n = number of radiating sources in the array.

NOTE: If secondary or side lobes in the linear broadside arrays are to be eliminated totally then the radiating sources must have current amplitudes proportional to the coefficients of binomial series.

The length of the array is directly proportional to the Directivity of the antenna array. Also the secondary or minor lobes appear, as length and Directivity increases.

It is high desirable that secondary lobes should be eliminated totally or reduced to minimum desired level. This work can be accomplished by arranging the arrays in such a way that ***radiating sources in the centre of broadside array radiated more strongly than the radiating sources at the edges.***

Secondary lobes can be eliminated entirely, by

(1) Spacing between the two consecutive radiating sources, should not exceed λ/2.

(2) The current amplitudes in the radiating sources (from outer, towards centre source) are proportional to the coefficients of the successive terms of binomial series.

Coefficients which corresponds to current amplitude of the sources - are obtained by putting n = 1, 2, 3 ...10, in the above equation. Where we find that these coefficients are from (PASCAL's triangle).

Number of Sources	Relative Current Amplitudes
n = 1	1
n = 2	1, 1
n =3	1, 2, 1
n = 4	1, 3, 3, 1
n =5	1, 4, 6, 4, 1
n= 6	1, 5, 10, 10, 5, 1
n =10	1, 9, 36, 84. 126, 126, 84, 36, 9, 1

NOTE:

(1) Elimination of secondary lobes takes place at the cost of directivity.

(2) Half power Beam width (HPBW) of Binomial array is more than that of uniform array of some length of the array.

For example

Radiating source n =5, spaced λ/2 apart HPBW of binomial and uniform arrays are respectively 23° & 31°.

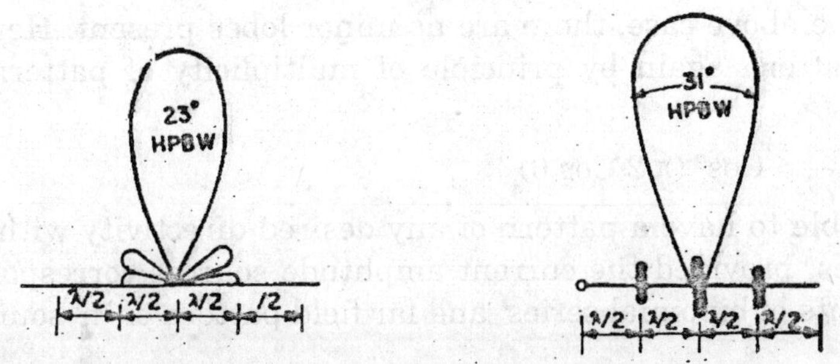

(a). Uniform Array. *(b). Binomial Array Amplitude ratio 1 : 4 : 6 : 4 : 1.*

Figure 3.25

Case: If another identical array of 2 point sources is superimposed on above array, then the array will take the shape of-

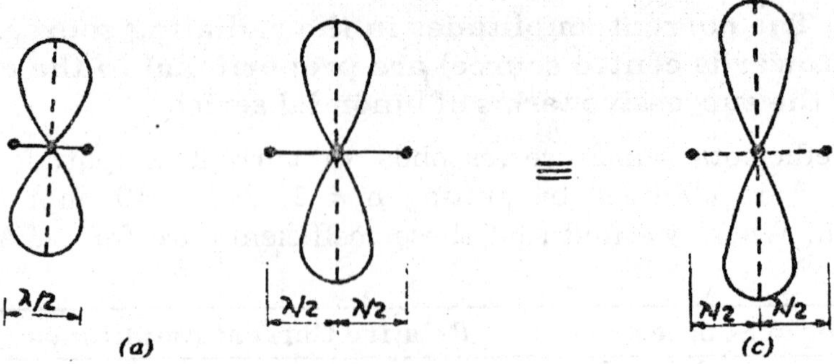

Figure 3.26

This superimposition of two sources at the centre increases its current amplitude to just double in comparison to sources at the edges.

Therefore, array now has three effective sources with amplitude ratio 1: 2: 1. Similarly if these three sources are superimposed with other identical array then an array of effective 4 sources with current amplitude Ratio 1 : 3 : 3 : 1 is obtained.

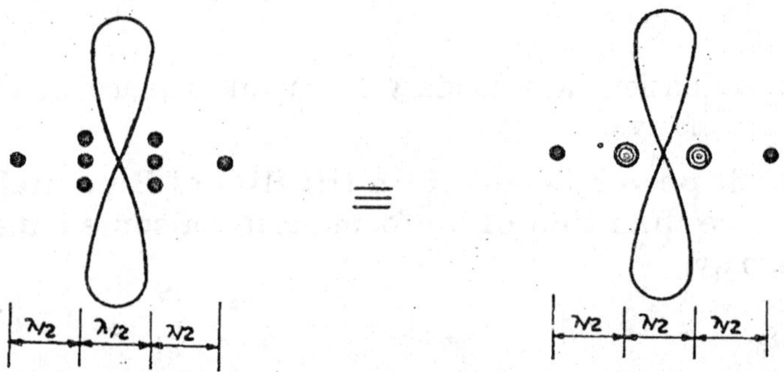

(a) 4 sources with amplitude ratio 1 : 3 : 3 : 1. *(b) The same Fig. shown superposition of centre sources.*

Figure 3.27

Here in the above case, there are no minor lobes present. Hence, the far-field pattern again by principle of multiplicity of pattern is given by-

$$E_{norm} = \text{Cos}^2 (\pi/2 \text{ Cos } \theta)$$

It is possible to have a pattern of any desired directivity without any minor lobes, provided the current amplitude sources corresponds to the coefficients of binomial series and far field pattern of `n' sources given by-

$$E_{norm} = \text{Cos}^{(n-1)} (\pi/2 \text{ Cos } \theta)$$

NOTE: Since minor lobes eliminations, in binomial array takes place at the cost of decrease in directivity in comparison with the same array (of equal amplitude sources). Therefore, in practice usually arrays of uniform size are designed as comparison between uniform & binomial arrays.

Disadvantage of Binomial arrays

(1) As HPBW increases Directivity of the array decreases.

(2) For design of large array, larger amplitude ratios of sources are required.

1. What is Pascal's triangle? How is it used in Binomial array? Prove using pattern multiplication principle that Binomial array is free from side lobe. **(Raj. Univ. 2003)**
2. Explain the importance of binomial array and discuss how it produces a pattern without secondary lobes. Discuss the effect of number of element of its directivity. **(Raj. Univ. 2002)**

3.7 GENERAL FORM OF AN ARRAY

In this an origin and coordinate system is selected, and then the `N' elements are positioned, each at location, which is clear from the given figure-

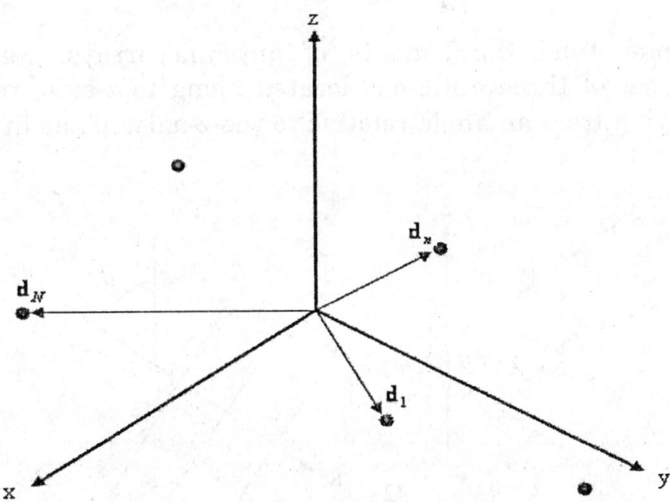

Figure 3.28

And the general form of an array is given by-

$$dn = [x_n \;\; y_n \;\; z_n]$$

Let X1, X2 ...XN represents the output from antennas. The output from these antennas are most often multiplied by a set of 'N' weights - $w_1, w_2 \ldots w_N$ - added together as in the given figure-

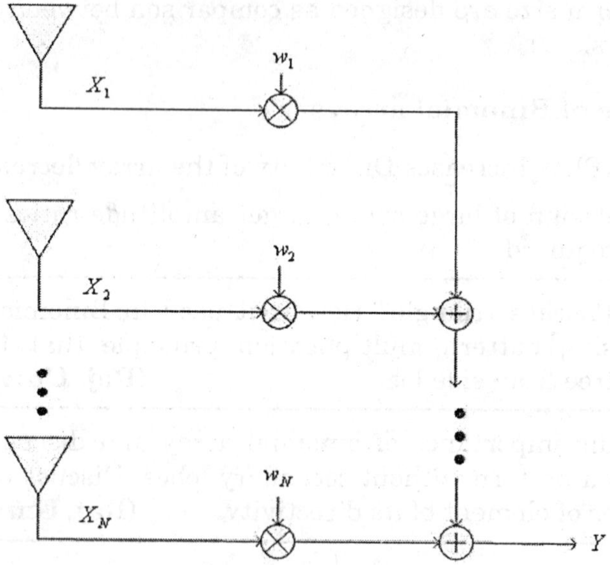

Figure 3.29

The output of an antenna array can be written succinctly as

$$Y = \sum_{n=1}^{N} w_n X_n$$

To understand the benefits of antenna arrays, we have to consider a set of three antennas located along to z-axis, receiving a signal arriving from an angle relative to the z-axis, `θ', as in the given figure –

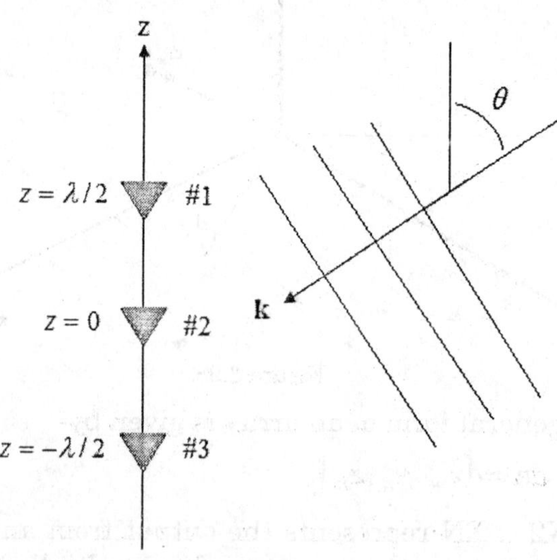

Figure 3.30

The antennas are spaced one half wavelength apart (centered at z=0). The E-field of the plane wave (assumed to have a constant amplitude everywhere) can be written as:

$$E\left(x, y.\ z\right) = e^{-j\frac{2\pi}{\lambda}\left(k_x\ x +\ k_y\ y + k_z\ z\right)}$$

$$= e^{-jlk\,(x.Sin\,\theta\ Cos\,\phi\ +\ y.Sin\,\theta\ Sin\phi\ +\ z.Cos\,\theta\,)} = e^{-jk.r}$$

Where, k = wave vector

(It gives the variation of the phase as a function of position).

The (x, y) coordinates of each antenna is (0, 0); only the z-coordinate changes for each antenna. Further, assuming that the antennas are isotropic sensors, the signal received from each antenna is proportional to the E-field at the antenna location. Hence for antenna `i', the received signal is-

$$X_i = e^{-j\frac{2\pi}{\lambda}\ Cos\ \theta\ z_i}$$

The received signals are distinct by a *complex phase factor*, which depends on the antenna separation and the angle of arrival on the plane wave. If the signals are summed together, the result is-

$$Y = e^{-j\frac{2\pi}{\lambda}\ Cos\,\theta\ z_1} + e^{-j\frac{2\pi}{\lambda}\ Cos\,\theta\ z_2} + e^{-j\frac{2\pi}{\lambda}\ Cos\,\theta\ z_3}$$

$$= \sum_{i=1}^{3} e^{-j\frac{2\pi}{\lambda}\ Cos\,\theta\ z_1} = \sum_{m=-1}^{1} e^{-jm\,\pi\ Cos\,\theta}$$

It is very interesting to note here that, if the magnitude of `Y' is plotted versus `θ' (the angle of arrival of the plane wave). The result is given by-

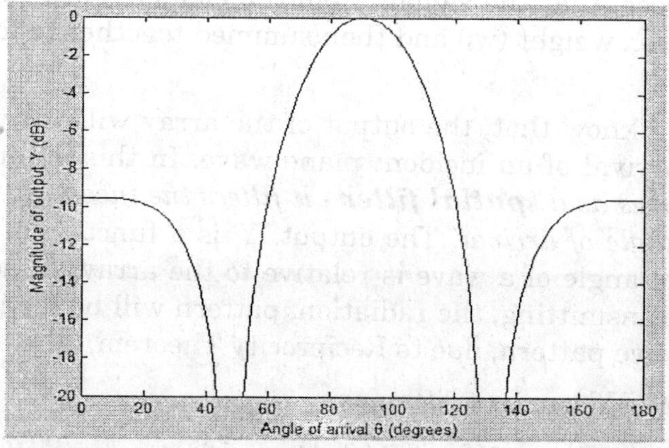

Figure 3.31

Here, the plot clears that, the array actually processes the signals better in some directions than others.

For instance, the array is most receptive when the angle of arrival is 90°. Similarly, the antenna array has zero output power (irrespective of the power of incident plane wave), when the angle of arrival is 45° or 135°. **Thus, a directional radiation pattern is obtained even though the antennas were assumed to be isotropic.**

NOTE: The same is true for the transmitting properties also, (as given for the receiver antenna), due to the Reciprocity Theorem.

It is very important to note here that-

(1) *The value and utility of an antenna array lies in its ability to determine (or alter) the received or transmitted power as a* ***function of the arrival angle.***

(2) *By choosing the* ***weights and geometry*** *of an array properly, the antenna array can be designed to cancel out energy from undesirable directions and receive energy most sensitively from other directions.*

3.8 THE ARRAY FACTOR

It is an important fundamental function of array theory. To derive it, let us, consider a set of `N' identical antennas oriented in the same direction, each with radiation pattern-

$$R\,(\theta,\,\phi)$$

Assuming that element `i' is located at position-

$$r_i = (x_i, y_i, z_i)$$

Suppose, that the signals from the elements are each multiplied by a complex weight (w_i) and then summed together to form the array output, 'y'.

As we know that, the output of the array will vary based on the angle of arrival of an incident plane wave. In this manner, *the array itself behaves as a* ***spatial filter*** *- it filters the incoming signals based on their angle of arrival.* The output `Y' is a function of (θ, φ), that is the arrival angle of a wave is relative to the array. In addition, if the array is transmitting, the radiation, pattern will be identical in shape to the receive pattern, due to Reciprocity Theorem.

Hence, `Y' can be written as

$$Y = R\,(\theta.\,\phi)\,w_1\,e^{-j\,k.r_1} + R\,(\theta,\phi)\,w_2\,e^{-jkr_2} + ... + R\,(\theta,\phi)\,w_N\,e^{-jkr_n}$$

Where, k is wave factor of the incidental wave.

The above equation can be factor simply as-

$$Y = R(\theta, \phi) \sum_{i=1}^{N} w_i \, e^{-jkr_i}$$

$$= R(\theta, \phi)\, AF$$

Therefore, $AF = \sum_{i=1}^{N} w_i \, e^{-jkr_i}$

The quantity `AF' is the Array Factor. It is a function of the positions of the antennas in the array and the weights used. By tayloring these parameters the array's performance may be optimized to achieve desirable properties. For instance, **the array can be steered (change the direction of maximum radiation or reception) by changing the weights.**

Using the steering vector, the AF can be written compactly as:

$$AF = w^T.\, v(k)$$

Where T = transpose operator.

NOTE: *If the elements are identical (array made up of all the same type of antennas), and have the same physical orientation (all elements face the same direction), then the radiation (or reception) pattern for an antenna array is simply the array factor multiplied by the radiation pattern R (θ, φ). This concept is known as* ***pattern multiplication.***

3.9 TYPES OF WEIGHTING METHODS

A weighting method is a means of selecting the weights that multiply the signals from the antenna.

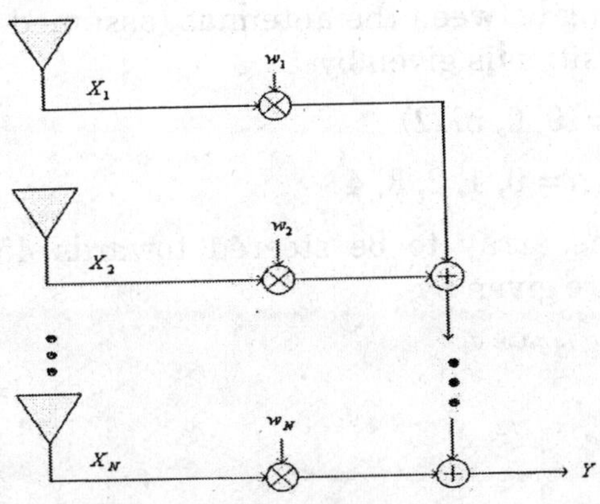

Figure 3.32

If a plane wave is incident upon an antenna array, as in the given figure below, the phase of the signal at the antennas will be a function of the angle of arrival of the plane wave. If the signals are then added together, they may add constructively or destructively depending on the phases.

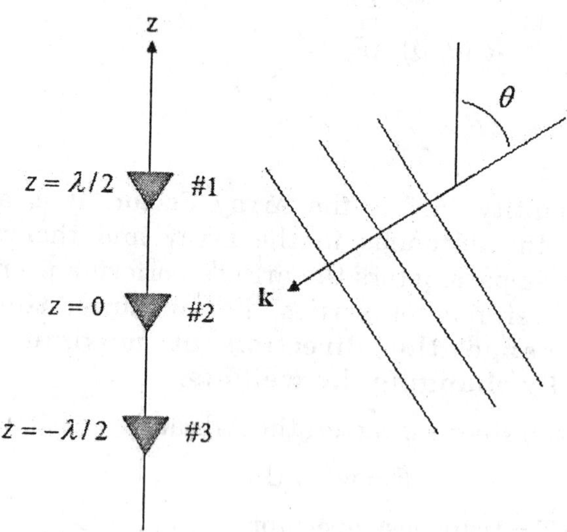

Figure 3.33

Let us discuss a case, where we want to receive signals from an angle of 45°. For this, let us first multiply the signal from each antenna by a complex phase (w_i), which cancels the phase change due to the propagation of wave. After this, when the signals from each antenna are added together to form the output of the array, they combines coherently. (Known as "**beam steering**").

Example

Suppose that we had n=5 element, linear array with one half wavelength spacing between the antennas (assumed to be isotropic for simplicity). Its position is given by-

$$d_n = (0, 0, n\lambda/2)$$

Where, now, n = 0, 1, 2, 3, 4

Again for the array to be steered towards 45° (θ_d = 45°), the weights applied are given by-

$$w_n = e^{jn\pi \, Cos \, \theta_d}$$

$$= e^{j\pi\sqrt{2}/2}$$

To be absolutely explicit about it, the weight vector can also be written as-

$$w = \begin{bmatrix} 1 \\ e^{j\pi\sqrt{2}/2} \\ e^{j\pi\sqrt{2}} \\ e^{j3\pi\sqrt{2}/2} \\ e^{j4\pi\sqrt{2}} \end{bmatrix}$$

The resulting Array Factor for this array becomes-

$$AF = w^r\, v(k) = \sum_{n=0}^{N-1} e^{jn\pi}\, \left(Cos\,\theta_d - Cos\,\theta\right)$$

Where, k = wave vector

v (k) = steering vector

N= 5.

The array response is the magnitude of the output, as a function of the incident angle of the plane wave. The array response is identical in form to the radiation pattern of this array (when the same weights are used), and is given by-

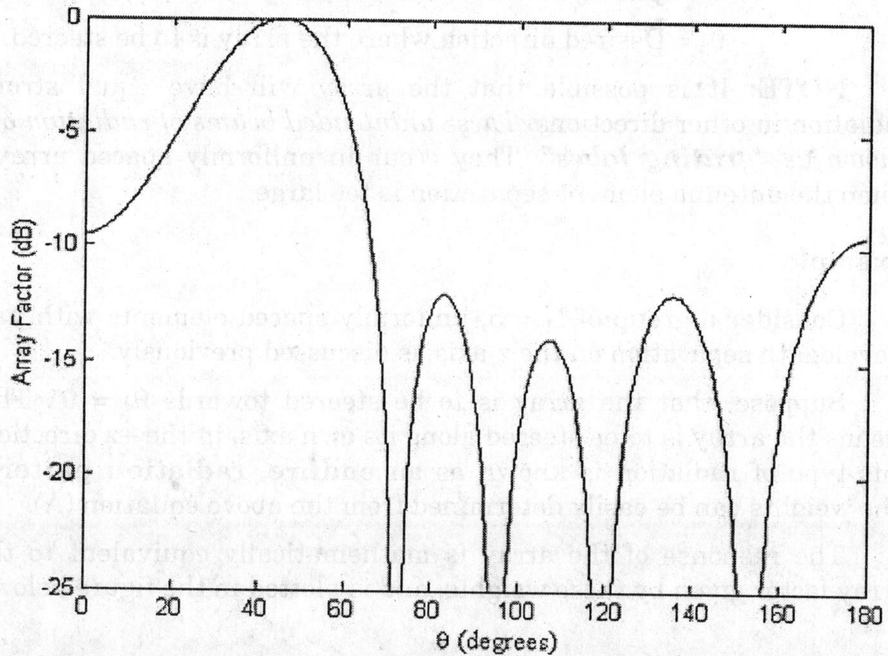

Figure 3.34

This plot captures the fundamental properties of the designed array. Here we can conclude that-

(1) The array has maximum radiation (or reception) in the direction of 45°, as designed.

(2) The array has nulls at roughly 72°, 94°, 120° and 153°. **The null directions are those in which the array completely blocks signals.**

(3) This array has an associated HPBW of roughly, 30°.

(4) The sidelobes are 12 dB below the peak of the main beam.

By employing the **"phased weighting scheme",** the array can be steered such that the direction of maximum reception is in a desired direction. The ease of implementation is responsible for its widespread use.

3.10 GRATING LOBES

We saw previously that, a uniformly spaced array with weights selected to be-

$$w_n = e^{jknd\, Cos\, \theta_d} \tag{A}$$

Where k = wave number

d = spacing between adjacent array element

θ_d = Desired direction where the array is to be steered.

NOTE: It is possible that the array will have equal strong radiation in other directions. *These unintended beams of radiation are known as **"grating lobes".*** They occur in uniformly spaced arrays, when the antenna element separation is too large.

Example

Consider a group of N = 5, uniformly spaced elements with half wavelength separation on the z-axis as discussed previously.

Suppose, that the array is to be steered towards θ_d = 0°. This means the array is to be steered along its own axis, in the +z direction; this type of radiation is known as an **endfire, radiation pattern.** The weights can be easily determined from the above equation (A).

The response of the array is mathematically equivalent to the array factor given by these weights, and is plotted in the figure below-

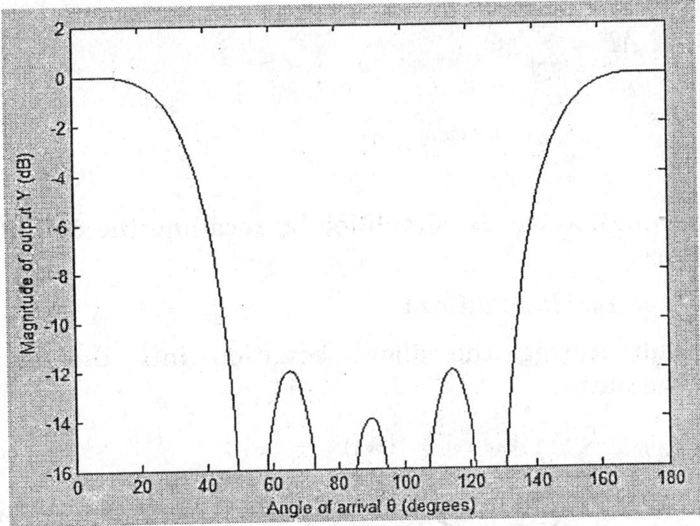

Figure 3.35

In the figure above, the array have a response which is maximum towards the desired direction (0° or + z - axis). In the same plot the lobe of maximum radiation towards 180°, is produced which is undesired, called *as grating lobe.* For uniformly spaced arrays, **grating lobes can be eliminated by decreasing the separation between the elements.**

After this, let us now look over at the properties of uniformly spaced arrays using the phased steering weighting scheme described by equation (A).

3.11 UNIFORM PHASED ARRAYS

Analysis

The general equation for the **array factor or array response** for an N-element uniformly spaced linear array with simple phased weights. When the array is steered towards direction θ_d, then these weights are given by-

$$w_n = e^{jknd \cos \theta_d}$$

Assuming that element `n' is at location given by-

$$r_n = (x_n, y_n, z_n) = (0, 0, nd)$$

This implies that, the *inter-element spacing* is constant and equal to **`d'**. When the antenna is steered towards `θ_d', with chosen weights according to the equation above the response of the array can be determined. By using the definition of the array factor, we have-

$$AF = \sum_{n=0}^{N-1} w_n \, e^{-jkr_n}$$

$$= \sum_{n=0}^{N-1} e^{jknd \, Cos \, \theta_d} \, e^{-jkr_n}$$

This equation can be simplified by recalling the definition of the wave vector-

$$k.\, r_n = k.\, n.\, d \, Cos \, \theta$$

On substituting the above equation into the array factor equation, we have-

$$AF = \sum_{n=0}^{N-1} e^{jknd(Cos \, \theta_d \, - \, Cos \, \theta)}$$

$$= \sum_{n=0}^{N-1} G^n$$

Where, G - dummy variable and is given by-

$$G = e^{jkd \, (Cos \, \theta_d \, - \, Cos \, \theta)}$$

As we know that, **the sum formula-**

$$\sum_{n=0}^{N-1} G^n = \frac{1-G^n}{1-G}$$

∴ the array factor can be rewritten by using the above identity as:

$$AF = \frac{1 - e^{jkNd \, (Cos \, \theta_d \, - \, Cos \, \theta)}}{1 - e^{jkd \, (Cos \, \theta_d \, - \, Cos \, \theta)}}$$

Here, considering the magnitude of the array factor only; we can factor out terms from the numerator and the denominator which simplifies the result with only the magnitude-

$$AF = \frac{-e^{jkNd \, (Cos \, \theta_d \, - \, Cos \, \theta)}/2 \left[e^{jkNd \, (Cos \, \theta_d \, - \, Cos \, \theta)/2} - e^{-jkNd \, (Cos \, \theta_d \, - \, Cos \, \theta)}\right]}{-e^{jkd \, (Cos \, \theta_d \, - \, Cos \, \theta)/2} \left[e^{jkd \, (Cos \, \theta_d \, - \, Cos \, \theta)/2} - e^{-jkd \, (Cos \, \theta_d \, - \, Cos \, \theta)/2}\right]}$$

Taking the magnitude of the above equation, we conclude that the multiplying complex exponentials have unity magnitude. It uses the following general formula for the `sine' function-

$$Sin \, x = \frac{e^{jx} - e^{-jx}}{2j}$$

The magnitude of the array factor reduces to-

$$AF = \frac{Sin\left[kNd\ \left(Cos\ \theta_d\ - Cos\,\theta\right)/2\right]}{Sin\left[kd\ \left(Cos\,\theta_d - Cos\ \theta\right)/2\right]}$$

Following part explains this equation (also the grating lobes), and extend the results to 2-D planar arrays.

3.11.1 TWO DIMENSIONAL PHASED ARRAYS

Let us concentrate our analysis on the two dimensional or planar arrays with uniform spacing and with phased weights, which are much simpler to understand than one dimensional array.

As, in one dimensional arrays, the weights are chosen to compensate for the phase propagation of a wave in a desired direction. The extension to two dimensional arrays is straight forward. For it let us assume, that we have an array in (x, y) plane, with positions given by-

$$d_{ab} = \left(a\lambda/2,\quad b\lambda/2,\,0\right) \qquad \text{For a, b} = 0, 1, 2$$

Resultant plot is given below-

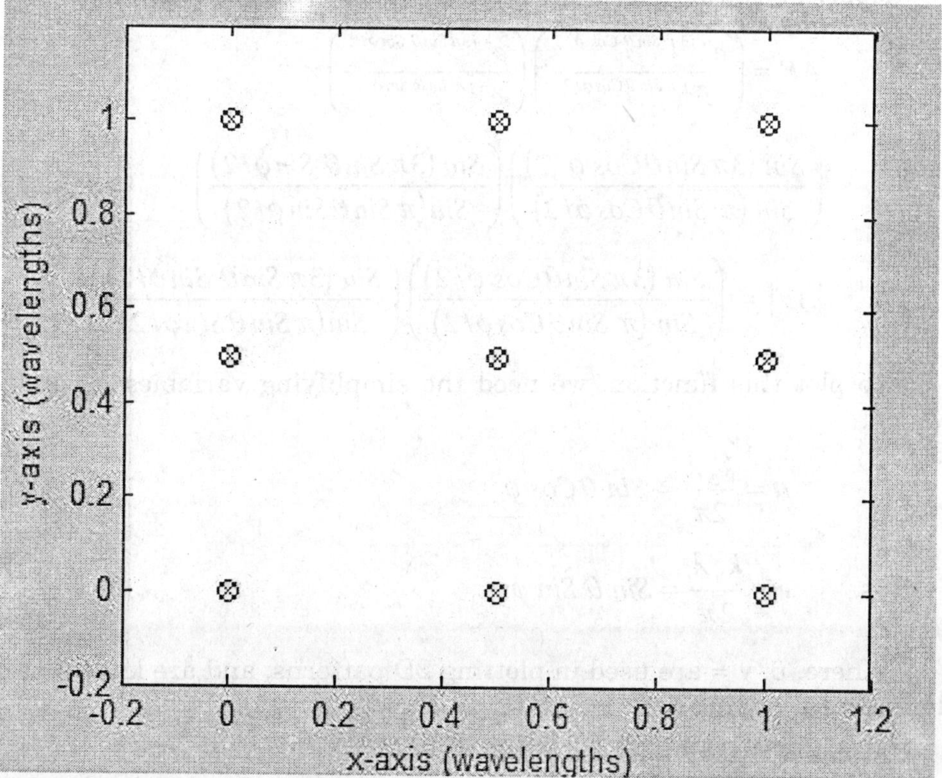

Figure 3.36 Array geometry for 2-D (planar) array

The phased weights for a two-dimensional planar array steered towards (θ_d, ϕ_d) are given by-

$$w_n = e^{jk d\, r_n}$$

$$= e^{j\left(k_{rd}\, x_n + k_{yd}\, y_n + k_{zd}\, z_n\right)}$$

$$= e^{j\, 2\pi/\lambda\, Sin\,\theta_d\, (Cos\, \phi_d\, x_n + Sin\, \phi_d\, y_n)}$$

In the above equation, the weights are desired to steer the array towards the wave vector in the desired direction, written as `kd'.

To illustrate this, we'll assume the array is steered to **broadside**-

$$(\theta_d, \phi_d) = (0,0)$$

In this case, all of the weights simply become equal to unity. The array factor becomes

$$AF = \sum_{b=0}^{2} \sum_{a=0}^{2} e^{-j\pi Sin\,\theta\, (a\, Cos\,\theta + b\, Sin\,\phi)}$$

Applying the sum formula twice, the above reduces to-

$$AF = \left(\frac{e^{-j3\pi\, Sin\,\theta\, Cos\,\phi/2}}{e^{-j\,\pi\, Sin\,\theta\, Cos\,\phi/2}}\right)\left(\frac{e^{-j3\pi\, Sin\,\theta\, Sin\,\phi/2}}{e^{-j\pi\, Sin\,\theta\, Sin\,\phi/2}}\right) \times$$

$$\left(\frac{Sin\,(3\pi\, Sin\theta\, Cos\,\phi/2)}{Sin\,(\pi\, Sin\theta\, Cos\phi/2)}\right)\left(\frac{Sin\,(3\pi\, Sin\,\theta\, Sin\,\phi/2)}{Sin(\pi\, Sin\,\theta\, Sin\,\phi/2)}\right)$$

$$|AF| = \left(\frac{Sin\,(3\pi\, Sin\theta\, Cos\,\phi/2)}{Sin\,(\pi\, Sin\theta\, Cos\phi/2)}\right)\left(\frac{Sin\,(3\pi\, Sin\,\theta\, Sin\,\phi/2)}{Sin(\pi\, Sin\,\theta\, Sin\,\phi/2)}\right)$$

To plot this function, we need the simplifying variables `μ' and `v'-

$$\mu = \frac{k_2\lambda}{2\pi} = Sin\,\theta\, Cos\,\phi$$

$$v = \frac{k_y\lambda}{2\pi} = Sin\,\theta\, Sin\,\phi$$

Where, u, v = are used in plotting 2D patterns, and are known as **directional cosines**

A plot of the AF above is shown-

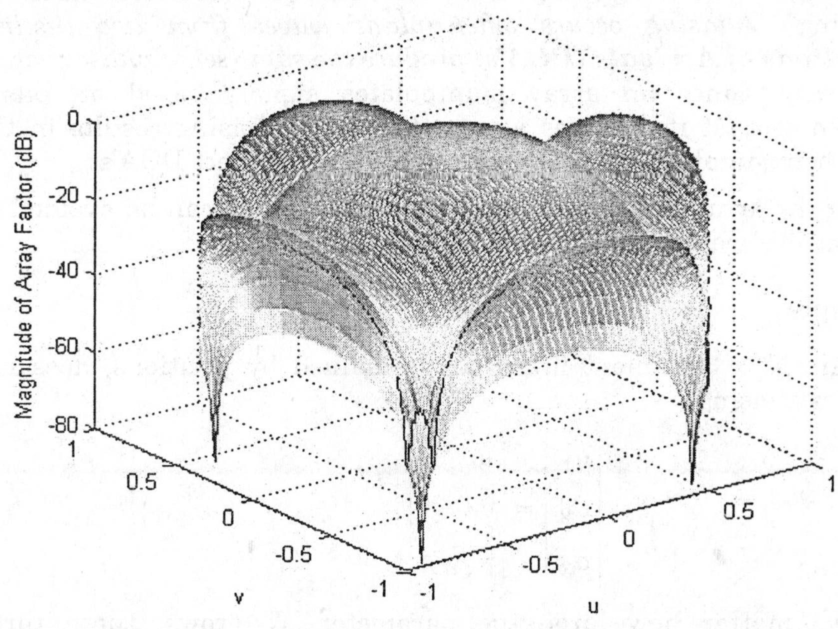

Figure 3.37

As expected, the array is maximum when (μ, v) = (0, 0) which corresponds to the desired direction (θd, ϕd) = (0, 0).

3.12 INTRODUCTION TO ANTENNA ARRAY GEOMETRY

We have already discussed the basis of antenna arrays and weighting method in array. The performance of an array output or reception pattern depends on-

(1) The number of elements in the array (generally more elements yields better performance)

(2) The weighting vectors.

(3) The geometrical (position) of the array.

There are few basic rules, related to the array geometry, but it is not, understood completely. In general, **if the inter element spacing between elements for an N-element array is increased, the beam width decreases.**

However, due to the grating lobes, increasing the size of the array will eventually produce grating lobes-undesirable directions of maximum radiations.

For uniformly spaced arrays (either 1-D, 2-D or 3-D arrays with a constant spacing between elements), the maximum spacing is a half-wavelength to avoid grating lobes. This effect is sometimes called as "aliasing". *Aliasing occurs when plane waves from two distinct* ***Directions of Arrivals (DOA's)*** *produce the same set of phases across the array.* Since an array manipulates signals based on phase differences that it observes across the array, aliasing results in the array being unable to distinguish signals from distinct DOA's.

If the array has non-uniform spacing, aliasing can be avoided (if the spacing are not multiplies of each other).

Example

An, N = 3 element linear array defined by positions, does not exhibits aliasing -

$$d = \begin{bmatrix} d_1 \\ d_2 \\ d_3 \end{bmatrix} = \begin{bmatrix} 0 \\ c\lambda \\ c\sqrt{2}\,\lambda \end{bmatrix}$$

No matter how large the parameter `C' grows, these three elements arrays will not exhibit aliasing. No two distinct directions of arrival will produce the same set of propagation delays across the array.

However, **the lack of aliasing does not imply that grating lobes will not exist.** In fact, if C is very large (>10), for any scan angle there will exist multiple grating lobes.

Non uniform arrays do not solve the grating lobe problem, but they do enable extra degree of freedom in designing the array.

In general, there is no quick method to determine if an array exhibits grating lobes. However, any array geometry can be quickly analyzed to determine if grating lobe occurs.

Next, let us look at the most popular 2D array: The hexagonal sampled array.

3.12.1 HEXAGONAL SAMPLED ARRAY

The basic geometry of a seven element hexagonal array is shown in Figure 3.38.

The hexagonal structure is such that each element is at fixed distance (C) away from its six nearest neighbors. In addition, any three adjacent elements form an equilateral triangle. This type of spacing is known as **hexagonal** because the outline of the outer six elements of the array in the above figure forms a hexagon; for arrays with more elements, the hexagons just repeat. Hence, *the hexagonal array can have as many elements as desired.*

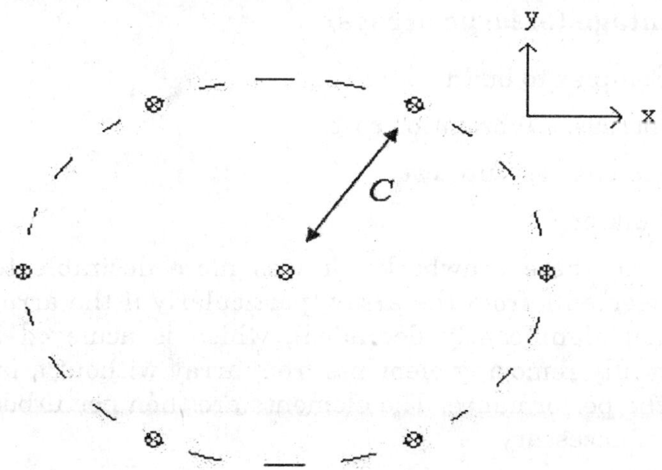

Figure 3.38 Hexagonal Two-Dimensional Antenna Array

Advantage of this array

1. In multidimensional digital signal processing, it is known that sampling with hexagonal lattice produces the least aliasing for band-limited signals.
2. As, the array has somewhat of a circular symmetry, so the response towards distinct azimuth angles do not vary by a large degree.
3. An **adaptive array developed by Raytheon** that attempts to block out interference (or jamming of signals) uses a hexagonal geometry.

NOTE: This array ends up being the optimal geometry for a large number of problems. For instance, for 2D arrays designed to have the minimum possible sidelobe level for a fixed beamwidth, the optimal geometry, with the optimal weights, becomes a hexagonal array, when N = 7.

In this case, the parameters `C' is chosen, such that the beamwidth meets the specified criteria and grating lobes do not occur.

3.13 THINNED ANTENNA ARRAYS: (Linear Arrays)

The earliest methods of optimizing array geometry around 1960, was **thinning arrays.** This method is the simplest method till now for this, large uniformly spaced array (linear or planar) elements are used.

Disadvantage (of large arrays)

(1) Complex to build

(2) Increased fabrication cost

(3) Increased setup cost

(4) Bulkier.

Due to, these drawbacks, it was more desirable to eliminate, antenna elements from the array (particularly if the array's performance is not significantly degraded); which is achieved by thinning systematically removing elements from array without a large degradation in the performance. The elements are then perturbed from their locations if necessary.

Example

Consider 20-elements uniformly spaced linear array (with half wavelength spacing), with positions given by-

$$d = \left[\pm\frac{\lambda}{4}, \pm\frac{3\lambda}{4}, \pm\ldots\ldots, \pm\frac{19\lambda}{4} \right]$$

To illustrate the concept of thinned arrays lets remove some of the elements and perturb the locations a little, we have a 12-elements linear array with positions given by-

$$d = \left[\pm\frac{\lambda}{4}, \pm\frac{3\lambda}{4}, \pm\frac{3\lambda}{2}, \pm\frac{5\lambda}{2}, \pm\frac{7\lambda}{2}, \pm\frac{21\lambda}{5} \right]$$

NOTE: Both arrays have approximately the same length. Assume that all the weights for the array are constant and set to one.

The resulting Array Factors for both arrays are plotted-

Some observations are immediately apparent after observing the figure 3.39.

(1) The **peak gain** of the array factor is decreased for the thinned array. It is because the array is made up of fewer elements.

(2) The **side lobe level** and the **beamwidth** are approximately the same for the two cases, except for large angular separation from the main beam.

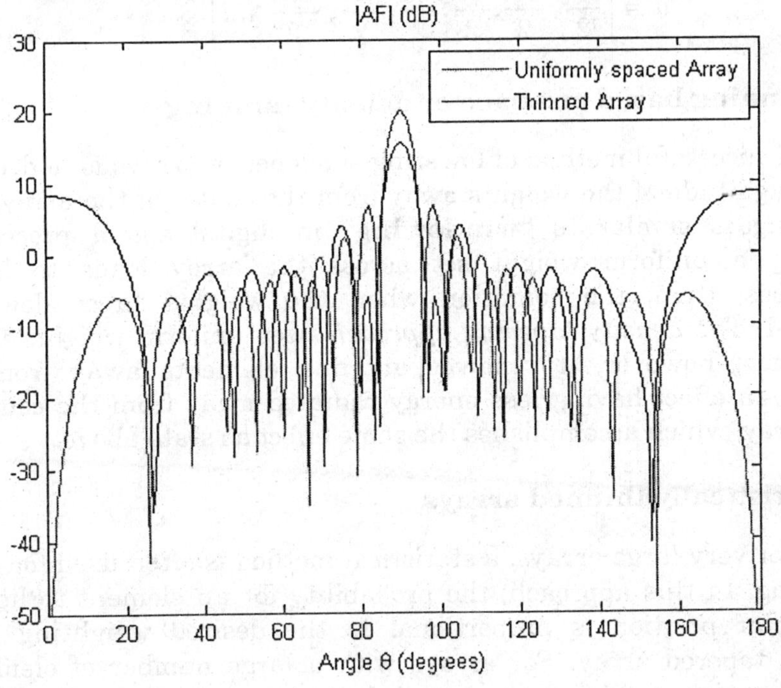

Figure 3.39

Conclusion

The performance of the full uniform array can be approximately achieved by using 40% fewer elements, which is often highly desirable.

NOTE: The large sidelobes in the thinned array are easily eliminated. As we already know that what is actually radiated by an antenna array is the product (pattern multiplication) of the Array Factor and the radiation pattern of the antennas that make up the array. Hence, by choosing an antenna which has low gain for small and large values of the polar angle 'θ' (such as in simple dipole antennas, where side lobes are instantly removed) we can construct thinned arrays.

Methods of array thinning are listed below

1. Thinning Based on an Empirical or Analytical Formula

Array thinning can be performed by using a set rule, which is advantageous because it requires no computation. For instance, array spacing can be designed to follow the prime number sequence, which leads to non uniform and sparse spacing-

$$d = \left[\frac{1\lambda}{2}, \frac{2\lambda}{2}, \frac{3\lambda}{2}, \frac{5\lambda}{2}, \frac{7\lambda}{2}, \frac{11\lambda}{2}, \ldots\right]$$

2. Thinning based on space or density tapering

A successful method of lowering sidelobes in arrays is to decrease the magnitude of the weights away from the center of the array. This tapering is similar to **"windowing"** in digital signal processing. Having a uniform weight set across the array leads to higher sidelobes, than, the condition when the weights taper down, in general. The *density tapering approach* uses uniform weights for all antennas; however, it removes antenna elements away from the center, in effect having less energy radiated away from the center of the array, which accomplishes the same effect as state above.

3. Statistically thinned arrays

For very large arrays, a statistical method is often used for array tapering. In this approach, the probability for an element to lie in a particular position is proportional to the desired weighting for a weight tapered array. For arrays with a large number of elements, this approach yields arrays that behaves accordingly and have low sidelobes.

4. Optimizing Algorithms

Finally, since computers are so computationally fast these days, thinning and placement optimization is often done via optimization algorithms. Examples include-

a. Genetic Algorithms (**GA**)

b. Particle Swarm Optimization (**PSO**) algorithm

c. Simulated Annealing (**SA**).

All of these methods employ some statistical optimization approach that guesses at the proper elements to remove, then removes them if this increases the performance of the array. This has been popular in the antenna literature over the past 15 years, primarily because it is simple to implement and can achieve interesting results.

3.14 ANTENNA MEASUREMENT

This section, is used to solve the common question, such as-

What is antenna testing?

When and where should we go for antenna testing?

To know how an antenna works?

To know that where a particularly designed antenna can be used?

One has to, has to, know about the fundamental parameters of the designed antenna. The most common and desired antenna measurements are its radiation pattern including gain and efficiency, impedance, bandwidth and polarization.

Basically there are 2 types of antenna measurements-

1. Impedance measurement

- Input impedance
- Mutual impedance

2. Pattern measurement

- Gain
- Radiation pattern
- Beamwidth
- Polarization
- Antenna Efficiency
- Antenna Noise

Here, in this part of the chapter we will be dealing with the topics-

- Required Equipments and their Ranges
- Radiation Pattern and Gain Measurement
- Phase Measurement
- Polarization Measurement
- Impedance Measurement and
- Scale Model Measurement

3.14.1 ANTENNA MEASUREMENT EQUIPMENT

For the testing purpose the test antenna (*primary antenna*) is considered to be a plane wave antenna. This approximation is achieved by using a source antenna with no radiation pattern and characteristics, in such a way that the field's incident upon the test antenna are approximately planar. The required equipment for antenna measurements include -

(1) **Source antenna/Transmitter:** This antenna is with known parameters, and is used to illuminate the test antenna. It is also called as the ideal / secondary antennas.

(2) **Receiver System:** It is used to determine, the total received power by the test antenna.

(3) **Positioning system:** This system is used to rotate the test antenna at desired angular steps (angle) relative to the source antenna, to measure the radiation pattern.

Block diagram of the above equipment is shown below -

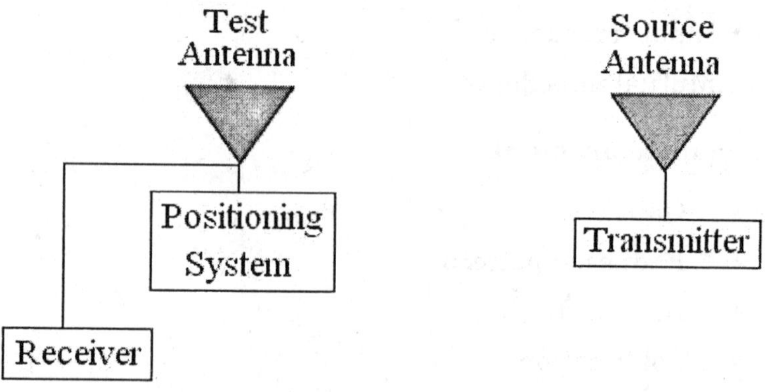

Figure 3.40 Diagram of required antenna measurement equipment

Now, let us briefly discuss the component of the above setup-

(1) SOURCE ANTENNA: It should radiate well at the desired/ set, test frequency. It must have the desired polarization and a suitable beamwidth for the given antenna test range. We generally go for horn antennas or for the dipole antennas with a parabolic reflector for this purpose.

(2) TRANSMITTING SYSTEM: This system should be able to transmit a constant known power. The output frequency should be tunable, and should be stable at this frequency.

(3) RECEIVING SYSTEM: This system is used to determine as though how much power is received from the test antenna. For this one can use a simple *bolometer,* (i.e. a device for measuring the energy of incident electromagnetic waves). The receiving system is very complex, with high quality amplifiers for low power measurements and more accurate as detection devices.

(4) POSITIONING SYSTEM: This system controls the orientation of the test antenna. To measure the radiation pattern of the test antenna as a function of angle, we need to rotate the test antenna, so that, the source antenna illuminates the test antenna from different angles. Hence, for the same purpose we go for the positioning system.

Now, to test any antenna, the above setup is replaced in special chambers.

Need of special chambers

By performing the testing of antenna in normal rooms, laboratory etc.; one faces the problem of **reflections** from walls, ceilings and the floor, which makes the reading inaccurate. Hence the **ideal location** to perform antenna measurements is somewhere in outer space, where no reflections can occur.

However, because space wave travel in an expansive behavior; therefore, there are two main types of ranges for these wave, those are-

(a) Free Space Ranges

(b) Elevated/Reflection Ranges.

Out of which Reflection ranges, are designed such that the reflections are added together in the test region to support a roughly planar wave; hence for the time, we focus on free space ranges (which are more commonly used due to their greater benefits).

(a) Free space ranges

These ranges are special antenna measurement locations, specially designed to simulate measurements that would be performed in space. Here, all the reflected waves from nearby objects and the ground, (undesirable) are suppressed as much as possible. The most popular free space ranges (space) are the **anechoic chambers.**

3.14.2 ANECHOIC CHAMBERS

These chambers are indoor antenna ranges. The wall, ceiling and floor of the chamber are lined with special electromagnetic wave absorbing material. *Indoor ranges are desirable because the test conditions can be much more tightly controlled than that of outdoor ranges.*

The material used, is often **jagged in triangular cone shapes** which are so designed, so that, the reflected waves from them tends to spread in random directions.

These reflections are added together with all the other random reflections **in-coherently and are thus suppressed further**. An anechoic chamber along with test equipment inside are shown in the Figure 3.41.

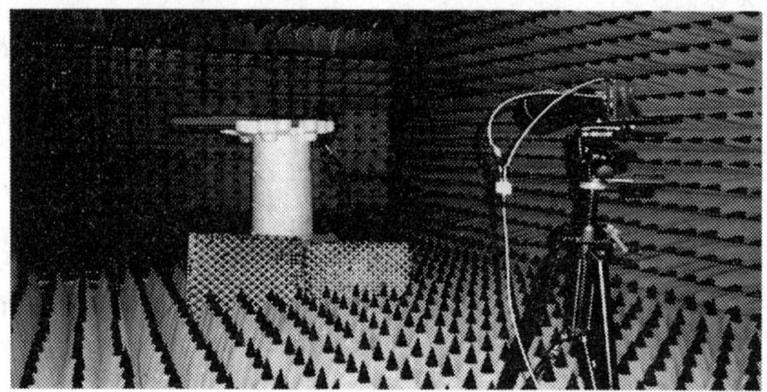

Figure 3.41

Drawbacks

(1) They are sometimes very large, to carry over to different places.

(2) Often antennas need to be of several wavelengths away from each other, to simulate far-field conditions. Therefore, it is desired to have anechoic chambers as large as possible but **cost** and **practical constraints** often limit their **size**.

NOTE: Some defense companies, which measure the Radar cross section of large air planes or other objects, are known to have anechoic chambers of the size of a basketball courts. Also, there are universities with chambers, of 3-5 m in length, width and height.

(3) Because of the size constraint, and **RF absorbing material**, there are best suitable only for UHF and higher frequencies i.e. above 300 MHz.

(b) Elevated/Reflection Ranges

Elevated ranges are the outdoor ranges. In this setup, the source and the antenna under test are mounted above the ground, that is on mountains, towers, building or wherever on fields.

This setup is for very large antennas or at low frequencies (VHF and below < 100 MHz) where indoor measurements are intractable.

The basic diagram of an elevated range is shown in the figure 3.42.

Hence, the source antenna is not necessarily at a higher elevation than the test antenna.

The **Line of Sight (LOS),** between the two antennas (illustrated by black ray in figure 3.41) must be unobstructed.

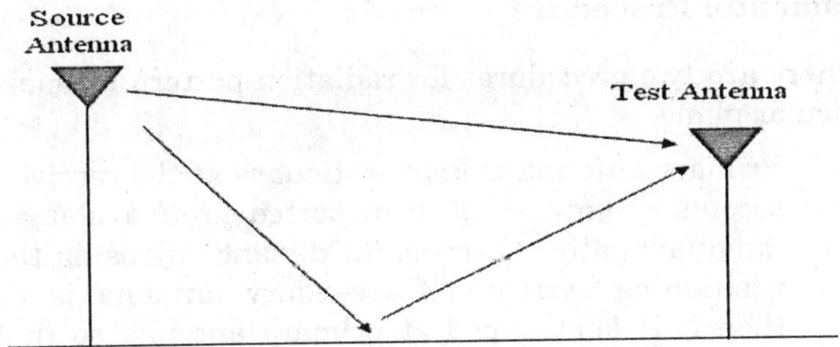

Figure 3.42 Illustration of elevated range

All the other reflections (such as red ray reflected from the ground) are undesirable).

For elevated ranges, once a source and test antenna locations are determined, the test operators, them determines where the significant reflections will occur, and attempt to minimize the reflections from these surfaces. Often the RF - absorbing material is used for this purpose, or other material that deflects the rays away from the test antenna.

3.14.3 RADIATION PATTERN MEASUREMENT

"Radiation pattern" – It can be well defined as the Field strength or the power density at a fixed distance from the antenna as a function of direction (3D - figure).

It needs measurement of field intensity all over the spatial (space) angles. Therefore, for Radiation pattern of antenna under test various space angles must be specified.

Radiation pattern (2D) –

i. X Y (Horizontal plane): It is used in - Broadcast Communication.

ii. X Z (vertical plane): It is used in – Radio astronomy

– Radar communication

– Space communication

The source antenna illuminates the antenna under test, with a plane wave, from a specific direction. The polarization and gain, for the fields radiated towards the test antenna, (i.e. of the source antenna) should be known.

Now, as we know from the *Reciprocity theorem*, that the radiation pattern from the test antenna is same for both the receiver and transmitter.

Fundamental Procedure

There are two procedures for radiation pattern measurements in a particular plane-

1. Primary antenna is kept stationary at the receive mode and secondary antenna is transported around along a circular (angular) path at a constant distance by using the antenna positioning system. (If secondary antenna is directional, then it is kept aimed at primary antenna so that primary antenna affects the results).

The field strength of secondary antenna with respect to primary antenna is recorded along the circle at different points. From these readings at number of points the plot of radiation pattern of prim. Antenna is made in Rectangular or Polar form.

Generally for the radiation pattern, spherical coordinate systems are used.

2. Both the antennas are kept in fixed position having a suitable spacing between them and secondary antenna's beam aimed at the primary antenna. Now primary antenna is rotated about a vertical axis and the secondary antenna is kept as the transmitter, so that field strength reading and the direction of primary antenna with respect to secondary antenna can noted down. The readings are taken at a number of points by stopping the rotation of primary antenna for recording the readings (or a continuous reading can also be taken if ``Pattern Recorder'' is available).

Arrangements for Radiation Pattern Measurements, is shown in the figure -

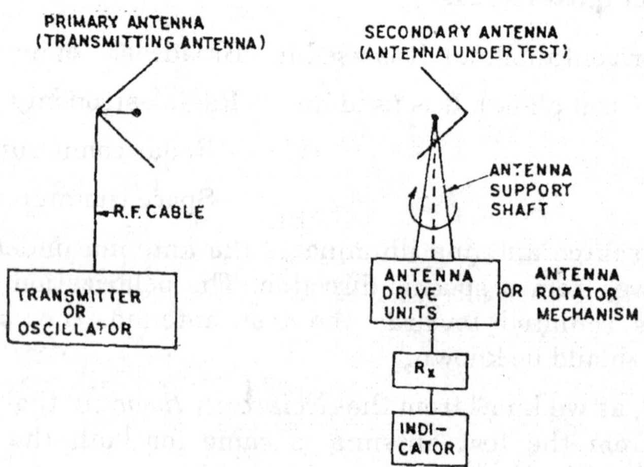

Figure 3.43

Here, the transmitting or the ideal antenna is the *Primary* antenna, Receiving or the antenna under Test is the *Secondary* antenna, and the *shaft* is used for rotating the secondary antenna. *Detector* is connected to the receiver and the *indicator* for indicating the relative magnitude of received field, these all makes the total system automatic.

Measurement example

To quote an example here, will make our vision more clear, as to how the radiations occur and can be measured for the given test antenna. Suppose the radiation pattern of a microstrip antenna is to be obtained.

Let the patch faces, towards the z-axis (i.e. normal to the surface of the path). The source antenna illuminates, the test antenna from +y - direction, as in the given figure-

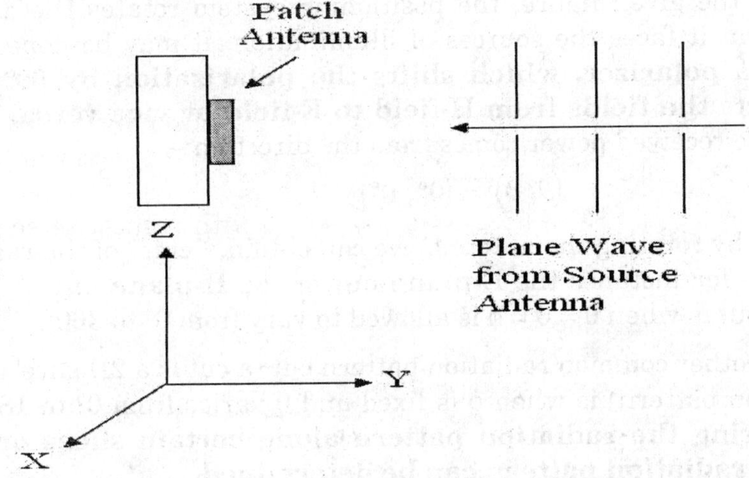

Figure 3.44 A patch antenna oriented towards the z-axis with a source illuminator from the +y direction

Hence, from the above figure, the received power for this case represents the power from the angle-

$$(\theta, \phi) = (90°, 90°).$$

We record this power. After this we, change the position of the test antenna only and record the reading again. Keeping, the power source once again in the same direction.

Suppose now, if we want to measure the radiation pattern normal to the patch's surface (straight above the patch). Than the measurement setup is changed as-

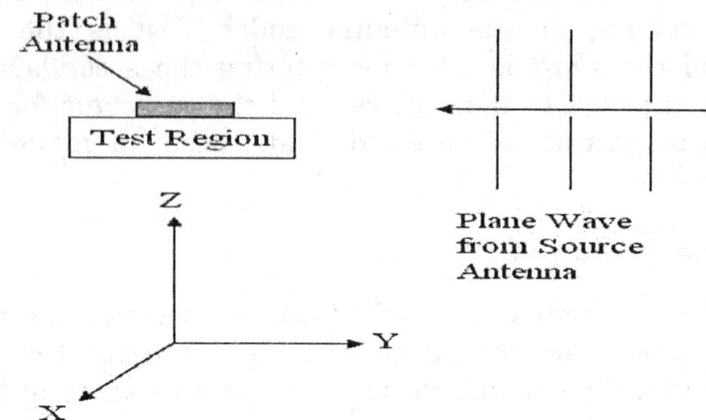

Figure 3.45 A patch antennas rotated to measure the radiation power at normal incidence

In the given figure, the positioning system rotates the antenna such that it faces the sources of illumination. It may be done by the use of a **polarizer, which shifts the polarization by 90° i.e. it converts the fields from H-field to E-field or vice versa.** In this case, the received power comes from the direction –

$$(\theta, \phi) = (0°, 0°).$$

So by rotating the antenna, we can obtain, ``cuts" of the radiation pattern- for instance the **E-plane cut** or the **H-plane cut**. A ``great circle" cut is when $\theta = 0 = \phi$ is allowed to vary from 0^0 to 360^0.

Another common radiation pattern cut (a cut is a 2D slice' of a 3D radiation pattern) is when ϕ is fixed and θ varies from 0^0 to 180^0. **By measuring the radiation pattern along certain slices or cuts, the 3D radiation pattern can be determined.**

NOTE: *The resulting radiation pattern will be correct only for a given polarization of the source antenna.*

For instance if the source is horizontally polarized and the test antenna is vertically polarized, the resulting radiation pattern will be zero everywhere. Hence the radiation pattern is sometimes classified as **H-pol** (Horizontal polarization) or **V-pol** (Vertical polarization).

NOTE: In addition, **the radiation pattern is a function of frequency.** As a result, the measured radiation pattern is only valid at the frequency; as that of the transmitting antenna or in other words to obtain a broadband signal from this transmitting antenna, the frequency of the receiver must be equalize to that of transmitter, to obtain the transmitted information signal.

Distance Requirement

1. In order to obtain accurate far field or Fraunhofer radiation pattern, the distance between primary and secondary antenna must be large.
2. *If the distance between the two antennas is very much small, then the near field or Fresnel pattern is obtained.*
3. For this it should be illuminated by a plane wavefront and the same wavefront is obtained only at infinite distance.

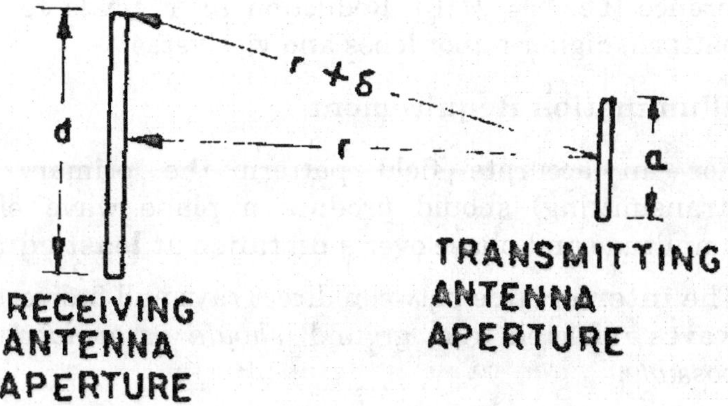

Figure 3.46

4. Phase difference between centre and at the edge of antenna under test should not exceed λ/16. Under this condition, the distance between Primary and secondary antenna should be-

$$r \geq \frac{2d^2}{\lambda}$$

Where, d = maximum linear dimension of any of the two antennas.

λ = wavelength

r = distance between transmitter and receiver.

The value of 'δ' may be calculated in terms of receiving aperture `d' and distance `r'.

$$(r+\delta)^2 = \left(\frac{d}{2}\right)^2 + r^2 \quad \because r \geq \delta,\ \delta \leq d. \text{ and } \delta - \textit{Phase difference error}$$

or $$r^2 + \delta^2 + 2r\delta = \frac{d^2}{4} + r^2$$

$\therefore \delta^2$ Can be neglected

$$2r\delta = \frac{d^2}{4}$$

$$\therefore \quad = r$$

$$= \frac{d^2}{8\,\delta}$$

This shows that **the minimum distance requirement depends on receiving antenna aperture'd' and wavelength 'λ'.** The value of r may be calculated by replacing ('δ' width) allowable phase difference (i.e. δ = λ/16). Reduction in 'r' tends to a broader radiation pattern, higher minor lobes and vice versa.

Uniform Illumination Requirement

1. For an accurate field pattern the primary antenna, (transmitting) should produce a plane wave of uniform amplitude and phase over a **distance at least equal to `r'.**
2. The **interference** between direct rays and indirect rays (i.e. waves reflected from ground) *should be avoided as far as possible.*
3. Besides **reflections** from surrounding objects like buildings, trees, etc *should also be avoided.*
4. Test should be conducted in **open** plain **area** and antenna should be *directional, installed on high*er towers or tops of high buildings.

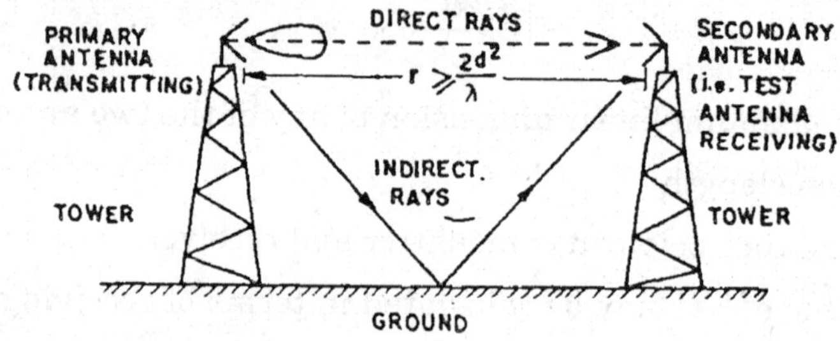

Figure 3.47

3.14.4 GAIN MEASUREMENT

In the previous section of radiation patter measurement, we learn how the radiation pattern of an unknown antenna can be plotted. For the pattern measurement, the peak value of gain for the given antenna was not initially known to us. We just measured the received power; which figure outs how directive an antenna is and the shape of the radiation pattern, which is actually a **"relative pattern".**

Here, in this section we focus our study on measuring the peak gain of an antenna; which tells us how much power could be received from a given plane wave.

Gain (G) can be defined as the relative ratio =

$$\frac{\text{Maximum Radiation Intensity (Test)}}{\text{Maximum Radiation Intensity (Reference) for same input power of both antenna}}$$

Directivity (D) is given by the ratio of the values from the same antenna (absolute value)

$$= \frac{\textit{Maximum Radiation Intensity}}{\textit{Average Radiation Intensity}}$$

Gain of antenna over isotropic lossless antenna is given by-

$$G_o = \alpha D$$

If matching is proper than $(\alpha = k)$

Where, G_o = *Gain w.r.t. isotropic antenna*

α = *Effectiveness Ratio* $(0 \leq \alpha \leq 1)$

D = Directivity and D > G (always)

k = Radiation efficiency

Measurement Techniques

I. Generalized

Gain can be measured by using the ``Friis transmission equation" along with a ``standard gain antenna".

NOTE: A **standard gain antenna** is a test antenna with an accurately known gain and polarization (generally linear). The most popular types of gain standard antennas are-

(1) Thin half wave **dipole** Antenna. (With peak gain nearly 2.15 dB).

(2) Pyramidal **Horn** Antenna (Most commonly used) (with peak gain range of 15-25 dB)

The separation between the source and test antenna is fixed, with constant frequency. Consider the test setup as given in the figure above, where a standard gain antenna is used in place of the test antenna, with the source antenna transmitting, a fixed amount of power (P_T). The gains of both of these antennas are accurately known.

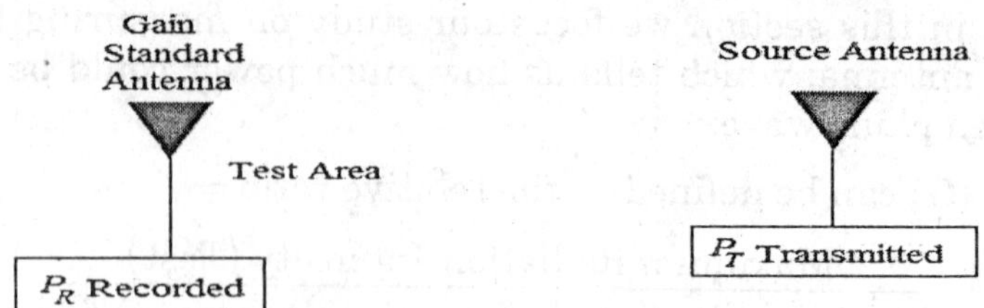

Figure 3.48 Record the received power from a gain standard antenna.

Now, from the Friis transmission equation, we know that the power received (P_R) is given by-

$$P_R = \frac{P_T G_T G_R \lambda^2}{(4 \pi R)^2}$$

If, now we replace the gain standard antenna with our test antenna, as in the figure below; the 'G_R' changes in the above equation (that is the gain of the receiving antenna).

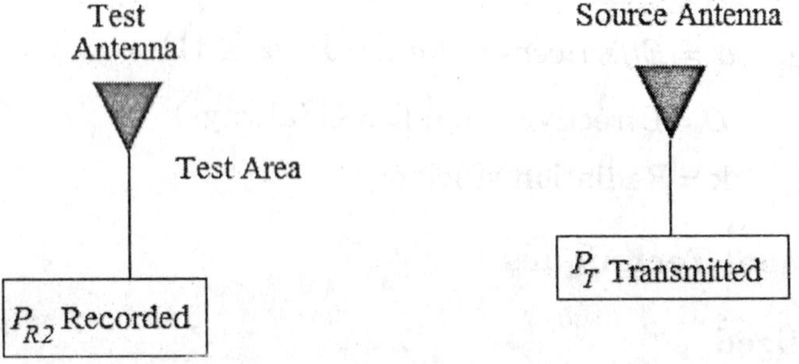

Figure 3.49 Record the received power with the test antenna (same source antenna)

Let, the received power from the test antenna is P_{R2}. Now **if the gain of the test antenna is higher than the gain of the "Standard gain antenna" then the received power will increase.**

Using this measurement technique, we can easily calculate the gain of the test antenna. 'G_T' by the formula-

$$G_T = G_G \frac{P_{R2}}{P_R}$$

Where G_G = gain of standard gain antenna.

P_R = Received power from the gain antenna under test.

P_{R2} = Power received with the test antenna.

The above equation uses **linear units (non-dB)**. If the gain is to be specified in decibels, (power received still in Watts), then the equation becomes-

$$[G_T]_{dB} = [G_G]_{dB} + 10 \log_{10} \left(\frac{P_{R2}}{P_R}\right)$$

II. Measurement of Gain by Direct Comparison Method

It is more suitably used for high frequency antenna gain measurement. **In this method, measurement of gain is done by comparing the signal strengths transmitted or received with unknown gain antenna and standard gain antenna.**

In place of primary antenna there are two antennas one the subject antenna under test and other standard antenna at constant distance so that coupling or interaction between two antennas could be avoided. Distance between two antennas should be $\geq 2d^2/\lambda$, so that the reflection between them should be minimized to the extent possible.

In the given figure-

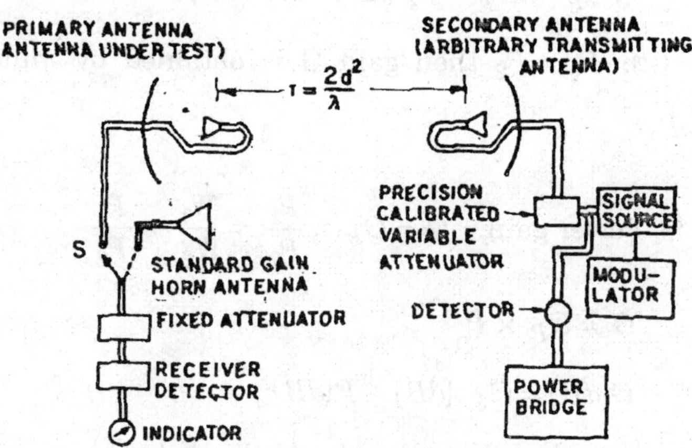

Figure 3.50

1. **Attenuator Pad-** It is used in order that the standard and subject antenna works at matched load condition with receiver at constant load. **Secondary Antenna and associated transmitter should be stable in respect of frequency** (that is, frequency should be constant throughout the experiment)

2. **Power Bridge –** It is used to keep frequency constant or to check the stability of the transmitter.

Procedure

(1) Standard antenna receiver is selected with the help of switch 'S' and the antenna is aimed at the secondary antenna in the direction of maximum signal intensity. The input of the transmitter antenna is then adjusted to a convenient level and corresponding reading at receiver is recorded. The readings of Attenuator and Power Bridge are also recorded (say W_1 and P_1 respectively).

(2) Now connect the subject antenna whose gain is to be measured in place of standard gain antenna. The attenuator dial is adjusted such that receiver indicates same previous reading as was with horn antenna. Let the readings in attenuator and power bridge now be, (say W_2 and P_2)-

CASE (1): If P1 = P_2 (no correction needed) and Power gain, $G_P = \dfrac{W_2}{W_1}$

and $$\log G_P = \log W_2 - \log W_1$$

or $$G_P\ (dB) = W_2\ (dB) - W_1\ (dB)$$

Case (2): $P_1 \neq P_2$ then gain G is obtained by multiplying G_P with $\dfrac{P_1}{P_2} = P$

$$\text{That is, Power gain} = G = G_P \times \frac{P_1}{P_2} = \frac{W_1}{W_2} \times \frac{P_1}{P_2}$$

or $$G = G_P \times P$$

or $$G(dB) = G_P\ (dB) + P(dB)$$

or $$G(dB) = W_2\ (dB) - W_1\ (dB) + P\ (dB)$$

By this method directivities may also be known if antenna Radiation efficiency is known. This is similar to the general method.

III. Use and Calibration of gain ``Standard"

There are two methods for the same-

(1) Two identical antennas are used.

(2) Three arbitrary antennas are used.

III. (1) The identical antennas having distance `r'

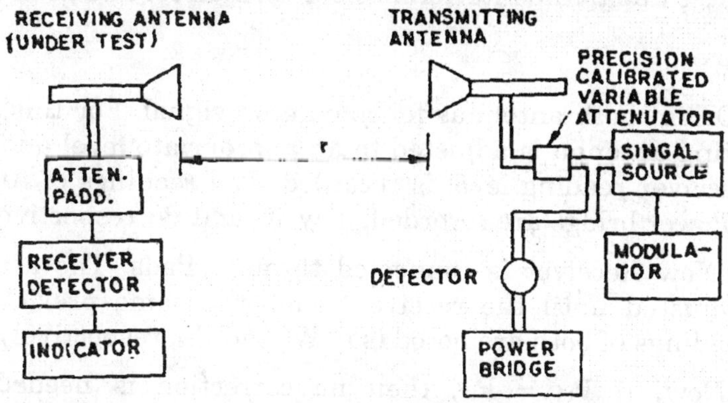

Figure 3.51

Where, W_t = Transmitted Power

W_r = Received Power

A_{et}, A_{er} = Effective Aperture of transmitting and receiving antennas.

As antennas are identical-

$$\therefore A_{et} = A_{er} = \frac{G_o \lambda^2}{4\pi}$$

$$\frac{W_r}{W_t} = \frac{A_{er} A_{et}}{\lambda^2 . r^2} = \left(\frac{G_o \lambda^2}{4 \pi}\right)\left(\frac{G_o \lambda^2}{4\pi}\right)\frac{1}{\lambda^2 r^2}$$

Where, r = distance between antennas

λ = wavelength.

$$\frac{W_r}{W_t} = \left(\frac{G_o \lambda}{4 \pi r}\right)^2$$

or
$$\sqrt{\frac{W_r}{W_t}} = \left(\frac{G_o \lambda}{4 \pi r}\right)$$

or
$$G_o = \frac{4 \pi r}{\lambda}\sqrt{\frac{W_r}{W_t}}$$

Now, due to effect of direct and indirect rays-

$$G_o = \frac{4\pi r}{\lambda F}\sqrt{\frac{W_r}{W_t}}$$

Where, F = Propagation constant,

= arise due to the interference between direct and indirect rays.

Procedure

(1) Orient both antennas for maximum signal. For this, input to transmitting antenna is adjusted to an appropriate level and corresponding receiver reading level is recorded. The readings of Attenuator dial and Power bridge are recorded, (say W_t and P_{t1}, respectively).

(2) Now Receiver is connected through Pads. The attenuation dial is adjusted until the receiver reads the same previous levels. Again Readings of both are noted (say W_r and P_{t2}, respectively)

(3) Now, if $P_{t1} = P_{t2}$, then no correction is needed due to transmitter level fluctuations and gain is calculated.

III. (2) this method is needed only when two identical antennas are not available

Efficiency

Efficiency of an antenna can be calculated by simply using the ratio of the peak gain to the peak directivity; that is-

$$\varepsilon = \frac{G}{D}$$

Where, the directivity is independent of the antenna gain and can be calculated from the measured radiation pattern.

Hence, once we have measured the radiation pattern and gain, efficiency can be calculated directly from the above equation.

1. Write short note on Antenna Gain measurement. **(Raj. Univ. 2004)**
2. Define an Antenna Gain Measurement technique if three different antennas of unknown gain are provided. **(Raj. Univ. 2006)**
3. Define gain and measure it. **(R.T.U. 2009)**

3.14.5 PHASE MEASUREMENT

For an antenna's radiation pattern to be completely specified we need both the magnitude of the power received or transmitted, along with the phase of the antenna. These measurements should be specified in two orthogonal directions in order to capture all the components (polarization) of the antenna.

For instance, suppose an antenna transmits at the frequency `f' with the fields travelling in the + y - direction. At any point, this field can be given by-

$$E = \hat{x} Ae^{jD} \; e^{j2\pi ft} + \hat{z} Be^{jF} \; e^{j2\pi ft}$$

Where A = the magnitude of x -component of the E-field

B = the magnitude of z -component of the E-field.

D = Phase of the x-component

F = Phase of the z-component (relative to the oscillation at frequency `f')

∴ The E-fields are orthogonal to the direction of travel in the far field region.

(1) If **D = F**, the components are in phase and the **polarization is linear.**

(2) If **D and F are separated by $\pi/2$ radians** (90° degree) and the amplitudes are equal, the E-field **is circularly polarized.**

Phase of an antenna is periodic and is defined in multiples of 360°. It is a relative quantity and thus, the reference must be provided during the measurement for the purpose of comparison. There are two types of measurements-

(i) Short distance (Near- field)

(ii) Long distance (Far- field)

(i) Near field distance

As we know that, the **phase is a relative quantity**; which means, it must be measured relative to some fixed reference. ∴ the easiest way to measure phase is the method with the setup shown in the figure.

In this method the test antenna is used as the source antenna, and another antenna is used to receive the fields. For this method to work, *the observation point is not too far from the test antenna, so that the source waveform feeding the test antenna can also be run into a **phase measurement box.*** This box compares the locations of the peaks and valleys of the received signals and determines the relative phase from this information. The receiver antenna is moved and then the process is repeated.

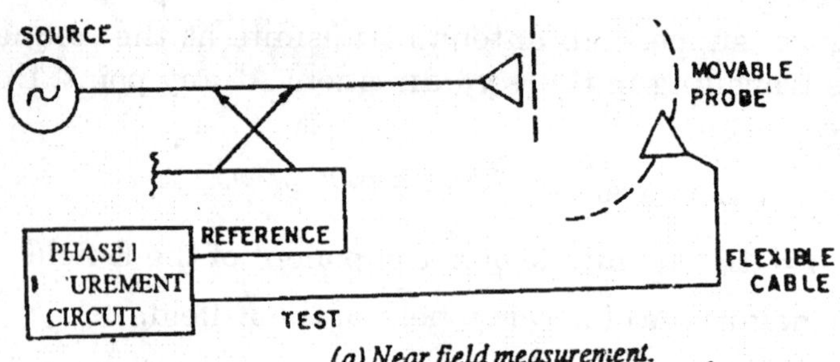

(a) Near field measurement.

Figure 3.52

The **probe antenna** should have ***good polarization,*** *so that it can pick up one component of the received field.* To get the other orthogonal component, it could simply be rotated or another probe antenna used.

(ii) Far field distance

Now, If the test **antennas are very far** from each other and the reference (source) waveform cannot be fed directly into the phase measurement circuit (this happens at low frequencies and large outdoor ranges, where many wavelengths becomes large with the distance), then a standard antenna with known phase characteristics is used to transmit a wave, which is used to compare with the received signal from the test antenna; as in the given figure-

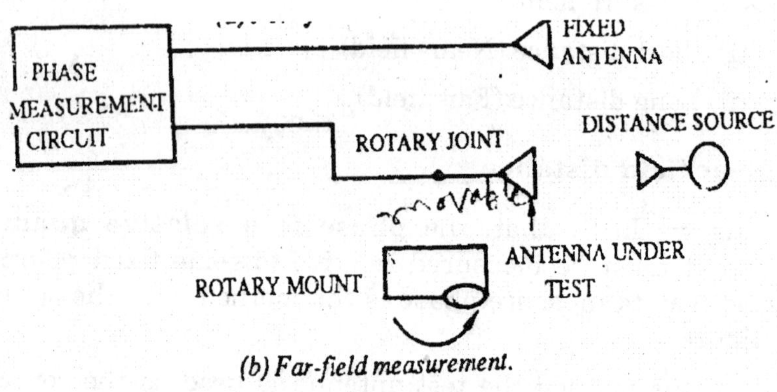

(b) Far-field measurement.

Figure 3.53

1. Write short note on phase measurement. **(Raj. Univ. 2003)**

3.14.6 POLARIZATION MEASUREMENTS

Polarization of electromagnetic field may be measured at one frequency as a function of space angles, that is polar angle 'θ' and azimuth angle 'φ' or at one angular position as function of frequency.

Measurement of polarization are desired where the dominant radiation is circularly or elliptically polarized.

Procedure: Polarization is measured by measuring the received signal voltage with linearly polarized receiving antenna (secondary antenna), since its polarization is rotated in the direction through 2 π or 360°.

If two maxima and two minima (or nulls) are obtained the field is **linearly polarized** in the **direction of maxima.**

The maxima and minima both will be 180° apart from each other or nulls will be 90° apart from the maxima. The directions of the nulls are measured accurately than the maxima.

If instead of nulls (zeros), minima or maxima are obtained the field is said to be **elliptically polarized-**

$$\text{Ratio of } \frac{\text{Maxima field intensity}}{\text{Minima field intensity}} = \text{Polarization Ratio}$$

When polarization ratio is measured in the axis of main beam of subject antenna, then it is called as **AR (Axis Ratio).**

If the field is constant as polarization of secondary antenna is rotated, then the field is called **circularly polarized** (special case of elliptical polarization when **AR = 1.**

Linearly polarized antennas such as half wave dipole or dipole array or dipole with reflectors may be used for polarization measurement rather any linearly polarized antenna will solve the purpose. Polarization measurements are also accomplished by using two fixed linearly polarized receiving antennas and measuring the ratio of the intensities received by each.

There are three methods used for measuring the polarization characteristics of EMW.

(1) Polarization Pattern Method: In which the polarization pattern with a linear antenna and the direction of rotation of E, are found.

(2) Linearly component method: In which the amplitudes (E_1 and E_2) of two perpendicular linearly polarized components of the wave and the relative phase angle 'δ' between them are measured.

(3) Circular component Method: In which the amplitudes (E_3 and E_4) of two circularly polarized components (of opposite rotation direction) of the wave and the relative phase angle 'δ' between them are measured.

The (1) and (2) methods are very much similar and are intended for general use. However, out of these three methods for measurement of elliptical polarization, the (3) method is a difficult method for accurate performance as it needs two identical circularly polarized antennas which is difficult especially for a wide band frequency.

Fundamental to an antenna's radiation pattern is its polarization. Hence, here in this section, the methods and techniques for determining the polarization of an antenna are discussed. (**NOTE:** polarization varies depending on the direction of radiation from an antenna. For instance, a circularly polarized antenna may be approximately circular only over a narrow beamwidth and linearly polarized away from the antenna's main beam (this is often the case for circularly polarized **patch antennas**).

To perform the measurement, the test antenna is used as the source. Next a linearly polarized antenna (typically a **half-wave dipole antenna**) is used as the receiver antenna. The linearly polarized receive antenna will be rotated, and the received power is recorded as a function of the angle of the receiver antenna.

In this manner, we can gain information on the polarization of the test antenna. This received information only applies to the polarization of the test antenna for the direction in which the power is received. For a complete description of the polarization of the test antenna, the test antenna must be rotated so that the polarization can be determined for each direction of interest.

The basic setup for polarization measurements is shown in Figure-

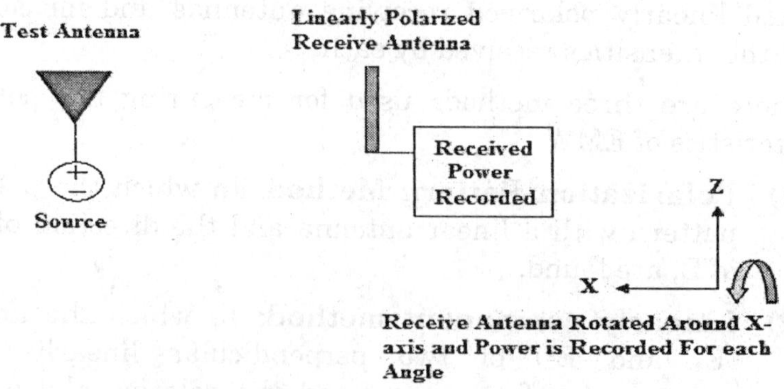

Figure 3.54 Setup for polarization measurements

The power is recorded for a fixed position (orientation) of the receive antenna, then it is rotated about the x -axis as shown in the same Figure, and the power is recorded again. This is done for a complete rotation of the linearly polarized receiver antenna.

CASE (A)

Again suppose that the test antenna is vertically linearly polarized, and that the receiver antenna is also vertically linearly polarized, and that the rotation angle zero has both antennas polarization matched. Then the output of our experiment, as a function of the rotation angle of the receiver antenna, would look something like the graph shown in Figure 3.55.

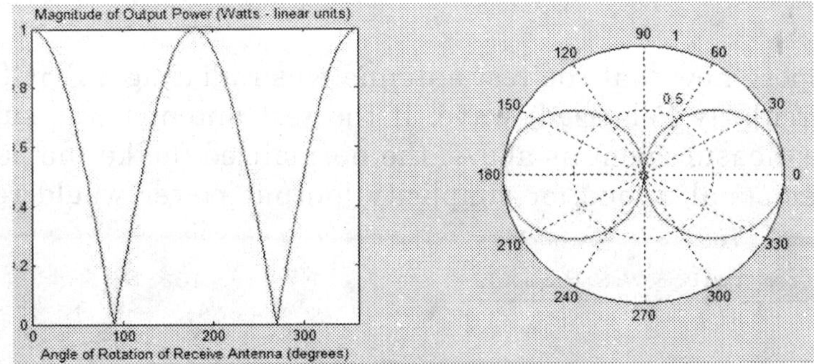

Figure 3.55 Output of measurement for linearly polarized test antenna. (Left), Rectangular Plot (Right) Polar Plot

The plots here give two views of the output. The left side gives an x-y plot of the output; while the right side gives a polar plot, which is helpful in visualizing the results.

NOTE: the result is periodic, when the receiver antenna is rotated 180^0, it is again vertically polarized so that the received power is identical.

CASE (B)

Suppose now the test antenna is horizontally polarized (again linearly polarized) but initially polarization did not matched to the received antenna. Then the resulting received power plots would resemble to the Figure 3.56.

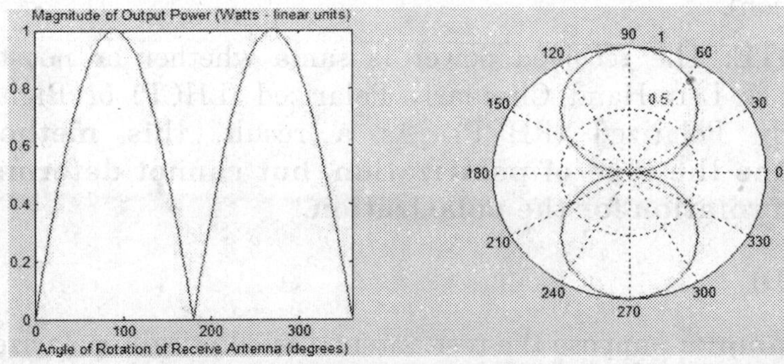

Figure 3.56 Output measurement of linearly polarized test antenna (Horizontal polarized)

In this case, it can be noticed the shape of the resulting measurement is same, but the peaks of the received power occur for different angles. As a result, when the test antenna is linearly polarized, the received power will resemble the shapes shown in the above Figure, and by determining the angle in which the received power is at a peak, we can determine the angle of the linear polarization.

CASE (C)

Suppose now that the test antenna was radiating a RHCP (Right Hand Circularly Polarized) wave. If the test antenna was subject to the same measurement as above, the normalized (make the peak output power equal to one for simplicity) output power would resemble that

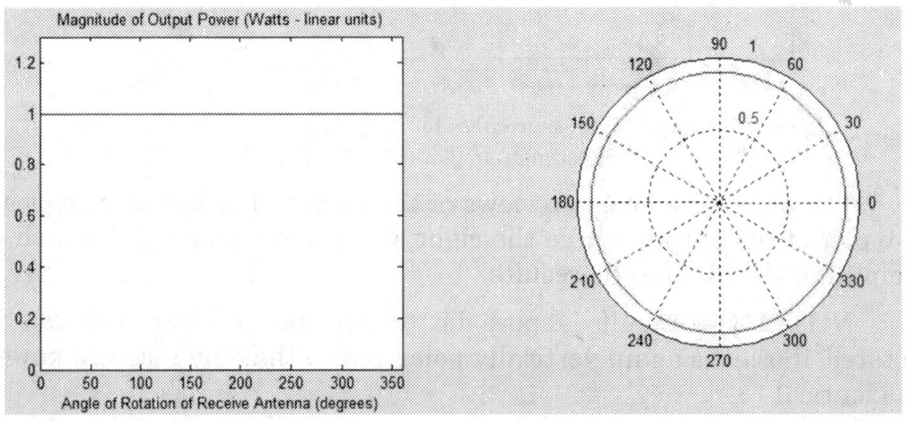

Figure 3.57 Output of Measurement when the Test Antenna is Circularly Polarized

Because a circularly polarized wave has equal amplitude components in two orthogonal directions, the received power is constant for a rotated linearly polarized antenna (feature of circular polarization is important where getting the orientation right is not a compulsion).

NOTE: The received power is same whether or not the test antenna is Left Hand Circularly Polarized (LHCP) or Right Hand Circularly Polarized (RHCP). As a result, **this method can determine the type of polarization, but cannot determine the sense of rotation for the polarization.**

CASE (D)

Example: Suppose the test antenna is elliptically polarized, with a tilt angle of 45^0 and an **axial ratio of 3 dB.** The E-field for this antenna is given by-

$$E = \hat{y}\left(1 - \frac{j}{2}\right) + \hat{z}\left(1 + \frac{j}{2}\right)$$

By using coordinate systems the resulting output is-

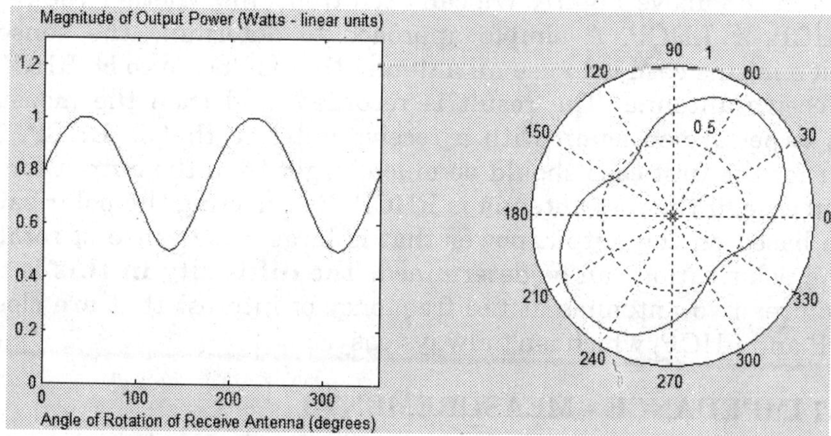

Figure 3.58 Output of Measurement when the Test Antenna is elliptically Polarized (axial ratio = 3 dB, major axis along 45 degrees)

Now, from the figure we conclude that, the tilt angle of the elliptically polarized wave by where the received power is at a peak - at 45^0. Again, the axial ratio can also be determined from the above plots. [The Ratio of the peak output (case 1.0 when the angle is 45^0) to the minimum power output (in this case 0.5, when the angle is 135^0) given the axial ratio (1/0.5 = 2 = 3 dB)].

NOTE: We still don't know the direction of rotation of the E-field (left or right).

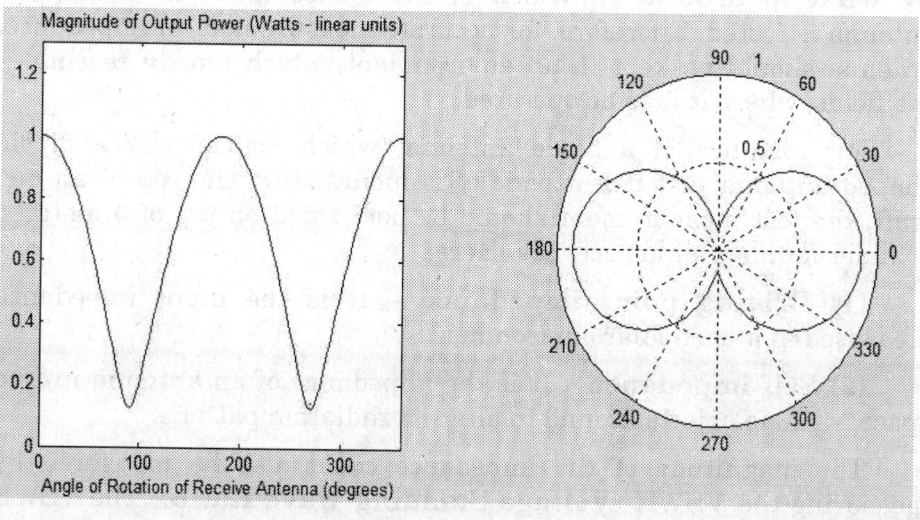

Figure 3.59 Output of Measurement when the Test Antenna is elliptically Polarized (axial ratio = 9 dB, major axis along 0 degrees)

CASE (E)

Finally, to determine the sense of polarization for the antenna under test, we suppose that the test antenna is RHCP, with output shown in the above Figure. We can't tell from this whether the result is RHCP or LHCP. A simple approach to determine the sense of rotation in this case is to use an antenna that is known to be RHCP as the receive antenna. The result is recorded, and then the measurement is performed again with a receive antenna that is LHCP. The result for the first case should be much larger than the output for the second case, if the test antenna is RHCP. By selecting the polarization sense based on the output power that is larger, the sense of rotation for the polarization can be determined. The **difficulty in this** is that is requires two antennas at the frequency of interest that are closely RHCP and LHCP, which isn't always easy.

3.14.7 IMPEDANCE – MEASUREMENTS

In this section, we will be discussing as to how to measure the impedance of an antenna. Impedance is a fundamental parameter of an antenna, which operates at RF frequencies. *If the impedance of an antenna is not "close" to that of the transmission line, then very little power will be transmitted by the antenna (if the antenna is used in the transmit mode), or very little power will be received by the antenna (if used in the received mode). Hence,* **without proper impedance, or matching network, antenna does not work efficiently.**

As we know that, the object placed near an antenna alters its, radiation pattern. As a result, its **input impedance is influenced by what is around** it, which is the environment in which the antenna is tested. Therefore, for accurate measurement of impedance, antenna should be kept in an environment, which mostly resembles the fields, where it is to be operated.

For instance, if a blade antenna (which is basically a dipole shaped antenna just like a paddle) is mounted on the top of an aircraft, the test measurement should be performed on top of a metallic cylinder for maximum accuracy. Here-

(1) Driving point impedance – It is the input impedance measured in a particular environment.

(2) Self impedance – It is the impedance of an antenna in free space, with no objects around to alter its radiation pattern.

The magnitude of the impedance could also be measured by measuring the **VSWR (Voltage Standing Wave Ratio)**. The VSWR is a function of the magnitude of the *reflection coefficient*, so as no phase information is obtained about the impedance (that is relative value of reactance divided by the resistance). VSWR gives a quick way of estimating about the reflected power by an antenna.

Generally, all the ways to measure impedance are functions of the reflected power from the antenna. **Impedance of an antenna is so important, because due to it, we get to know how much power is transferred to or from the antenna.**

One method we can use in this case is that of **Scale Model Measurements**. In this technique, a ***scaled model*** *(typically a much smaller physical model of the actual structure)* is used to represent the platform on which the antenna operates. With this technique, we get proper measurements from a smaller model.

The frequency is one of the fundamental properties of the signal, which can be represented as a combination of frequencies.

Also, all of electromagnetic and antenna theory can be completely described by Maxwell's Equations.

Examples

1. Hence, to understand this scale mode let us know the properties of a monopole antenna at f = 300 MHz (wavelength is 1 meter). Suppose this antenna is mounted on an airplane that is 30 meters long, so that the airplane is 30 wavelengths long. Now suppose we build a scale model of our airplane that we can fit in our **anechoic chamber**, and this airplane model is 3 meters long. If we want to get the electromagnetic waves to behave the same way as they do on a real airplane at 300 MHz, we need to have the scale model be 30 wavelengths long.
2. Hence, if we operate at f = 3000 MHz (3 GHz, where the wavelength is 0.1 meters), the model is now 30 wavelengths long.
3. If we scale the monopole antenna by the same factor (10), then our measurements at 3 GHz on the scaled model theoretically will be identical to measurements performed on the actual airplane at 300 MHz.

When a single frequency E-field wave, is obstructed by any obstacle, its behavior here depends only on the size of the obstacle (in wavelengths) i.e. plane waves here behaves similar to the perfectly conducting circular plate.

The quality of the results depends on the quality of the model. This method is a valid and often practiced method of antenna measurements in the-

- Aerospace
- Defense world

Figure 3.60

The following table is provided for understanding proper scaling. In the table, the model is assumed to be scaled down by a factor of 'n'. For instance, if n = 100 (in previous example n=10), then an L = 100 meter real-world object would be represented by a scaled model that is **L/n = 1 meter** in size.

Table 3.1. Relationship among Quantities in Scaled Measurements

QUANTITY	SCALED RELATIONSHIP
Length (L)	L/n
Frequency (f)	f*n for $T_D < T_R$
Permittivity (ε)	ε
Permeability (μ)	μ
Conductivity (σ)	σ*n
Impedance (Z)	Z
Gain (G)	G
Radar Cross Section (A)	A*n^2
Capacitance (C)	C/n
Inductance (L)	L/n

NOTE

1. The above table is a result work through mathematics involved in Maxwell's equations.
2. Some properties like – permittivity, permeability, Antenna impedance Gain, do not need scaling.
3. Permeability and permittivity are used where modeling is done for dielectric or magnetic materials.

As, the real world aircraft has a shell that has high conductivity (metals). Therefore, care is taken that the conductivity of the model is kept as high as possible; which is achieved by polishing high quality metals on the scaled model.

There are two types of impedance measurement-

1. IMPEDANCE BRIDGE METHOD (LF for below 30MHz)

2. SLOTTED LINE (HF for above 1000 MHz)

Between (30 MHz – 1000 MHz) either of the two methods can be used.

Impedance $\quad Z = \frac{V}{I}$ (generalized)

or $\quad Z = R + jX$

Where, R = Resistive

X = Reactive (complex)

$\therefore \theta$ = phase difference between 'I' and 'V'

$$\theta = \tan^{-1}\left(\frac{X}{R}\right)$$

Where, **'V' and 'I' are in phase when X = 0**

At low Radio frequencies 'V' and 'I' can be measured by voltmeters and Ammeters respectively. Phase angle is measured by CRO by using Lissaju's figure, (which is a common method at low frequency).

(I) Impedance Bridge method (for low frequency):

For this method we use a Wheat stone Bridge. It is used to measure unknown impedance (R, L or C) by comparing the value by known impedance.

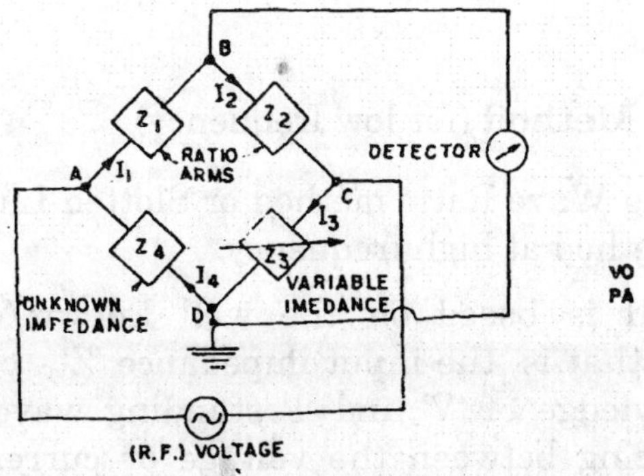

(a) Wheat stone bridge for impedance measurement.

VO
PA

Figure 3.61

Where, Z_1 Z2 Z_3 and Z_4 = Four Arms

Z_1Z_2 = Ratio Arms (Balanced)

Z_3 = Variable arm Impedance which is varied to get null in the detector.

Z_4 = unknown impedance (under measurement).

When the Bridge is balanced by Z_3, no potential difference exists between point B and D and meter in the detector circuit gives null. Under this condition-

$$\frac{Z_1}{Z_2} = \frac{Z_4}{Z_3}$$

Bridge is balanced not for impedance magnitude but also for phase balance-

$$\frac{Z_1 \angle\theta_1}{Z_2 \angle\theta_2} = \frac{Z_4 \angle\theta_4}{Z_3 \angle\theta_3}$$

$$\therefore Z_1\ Z_3\ (\angle\theta_1 + \angle\theta_3) = Z_2\ Z_4\ (\angle\theta_2 + \angle\theta_4)$$

There are two balance conditions-

$$\therefore Z_1\ Z_3 = Z_2\ Z_4$$

$$\therefore (\angle\theta_1 + \angle\theta_3) = (\angle\theta_2 + \angle\theta_4)$$

For magnitude and phase angle balance

$$\therefore \textit{Unknown impedance } (Z_4) = \frac{Z_1\ Z_3}{Z_2}$$

As **D is grounded** ∴ it is suitable for low frequency grounded vertical antenna. Bridge is balanced with unknown impedance terminal short circuited (or open circuited). Actual indicating device is d. c. meter.

(II) Slotted line Method (for low frequency)

The Standing Wave Ratio method or Slotted Line Method is used for Impedance method at high frequency.

This method is based on the well known characteristics of traveling waves, that is, the input impedance 'Z' - can be determined by the and knowledge of 'V' and 'I' standing wave Ratio (ISWR or VSWR), the spacing between the voltage or current minimum and reference point at which the impedance is desired.

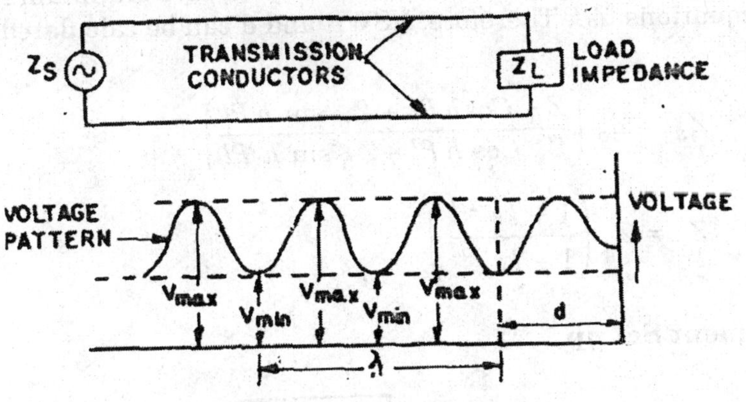

nt. *(b). Standing wave along the line.*

Figure 3.62

∴ **Slotted line method is also called as standing wave ratio method.**

It can determine two things

1. Impedance (Z)
2. Spacing (d)

$$Z_L = Z_o \left(\frac{1+k}{1-k}\right)$$

$$VSWR = \left|\frac{V_{max}}{V_{min}}\right| = \frac{|V_i| + |V_r|}{|V_i| - |V_r|} = \frac{1 + \left|\frac{V_r}{V_i}\right|}{1 - \left|\frac{V_r}{V_i}\right|}$$

As k = Reflection coefficient $= \frac{V_r}{V_i}$

$$\therefore VSWR = \frac{1+|k|}{1-|k|}$$

or $$k = \frac{VSWR - 1}{VSWR + 1} = \frac{S-1}{S+1}$$

$$\theta = \pi - 2\beta d = \pi\left(1 - \frac{4d}{\lambda}\right) \text{ radian as } \beta = \frac{2\pi}{\lambda}$$

Where,'d' - spacing between load end and first minimum from the above equations 'Z_L'. Therefore, VSWR and d can be calculated, (if k is known)-

$$Z_s = Z_o \left(\frac{Z_R \; Cos\,h\,Pl + Z_o \sin h\,Pl}{Z_o \; Cos\; h\,Pl + Z_R \sin\, h\; Pl} \right)$$

$$Z_s = Z_o \left(\frac{1 + ke^{-2pl}}{1 - ke^{-2pl}} \right)$$

Experiment Set up

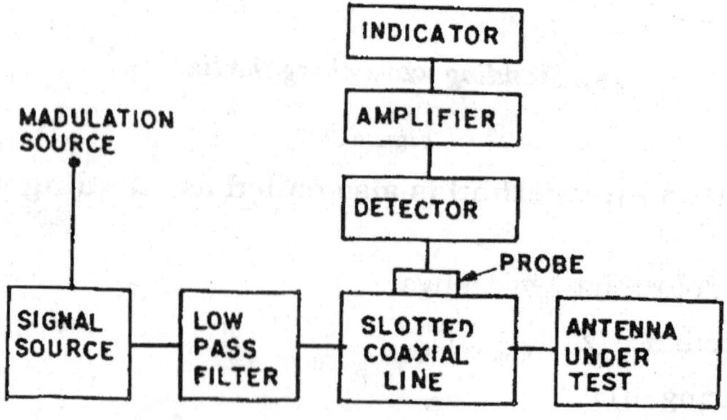

Figure 3.63

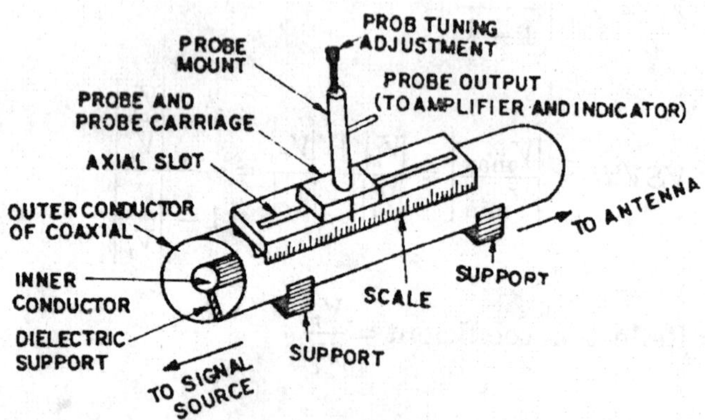

Figure 3.64

A transmission line system terminated in an antenna if not perfectly matched to the feeding transmission line, the incident and reflected waves and consequently produce standing wave will be set up along transmission line.

A part of transmission line is replaced by axial slotted line, VSWR or impedance measuring set up.

Here the transmission line consists of a coaxial cable which along with moves the travelling carriage carrying a probe moves. The detector is combination of Crystal and micro ammeter which is used as measuring device ('V' or 'P').

A signal source (Transmitter or Oscillator) is connected to left end and right end is connected to **unknown Impedance** (antenna impedance). **The standing wave pattern is obtained by moving the probe along the carriage and observing the resulting variation in crystal detector output.**

In fact device measure electric field intensity but since it is proportional to 'V' between conductors. ∴ Standing wave indicator is assumed to be 'V' measuring device. The probe in the slotted coaxial cable is moved and the two consecutive points V_{max} and V_{min} are noted. Their ratio is VSWR.

The probe is inserted deeply into axial slot line in order to sample the standing wave pattern. In detectors (standing wave), **a HPF or LPF is used to avoid harmonics** spurious signal sources and unwanted signals between slotted and signal source.

Modulation source is for modulating signal source with a square or pulse wave.

1. Describe the impedance bridge method and slotted line technique for impedance measurement. **(Raj. Univ. 2008, 2006)**
2. Explain measurement of antenna impedance by bridge and slotted line method. **(Raj. Univ. 2007)**
3. Which method is used for impedance measurement? Explain this. **(R.T.U. 2009)**

3.14.8 DIRECTIVITY MEASUREMENT

It can be defined as the ratio of

$$= \frac{\textit{Maximum Radiation Intensity}}{\textit{Average Radiation Intensity}}$$

$$D = \frac{4\,\pi \times \textit{Maximum Radiation Intensity}}{\int\limits_{\phi=0}^{2\pi} \int\limits_{\theta=0}^{\pi} \textit{Radiation Intensity} \times \sin\theta \; d\,\phi \; d\,\theta}$$

Directivity can be obtained if we know the radiation pattern of antenna.

The particular quantity of the radiation pattern to be used is the **radiation intensity** which is defined as the power radiated from the antenna per unit solid angle in the given direction.

Total power Radiated = $4\pi \times$ *Average Radiation Intensity*

Where, θ = Polar angle of spherical co-ordinate system

ϕ = Azimuth angle of spherical co-ordinate system of Radiation Pattern

There are two most commonly used methods :

(1) Orange Slice Method

For discrete value of ϕ, a set of pattern is obtained by measuring radiation intensity versus θ. Each pattern is multiplied continuously by $\sin\theta$, the weighting factor and then integrated. The integrated values for several patterns are then added according to the above equation.

(2) Conical cut Method

For a number of discrete values of θ, a set of pattern is obtained by measuring radiation intensity versus ϕ. Each pattern is integrated and each of these values are multiplied by the appropriate $\sin\theta$ the weighting factor and then added.

Radiation patterns which are measured for determining directivity include the effect of cross polarization. Radiation intensity involves both power radiated in normal and cross polarization (Cross Polarization is large in direction away from peak of the pattern). ∴ **Orthogonal polarization is used.**

1. Define the following terms regarding Antenna:

(i) Antenna measurements

Q.1. In the radiation pattern of a 3-element array of isotropic radiators equally spaced at a distance of $\lambda/4$, it is required to place a null at an angle of 33.56° off the endfire direction. Calculate the progressive phase shifts to be applied to the elements. Also calculate the angle at which main beam is placed for this phase distribution. **(Raj. Univ. 2005)**

Ans: The geometry of array is given below –

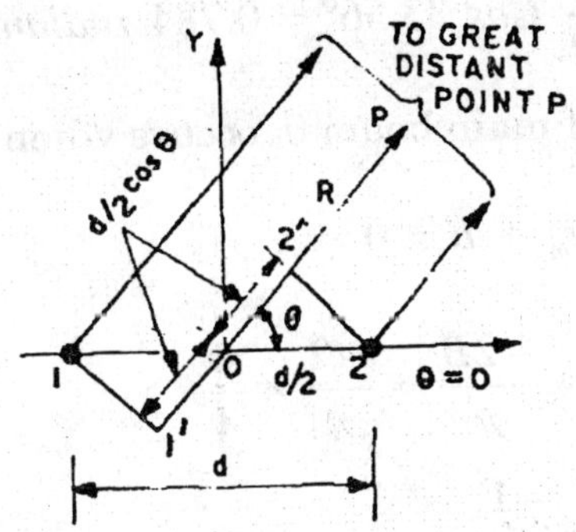

Figure 3.65

Given: 3 element array, d=λ/4 and θ=33.56°

The phase difference is given by-

$$\psi = \beta d \; Cos \; \theta = \frac{2\pi}{\lambda} \times \frac{\lambda}{4} \; Cos \; \theta$$

$$= \frac{\pi}{2} \; Cos \; \theta$$

$$\text{(i)} \quad \therefore E_t = E_o \left[e^{-j\left(\frac{\pi}{2} Cos \, \theta + \alpha\right)} + 1 + e^{j\left(\frac{\pi}{2} Cos \, \theta + \alpha\right)} \right]$$

$$= E_o \left[1 + 2 \; Cos \left(\frac{\pi}{2} \; Cos \; \theta + \alpha \right) \right]$$

For endfire antenna the null occurs at θ = 90°

$$\therefore 1 + 2 \; Cos \left(\frac{\pi}{2} \; Cos \, \theta_n + \alpha \right) = 0$$

After solving, *with* $\theta_n = 33.56°$, the above equation becomes-

$$1 + 2 \; Cos \left(\frac{\pi}{2} \; Cos \; 33.56° + \alpha \right) = 0$$

$$\text{or} \quad Cos \left(\frac{\pi}{2} \; Cos \; 33.56° + \alpha \right) = \frac{-1}{2}$$

$$\frac{\pi}{2} \; Cos \; 33.56° + \alpha = \frac{2\pi}{3}$$

$$\therefore \alpha = \frac{2\pi}{3} - \frac{\pi}{2} Cos\ 33.56° = 0.784\ radians\ = 45°.$$

(ii) The position of main beam θ_n occurs when-

$$\frac{\pi}{2} Cos\ \theta_m + B = 0$$

$$Cos\ \theta_m = \frac{-2B}{\pi} = \frac{-2}{\pi} \times \frac{\pi}{4}$$

$$= \frac{-1}{2}$$

$$\therefore \theta_m = 120°$$

Q.2. Consider an array of two non-directional radiators with spacing (d - 0.5 λ). Determine the direction of maximum radiation when radiators are energized in phase by high frequency currents, of equal amplitude and frequency. Calculate the phase shift required for turning the direction of maximum radiation by 90° keeping the separation `d' unchanged. **(Raj. Univ. 2004, 2002)**

Ans:

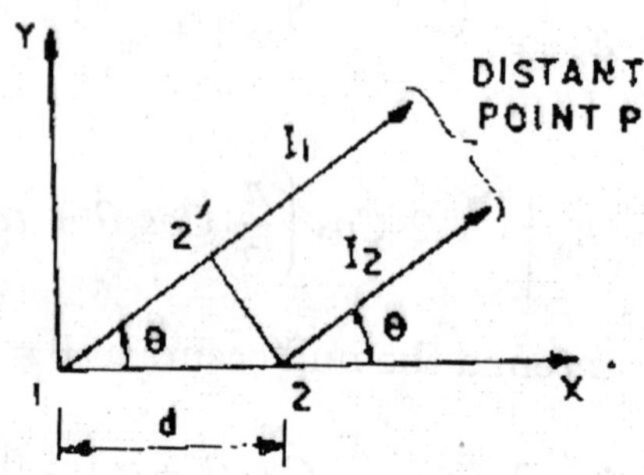

Figure 3.66

$$E_n = Cos\left(\frac{\psi}{2}\right) = \cos\left(\frac{\pi}{2} \cos\theta\right) = \pm 1$$

$$\frac{\pi}{2} \cos\theta_{max} = \pm n\pi \qquad\qquad Put\ n = 0$$

$$Cos\ \theta_{max} = 0$$

$$\theta_{max} = \frac{\pi}{2}\ or\ \frac{3\pi}{2}$$

For turning the direction of maximum radiation by 90° then-

$$\theta_{max} \text{ occurs at } 0, 2\pi$$

$$\cos \theta_{max} = \pm 1$$

$$\frac{\pi}{2} \cos \theta_{max} = \pm\frac{\pi}{2}$$

Since for maxima, $E_N = \pm 1$

$$\sin\left(\frac{\pi}{2} \cos \theta_{max}\right) = \sin\left(\pm\frac{\pi}{2}\right) \qquad \therefore E_N = \sin\left(\frac{\pi}{2} \cos \theta_{max}\right)$$

$$= \pm 1$$

This shows that two point sources are out of phase and of equal amplitude with separation 'd' unchanged.

Q.3. An elemental antenna is 1 meter long and a uniform r. m. s. current of 1 ampere is flowing through it. If the signal frequency is 100 kHz, then calculate its total radiated power and radiation resistance. **(Raj. Univ. 2000).**

Ans:

Given: length = 1 meter

I_{rms} = 1 Ampere

f = 100 kHz

To find: W and R_r =?

$$\text{Where, } \lambda = \frac{c}{f} = \frac{3 \times 10^8}{100 \times 10^3} = 3000\ m$$

(i) Total radiated power (W)

$$W = \left[80\ \pi^2 \left(\frac{l}{\lambda}\right)^2 I_{rms}^2\right] . \text{ Watts}$$

$$= \left[80\ \pi^2 \left(\frac{1}{3000}\right)^2 (1)^2\right] \text{ Watts}$$

$$= \left[\frac{80\ \pi^2}{9 \times 10^6}\right] \text{Watts} = 8.76 \times 10^{-5} \text{ Watts}$$

$$\therefore W = 8.76 \times 10^{-5} \text{ Watts}$$

(ii) Radiation Resistance (R_r)

$$R_r = \left[80\,\pi^2 \left(\frac{l}{\lambda}\right)^2\right] \Omega$$

$$= \left[80\,\pi^2 \left(\frac{1}{3000}\right)^2\right] \Omega$$

$$= \left[\frac{80\,\pi^2}{9 \times 10^6}\right] \Omega$$

$$= 8.76 \times 10^{-5}\ \Omega$$

$$R_r = 8.76 \times 10^{-5}\ \Omega$$

Q.4. Calculate the beam width between nulls for a broadside array with 10 half wave dipole elements spaced $\lambda/2$. **(Raj. Univ. 2000)**

Ans. **Given:** $d = \frac{\lambda}{2}$, n = *no. of arrays* = 10

To find: BW FN

Where, $D = 2N\left(\frac{d}{\lambda}\right)$

$$= 2 \times 10 \left(\frac{\lambda}{2\lambda}\right)$$

$$D = 10$$

(i) $\text{BWFN} = \frac{114.6}{L/\lambda}$ as $D = \frac{2L}{\lambda} = 10$

$\therefore\ BWFN = \frac{114.6}{5\lambda/\lambda}$ $\therefore\ L = 5\lambda$

$$= \frac{114.6}{5}$$

$$= 22.92$$

BWFN = 22.92 Ans.

SHORT QUESTIONS

1. Describe the physical arrangement of a ground screen.
2. What is the purpose of antenna stubs?

3. What is the maximum number of elements ordinarily used in collinear array?
4. Why is the number of elements used in a collinear array limited?
5. How can the frequency range of a collinear array be increased?
6. How is the directivity of a collinear array affected when the number of elements is increased?
7. What is the primary cause of broadside arrays losing efficiency when not operating at their designed frequency?
8. When more than two elements are used in a broadside array, how are the elements arranged?
9. As the spacing between elements in a broadside array increases, what is the effect on the major lobes?
10. What are some disadvantages of the end-fire array?
11. Where does the major lobe in the end-fire occur?
12. To maintain the required balance of phase relationships and critical feeding, how must the end-fire array be constructed?

ANSWER

1. It is composed of a series of conductors arranged in a radial pattern and buried 1 to 2 feet below the ground.
2. To produce desired phase relationship between connected elements.
3. Four
4. As more elements are added, an unbalanced condition in the system occurs which impairs efficiency.
5. By increasing the lengths of the elements of the array.
6. Directivity increases.
7. Lower radiation resistance.
8. Parallel and in same plane.
9. They sharpen
10. Extremely low radiation resistance, confined to one frequency, and affected by atmospheric conditions.
11. Along the major axis.
12. Symmetrically.

UNIT 4
RADIO WAVE PROPAGATION- I

CONTENTS

RADIO WAVE PROPAGATION I

4.1 INTRODUCTION

In earth's environment, electromagnetic waves propagate in space that not only depend on their own properties but also dictated by the environment itself. The real environment in which radio waves are propagated may have obstacles, discontinuities and propagation medium changes. The only environment of importance for practical radio wave propagation is that of earth and its immediate surroundings. Radio has invaded ``out space" too in recent years. In many respects the concept of free space propagation is realized in space for from the earth. **Free space is the space which does not interfere with normal radiation and propagation of radio waves.**

OR

In *free space*, no magnetic or gravitational fields or solid bodies or ionized particles are assumed to exist.

The power radiated from transmitter is spread over a large area and so power available at most of receiving antenna is only small fraction of radiated power. Transmitter loss (Ratio of Radiated to Received Power) is $10^{15} - 10^{20}$ (150-200 dB) in some cases and so the transmitter loss between transmitting and receiving antennas determine whether the received signal will be useful or not. This energy can be transmitted over a distance by 3 modes of propagation.

1. Ground or surface wave propagation.
2. Sky wave or Ionospheric propagation.
3. Space wave propagation.

Propagation of waves are used in-

1. Radio communication for transmission of intelligence over short and long distances.

2. RADAR
3. Radio direction finding
4. Control of machine from distance etc.

4.2 ELECTROMAGNETIC OR RADIO WAVES

Electromagnetic waves are just like oscillations which propagate with the velocity of light (C = 3 x 10^8 m/s) in free space. EM waves consist of moving field of electric and magnetic forces. The lines of forces are at right angle. These forces are at right angle to direction of propagation. **The direction of *electric field, magnetic field* and *propagation* are mutually perpendicular in EM waves i.e. "Oscillation" are perpendicular to the direction of propagation of waves.**

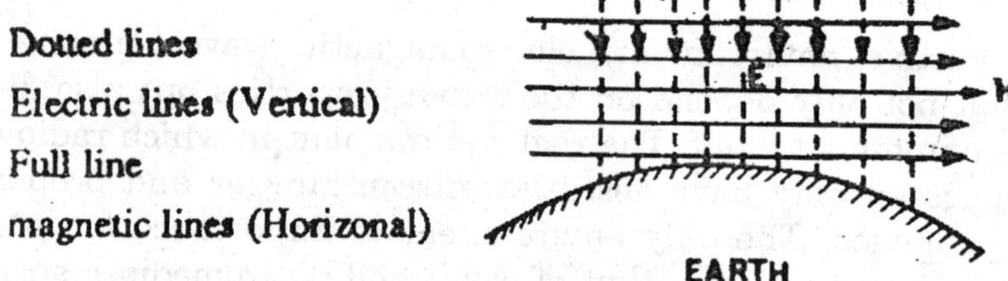

Figure 4.1 Direction of E and H field

In the figure, electric lines perpendicular to earth and magnetic lines horizontal. However, they can be of vice versa case also in orientation. Arrows indicates instantaneous direction of field for a wave traveling towards reader.

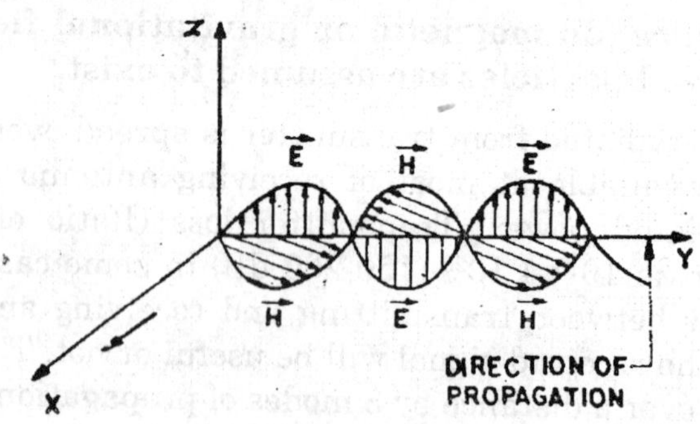

Figure 4.2. E-M wave

The orientation of Electric field w.r.t. earth gives the polarization. In the given figures the wave E-field is vertically polarized. The starting polarization is determined by orientation of antenna i.e. if antenna is vertical or horizontal, then it will produce the same polarized wave. **The wave usually maintains its polarization during the travel as it was generated.** Till then no obstacles or interference in free space is assumed to be present so EM waves spread equally in all directions from a point source. Thus the wave front is spherical as shown in fig and actual wave front is shown.

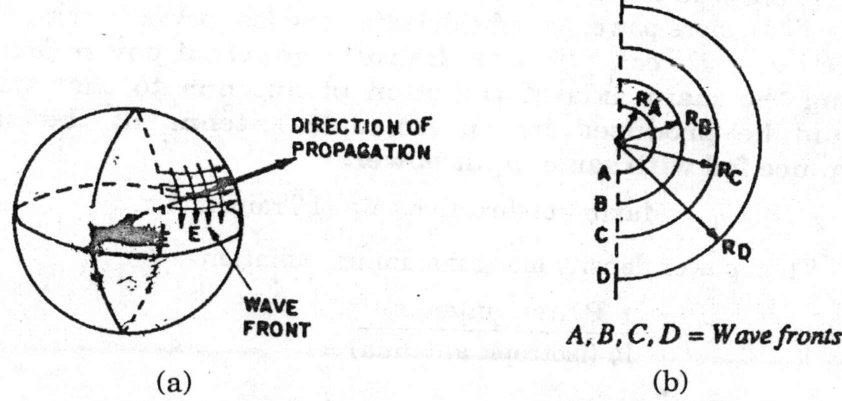

Figure 4.3. (a) Spherical wave front representation, (b) wave fronts and rays from a point source in free space

Sometimes to understand, even ray are also imagined which radiates from point source in all direction and they are perpendicular to plane of wave front. **Optical phenomena of reflection, refraction, diffraction and polarization are also observed and followed by EM waves.**

4.3 FUNDAMENTAL EQUATION FOR FREE SPACE PROPAGATION

The ratio of Radiated to Received power is called as the *Radio transmission loss*. The main concept in estimating radio transmission loss is loss expected in free space i.e. in region of all objects that might absorb or reflect radio energy. This concept is inverse square law in optics applied to radio transmission for free space fundamental equation:

1. Consider an average power = W_T

(Assumed to be radiated equally in all direction)

Assumption: Isotropic radiator is placed in free space, homogenous and non-absorbing medium dielectric constant of Isotropic radiator is unity. These radiations will spread out spherically as travels away from source. The power density in wave will be-

(Power per unit area of wave front) =>

$$\boxed{P_r = \frac{W_T}{4\pi d^2} \ \text{W/m}^2} \qquad (1)$$

Where $4\pi d^2$ = surface area of sphere

d = radius centered on the source.

2. Since all the practical antenna have directional characteristics i.e. radiate more power in some direction and loss power in some other direction, ∴ ***Directivity gain*** **the ratio of actual power density along the main axis of radiation of antenna to that which would be produced by an isotropic antenna at the same distance fed with same input power.**

G_T = Maximum directive gain of Transmitter.

Then power density along maximum radiation

$$G_T = \frac{P_o \text{ (test antenna)}}{P_r \text{ (isotropic antenna)}}$$

$$P_o = G_T \cdot P_r$$

or $$P_D = G_T \frac{W_T}{4\pi d^2} \quad (2)$$

3. Now a receiving antenna may be positional so that it collects maximum power from the wave. Let `W_R' be power delivered by antenna to load (receiver) under method condition. When so positioned the antenna can be considered to have an effective aperture area (Ae) so that-

$$\boxed{W_R = P_D \cdot A_e} \quad (3)$$

or

$$\boxed{W_R = \frac{G_T W_T A_e}{4\pi d^2}} \quad (4)$$

But for any antenna maximum directivity gain and effective area

$$\Rightarrow G = \frac{4\pi}{\lambda^2} \cdot A_e$$

Or $$\frac{A_e}{G} = \frac{\lambda^2}{4\pi} \quad (5)$$

Where, λ = wavelength being radiated.

let G_R = Maximum directivity gain of Rx. antenna.

$$\boxed{A_e = G_R \left(\frac{\lambda^2}{4\pi}\right)} \quad (6)$$

Equation (4) and (6)

$$W_R = \frac{G_T\, W_T}{4\pi d^2} \cdot G_R \left(\frac{\lambda^2}{4\pi}\right)$$

or $$\boxed{\frac{W_R}{W_T} = \frac{G_T \cdot G_R}{4\pi d^2}\left(\frac{\lambda^2}{4\pi}\right) \text{ watts}} \qquad (7)$$

This is the fundamental equation for free space transmission or propagation (FRIIS FREE SPACE EQN.).

Its alternative forms are-

$$\boxed{\frac{W_R}{W_T} = G_T \cdot G_R \left(\frac{\lambda}{4\pi d}\right)^2} \qquad (8)$$

$$\boxed{W_R = W_T \frac{G_T \cdot G_R}{\left(\frac{4\pi d}{\lambda}\right)^2}} \qquad (9)$$

Where, W_T = Radiated power

W_R = Received power

G_T = Maximum directivity gain (tx)

G_R = Maximum directivity gain (Rx)

λ = Wavelength (m) = f/c

d = distance (m)

4. From Equation (7) larger the denominator, smaller the value of received power W_R and hence **it is desirable-**

$$\boxed{L_S = \left(\frac{4\pi d}{\lambda}\right)^2} \qquad (10)$$

as the **spatial (related to space) attenuation** which is distinct from dissipative or absorption losses.

5. To form Equation (7) in logarithmic form

E (dB μv/m) = W (dB m W)

E (dB μ) = W (dB m)

Now Equation (7) is

$$10\log_{10}\left(\frac{W_R}{W_T}\right)=10\log_{10}(G_T)+10\log_{10}(G_R)+10\ \log_{10}\left(\frac{\lambda}{4\pi d}\right)^2$$

Or

$$\boxed{W_R\,(\text{dBW})=W_T\,(\text{dBW})+G_T\,(\text{dB})+G_R\,(\text{dB})-L(S)} \quad (11)$$

$$\text{Where, } L(S) = 10\log_{10}\left(\frac{\lambda}{4\pi d}\right)^2$$

$$= 20\log_{10}\left(\frac{\lambda}{4\pi d}\right)$$

$$\boxed{L(S)\,\text{dB} = 32.45 + 20\ \log_{10} f + 20\ \log_{10} d} \quad (12)$$

$$\text{w.r.t. } \lambda = \frac{C}{f} = \frac{3\times10^8\ (\text{m/s})}{f\,(\text{H}_Z)} = \frac{3\times10^8}{f\times10^6\ (\text{MH}_Z)}$$

$$\boxed{\therefore\ \lambda = \frac{300}{f\,(\text{MH}_Z)}} \quad (13)$$

Now equation (8) is by using the value of λ

$$\left(\frac{W_R}{W_T}\right) = G_T G_R\left(\frac{\lambda}{4\pi d}\right)^2 = G_T G_R\left(\frac{C}{4\pi\,df}\right)^2$$

$$= \frac{G_T G_R}{16\pi^2}\left(\frac{300\times300}{d^2\times((10)^3)^2\times f\,(\text{MH}_Z)}\right)$$

$$= G_T\,G_R\left(\frac{9\times10^4}{16\times9.8596\times10^6\ (df\ )^2}\right) = G_T G_R\left(\frac{9\times10^{-2}}{157.7536}\right)\frac{1}{(df\ )^2}$$

or

$$\boxed{\frac{W_R}{W_T} = G_T G_R\ \frac{0.5\times10^{-3}}{(df)^2}} \quad (14)$$

Expressing power ratio in decibels

$$10\log_{10}\left(\frac{W_R}{W_T}\right) = 10\log_{10} G_T + 10\log_{10} G_R + 10\log_{10} 0.57 +$$

$$10\log_{10} 10^{-3} - 10\log_{10} d^2 - 10\log_{10} f^2$$

$$= G_T\,(\text{dB}) + G_R\,(\text{dB}) - \{32.45 + 20\log_{10} d + 20\log_{10} f\}$$

(i) $$\boxed{\left(\frac{W_R}{W_T}\right)(\text{dB}) = G_T\ (\text{dB}) + G_R\ (\text{dB}) - L(s)\ (\text{dB})}$$

where, (ii) $$\boxed{L(s)\ \text{dB} = 32.45 + 20 \log_{10} d + 20 \log_{10} f}$$

$d \rightarrow$ km and $f \rightarrow MH_Z$

If here are microwave frequencies, the loss will be greater because of atmospheric absorption.

Often it is required to know the electric field strength of wave at Rx. antenna. This is obtained by ***Poynting Theorem***.

$$\boxed{P_D = E \times H}$$

or $$\boxed{\frac{E}{H} = \eta = 120\,\pi} \qquad (16)$$

Hence $$P_D = E.\frac{E}{\eta} = \frac{E^2}{\eta}$$

$$E = \sqrt{P_o . \eta}$$

$$= \sqrt{\frac{G_T\, W_T}{4\,\pi\, d^2} .\ 120\,\pi} \quad \text{by equation (2)}$$

$$E = \sqrt{\frac{30\, G_T\, W_T}{d^2}} \quad \text{V/m} \qquad (17)$$

This is fundamental equation which field strength at receiving antenna for free space - propagation conditions. A receiving antenna therefore, has effective length or height l_e analogues to effective area and hence open circuit emf of Rx. antenna is given by

$$V_{oc} = E\ .\ l_e$$

Equation (17) is also written sometimes in terms of field strength E_o at unit distance i.e. 1 km. Thus at d = 1 km we have-

$$\boxed{E = E_\theta = \sqrt{30\, W_T\, G_T} \quad \text{V/m}} \qquad (18)$$

Equation (18) and (17)

$$\boxed{E = \frac{\sqrt{30\, W_T\, G_T}}{d} = \frac{E_\theta}{d} \quad \text{V/m}} \qquad (19)$$

$$\boxed{E = E_\theta \ \frac{1}{d\,(m)} \ V/m} \tag{20}$$

Further Equation 17 for (1 km and $P_T = 1\,kW$)

Then,

$$E_\theta = \frac{\sqrt{30\,W_T\,G_T}}{d} = \frac{\sqrt{30\times 1000\times 1\times W_T}}{1000} = \frac{100\times\sqrt{3\times W_T}}{1000}\ V/m$$

$$= \frac{1.732\,\sqrt{W_T}}{10} \cong 0.173\,\sqrt{W_T}\ V/m$$

$$E_\theta \cong 0.173\,\sqrt{W_T}\ mV/m \tag{21}$$

if $\quad W_T = 1\,kW$

$$E_\theta = 173\,mV/m$$

Hence, it can be written in general, as

$$\boxed{E_\theta\,(mV/m) = \frac{173\,\sqrt{W_T\,(kW)}}{d\,(km)}} \tag{22}$$

4.4 MODES OF PROPAGATION

The radio waves from Tx. antenna may reach to Rx. antenna using any of the following modes of propagation depending on the factors like-

- frequency of operation
- Distance between Tx. and Rx antenna etc.

4.4.1. Ground wave or surface wave propagation (up to 2 MHz)

- Also called as Norton's surface wave.
- It is of practical importance at broadcast and low frequencies i.e. for medium waves, long waves and for very long waves. **The ground wave is always guided along surface of earth just as an EM wave is guided by a waveguide or Tx. live.** Surface wave permits the propagation around the curvature of earth.
- **This mode of propagation exists when the transmitting and receiving antennas are close to the surface of earth. It is also supported at its lower edge by presence of ground.**

- They are produced usually by vertical antennas i.e. *vertically polarized.*

  T_x = Transmitting antenna

  R_x = Receiving antenna

  Path A = ground wave propagation

  Path B = sky or Ionospheric propagation

  Path C = space wave propagation.

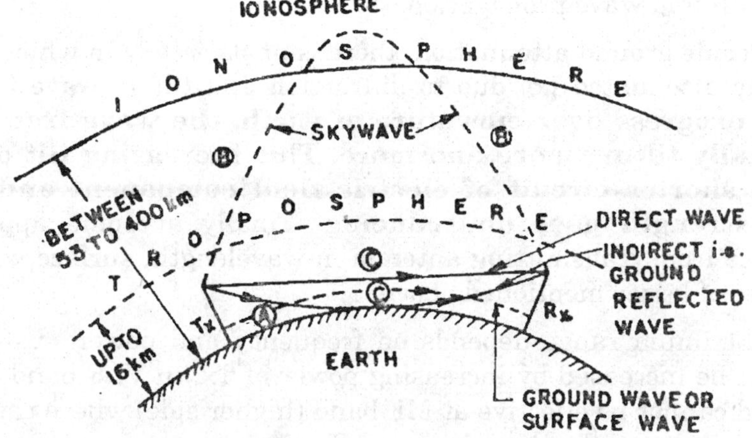

Tx = Transmitting antenna.
Rx = Receiving antenna.
Path A = Ground wave propagation.
Path B = Sky or ionospheric propagation.
Path C = Space wave propagation.

Figure 4.4 Possible propagation paths from transmitting antenna to receiving antenna

- **Any horizontal component of electric field in contact with earth is *short-circuited* by the earth. The gnd. wave propagation along surface of wave, induce charges in earth. These chases travel with wave and hence constitute a current while carrying this induce current earth behaves just as a leaky capacitor and so earth can be represented as resistance in shunt with a capacitor.** This behavior of earth as a conductor may be described in terms of conductivity and dielectric constant `K'.

When surface wave travels over the surface of earth energy is abstracted from surface wave to supply the losses in earth. Thus while passing over the surface of earth, the surface wave losses some of its energy by absorption. This energy less, is however, replenished to a certain extent by energy diffracted downward from upper portion of wave front present somewhat above immediate surface of earth ∴ This mode of propagation suffers varying amount of attenuation while propagating along the curvature of earth, depending upon frequency surface irregularities, permittivity and conductivity. Earth's (attenuation) as (f) and so this mode is suitable for low and medium f i.e. up to 2MHz only ∴ **at high frequency, attenuation by ground is more than at low-frequency over same ground**. It is also called as medium wave propagation. Its use is invariable in *local broadcasting.* All the broadcast signals, which received during day time, are due to ground wave propagation.

Beside ground attenuation, there is another way in which surface wave is attenuated i.e. due to diffraction and tilt in wave front. **As wave progress over curvature of earth, the wave fronts start gradually tilting more and more. This increasing tilt of wave cause shorter circuit of electric field component and hence field strength goes on reducing**. Finally at some appreciable distance from transmitting antenna in wavelength, surface wave dies because of losses mentioned above.

Maximum range depends on frequency and power. ∴ Range of Tx. can be increased by increasing power of Tx. in VLF band but this method cannot be effective at MF band (higher side) where the tilting, which is due to diffraction, in more effective.

Field strength

$$\boxed{E = \frac{120\ \pi . h_t . h_r . I_s}{\lambda d.}\ V / m}$$

For flat earth

$$\boxed{Eg = \frac{Eo.A}{d}}$$

Unit distance field strength E. depends upon

(1) Power radiation of transmitting antenna.

(2) Directivity in vertical and horizontal planes.

Radiated power = 1 kW. = P.

$$E_o = \frac{300\ \sqrt{P}}{d.} V / m = \frac{300\ \sqrt{1}}{1000} V / m = 300\ mV / m$$

d = distance in km

Short vertical unipole antenna, E_o at distanced on a hypothetical flat perfectly conducting earth is

$$\boxed{E_o = \frac{\sqrt{90}\ P}{d}\ V/m}$$

Ground wave attenuation factor

$$\boxed{A \cong \frac{2 + 0.3\ P}{2 + P + 0.6\ P^2}}$$

P = numerical distance

Q. Write a brief note on surface wave propagation. **[Raj. Univ. 1999]**
Q. Explain propagations with ranges; polarizations: i. Ground wave. **(R.T.U. 2009)**

4.4.2. Sky wave or Ionospheric wave propagation (Between 2 - 30 MHz)

The sky waves are having practical importance at medium and high frequencies (i.e. at medium and short waves) for very long distance radio communications. In this mode, **electromagnetic wave reaches the Rx. point after reflection from ionized region in upper atmosphere called ionosphere**. It is situated between 50 km - 400 km above earth surface - under favorable conditions (Illustrated as in path (B)). Ionosphere acts as ***reflecting surfaces***. It is able to reflect back the EM waves of frequencies between 2t 30 MHz EM waves of more than 30 MHz frequency are not reflected back from ionosphere rather they ***penetrate***. Mostly sky wave propagation is favorable for frequencies between 2 to 30 MHz, so this mode of propagation is also called as ***short wave propagation***.

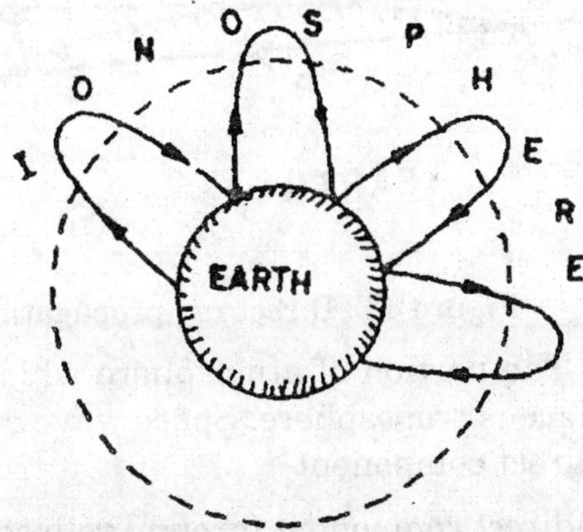

Figure 4.5 Multiple reflections of radio waves from ionosphere

Sky waves propagate after reflecting from the ionosphere, so it is called as ***Ionospheric propagation***.

Long distance point to point communication is possible with sky wave propagation. So it is also called as ***point to point propagation Communication***.

Extremely long distance i.e. round the globe communication is also possible by multiple reflections of sky waves. As in a single reflection from the ionosphere the radio waves cover a distance **not more than 4000 km.**

The signals received due to sky wave propagation are, however, subjected to *fading* in which signal strength varies with time. It is due to a large no. of waves follow a different number of paths at receiving point ∴ fading occurs.

Q. Explain sky wave propagation?	**[Raj. Univ. 2008]**
Q. Explain propagations with ranges; polarizations: i. Sky wave	**(R.T.U. 2009)**

4.4.3. Space wave Propagation - (above 30 MHz)

It is used at VHF bands (between 30 MHz - 300 MHz), VHF and microwaves and communications like T.V., Radar, FM etc. These utilize this mode of propagation. **In this mode of propagation EM-waves from Tx. antenna reach the Rx antenna by two methods either directly or after reflections from ground in earth's *tropospheric region.***

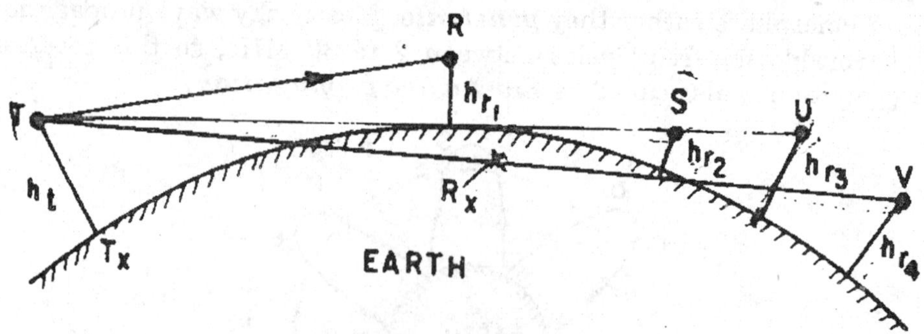

Figure 4.6. Direct ray propagation

Troposphere - The portion of atmosphere which extends up to 16 km from earth surface is troposphere. Space wave consists of at least 2 components – (a) Direct component

(b) Indirect component (ground reflections).

Direct components are the components in which, wave reaches directly from transmitting antenna to receiving antenna and indirect component are the components in which, waves reaches Receiving antenna after reflection from the ground, here phase change of 180° takes place. **Both these waves travel in same phase from T_x, but may reach the receiving antenna either in or out of phase as two waves travel different path length.** ∴ Strength of resultant wave at R_x. may be stronger or weaker depending or in phase or out phase respectively. This signal at receiver is vector addition of both types of waves. Because these waves propagates through troposphere so called as *Tropospheric propagation.*

Space wave propagates mainly at VHF band, as at such frequencies sky wave and ground wave propagation both FAIL." Beyond 30 MHz sky wave propagation fails as wavelength becomes too short to be reflected from ionosphere and ground waves are propagating does to the antenna only, as attenuation is very high. So just after few 100 feet ground waves also die due to attenuation and wave tilt.

This mode is called as ``line of sight" propagation because at VHF, UHF and microwave frequencies.

Limitations - (1) Line of Sight distance

(2) Curvature of earth (as shown in figure)

Actually, in actual practice space wave propagates even slightly beyond the line of sight distance due to refraction in atmosphere of earth. In LOS distance Tx. antenna and Rx. antenna can usually in front of (see) each other.

In fact, LOS i.e. range of communication can also be increased by *increasing the heights of Tx or Rx antennas* (as in U antenna). The curvature of earth and height of Tx and Rx antennas determines maximum range of communication through direct waves.

See the figure, the height of Tx. antenna `ht.' and Rx. antenna `h_{r1}', the direct wave communication Range is T_R.

- When the range is increased i.e. Rx antenna is moved away from R than a point is reached. When LOS distance from T to S increases. Then say Ts represents maximum range of LOS up to which communication is possible with Tx and Rx. height h_t and h_{r2}.
- This LOS distance can be increased if heights of eight antennas (Tx or Rx) is increased further (say h3) (point U) which means range has increased from Ts to T_U.

- If Rx antenna is moved to a distance which is not in LOS distance like V point of some height hr3, then direct wave signal reception is not possible.

This is used in space communication or specially satellite Communication, Which has facilitated trans-oceanic propagation of microwave with potentiality of large B.W. (means-Radio traffic between ground station and satellite)

Q. What is LOS propagation? How does it differ from ground wave and sky wave propagation? **[Raj. Univ. 2003]**

Q. What is space wave propagation? **[Raj. Univ. 2008]**

Q. How do the earth's magnetic field and solar active affect line of sight and ionospheric propagation? **[Raj. Univ. 2003].**

Q. Explain propagations with ranges; polarizations:

i. Space wave **(R.T.U. 2009)**

Q. A VHF communication is to be established with 35 W transmitters at 90 MHz. Find the distance upto which line of sight communication may be possible if the height of transmitting and receiving antenna are 40 m and 25 m respectively. Also determine the field strength at receiving end. **(R.T.U. 2009)**

4.4.4. Tropospheric scatter propagation or forward Scatter Propagation (UHF and microwaves) (above 300 MHz)

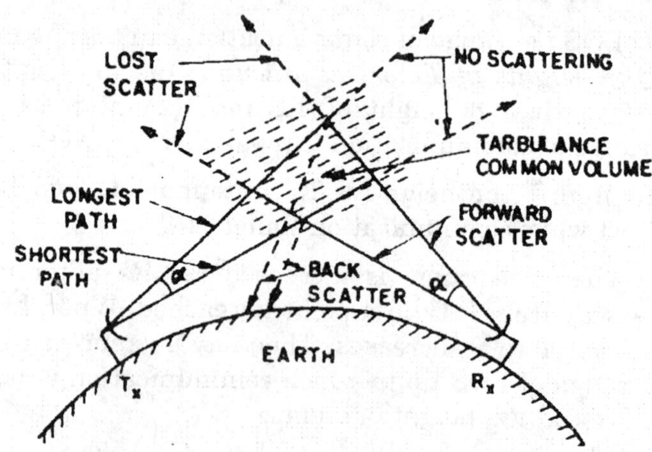

Figure 4.7. Tropospheric scatter propagation

It is used at VHF, UHF and microwaves. **UHF and microwaves signals propagate much beyond the LOS propagation through**

forward scattering in tropospheric irregulations. It uses certain properties of troposphere and is also known as *Troposcatter*.

This has lead to the discovery of ionosphere scatter propagation for signal frequencies in lower end of VHF band ∴ it has been established in recent years that it is possible to achieve a very reliable communication over communication range of 160 km - 1600 km by using high power Tx and high gain antennas i.e. reliable scatter propagation is possible in VHF and UHF bands.

The name scatter propagation (**beyond the horizon propagation**) is given due to mechanism involved in the phenomenon.

Two theories

(1) The first mode is Ionospheric. It is believed to be resulted from the scattering of radio waves from lower E-layers of ionosphere.

(2) The second mode is tropospheric. It is thought to be result of scattering from either blobs or fine layers in the troposphere.

It is suggested that Ionospheric scatter might be due to blobs or fine layers at lower edge of E-layer or it could be from ionized trails of myriads of small meteors which bombard earth from the outer space.

- Ionospheric communication in range. (1000 km - 2000 km)
- frequency Range (25 MHz - 60 MHz)
- decrease of use beyond 60 MHz

But at the same time tropospheric scatter propagation appears to be effective starting from 100 MHz to at least 10 GHz i.e. 10,000 MHz

Because of greater attenuation of signals along path, forward scatter propagation is mainly useful for point to point communication, radio and television relay links where it is possible to use extremely high gain antennas and high power transmitters.

- Two directional (Tx and Rx) (according to figure description) antennas are so pointed that their beams intersect midway between them above the horizon.
- If one is UHF Tx. antenna and other UHF Rx. antenna. Then sufficient radio energy is directed towards in receiving antenna to make this a useful communication system.
- The blobs of air masses or eddies in tropospheric scatter radio waves are generated due to turbulence. This happens when they are situated in common volume facing Tx and Rx antenna beams.

- **When wavelength is more (frequency low) than the eddies, the scattering dominates into the cone of angle α. This angle should be very small.**
- The best and typical used frequencies are centered on 900 MHz, 2000 MHz, 5000 MHz
- Both Ionospheric and tropospheric scatters sometimes produce undesirable noise and fading which may be minimized to certain extent by diversity reception.

> **Q.** What is tropospheric scattering? What is the frequency range for it? Why are such ranges only applicable for tropospheric scattering? **[Raj. Univ. 2003, 2001]**
>
> **Q.** Write short note on tropospheric scatter communication **[Raj. Univ. 2004, 2005]**
>
> **Q.** What do you understand by tropospheric scatter? What frequency band is used for such propagation? What are its main advantages and limitations? **[Raj. Univ. 2002, 1999]**

4.5 STRUCTURE OF ATMOSPHERE

As the medium between transmitting and receiving antennas play an important role, ∴ it is necessary to study the medium above the earth, by which radio waves propagates before further study on modes of propagation is taken

(a) Structure of troposphere

(b) Structure of Ionosphere

(c) Outer Atmosphere

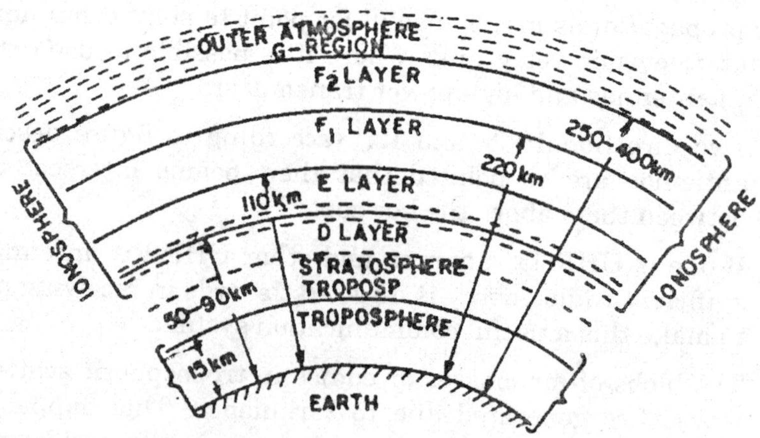

Figure 4.8. Medium above earth surface

4.5.1 Structure of Troposphere

Medium above earth surface

(a) Troposphere up to 15 km

(b) Ionosphere (50 - 400 km)

(c) Outer atmosphere (>400 km)

- Troposphere is the portion of earth atmosphere which extends up to a highest of 8-10 km. from the earth's surface at polar latitude, 10-12 km at moderate latitude and up to 16-18 km at the equator.
- On an average troposphere is said to be extending up to a height of 15 km from the earth surface.
- **Thus actual height is at least at the poles and maximum at equators and composition of Troposphere remains approx. some as at the surface of earth.** Entire belt is called as *troposphere or **"Region of change"***

In troposphere, the percentage of gas components remains almost constant with increase of height, yet the water vapour components sharply decrease with height one more important property of troposphere is that temperature decreases with increase of height and falls to a minimum temperature of - 68°F.

After the 'top of troposphere', **tropopause starts** and it ends at the beginning of **stratopause, or region of calm**. Above critical height called Tropopause, the temperature remains constant through the narrow belt and begins to increase afterwards.

Stratosphere-No intermixing due to air currents take place and composition of atmosphere varies with height and stratification begins.

$$\text{Ref. index } (\mu) = 0.0003\% \ \textit{greater than unity (for troposphere)}$$

$$N = (\mu - 1) \times 10^{-6}$$

N - Excess refractive index over unity, in millionths.

4.5.2 Structure of Ionosphere

The upper most part of atmosphere where ionization is appreciable is known as *ionosphere*. The upper most part of earth's atmosphere absorbs large quantities of radiant energy from seem. It not only heats the atmosphere but also produces ionizations i.e. formation of positive and negative ions occurs. **The ionized region so consists of *free electrons, positive ions* and *negative***

***ions.* The most important ionizing agents are *UV - rays, x - rays, B-rays, cosmic rays* and *meteors*.** As in gas under very low pressure it is possible to knock off one or two e out of a gas molecule. When a molecule donate one e it become positive by charged ion and the molecule which accept it becomes negatively charged ions. These ions can easily be effected (i.e., attracted or repelled) under electric forces. e can be knocked off not only by fast moving e but also by certain type of radiations such as ultra violet and cosmic rays.

High atmosphere, conditions are very much favorable for ionization as pressure is less and sun is constantly radiation UV Rays. When UV and cosmic rays reach the upper atmosphere they cause a large proportion of air particles to become ionized.

Further the ions, electrons and atoms in a gas are continuously in motion so frequent collisions occur between them and consequently *process of recombination* continued all the time. So once a molecule is ionized, does not remain ionized indefinitely. **The *time of recombination* depends on distance between 2 particles of gas.**

In lower part of earth atmosphere collisions are very high and hence air molecules does not remain ionized for a longer time. Besides the UV-rays from the sum are greatly absorbed by upper parts of atmosphere and hence there is little ionization in lower part of earth's atmosphere and very little ionization below about 50 km.

Above the height of 400 km air particles present are so less that density of ionization is again very loss. However, considerable ionization, exist in the intermediate height (i.e. between 50 km to 400 km) and this region is ∴ has most influence on sky wave propagation.

As sky waves of different frequencies are found to return to earth from different heights, it means ionosphere is not having one but *several layers*. The different layers in ionosphere are due to the **fact that different gases in earth's atmosphere are ionized at *different pressures* (i.e. different height) and there are many *ionizing agent* to do the ionization.**

So due to different ionizing agent and different physical properties of atmosphere at different heights, the ionization in the ionosphere stratified and the levels at which e density reaches a maximum are called *layers.* The number of layers, their heights and the amount of sky wave that can be bent by them, will vary from day to day, time to time, season to season.

- For each layer there is critical frequency, above which is radio wave is sent vertically upward, will not return to earth but will penetrate and move outside the atmosphere.

These principle layers – E

– F1

– F2

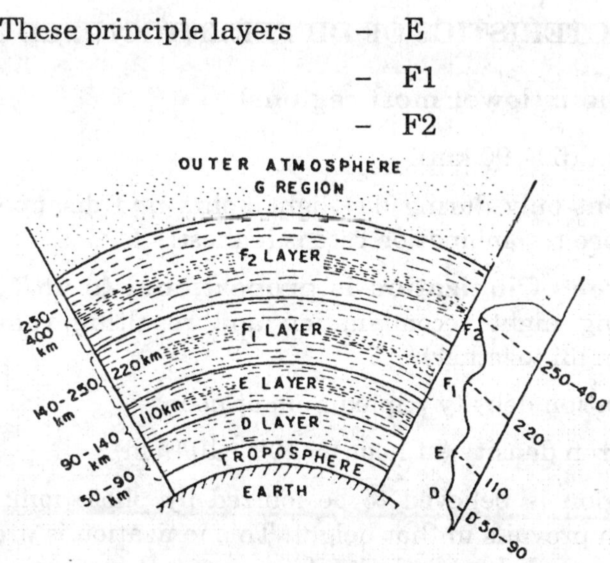

Figure 4.9. Typical height of Ionospheric layers above the earth

Beside these regular layers there is a region below E, called D-Region. It lies between 50-90 km and is responsible for much of day time attenuation of high `f' waves.

During night = **the f1 and f2 layers combined and from one layer called f layer and D-region disappears**. Thus in night only 2 principal layers exists i.e. E and F layers. E-layer is found 110 km, but may vary 90-140 km. Similarly F2 lies at 200 with little diurnal or seasonal variations. (250-400 km)

D, E, F1: Sir E Appleton

F: Appleton layer

E: Kennelly Heaviside layer

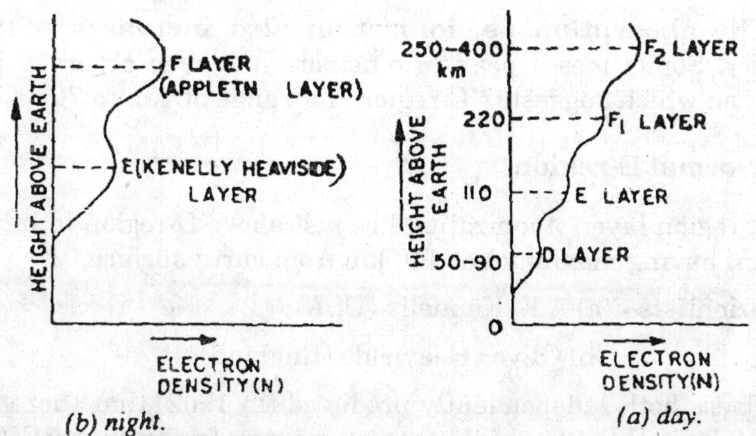

Figure 4.10 Electron density Vs height above earth in day and night

4.6 CHARACTERISTICS OF DIFFERENT IONIZED REGIONS

4.6.1 D-Region: (lower most regions)

- Height (50 - 90 km)
- Present only during day light hours and disappears at night because recombination rate is highest.
- **Degree of ionization is proportional to altitudes of sun** (during night recombination and resulting vanishing of D-region all together).
- Ionization density = maximum at noon.
- Electron density (at noon) = 10^{14} - 10^{16}/cm^3
- D-region is believed to be ionized by photo ionization of No. which prevails in that height. This ionization is produced by Lα (**Lyman alpha**) radiation from sum. D-region is due to photo ionization of 0xygen (O_2) molecule at its first ionization potential.
- It is capable of reflecting VLF signals and LF (long waves) and absorbs to certain extent MF and HF signals. Its **critical frequency is about 100 KHz** (but it can't be observed due to very high selective absorption).
- D-layer is also known as absorbing layer for short wave signals (i.e. HF). **As the density of e in this region is not sufficient to effect appreciable bending of radio waves and hence they suffer attenuation while passing.**
- Diurnal variation = Maximum = at noon

  Seasonal variation = maximum = in summer

The absorption i.e. ionization also increases with *solar activity*. Sometimes a peak in e density has been observed in lower `D' region which suggests ``C-region'' for range 50 km to 70.

4.6.2 Normal E-region

E region layer of ionization lies just above D-region in height (90-140 km) having maximum at 110 km from earth surface.

Scientists - (a) A.E. Kennelly (USA)

(b) Oliver Heaviside (England)

These both independently predicted (in 1902) that there must be densely *ionized layer* **which acts as mirror for turning EM waves back to earth reflection or refraction**. This is why it is also called as ``**KENELLY HEAVISIDE region or layer)**" (It is misleading as it don't have a boundary).

- Layers occur only during day light hours + constant.
- Maximum density at average height of 110 km. **which has appreciable** effect on the direction of propagation of radio waves.
- During might hours E-region remains weekly ionized.
- Day to Day variations are less.
- Since obliquity of sum reduces ionization and hence e density is low in winter and high in summers comparability.
- electron density in day time = 10^5 to 4.5×10^5
- electron density at night = 5×10^3 to 10^4
- **It is maximum at noon in summer and increases with increased solar activity (i.e. sun spot cycle)**
- Critical frequency 3 MHz - 5 MHz (noon/low latitudes)
- **E-Region is formed by ionization of all gasses by soft X-ray radiations.**
- *Used for* = Long distance communication.
- Main function = **to reflect some HF waves in day hours.**

4.6.3 Sporadic (very irregular) E region (Es)

- As the presence of this region is very much irregular hence it is termed as *sporadic* E-layer.
- It occurs inform of clouds, varying in size from about one km to several 100 km across.
- **Its presence is purely regional and its occurrence and intensity of ionization has no connection with sun radiation.** The occurrence of sporadic E-layer is quite unpredictable and it is may be observed both in day and night hours, and in any season of year.
- E_s is very thin layer of *high ionization density* (10 times of normal E - layer)
- Height Range = 90 km - 130 km with normal E - layer.
- It is stable; height may differ by not more than 5-10 km.
- At polar Region - E_s occurs **only at night**

  No Seasonal variations.

- At Equatorial -At day Hours.

  Possibility of Es occurrence increases with geomagnetic latitude.

- Sometimes they are produced by meteoric ionization, then at other times due to vertical transport of ion clouds i.e., *formation of Es is by the infiltration of charged particles* from the overlying F2 layer due to turbulent motion of air masses.
- *Thunder storm or geomagnetic disturbances* are also the cause for Es formation.
- Not for long distance communication but sometimes gives unexpected good results.

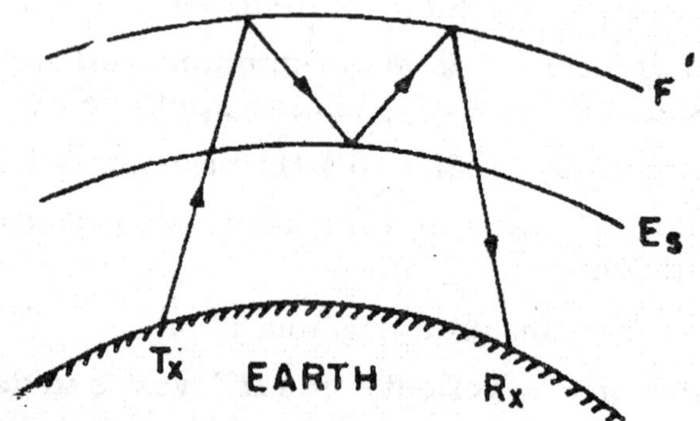

Figure 4.11. M-Type reflection from Es and F layer

- Mostly Ionospheric *reflections* from Es and from higher f-regions are observed simultaneously if layers are smaller compared with wavelength. Es region long distance scatter propagation of V.H.F. signals.
- M-type reflection occurs, i.e. signal path is like latter M.

4.6.4 F_1, F_2 and F-region or Appleton Regions

- Region lying in range of 140-400 km from earth surface is called as F-region.
- Average height = 270 km.
- Uppermost layer.
- **It always remains ionized irrespective of hours, day or seasons of the years.**
- It was in 1925, Appleton showed, a more densely ionized layer than E Regions. For this he received echo effects by sending short duration HF signals projected vertically upward.

- **Ionization persists throughout dark hours of night.**
- **F-region facilitates long distance sky wave propagation of radio signals during night hours. This existence is due to -**
  (1) Being top most layer, highly ionized and hence some ionization remains even after sun set.
  (2) As Ionization density is high, the actual air density is not much and so most of molecules of this layer are ionized.
- During day, sometimes after sunrise, the F-region is splited up into 2-layers called F_1 and F_2 in low latitude stations throughout the year and in high latitude station only in summer.
- **Fl - layer**
  - 140 - 250 km.
  - Average height = 220 km.
  - Behave as E-Region (normal) follows the **Chapman's law of variation.**
  - Critical frequency = 5 MHz - 7 MHz
  - electron density = 2×10^5 to 4.5×10^5
  - It shows great variations in height and its density is lower in winter then in summer.
  - It is formed by ionization of O_2 atoms.
  - Although some waves are reflected from F_1 but most it and in reflected from F_2 layer.
- **F2 layer**
  - height = 250-400 km
  - In day, highest e-density of all the Ionospheric layers is observed.
  - It **falls to 300 km at night** where it combines with F1 layer.
  - electron density = 3×10^5 to 2×10^6
  - **Being highest in height, the air density is so low in this region that ionization disappears very slowly.**
  - Formed by ionization of UV, x-rays and probably corpuscular radiations.

- Critical frequency = 10 MHz o still more at low altitude stations.
- Ranges 5 MHz to 12 MHz
- **It is effected by earth's magnetic field, atmospheric (i.e. Ionospheric tides and winds), Ionospheric storms and other geomagnetic disturbances.**
- Ionization density changes with solar activity and change from sun spot minimum to sum spot maximum is highest for this layer.
- **Does not follow** - Chapman's law
- large no of irregulations as it is located at maximum height
- **Splitting up to f1 and f2 is produced by tidal effects and increase in temperature with increase in height. ∴ It is most important reflecting medium for high frequency radio waves.**

4.6.5 Outer Atmosphere or G-Region

The upper limit of ionosphere is not known but region further away from 400 km is known as G-region. The outer region of ionosphere is occupied by radiation belts girdling the earth and consisting of charged particles trapped due to the terrestrial magnetic field, having the shape of magnetic lines of force.

4.7 SKY WAVE PROPAGATION

- Propagation of space and ground waves are limited by the *curvature of earth* and hence these modes of propagation fail for communication over large distances (Which is then performed by sky wave propagation (1000's Km) (Ionospheric propagation).
- **Sky waves are reflected by some of ionized ionospheric layers and return back to the earth either in *single hop* or in *multiple hops* of reflections.** (Thus for sky wave of suitable frequency it is possible to cover any distance round the earth).
- Radio wave of frequency 2 MHz to 30 MHz (i.e. H.F. signals or short waves) is reflected from ionosphere but during day time the lower frequencies of 2-30 MHz are *highly attenuated* and hence long distance communication and broadcasting is performed in frequency range of 10 MHz - 30 MHz

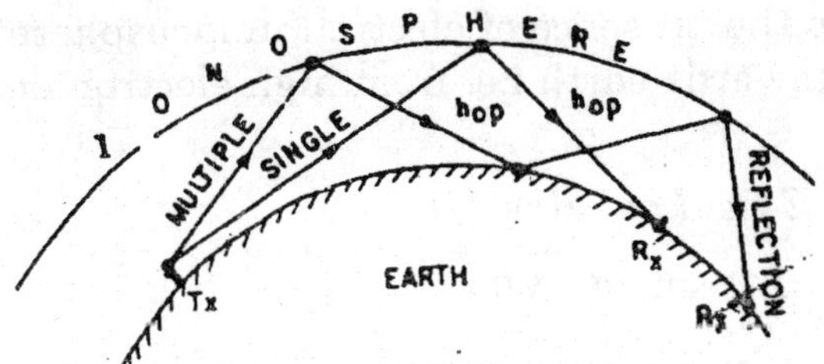

Figure 4.12. Single and multiple hops reflections for long distance communication

As at height higher frequencies around 30 MHz is not at all reflected back to earth, so during *night some lower frequencies* are utilized for long distance or broadcasting.

- Further sky waves follows different paths with in ionosphere and at receiving point, the received signal is vector sum of all, ∴ Fading occurs that is minimized by A.V.C. or Diversity reception.

> **Q.** Explain the formation of ionosphere. What are the various layers of the ionosphere? With the help of a neat diagram show their respectively electron densities and heights above the ground. **[Raj. Univ. 2004]**

4.7.1 Propagation of Radio Waves through the Ionosphere (Neglecting Earth Mag. field Theory of Eccles and Larmor) or Expression for the *Refractive Index of the Ionosphere*

In an ionized medium, having free electrons and ions, when the radio waves pass through it, set these charged particles in motion. So the mass of ions are much heavier than the electrons so their motions are negligible for all practical purposes. The radio waves passing through the ionosphere influenced by electrons only and E-field of RW set electron of ionosphere in motion. These electrons then vibrate simultaneously along paths parallel to E-field of R.W. and vibrating electron represent on *A.C. current proportional in velocity of vibration.* Now the effect of earth's magnetic field of vibrations of ionospheric electron lags behind E-field of wave, thus security e- *current* is inductive. The actual current which is flowing through a volume of space in ionosphere, consists of components e.g. the usual *capacitive current (which loads the voltage by 90°) and the e-current (which lags the voltage by 90°)* hence subtracted from capacitive current.

So free electron in space decreases the current and so the dielectric constant of space is also reduced below the value that would be in the absence electron. The reduction in dielectric constant due to the presence of electron in ionosphere causes the path of R.W. to tilt towards earth i.e. from high electron density to lower e-density.

(1) Let an *E-field* of value

$$E = E_m \sin \omega t \text{ V/m} \tag{1}$$

is acting across a cubic meter of space in ionosphere,

Where ω = angular velocity

E_m = Maximum amplitude

(2) Force exerted by E-field on each electron

$$F = -eE \text{ Newton} \tag{2}$$

Where e = charge of an electron in coulomb.

(3) Let us assume that there is no collision, the electron will have an instantaneous velocity in m/sec. in direction opposite to field.

$$\text{Force} = \text{Mass x Acceleration}$$

$$= mA = \frac{m\,dV}{dt} \tag{3}$$

Equating (2) and (3)

$$-Ee = \frac{m\,dv}{dt}$$

Where, m = mass of e-

$$\frac{dv}{dt} - \text{Acceleration}$$

or $$\frac{dv}{dt} = \frac{-Ee}{m}$$

or $$dv = \frac{-Ee}{m}.dt$$

On integrating both the sides

$$\int dv = -\int \frac{Ee}{m} dt$$

$$\int dv = \frac{-e}{m} \int E\,dt$$

$$= \frac{-e}{m} Em \sin \omega t$$

$$v = \frac{+eE_m \cos \omega t}{m\,w}$$

or

$$\boxed{V = \left(\frac{e}{m\omega}\right) E_m \cos \omega t} \tag{4}$$

Constants of integration are set to zero.

(4) If N be the number of e- per cubic meter, then instantaneous electric current (i_e) constituted by these N - electron moving with instantaneous velocity V, is

$$ie = -Nev \text{ amp/m}^2$$

$$= -Ne\left(\frac{e}{m\omega}\right) E_m \cos \omega t$$

or

$$\boxed{i_e = -\left(\frac{Ne^2}{m\omega}\right) E_m \cos \omega t} \tag{5}$$

It shows that `i_e' lags behind E by 90°.

(5) Besides this inductive current or conduction current component obtained by ionization of air i.e. presence of e and its motion, there is usual capacitive current (i_e) (or displacement current exists in a unionized air). The capacitance of unit volume is

$$k_o = 8.854 \times 10^{-12} \text{ F/m}$$

Hence the capacitive or displacement current through this capacitance is

$$i_c = \frac{dD}{dt} = \frac{d}{dt}(k_o E) \tag{6}$$

$$= k_o \left[\frac{d}{dt}(E_m \sin wt)\right]$$

$$\because D = \varepsilon_o F = K_o F \qquad as\ k_o = constant$$

$$\boxed{i_c = k_o E_m \cos \omega t\,(\omega)} \tag{7}$$

(6) Total current `i' that flows through m^3 of ionized medium is

$$i = i_c + i_e \tag{8}$$

$$i = k_o E_m \omega . \cos \omega t - \frac{Ne^2}{m\omega} E_m \cos \omega t$$

$$i = E_m \cos \omega t\,(\omega)\left[k_o - \frac{Ne^2}{m\omega^2}\right] \tag{9}$$

(7) On comparing (7) and (9). The effective dielectric constant k of ionosphere (i.e. ionized space)

$$k = k_o - \frac{Ne^2}{m\omega^2}$$

$$= k_o \left[1 - \frac{Ne^2}{m\omega^2 k_o}\right]$$

Hence the relative dielectric constant w.r.t. vacuum or air

$$k_r = \frac{k}{k_o} = 1 - \frac{Ne^2}{m\omega^2\, k_o}$$

Thus relative ref. index (μ) of ionosphere w.r.t. vacuum or air (i.e. in ionized air)

$$\mu = \sqrt{k_r} = \sqrt{\frac{k}{k_o}} = \sqrt{1 - \frac{Ne^2}{m\omega^2\, k_o}}$$

$$\because v = \frac{C}{\sqrt{k_r}} = \frac{C}{\mu} \qquad \therefore \mu = \sqrt{k_r} \qquad (10)$$

If we put the values

i.e.

(i) $m = 9.107 \times 10^{-31}\ kg.$

(ii) $e = 1.602 \times 10^{-18}\ coloumbs$

(iii) $k_o = 8.854 \times 10^{-12}\ F/m$

$$= \frac{1}{36\ \pi \times 10^9}\ F/m$$

(iv) $\omega = 2\,\pi\, f.$

We get expression for reference index of ionosphere as

$$\boxed{\mu = \sqrt{1 - \frac{81N}{f^2}}} \qquad (11)$$

Where N = No of e^- per m^3 or ionic density

f = frequency in H_z

If N is per cm^3 then f is in k Hz

(11) Shows that ref. index of ionosphere is less than one where that of un-ionized, medium is one. So presence of electron in ionosphere reduces the ref. index of air & reduction is higher if electrons are more.

Deriving this equation, it is assumed that electrons do not undergo any in elastic collision during their motion and there is no dissipative loss of energy.

4.7.2 Mechanism of R.W. bending by the ionosphere

The bending of R.W. at ionosphere can readily be understood with the help of ref. index formula of the ionized medium i.e.

$$\mu = \sqrt{k_r} = \sqrt{1 - \frac{81N}{f^2}}$$

Where N = ionic density (in M^{-3})

f = frequency (Hz)

Equation (11) shows that real value of refractive index of ionosphere is always less than unity and the deviation of μ from unity becomes greater, if the **ionic density is higher and frequency is lower. If $f^2 < 81$ N = Ref. index becomes imaginary it means under such condition the R.W are attenuated at this frequency and ionosphere is not able to transmit or bend the R.W.**

The bending of R.W. from the ionosphere is governed by ordinary optical laws. By Snell's law, the angle of incidence (i) and refraction (r) at any point is given by

$$\mu = \frac{\sin i}{\sin r}$$

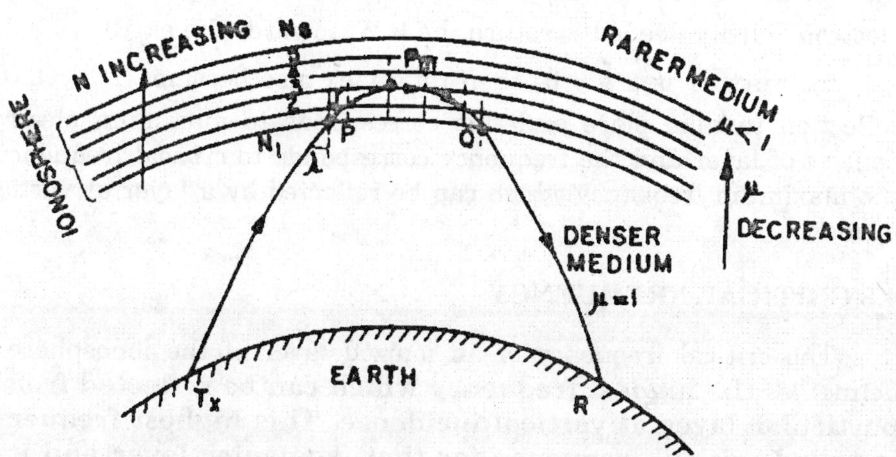

Figure 4.13. Reflection of R.W. in ionosphere

$\because \mu < 1$ For the ionosphere so sin i < sin r i.e. angle of refraction will go on deviating by the normal as the wave will encounter in comparatively rare medium.

If successive layers of the ionosphere are of higher electron density i.e.-

$$N_6 > N_5 > N_4 > N_3 > N_2 > N_1$$

$$\therefore \mu \text{ goes on decreasing}$$

$$\mu_1 > \mu_2 > \mu_3 > \mu_4 > \mu_5 > \mu_6$$

So As if a wave enters at, say point P, will be deviating more and more and a point will reach where it travels parallel to earth (at P_m). Here the angle of refraction is 90° and the point P_m is the highest point in the ionosphere reached by R.W. If μ_n, be refractive index and N_m be maximum electron density at point P_m then

$$\mu_m = \sin i_m$$

$$\because \sin r = \sin 90^\circ = 1$$

The point P_m is usually called as point of reflection although it is actually a point of refraction. At this point total internal reflection takes place and the wave gets bent toward earth and ultimately returns to earth.

Hence the R.W. once enter at point P, leave the ionosphere at point Q after slight penetration into the ionosphere and thus R.W. are reflected back to earth after successive refraction in ionosphere.

(i) Equation (12) suggests that as smaller the angle i, as smaller the refractive index (μ_m), which implies that higher should be electron density needed to return the R.W. towards the earth.

(ii) Further is $\angle i = 0$, then ref. index also becomes $\mu_m = 0$ for reflection to take place and this corresponds to maximum electron density of layer and the frequency corresponds to critical frequency - the maximum frequency which can be reflected by a layer at vertical incidence.

4.8 CRITICAL FREQUENCY

The critical frequency of an ionized layer of the ionosphere is defined as **the highest frequency which can be reflected from a particular layer at vertical incidence. This highest frequency is called critical frequency for that particular layer and it is different for different layers. It is usually denoted by f_o or f_c.**

Critical frequency of the particular regular layer is proportional to square root of maximum e⁻ density in layer as shown below. From equation (11) and (12), we can write-

$$\mu = \frac{\sin i}{\sin r} = \sqrt{1 - \frac{81\,N}{f^2}}$$

By definition, at vertical incidence

Angle of incidence, $\angle i = \theta; N = N_{\max}$ *and* $f = f_c$

As angle of incidence start decreasing and reaches to zero (i.e. vertical incidence) the e⁻ density starts increasing and reaches to maximum electron density (N_m). The highest frequency which can be reflected back by the ionosphere is one for which ref. index `μ' becomes zero.

$$\therefore \mu = \frac{\sin \theta}{\sin r} = \sqrt{1 - \frac{81\,Nm}{f_c^2}} = 0$$

$$1 = \frac{81\,Nm}{f_c^2}$$

or $$f_c = \sqrt{81\,(N_m)}$$

or $$f_c = 9\sqrt{N_m} \tag{13}$$

Where, f_c is expressed in MHz and N_m is expressed in per cubic meter. So if the maximum electron density N_m is known the critical frequency can be calculated by equation (13). Surely critical frequency is the highest frequency which can be reflected by a particular layer **at vertical incidence but it is, not the highest frequency that will get reflected for any other angle of incidence.** The frequency which can be reflected from a layer is a function of angle of incidence i) and is called MUF (maximum usable frequency)

So critical frequency gives an idea that Radio Wave o frequency equal to or less than the critical frequency will certainly be reflected back by the ionospheric layer irrespective of angle of incidence. Radio waves of frequency higher than critical frequency will also be returned to earth only when the angle of incidence i) is sufficiently glancing so that equation (11) is satisfied at the frequency involved, otherwise the wave will penetrate the layer concerned. It may be reflected back by a higher layer Thus for a wave of frequency greater than critical frequency which is going to be reflected, the condition is

$$\sin i > \mu_m$$

$$> \sqrt{1 - \frac{81\,N_m}{f^2}}$$

$$\sin i > \sqrt{1 - \frac{f_c^2}{f^2}}$$

Where $$fc = \frac{81\,N_m}{f^2} = \frac{N_m\,e^2}{m\,4\,\pi^2 f^2 k_o}$$

4.9 VIRTUAL HEIGHT

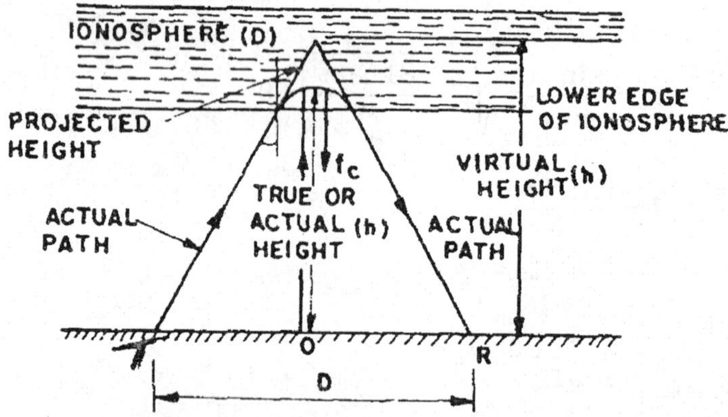

Figure 4.14. Virtual and actual heights of ionized layer

It can be understood with help of the given diagram. Figure shows that wave is refracted by the layer, it is tilted down gradually rather sharply. The actual path of wave in ionized layer is a curve and is due to refraction of wave, as happens in case of refraction from the prism. As it is more convenient to think about the wave being reflected rather the wave being refracted.

∴ the path can be assumed straight lines TD and RD as shown. This assumption is made in measurement of height of layer.

OD -> virtual height of ionized layer as it is not true height.

Virtual height is always greater than the actual height

- If the virtual height of layer is known, then it is easy to calculate the angle of incidence required for the wave to return to earth at desired point.

Definition : ``The height to which a short pulse of energy sent vertically upward and travelling with the speed of light would reach taking the same two ways travel time as does the actual pulse reflected from the layer.

Note: In the measurement of virtual height the tx point and the receiving point (R) are usually placed very close together so that the wave sent nearly vertically upward.

The most common method of height virtual measurement is in which the transmitted signal consists of pulses of RF energy of small duration. The receiver which is located close to Tx, pick up both the direct and reflected signals. The gap between these signals on the time axis of CRO gives a measurement of height of layer. The actual height (h) is smaller than virtual height **because the interchange of energy takes place between the wave and the electron of ionospheres causes the velocity of propagation to be reduced.**

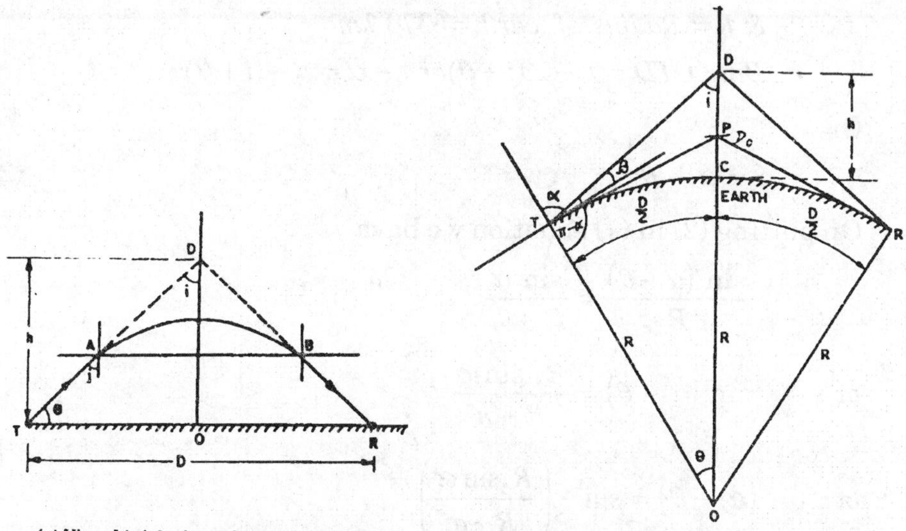

(a) Virtual height determination for flat earth. *(b) Virtual height determination for curved earth.*

Figure 4.15. Virtual height determination

The virtual has highest advantage of being easily measured, and it is very useful in Tx path calculations.

For flat earth assumptions and assuming that ionospheric conditions are symmetrical for incident and reflected waves, Tx. path distance TR is obtained from

$$\tan \beta = \frac{Do}{To} = \frac{h}{T_R/2}$$

or $$\frac{T_R}{2} = \frac{h}{\tan \beta}$$

$$\therefore T_o = \frac{T_R}{2}$$

$$\boxed{\therefore T_R = \frac{2h}{\tan\beta} = D} \qquad (1)$$

When curvature of earth is accounted for, then the Tx path distance may be calculated from the geometry from this figure. From ΔTOD

$$\frac{\sin i}{R} = \frac{\sin(\pi - \alpha)}{R+h} = \frac{\sin\alpha}{R+h}$$

$$or \ \frac{\sin i}{R} = \frac{\sin\alpha}{R+h} \qquad \because \sin(\pi-\alpha) = \sin\alpha.$$

$\& R = Radius\ of\ earth = 6370\ km.$

$But \angle OTD = \pi - \angle(i+\theta) or \pi - \alpha = \pi - (i+\theta)$

Or

$$i = \alpha - \theta \qquad (2)$$

On putting (2) in (1) equation we have

$$\frac{\sin(\alpha - \theta)}{R} = \frac{\sin\alpha}{R+h}$$

or $$\sin(\alpha - \theta) = \frac{R.\sin\alpha}{R+h}$$

or $$(\alpha - \theta) = \sin^{-1}\left(\frac{R.\sin\alpha}{R+h}\right)$$

or

$$\theta = \alpha - \sin^{-1}\left(\frac{R\sin\alpha}{R+h}\right) \qquad (3)$$

Further from the figure

$$90° = \alpha + \beta$$

Or

$$\boxed{\alpha = 90° - \beta} \qquad (4)$$

Hence putting - (4) in terms of angle of elevation θ, can be written as

$$\theta = (90^\circ - \beta) - \sin^{-1}\left(\frac{R \sin \alpha}{R + h}\right)$$

$$= (90^\circ - \beta) - \sin^{-1}\left(\frac{R \sin (90 - \beta)}{R + h}\right)$$

$$\boxed{\text{or } \theta = 90^\circ - \beta - \sin^{-1}\left(\frac{R \cos \beta}{R + h}\right) \text{ Radians}} \quad (5)$$

Now we k.t $\quad Angle = \dfrac{Arc}{Radius}$

$$\theta = \frac{Arc \ \ Tc}{R} = \frac{D/2}{R} = \frac{D}{2R}$$

or $\quad D = 2R . \theta = 2 R \left[(90^\circ - \beta) - \sin^{-1}\left(\frac{R \cos \beta}{R + h}\right)\right] \quad (6)$

Q.1. Explain virtual height and prove it.	**(R.T.U. 2009)**

4.10 IONOSONDE

Measurement of virtual height is done by its means. The basic method is to transmit vertically upward a PM-radio wave with pulse duration of about 150 m - sec. The reflected signal is received close to the tx-point and the time required T, for round trip measurement. The virtual height is then given by

$$\boxed{h = \frac{CT}{2} = \text{virtual height}}$$

Where, C - Velocity of light = 3 x 10^8 m/sec.

4.11 MAXIMUM USABLE FREQUENCY (MUF)

Critical frequency is maximum frequency of R.W. which is returned from a ionized layer at vertical incidence. In case the frequency of R.W. exceeds the critical frequency, the influence of the ionospheric layer on the path of propagation (i.e. communication) depends on the angle of incidence at the ionosphere. So, MUF is also a limiting frequency which can be reflected back to earth but this time for specific angle of incidence rather than the vertical. The highest possible value of frequency for which reflection takes place for a given distance of propagation, is called as MUF for that distance, and for given ionospheric layer.

If the wave frequency is higher than this, then wave crosses ionized layer and does not reflect back to the earth.

Definition (2) frequency which gives receiving point corresponds to a distance from the transmitter equal to skip distance.

Definition (3) MUF is the highest frequency which can be used for sky wave communication between 2 given points on earth.

MUF is greatest frequency which can be used for sky wave communication between 2 given points on the earth. There is different value of MUF for each pair of points on globe.

MUF Range = 8 - 35 MHz

After unusual solar activity MUF may be as high as 50 MHz. Now highest working frequency between two particular points on earth is obviously a bit less than MUF.

For Sky wave to return to earth, angle of refraction i.e.-

$\angle r = 90^\circ$ *which* $= N = N_{max}$ $f = f_{max}$ *i.e.* Maximum values of same

$$\mu = \frac{\sin i}{\sin 90^\circ} = \sqrt{1 - \frac{81\, N_m}{f_{muf}^2}} \quad \textit{Applying condition of MUF}$$

$$\mu = \sin i = \sqrt{1 - \frac{81\, Nm}{f_{muf}^2}} \quad \textit{But } f_c^2 = 81\, Nm$$

$$\therefore \sin i = \sqrt{1 - \frac{f_c^2}{f_{muf}^2}}$$

or $$\sin^2 i = 1 - \frac{f_c^2}{f_{muf}^2}$$

or $$\frac{f_c^2}{f_{muf}^2} = 1 - \sin^2 i = \cos^2 i$$

$$f_{muf}^2 = f_c^2 \cdot \sec^2 i$$

$$\boxed{f_{muf} = \sec (i).\, f_c}$$

Conclusion

$f_{muf} > f_c$ By the factor Sec i. **This is called as SECANT LAW** and gives maximum frequency which can be used for sky wave communication for a given angle of incidence i) between 2 points on earth.

This equation is valid till a distance of 1000 km. However, as the gap between 2 points (Tx and Rx) is increased a limit occurs due to curvature of earth (fig) where the path of wave is tangent to surface of earth at these 2 points. The angle i) corresponding to its limiting distance is about 74° for F- layer. Hence MUF is given by (for this case)-

$$F_{muf} = \sec 74.\ f_c$$

$$\boxed{f_{muf} = 3.6\ f_c}$$

This is the maximum f which can be reflected from ionosphere.

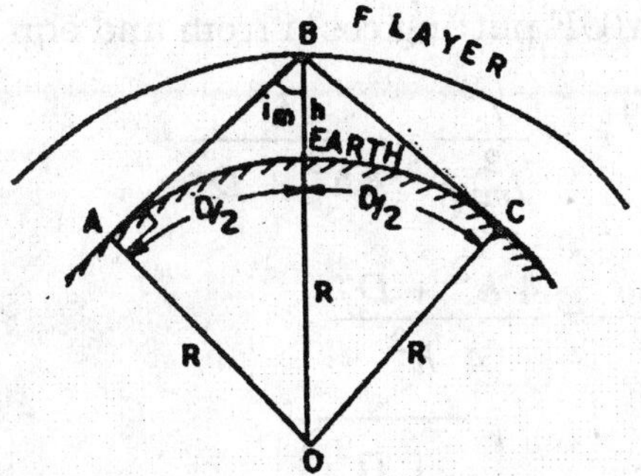

Figure 4.16. Maximum angle of incidence of wave

4.11.1 Calculation of MUF

Case I - Thin layer (or flat Earth)

The ionized layer may be assumed to be thin layer with sharp ionization density gradient, which gives mirror image reflection of R.W.

For the shorter distance of communication (say up to 500 km) the earth can be assumed to be flat.

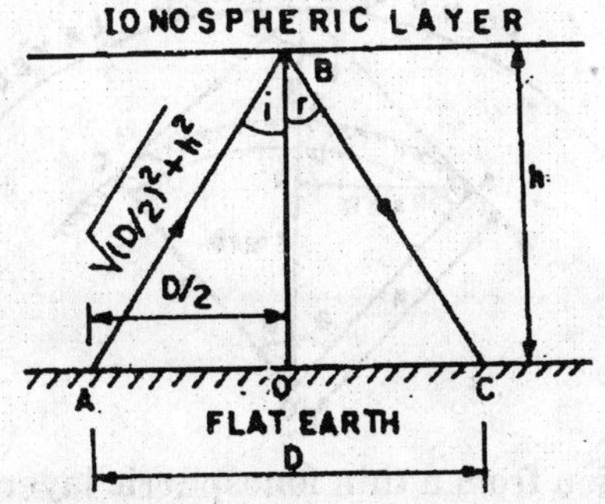

Figure 4.17. Reflection from a thin layer on flat earth

From figure = $\cos i = \frac{BO}{AB} = \frac{h}{\sqrt{h^2 + (D/2)^2}}$

$$= \frac{2h}{\sqrt{4h^2 + D^2}} \quad (1)$$

h - Height of Layer
D - Propagation distance AC

(1) $$\mu = \sin i = \sqrt{1 - \frac{81\, Nm}{f_{muf}^2}} = \sqrt{1 - \frac{f_c^2}{f_{muf}^2}}$$

Where fc = critical frequency

f_{muf} = MUF putting $\cos^2 i$ from and eqn. (1)

$$\cos^2 i = \frac{f_c^2}{f_{muf}^2} = \frac{4h^2}{4h^2 + D^2}$$

or $$\frac{f_{muf}^2}{f_c^2} = \frac{4\,h^2 + D^2}{4\ h^2}$$

$$\frac{f_{muf}}{f_c} = \sqrt{1 + \left(\frac{D^2}{4h^2}\right)}$$

$$f_{muf} = f_c\sqrt{1 - \left(\frac{D}{2h}\right)^2}$$

This is expression for maximum UF in terms of critical frequency (fc), propagation distance D and height of layer h.

Case II. Thin layer (curved Earth)

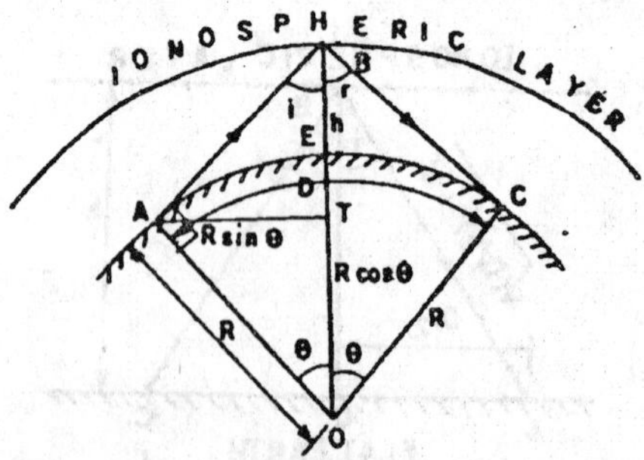

Figure 4.18. Reflection from a thin ionospheric layer but on curved earth according its curvature

If the curvature of earth is taken into account, the reflecting region is considered to be concentric with earth. In Figure 4.18, the transmitted wave leaves the Tx tangentially to earth. Let 2 θ be the angle subs tented by Tx distance D at the centre of O (earth).

$$\text{Then - Angle} = \frac{Arc}{Radius}$$

$$\therefore\ 2\,\theta = \frac{D}{R}$$

or

$$\boxed{D = 2R\theta}$$

Now At = R sin θ and OT = R cos θ

$$\therefore\ BT = OE + EB - OT$$

$$= \text{h} + \text{R} - \text{R}\cos\theta$$

$$\text{AB} = \sqrt{AT^2 + BT^2}$$

$$= \sqrt{(R\sin\theta)^2 + (h + R - R\cos\theta)^2}$$

$$\text{Hence, } \cos i = \frac{BT}{AB} = \frac{h + R - R\cos\theta}{(R\sin\theta)^2 + (h + R - R\cos\theta)^2}$$

$$\text{from equation } = f_{muf}^2 = f_c^2 \,.\, \sec^2 i$$

$$\cos^2 i = \frac{f_c^2}{f_{muf}^2} = \frac{(h + R - R\cos\theta)^2}{(R\sin\theta)^2 + (h + R - R\cos\theta)^2}$$

The curvature of earth limits both the MUF and skip distance D and the limit is obtained when waves leave the transmitter at grazing angle (implies $\angle OAB = 90°$).

Thus when D is maximum (i.e. max. skip distance to be seen next), θ is maximum, given by

$$\cos\theta = \frac{OA}{OB} = \frac{R}{R + h} \qquad (3)$$

However, actual value of θ is very small, so equation (3) can be expended

$$\cos\theta = \frac{R}{R\left(1 + \dfrac{h}{R}\right)} = \left(1 + \frac{h}{R}\right)^{-1}$$

$$\cos\theta = \left(1 - \frac{h}{R} + \ldots\right) \qquad \because \frac{h}{R} << 1$$

$$1 - \frac{\theta^2}{2} = 1 - \frac{h}{R}$$

or $$\theta^2 = \frac{2h}{R}$$

Hence from equation $= 2\theta = \dfrac{D}{R}$

$$D^2 = 4R^2\ \theta^2$$

$$= 4\ R^2 \cdot \frac{2h}{R}$$

$$= 8\ \text{h R.}$$

$$\boxed{\therefore h = \frac{D^2}{8\,R}} \qquad (4)$$

From equation $\cos\theta = \left(1 - \dfrac{h}{R} + \ldots\right)$

$$\boxed{\cos\theta = \left(1 - \frac{D^2}{8R^2}\right) \quad \textit{and as } \theta \textit{ is small}} \qquad (5)$$

$$\boxed{\sin\theta \approx \theta = \frac{D}{2\,R}} \qquad (6)$$

on Putting equation (5) and (6) in equation (2)

$$\frac{f_c^2}{\left(f_{muf}^2\right)_{\max}} = \frac{\left\{h + R - R\left(1 - \dfrac{D^2}{8R^2}\right)\right\}^2}{R^2 \cdot \dfrac{b^2}{4\,R^2} + \left[h + R - R\left(1 - \dfrac{D^2}{8\,R^2}\right)\right]^2}$$

$$= \frac{\left(h + D^2/8R\right)^2}{\dfrac{D^2}{4} + \left(h + \dfrac{D^2}{8R}\right)^2}$$

$$\text{or} \quad \frac{\left(f_{muf}\right)_{\max}}{f_c} = \frac{\sqrt{D^2/4 + \left\{h + D^2/8R\right\}^2}}{\sqrt{\left(h + D^2/8R\right)^2}}$$

$$\boxed{\left(f_{muf}\right)_{\max} = f_c \frac{\sqrt{D^2/4 + \left(h + D^2/8R\right)^2}}{\sqrt{\left(4 + \dfrac{D^2}{8R}\right)^2}}} \qquad (7)$$

From equation (4)

$$\boxed{\left(D_{skip}\right)_{\max} = \sqrt{8hR}} \qquad (8)$$

As D is nothing but skip distance

$$D = 2\left(h + \frac{D^2}{8\,R}\right)\sqrt{\left(\frac{f_m}{f_c}\right)^2 - 1}$$

4.12 LOWEST USABLE FREQUENCY: (LUHF/LUF)

The absorption of (HF) RW in D-region of ionosphere is proportional to the inverse square of frequency. The sensitivity of HF receiver is normally limited by external noise which increases as frequency is reduced. So there is a frequency limit below which the SNR fails to reach an acceptable value for the service required.

LUF depends upon = engineering characteristics of link viz. transmitted power.

In addition to absorption limitations, the signal can reduce its energy after it has been transmitted by several other mechanisms for e.g.:

(i) Free space propagation i.e. spatial spreading of energy.

(ii) Polarization change caused by earth's magnetic field.

(iii) Scatter process

(iv) Focusing and Defocusing caused by ionospheric curvature.

These processes complicate evaluation of expected performance of the circuit.

For a given distance of Tx which has limited transmitter power, the lowest usable high frequency for sky wave propagation is limited mainly due to following 2 phenomena:

(1) Sky wave absorption increases with decreasing frequency and so reducing the received field intensity for the lower frequencies.

(2) The atmospheric noise as also mainly of man-made electrical noises increases with decreasing frequency in short wave or high f range. It results, with lowering of Tx frequency a situation arises where received noise field is stronger then desired signal intensity.

As the MUF limits maximum permissible frequency for sky wave propagation in a given path, the LUF gives lowest permissible frequency. For lower `f' of Tx the received sky wave signal gets lost in the background noise and no communication is possible. The LUF is limited by absorption in D-layer during day light hours. Whereas during night, at lower frequencies the layer is limited by the maximum noise. The value of day time LUF is normally much higher than the night time and further increases during SID's. The value of LUF can be calculated from the measurement of noise level at receiving site and the estimated value of sky wave absorption in given propagation path.

4.13 SKIP-DISTANCE

Radio wave radiated horizontally by transmitter near earth's surface is quickly absorbed due to large ground loses and so only short distance communication is carried out by this horizontal radiations of ground or surface wave. Radio wave radiated at high angle may not tilt sufficiently at the ionospheric layers to return to earth at all that's why escapes rather penetrate the layer. By this way radio wave radiated at shallow angle (i.e. angle between horizontal and high angle) just great enough to escape absorption by earth, will enter the lower layer, suffer attenuation, be tilted at upper layer and return to earth.

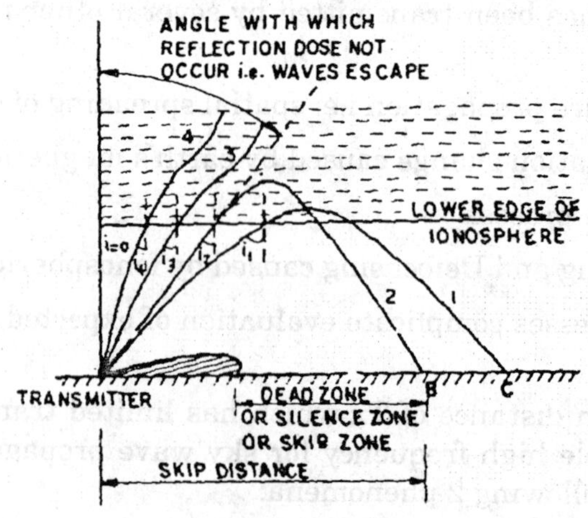

Figure 4.19. Skip distance

Definition: The distance at which surface wave becomes negligible and the distance at which first wave returns to earth from the ionospheric layer, here is a zone which is not covered by any wave i.e. neither ground nor sky) this is called skip zone or area and distance across it is skip distance.

Definition: It is very usual to consider skip distance from transmitter to the point where first sky wave is received, as range of surface wave is always small.

Hence it can be redefined as-

(1) Minimum distance from the Tx at which a sky wave of a fixed frequency is returned to earth by the ionosphere. It is represented by D.

(2) The minimum distance from the Tx to a point where sky wave of a given frequency is first received.

(3) Minimum distance in which a sky of given frequency fails to be reflected back.

(4) Minimum distance for which sky wave propagation just takes place and no sky wave propagation is possible for points nearer than this distance

$$frequency\ \ increases\ \ =\ \ skip\ \ distance\ \ increases$$

- Frequency less than critical frequency of layer skip distance is zero.
- The effect of ionosphere is dependent on the angle of incidence at ionosphere in which waves of diff. angle of incidence is shown.
- -As the angle of incidence decreases (↓), the distance from transmitter, at which the ray returns, first decreases.

It is continues till an angle of incidence is reached at which the distance becomes minimum. The minimum distance is called skip distance D.

With furthers decrease in angle of incidence the wave penetrates the layer and does (3 and 4) not reverted to earth. **Skip distance is the distance skipped over by sky wave.** This happen because

(1) As the angle of incidence `i' is large.

$$\mu = \sin i = \sqrt{1 - \frac{81N}{f^2}}\ \ is\ satisfied\ with\ \ small\ e^{-density}$$

Where $\mu = .$ *slightly less than unity & hence wave returns after slight penetration into layer*

As the $\angle i$ is further decreased (2) sin i decreases and so the μ and N becomes comparatively more ∴ wave penetrate still more before it reaches to earth.

Lastly when $\angle i$ is small enough, so that μ = sin i can not be satisfied even by maximum electron density of layer, then wave penetrates (3 and 4)

The freq. which makes a given distance corresponds to skip distance, is MUF for those two pts. If Rx. is placed with skip distance no signal is heard unless ground wave is strong enough as at A.

For given freq. of propagation $f = f_{mum}$

∴ Skip distance can be calculated

$$\frac{f_{muf}}{f_c} = \sqrt{1 + \left(\frac{D}{2h}\right)^2}$$

$$\left(\frac{f_{muf}}{f_c}\right)^2 - 1 = \left(\frac{D_{skip}}{2h}\right)^2$$

$$\therefore D_{skip} = 2\,h\sqrt{\left(\frac{f_{muf}}{f_c^2}\right)^2 - 1}$$

> 1. Explain Define short questions:
>
> i. Skip Distance. **(R.T.U. 2009)**

4.14 OPTIMUM WORKING FREQUENCY (OWF) (and day and night frequency)

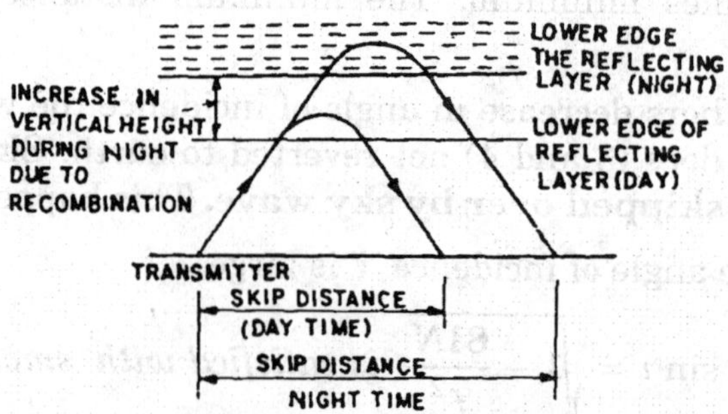

Figure 4.20. Optimum Working Frequency

- For radio communication, for satisfactory reception the frequency should be less than MUF absorption of waves by ionosphere be small.
- Due to presence of free electron which gives rise to certain conductivity and this effect is important at lower edge of the ionosphere.
- Absorption $\alpha \dfrac{1}{f^2}$
- **So highest frequency gives strongest sky wave signal at T_x. ((≈ *MUF*)**
- Optimum frequency is nearly 15% of mean value of daily variation (monthly observation) so to use a frequency 85% of predicted MUF.
- **∴ There is a frequency called OWF or optimum Traffic frequency (OTF) which is (50-85)% of MUF is used to accommodate a no. of channels i.e. (OWF = 85% MUF)**
- As MUF changes from time to time, so the same with OWF.
- But it is not possible to change the frequency from time to time ∴ it is necessary to the at least 2 frequency, one for day and night each.
- Third frequency - is used for transition period (sometimes)
- Vertical Height (Night) > Vertical height (day)

  ∴ Skip distance increases at night.

Example 1. A transmitter radiates 35 watts of power at a wavelength of 6 cm. Calculate the power received by an antenna of 150 km if the gain of the transmitting and receiving antennas is equal and has a value of 40 dB.

Solution:

Given, $(W_T)\text{ dB} = 10\log_{10} W_T$

$$= 10\log_{10} 35$$

$$= 10 \times 1.544 = 15.44$$

$$W_R = ? \text{ and } G_T = G_R = 40\text{ dB}$$

$$\lambda = 6\text{ cm} = \frac{6}{100}\text{ meters and } \therefore \lambda = \frac{300}{f\text{ (MHz)}}$$

$$d = 150\text{ km}$$

Therefore, $\left(\dfrac{W_R}{W_T}\right)_{dB} = G_T\,(dB) + G_R\,(dB) - L_S\,(dB)$

$$L_S\,(dB) = 32.45 + 20 \log 150 + 20 \log_{10} 5000$$
$$= 32.45 + 43.52 + 73.98$$
$$= 149.949$$

Then $W_R\,(dB) = W_T\,(dB) + G_T\,(dB) + G_R\,(dB) - L_S\,(dB)$

$$= 15.44 + 40 + 40 - 149.949$$
$$= 95.44 - 149.949$$
$$= -54.5094 \text{ dB}$$

As, the transmitting power is given in watts we may calculate power received also in watts as follows

$$W_R\,(dB) = 10 \log_{10} W_R$$
$$-54.5094 = 10 \log_{10} W_R$$
$$W_R = \text{antilog}\,(-54.5094)$$
$$= 3.540 \times 10^{-6}$$
$$= 3.54\ \mu \text{ watt}$$

Ans.

Example 2. A communication link is to be established between two stations using half wavelength antenna for maximum directive gain. Transmitter power is 2 kW, distance between transmitter and receiver is 200 km. What is the maximum power received by the receiver, frequency of operation is 150 MHz?

Solution: The maximum received power by antenna is given by

$$\frac{W_R}{W_T} = G_T G_R \left(\frac{\lambda}{4 \pi d}\right)^2$$

Where, W_R = Received power by antenna

W_T = Transmitted power

G_R = Directivity of receiver = 1.64

G_T = Directivity of transmitter = 1.64

$$\lambda = \text{wavelength} = \frac{300}{f\,(\text{MHz})} = \frac{300}{150} = 2 \text{ meters}$$

$$W_T = 2 \times 10^3 \text{ watts}$$

d = 200 km = 200 x 10^3 meters

$$(W_R)_{max} = W_T G_T G_R \left(\frac{\lambda}{4\pi d}\right)^2$$

$$= 2\times10^3 \times(1.64)^2 \left(\frac{2}{4\pi\times200\times10^3}\right)^2$$

$$= \frac{8\times10^3 \times(1.64)^2}{16\times(3.14)^2 \times(200)^2 \times10^6}$$

$$= \frac{21516.8}{6.3110144\times10^{12}}$$

$$= 3.4099\times10^{-9} \text{ watts}$$ **Ans.**

Example 3. What is the maximum power that can be received over a distance of 1.5 cm in free space with a 1.5 GHz circuit consisting of the transmitting antenna with a gain of 30 dB? Transmitted power is 200 w.

Solution: Given

$$G_T = 25\,\text{dB} = 316.228$$

$$G_R = 30\,\text{dB} = 1000$$

$$\lambda = \frac{3\times10^5}{1.5\times10^9} = 0.2\,\text{m}$$

The maximum power received will be

$$P_R = P_T G_T G_R \left(\frac{\lambda}{4\pi R}\right)^2$$

$$= 200\times316.228\times1000\times\left(\frac{0.20}{4\pi\times1.5\times10^3}\right)^2$$

$$= 7.120\,\text{mW}$$ **Ans.**

Example 4. Microwave link is 50 Km long. It operates at 2GHz with antenna gains of 20 dB at either end. What is the voltage available at the terminals of a matched load 50W receiver if the transmitted power is $P_i = 1\,\text{w}$? Ignore problems of fading and assume an unobstructed line of sight path.

Solution: Given $G_R = G_T = 20\text{dB} = 100$

$$\lambda = \frac{3\times10^5}{2\times10^9} = 0.15\,\text{m}$$

The received power is given by

$$P_R = P_T G_T G_R \left(\frac{\lambda}{4\pi R}\right)^2$$

Putting the values of parameters we get

$$P_R = 1\times100\times100\times\left(\frac{0.15}{4\pi\times50\times10^3}\right)^2$$

$$= \frac{15\times15}{16\pi^2\times25\times10^8} = 0.05699\times10^{-8}\ \text{w}$$

Since load implied is 50 ohm, we can write for the terminal voltage V the following relation

$$\left(\frac{V}{R}\right)^2 R = P_R$$

$$V = \sqrt{P_R\times R}$$

$$= \sqrt{0.05690\times10^{-8}\times50}$$

$$= 1.688\times10^{-4}$$

$$= 168.8\times10^{-6}\ \text{V}$$

$= 168.8\ \mu$ volt. **Ans.**

Example 5. Two identical 1.22 m diameter parabolic dish antenna with typical effective antenna areas are used to transmit and received at a 4 GHz over a 96 km free space path. What transmitter power is required for a received signal of –90dBm?

Solution: The received power is given by

$$P_R = P_T G_T G_R \left(\frac{\lambda}{4\pi d}\right)^2$$

Since two antennas are symmetrical $G_T = G_R = G$

Thus, $$\frac{P_R}{P_T} = G^2\left(\frac{\lambda}{4\pi d}\right)^2$$

$$= \left(\frac{4\pi A}{\lambda^2}\right)^2 \left(\frac{\lambda}{4\pi d}\right)^2 = \frac{A^2}{\lambda^2 d^2}$$

$$P_T = P_R \frac{\lambda^2 d^2}{A^2}$$

For this case $\lambda = \frac{3\times10^8}{4\times10^9} = 0.075\,\text{m}$

$$A = \frac{\pi D^2}{4} = \frac{\pi\times(1.22)^2}{4} = 1.1689\,\text{m}^2$$

Also, $\quad -90\,\text{dBm} = 10\log\left(\frac{P_R}{10^{-3}}\right)$

or $\quad P_R = 10^{-3}\times10^{-9} = 10^{-12}$ watt

Thus $\quad P_T = \frac{10^{-12}\times(0.075)^2\times(96\times10^3)^2}{(1.1689)^2}$

$$= \frac{10^{-6}\times(0.075)^2\times(96)^2}{(1.1689)^2} = \frac{51.84\times10^{-6}}{1.36648} = 37.9369\,\mu\text{W}$$ **Ans.**

Example 6. Find the electric field strength at 10 km from a 1.5 MHz transmitter having E_D = 2500 mw/m over a ground path with $\sigma = 3\,\text{ms/m}$ and $s_r = 7$.

Solution: Numerical distance

$$P = \frac{0.582\,d_{km}\,f^2\,(\text{MHz})}{\sigma\,(\text{ms/m})} = \frac{0.582\times10\times(1.5)^2}{3}$$

$$= \frac{0.582\times22.5}{3} = 0.582\times7.5 = 4.365$$

But $\quad A = \frac{2+0.3P}{2+P+0.6P^2} = \frac{2+0.3\times4.365}{2+4.365+0.6\times(4.365)^2}$

$$= \frac{2+1.3095}{6.365+0.6\times19.0532}$$

$$= \frac{3.3095}{6.365+11.4319} = \frac{3.3095}{17.7959} = 0.185959$$

$$\simeq 0.186$$

and $E_v = 2500$ mv/m

$$E_s = \frac{E_0 A}{d} = \frac{2500 \times 0.186}{10} = 250 \times 0.186$$

$E_s = 46.5$ mv/m **Ans.**

Example 7. A 50 kW transmitter employs a vertical grounded half-wave antenna with a directivity of 1.41 as compared to a short dipole. If the transmitter frequency is 1 MHz, find the field $\varepsilon_r = 15$ and $\sigma = 10^{-3}$ mhos/m

Solution: Given $f = 1$ MHz $\sigma = 10^{-3}$ mhos/m;

$\varepsilon_r = 15$; $P = 50$ Kw

$$x = \frac{1.8 \times 10^{12}\ \sigma}{f} \text{ mho/m}$$

$$x = \frac{18 \times 10^{12} \times \sigma}{1 \times 10^{6}} \text{ mhos/m}$$

$$x = \frac{1.8 \times 10^{3}}{100} \text{ mhos/m} = 18 \text{ mho/m}$$

$$b = \tan^{-1}\left(\frac{\varepsilon_r + 1}{x}\right) = \tan^{-1}\left(\frac{15+1}{18}\right) = \tan^{-1}(0.8888)$$

$= 41.62°$

$$P = \frac{\pi}{x} \cdot \frac{d}{\lambda} \cos b;$$

$$d = \frac{100}{3\sqrt{f}} = \frac{100}{\sqrt{1}} \text{ km} = 100 \text{ km};$$

$$\therefore \lambda = \frac{300}{1} = 300 \text{ m}$$

$$= \frac{3.14}{18} \cdot \frac{100 \times 10^{3}}{300} \cos 41.62 = \frac{3.14}{54} \times 10^{3} \times 0.7482$$

$$= \frac{2.349348 \times 10^{3}}{54} = 43.50$$

$$A = \frac{2 + 0.3\,P}{2 + P + 0.6\,P^2} = \frac{2 + 0.3 \times 43.5}{2 + 43.5 + 0.6 \times (43.5)^2}$$

$$= \frac{2+13.05}{45.5+0.6\times1892.25}$$

$$= \frac{15.05}{45.5+1135.35} = \frac{2+15.05}{1180.85} = 0.012745$$

Hence, Eg. $\frac{E_v A}{d} - \frac{300\sqrt{P}}{d}$ A

As the directivity is 1.41 so,

$$Eg = \frac{300\sqrt{50}}{100} \times 0.012745\times 1.41$$

$$= 3\times \sqrt{50} \times 0.0179704$$

$$= 0.0539\times7.07 = 0.3811 \text{ mw/m}$$ **Ans.**

Example 8. A television transmitting antenna mounted at a height of 120 m radiates 15 km of power equally in all directions in azimuth at a frequency of 50 MHz. Calculate (i) maximum line of sight range (ii) the field strength at a receiving antenna mounted at a height by 16 m at a distance of 12 km and (iii) the distance at which the field strength reduces to 1 mv/m.

$$LOS = 4.12\left[\sqrt{h_1}+\sqrt{h_f}\right]$$

Solution (i): $LOS = 4.12\left[\sqrt{120}+\sqrt{16}\right]$

$$= 4.12\ [10.9544 + 4]$$

$$= 4.12 \times 14.9544 = 61.61 \text{ km}$$ **Ans.**

(ii) $E_R = \frac{88\sqrt{P}\,h_1 h_r}{\lambda d^2} = \frac{88\sqrt{15000}\times 120\times16}{6\times(12\times10^3)^2}$

$$= \frac{88\times122.474\times20\times16}{144\times10^6} = 23.95 \text{ mv}$$ **Ans.**

(iii) $d^2 = \frac{88\sqrt{P}\,h_r h_f}{\lambda E_R} = \frac{88\sqrt{15000}\times120\times16}{6\times1\times10^{-3}}$

$$d^2 = 88\times122.474\times20\times16 = 3448867.8\times10^2$$

or $d = \sqrt{3448867.8\times100}$

$$d = 587270 \text{ x } 10$$

$$d = 58727.0 \text{ km}$$ **Ans.**

Example 9. Find the maximum possible distance along the surface of earth for tropospheric transmission. The height of transmitting antenna is 100 ft that of receiving antenna is 50 ft.

Solution: $d = \sqrt{2}\left[\sqrt{h_f} + \sqrt{h_r}\right]$

$$d = 1.4142\ [\sqrt{100} + \sqrt{50}\,]$$

$$= 1.4142 \times 17.07$$

$$d = 24.1418 \text{ meter}$$ **Ans.**

Example 10. A part of microwave link can be assumed to be a free space conditions. The antenna, gains are each 40 db. The frequency is 10 GHz and the path length is 80 km. calculate the transmission path loss & the received power for a transmitted power of 10 watts.

Solution: Given that

$G_T = 40$ db

$G_R = 40$ dB

$f = 10$ GHz $= 10{,}000$ MHz

$d = 80$ km.

$W_T = 10$ W

$$L_s\,(dB) = 20 \log_{10} f + 20 \log_{10} (d) - 147.56$$

$$= 20 \log_{10} 10 \times 10^9 + 20 \log_{10} 80000 - 147.56$$

$$= 200 + 18.06 + 80 - 147.56$$

$$= 150.512 \text{ dB}$$ **Ans.**

$$(W_R)\,dB = G_T\,dB + G_R\,dB - L_s\,dB + W_T\,dB$$

$$= 40 + 40 - 150.512 + 10$$

$$= -70.512 + 10 = -60.512$$

$[W_T\,dB = 10 \log_{10} 10 = 10\text{ dB}]$

Converting dB in watts

$W_R = 0.8908$ Watts **Ans.**

Example 11. The antenna of gain 2.2 dB of a VHF transmitter 5 KW at 62.25 MHz is located at a height of 40 meters above the surface of earth. If the height of the receiving antenna of gains 1.6 dB is 10 meters only. What is the ultimate maximum distance up to which a LOS communication may be possible? Assume a standard atmosphere. What will be power received at this maximum distance.

[Raj. Univ. 1993, 1998, 2005]

Solution: According to question following parameters are given:

(i) Gain of transmitting antenna G_t = 2.2 dB

(ii) Power of transmitting antenna W_t = 5 KW

(iii) Carrier frequency = 62.25 MHz

(iv) Height of transmitting antenna h_t = 40 meters

(v) Height of receiving antenna h_r = 10 meters

(vi) Gain of receiving antenna G_r = 1.6 dB

Using the concept of effective earth radius and equation below for standard atmosphere-

$$d = 4.12\left[\sqrt{h_r} + \sqrt{h_t}\right] km$$

$$d = 4.12\left[\sqrt{40} + \sqrt{10}\right]$$

$$d = 39.085 \text{ km}$$ **Ans.**

Now using formula for free space loss

$$Ls\,dB = 20\log_{10} f + 20\log_{10} d - 147.56$$

$$L_s\,dB = 20\log_{10}\left(62.25\times10^6\right) + 20\log_{10}\left(39.085\times10^3\right) - 147.56$$

$$L_s\,dB = 155.88 + 91.84 - 147.56 = 100.16\,dB$$ **Ans.**

and W_t in dB is

$$W_t\,dB = 10\log_{10} W_t \text{ in watt} = 10\log_{10} 5\times10^3 = 36.99\,dB$$

$$W_{t\,dB} = W_{tdB} + G_{tdB} + G_{rdB} - L_s\,dB$$

$$W_r = (36.99 + 2.2 + 1.6 - 100.16)\text{ dB}$$

$$W_r = -59.37\text{ dB}$$ **Ans.**

Converting dB in watts

$$W_r = 1.15\ \mu W$$ **Ans.**

Example 12. A television transmitting antenna mounted at a height of 100 meters radiates 5 kW of power at a frequency of 60 MHz. Power is radiated equally in all directions in azimuth. Calculate (a) the maximum line of sight range (b) the field strength at a receiving antenna mounted at height of 16 meters at a distance of 5 km. (c) Calculate the distance at which the field reduces to 2 mV/meter.

[Raj. Univ. 1993, 2002]

Solution: (a) The maximum range is defined as

$$ran, e = 3570\left[\sqrt{h_t} + \sqrt{h_r}\right] meters$$

$$= 3570\left[\sqrt{100} + \sqrt{16}\right] meters$$

$$= 3570 \times 14\ metres$$

Ra ıge = 49,980 meters **Ans.**

(b) N(w λ = 5 meters.

At 5 k n, the field strength will be given as

$$E = \frac{4\pi h, h_t}{\lambda d^2} E_o$$

Wh :re E_o is defined in terms of earth's conductivity and other certain features. Normally

$$E_o = \frac{8\sqrt{\rho}}{d}$$ For a hill so substituting this value in above equat on yields

$$E = \frac{4 \times 3.14 \times 8 \times \sqrt{5 \times 10^3} \times 100 \times 16}{5 \times \left(5 \times 10^3\right)^2} Volt/meter$$

$$= \frac{3.14 \times 32 \times 16 \times \sqrt{5 \times 10^3} \times 100}{5 \times 25 \times 10^6}$$

$$= \frac{3.14 \times 32 \times 66 \times \sqrt{0.5} \times 20}{10^6} = \frac{128624 \times 0.71}{10^6}$$

E = 90 mV/meter **Ans.**

(c) $$E = \frac{4\pi h + h_t}{\lambda d^2} E_o$$

$$E_o = \frac{8\sqrt{\rho}}{d}$$

$$2 \times 10^{-3} = \frac{3.14 \times 32 \times 16 \times \sqrt{5 \times 10^3} \times 100}{5d^2}$$

$$d^2 = \frac{3.14 \times 32 \times 16 \times \sqrt{5 \times 10^3} \times 100}{5 \times 2 \times 10^{-3}}$$

$$= \frac{3.14 \times 32 \times 16 \times \sqrt{0.5} \times 10000}{10^{-2}}$$

$$= 1607.68 \times \sqrt{50} \times 10^5$$

$$d^2 = 11.4 \times 10^8 \; meters$$

$$d = 3.34 \times 10^4 \; meters.$$ **Ans.**

Example 13. Calculate the maximum wavelength at which propagation is possible by means of a ground based duct 150 ft height when $\Delta m = 36$. **[Raj. Univ. 2000]**

Solution: The maximum wavelength in = λ_{max}, d = duct height

$$\lambda_{max} = 0.084\, d^{3/2}$$

$$= 0.084\, (150)^{3/2}$$

$$= 154.3178$$ **Ans.**

Example 14. Two planes 15 km apart are in radio communication. The transmitting plane delivers 500 watt. Its antenna gain being 10 in the direction of other plane power absorbed in 2 micro watts by the receiving antenna of the second plane. Find the effective area assuming that the ground and ionospheric reflections are negligible.

Solution: Given

d = 10 km

$G_T = 10$

$W_T = 500$ watts

$W_R = 2 \times 10^{-6}$ watts

Ae = ?

$$P_D = \frac{W_T\, G_T}{4\, \pi d^2}$$

$$W_R = P_D \,.\, A_e = \frac{W_T\, G_T}{4\, \pi\, d^2}\, A_e$$

$$A_e = \frac{W_R}{W_T} \times \frac{4\, \pi d^2}{G_T}$$

$$= \frac{2 \times 10^{-6} \times 4\, \pi \times \left(10 \times 10^3\right)^2}{500 \times 10}$$

$$= \frac{2 \times 4\, \pi}{50} = \frac{25.12}{50} = 0.5024\, m^2$$ **Ans.**

Example 15. Find the basic path loss for a communication from the moon the earth operating at 300 MHz Assume distance between moon and earth is 384000 km.

Solution: Path loss = 32.45 + 20 $\log_{10} f_{HZ}$ + 20 $\log_{10}$ d km

$= 32.45 + 20 \log_{10} 3000 + 20 \log_{10} 38400$

$= 32.45 + 20 \times 3.4771 + 20 \times 5.5843$

$= 32.45 + 69.542 + 111.686$

$= 213.678$ dB **Ans.**

Example 16. VHF communication is to be established with 100 watt transmitter at 150 MHz. Find the distance up to which line of sight communication may be possible if the height of the transmitting and receiving antennas are 25 m and 16 m respectively. Also calculate the field strength at the receiving end. [Raj. Univ. 2005]

Solution: Line of sight $= 4.12 \left(\sqrt{h_t} + \sqrt{h_r}\right)$

$= 4.12 \left(\sqrt{25} + \sqrt{16}\right)$

$= 4.12 (5 + 4)$

$= 37.08$ km **Ans.**

$$E_r = \frac{88 \sqrt{P}\, h_t h_r}{\lambda d^2}$$

$$\lambda = \frac{c}{f} = \frac{3 \times 10^8}{150 \times 10^6} = 2$$

$$= \frac{88 \sqrt{100} \times 25 \times 16}{2 \times (37.08)^2 16} = \frac{35200 \sqrt{100}}{2749.85 \times 10^6}$$

$E_r = 128 \times 10^{-6}\ V$ **Ans.**

Example 17. Two aircrafts are flying at altitudes of 3000 m and 5000 m respectively. What is the maximum possible distance along the surface of earth over which they can have effective point to point microwave communication? Radius of earth is 6.37×10^6 meters.

[Raj. Univ. 1997, 2003]

Solution: According to question following parameters are given:

(i) Height of transmitting aircraft h_t = 3000 m

(ii) Height of receiving aircraft h_r = 5000 m

(iii) Radius of earth = 6.37×10^6 m

After taking the consideration of effective earth radius, the maximum possible distance on the earth surface is given by

$$d = 4.12\left[\sqrt{h_t} + \sqrt{h_r}\right] km$$

$$d = 4.12\left[\sqrt{3000} + \sqrt{5000}\right] km$$

d = 516.989 Km **Ans.**

Example 18. Two space craft's are flying at heights of 10 km and 15 k above the earth surface. The distance of LOS path is 916.59 km Find the effective earth radius and verify your answer if actual earth radius is 6.37×10^6 meters. **[Raj. Univ. 1993, 2003]**

Solution: According to equation following parameters are given:

(i) Height of transmitting spacecraft, h_t = 10 km

(ii) Height of receiving space craft, h_r = 15 km

(iii) Distance of LOS communication, d = 916.59 km

(iv) Actual earth radius, r = 6.37×10^6 meters

We know the formula for distance d with effective earth radius.

$$d = 4.12\left[\sqrt{h_t} + \sqrt{h_r}\right] km$$

or using equation

$$d = \sqrt{2r'}\left[\sqrt{h_t} + \sqrt{h_r}\right]$$

$$916.59 \times 10^3 = \sqrt{2r'}\left[\sqrt{10 \times 10^3} + \sqrt{15 \times 10^3}\right]$$

$$\sqrt{2r'} = \frac{916.59}{222.47} \times 10^3 = 4.12 \times 10^3$$

$$r' = 8.4872 \times 10^6 \; meter$$ **Ans.**

We know the effective earth radius is 4/3 times of actual earth radius them

$$r' = \frac{4}{3} r$$

$$r = r' \times \frac{3}{4} = 6370 \; km$$

Answer is verified.

Example 19. A pulse of a given frequency transmitted upward is received back after a period of 6 m sec. Find the virtual height of the reflected layer. **[Raj. Univ. 1997, 2004]**

Solution: Since $h = \frac{CT}{2}$ *meters*

$$T = 5m \text{ sec} = 5 \times 10^{-3} \text{ sec.}$$

$$C = 3 \times 10^{8} \; m/\text{sec}$$

$$h = \frac{3 \times 10^{8} \times 5 \times 10^{-3}}{2} = 7.5 \times 10^{5} \; meters. \qquad \textbf{Ans.}$$

Example 20. A television transmission antenna mounted at a height 144 m radiates 10 kW of power equally in all directions in azimuth at a frequency of 50 MHz. Calculate maximum line of sight range, height of receiver is 16 meters. [Raj. Univ. 1997, 2004]

Solution: $Loss = 4.12\left(\sqrt{h_t} + \sqrt{h_r}\right)$

$$= 4.12\left(\sqrt{144} + \sqrt{16}\right)$$

$$= 4.12\,(12 + 4)$$

$$= 4.12 \times 16$$

$$= 65.92 \text{ km} \qquad \textbf{Ans.}$$

Example 21. If the height of transmitting antenna h_t is 40 meters, height of receiving antenna h_r is 20 meters. Calculate the radio wave horizon. If the height of h_t and h_r are doubled and halved respectively, what impact would it have on radio wave horizon?

[Raj. Univ. 2001, 2005]

Solution: The line of sight distance is given by

$$d = 4.12\left[\sqrt{h_t} + \sqrt{h_r}\right] km$$

$$= 4.12\left[\sqrt{40} + \sqrt{20}\right] km$$

$$= 4.12\,[6.324 + 4.472]$$

$$= 44.48 \text{ km}$$

If the h_t height is doubled then h_t becomes = 80 m

If the h_r height is halved then h_r becomes = 10 m

Then the line of sight or radio of horizon distance will be

$$d = 4.12\left[\sqrt{80} + \sqrt{10}\right] km$$

$$4.12\ [8.944 + 3.162\ ]\ km$$

$$= 49.877 \text{ km}$$ **Ans.**

Therefore, the radio of horizon distance is extended.

Example 22. Calculate the maximum usable frequency for an electromagnetic wave incident at an angle of 45° to be propagated through ionospheric layer having electron density of 4 x 10^4 electron m^3.

[Raj. Univ. 1999]

Solution: Since

$$\sin^2 t = 1 - \frac{f_c^2}{f_{MUF}^2}$$

Now $$f_c = 9\sqrt{N_{\max}}$$

$$= 9 \times \sqrt{4 \times 10^4}\ Hz$$

$$= 9 \times 2 \times 10^2\ Hz$$

$$= 1800 \text{ Hz}$$

or $$\sin^2 45° = 1 - \frac{(1800)^2}{f_{MUF}^2}$$

or $$\frac{(1800)^2}{f_{MUF}} = 1 - \sin^2 45° = 0.5 \qquad \left[\sin 45° = \frac{1}{\sqrt{2}}\right]$$

Therefore $$f_{MUF}^2 = 2 \times 1800 \times 1800$$

or $$f_{MUF} = 2545.58\ Hz$$ **Ans.**

Example 23. A VHF communication is to be established with 35 Watt transmitter at 90 MHz. Find the distance up to which line of sight communication may be possible if the height of the transmitting and receiving antennas are 40 m and 25 m respectively. Also determine the field strength at the receiver end. **[Raj. Univ. 1999]**

Solution: Given P = 35 Watt, h_t = 40 m, h_r = 25 m, f = 90 MHz

$$d = 4.12\left[\sqrt{h_t} + \sqrt{h_r}\right]\ km$$

And field $$E = \frac{88\sqrt{P}\ h_t h_r}{\lambda d^2}\ V/m$$

So $d = 4.12\left[\sqrt{40} + \sqrt{25}\right] = 46.65\ km\ Ans.$

Now $\lambda = \dfrac{300}{90\ (mHz)} = \dfrac{10}{3}\ metre$

$$E_r = \frac{88 \times \sqrt{35} \times 40 \times 25}{\frac{10}{3} \times \left(46.65 \times 10^3\right)} = \frac{88 \times 3 \times 100\ \sqrt{35}}{(46.66)^2} \times 10^{-6}\ V/m$$

$= 71.70\ \mu V/m$ **Ans.**

Example 24. Calculate the value of the factor by which the horizon distance of a transmitter will be modified if the gradient of refractive index of air near the ground is $-0.5 \times 10^{-6-}\ m^1$. The radius of the earth is given to be 6370 km. **[Raj. Univ. 1996]**

Solution: Since

$$k = \frac{1}{1 + r\dfrac{dx}{dh}} \quad where\ x\ is\ refractive\ index$$

Given, $r = 6370\ km\ \dfrac{dx}{dh} = -\ 0.05 \times 10^{-1}$

$$k = \frac{1}{1 - 0.05 \times 10^{-6} \times 6.37 \times 10^6}$$

$$= \frac{1}{1 - 0.3185} = \frac{100}{68} = \frac{25}{17}$$

Therefore the horizon distance is modified by

$$r' = kr = \frac{25}{17} r$$

(Where r is the radius of earth)

Example 25. A television transmitter antenna has a height of 169 m and the receiving antenna has a height of 16 m. What is the maximum distance through which the TV signal could be received by space wave propagation? What is the radio horizon in this case?

Solution: $d = 4.12\left(\sqrt{h_t} + \sqrt{h_r}\right)$

$= 4.12\left(\sqrt{169} + \sqrt{16}\right)$

$= 4.12 \times 17 = 70.04\ km$ **Ans.**

Radio Horizon $= \sqrt{2r\ h_t}$

$$= \sqrt{2 \times \frac{4}{3} \times \frac{6370}{1000} \times h_t}$$

$$= \sqrt{16.9855 \times h_t}$$

$$= 4.12\sqrt{h_t}$$

$$= 4.12\sqrt{169}$$

$$= 4.12 \times 13 = 53.56\ km.$$ **Ans.**

Example 26. Distance between transmitting and receiving antenna of a microwave link is 30 kms. Assuming the frequency of operation as 10 GHz, both antennas having gain of 13 db each, calculate the required transmitter power is dbμ, if the minimum required power at receiver is -10 dB. **[Raj. Univ. 2001]**

Solution: $G_T = G_r = 13$ dB

$$f = 10\text{ GHz} = 10 \times 10^3\ MHz = 10^4\ MHz$$

$$d = 30\text{ km},\ W_R = 10\text{ dB}$$

$$\text{Path loss} = 32.45 + 20\log_{10} f_{MHz} + 20\log_{10} d_{km}$$

$$= 32.45 + 2\log_{10} 10^4 + 20\log_{10} 30$$

$$= 32.45 + 80 + 29\text{-}542$$

$$L_s = 141.99\text{ dB}$$

$$W_R = W_T + G_T(\text{dB}) + G_R(\text{dB}) - L_s(\text{dB})$$

$$-10 = W_T + 13 + 13 - 141.99$$

$$W_T = 105.99\text{ dB}$$

$$W_T\ (\text{in Watt}) = 3.97 \times 10^{10}\text{ W}$$

$$(W_T)_{dB\mu} = 10\log_{10}\left(\frac{3.97 \times 10^{10}}{10^{-6}}\right)$$

$$= 165.99 = 16\text{ dB}\mu$$ **Ans.**

Example 27. A transmitting antenna radiates 10 watts of power at 5 GHz frequency. If the received power at a distance 25 km is 13 dBμ. Calculate the effective aperture of receiving antenna assuming a transmit antenna gain of 6 dB. **[Raj. Univ. 2000]**

Solution: Given $P_1 = 10$ watt,

$$\lambda = \frac{300}{5000} = \frac{3}{50}\ m$$

$$d = 25\times10^3\ m,\ 13\ dBm = 19.952\mu\ watt.$$

6 dB = 3.981 (in units)

Therefore, the gain of receiving antenna can be written as

$$G_R = \frac{P_r \times (4\pi d)^2}{P_t G_i \lambda^2} = \frac{19.95\times10^{-3}\left(50\times4\times3.14\times25\times10^3\right)^2}{9\times10\times3.981}$$

$$G_R = \frac{4\,\pi\,A_{eff}}{\lambda} = 13724847.19\times10^3 = 101.37\,dB$$

$$\text{and } A_{eff} = \frac{\lambda}{4\pi}\times13724847.19\times10^3 = 65564.55\ squar\,meter \quad \textbf{Ans.}$$

Example 28. Two space craft's A and B are separated by 3000 km. Each has a paraboloidal reflector antenna of 0.90 m diameter operating at a frequency of 2 GHz with an aperture efficiency of 65% (i.e. K = 0.65). Calculate the path loss and antenna gain in dB. What should be the transmitted power is dBμ from space craft A if the receiver on spacecraft B requires - 90 dBμ. **[Raj. Univ. 2004]**

Solution: The gain of paraboloidal aperture efficiency is given by

$$G = \frac{K.4\pi A}{\lambda^2}\ where\ K\ is\ aperture\ (K = 0.65)$$

Given, $D = 0.90\text{ m},\ \lambda = \frac{300}{2000} = \frac{3}{20}m,$

$$A = \frac{\pi D^2}{4}$$

Substituting the values,

$$G = K.\frac{4\pi}{\lambda^2}\times\frac{\pi D^2}{4} = 0.65\times(3.14)^2\times\left(\frac{0.90\times20}{3}\right)^2$$

Since the receiver receives - 90 dBμ power therefore the transmitted power is given by the formula -

$$P_t = \frac{P_r\times(4\pi d)^2}{G_T\,G_R\,\lambda^2}$$

Where d is the distance between two spacecrafts.

$$\lambda = \frac{3}{20}\ meter; d = 3\times10^6\ meter;\ P_r = 90\ dB\mu = 10^{-9}\ \mu w$$

so, $P_t = \dfrac{10^{-9}\ \mu w \times 16 \times \pi^2 \times \left(3 \times 10^4\right)^2 \times 400}{(230.71)^2 \times 9}$

so, P_t in dBμ $P_t = 10 \log 1185.51 = 30.7$ dBμ

The path loss can be calculated by-

$$(L)_{dB} = (LOSS)_{dB} = 32.45 + 20\log_{10} f(MHz)$$

$$+ 20 \log_{10} (km)$$

$$= 32.45 + 20 \log_{10} 2000 + 20 \log_{10} 3000$$

$$= 32.45 + 66.02 + 69.54 = 168.01 \text{ dB}$$ **Ans.**

Example 29. Two spacecrafts are separated by 3000 km. Each has a paraboloidal reflector antenna of 0.85 m diameter operating at a frequency of 2 GHz with an aperture efficiency of 64%. If the spacecraft A's receiver requires 1 PW for a 20 dB signal to noise ratio, what transmitter power is this signal to noise ratio?

[Raj. Univ. 1997, 2000]

Solution: At a distance R from the transmitting antenna B, the power density -

$$W = I_B \frac{P_B \times G_B}{4\pi R^2}$$

Where P_B is the power transmitted by B, is watts e_B is the transmitting efficiency of B and the gain of antenna B -

$$G_B = \frac{4\pi}{\lambda^2} \cdot \frac{\pi D^2}{4} \quad \textit{where D is diameter.}$$

At the receiving antenna A, the power received -

$$P_A = w \times e_A \times \frac{\pi D^2}{4} \quad \textit{Where } e_A \textit{ is the efficiency of A.}$$

$$P_A = \frac{e_B}{4\pi} \cdot \frac{P_B}{R^2} \cdot \frac{4\pi}{\lambda^2} \cdot \frac{\pi D^2}{4} \times e_A \times \frac{\pi D^2}{4}$$

$$e_A = e_B = 0.64, P_A = 1 \times 10^{-12} W,$$

$$R = 3000 \times 10^3 m$$

$$\lambda = \frac{3 \times 10^8}{2 \times 10^9} = 0.15\ m, D = 0.085 m$$

$$10^{-12} = \frac{0.64 \times P_B}{4\times\left(3\times10^{6}\right)^2} \times \frac{4\pi^2 \times 10.85 \times 0.64}{10\times 0.15^2}$$

$$= 6.513 \times 10^{-13}\ P_B$$

$$\Rightarrow P_B = 1.535\ watt$$

Ans.

Example 30. What is the maximum power received at a distance of 10 km over a free space? A 1000 MHz circuit consists of a transmitting with a 25 dB gain and a receiving antenna with a 20 dB gain w.r.t. isotropic antenna. Input power to transmitting antenna is 150W.

[Raj. Univ. 1997, 2001]

Solution: According to question following parameters are given -

(i) Distance d = 10 km = 10×10^3 *meter*

(ii) Gain of transmitting antenna G_T = 25 dB

(iii) Gain of receiving antenna G_r = 20 dB

(iv) Power of transmitting antenna, W_T = 150 W

(v) Carrier frequency f = 1000 MHz = 10^9 Hz

First of all we convert the 150 W values in dB and calculate the free space loss according to equation (1.252) the path loss $\left(\frac{4\pi R}{d}\right)^2$ *in dB.*

$$W_T\ dB = 10\log_{10} 150 = 21.76 dB$$

$$L_s dB = 20\ \log_{10}\ (f) + 20\log_{10}\ (d) - 147.56$$

$$L_s\ dB = 20\log_{10}\ \left(10^9\right) + 20\ \log\left(10^4\right) - 147.56$$

$$L_s\ dB = 180 + 80 - 147.56 = 112.44\ dB$$

We can also calculate this - 45.68 dB in watts.

10 $\log_{10}$ (W_r in watts) = - 45.68

or W_r = 27.03 m watt

Ans.

Example 31. Calculate the free space loss in a satellite communication where the satellite is at a height of 36,000 km above the earth surface. Given G_T = 15 dB, G_R = 45 dB and f = 4 GHz. What is the power received if the power radiated is 200 watts.

[Raj. Univ. 2005, 2004]

Solution: According to the question following parameters are given -

(i) Gain of transmitting antenna, G_T = 15 dB

(ii) Gain of receiving antenna, G_r = 45 dB

(iii) Distance between T_x and R_x antenna, d = 36000 km

(iv) Carrier frequency = 4 GHz = 4×10^9 Hz

Using the formula for free space loss -

$$L_s\, dB = 20 \log_{10} f + 20 \log_{10} d - 147.56$$

$$L_s\, dB = 20 \log_{10} \left(4 \times 10^9\right) + 20 \log_{10} \left(36 \times 10^6\right) - 147.56$$

$$L_s\, dB = 192.04 + 151.12 - 147.56$$

L_s = 195.6 dB **Ans.**

Further, convert radiated power in dB.

W_t dB = 10 $\log_{10}$ 200 = 23.01 dB

Now, applying formula for receiving power -

W_r dB = W_rdB + G_t dB + G_rdB – L_sdB

W_r dB + 23.01 + 15 + 45 – 195.6

W_r = – 112.59 dB

Negative sign shows that there is loss in system.

Then convert - 112.59 dB in Watts.

W_rdB = 10 $\log_{10}$ W_r in Watts

– 112.59 = 10 $\log_{10}$ W_r

W_r = 5.50×10^{-12} W = 5.50 P watt.

Example 32. A transmitter radiates 20 watts of power at a wavelength of 4 meter. Calculate the power received by an antenna at a distance of 100 km if the gain of the transmitting and receiving antennas is equal and have a value of 30 dB. **[Raj. Univ. 2005, 2002]**

Solution: According to question following parameter are given -

(i) Wavelength of carrier $\lambda = 4cm = 4 \times 10^{-2}\ meter$

(ii) Distance between T_x and Rx antenna, d = 100 km,

(iii) Gain of transmitting antenna G_r = 30 dB

(iv) Gain of receiving antenna, by = 30 dB

(v) Transmitting power, W_t = 20 watts.

First all convert the 20 watt is dB

$W_t\,dB = 10 \log_{10} W_t$ in watt

$W_t\,dB = 10 \log_{10} 20 - 13.01$ dB

And then calculate the carrier frequency -

$$f = \frac{C}{\lambda} = \frac{3\times10^8}{4\times10^{-2}} = 75\times10^8\ Hz$$

Now using the formula for free space -

$$L_s\,dB = 20\log_{10} f + 20\log_{10} d - 147.56$$

$$L_s\,dB = 20\log_{10}\left(75\times10^8\right) + 20\log_{10}\left(100\times10^3\right) - 147.56$$

$$L_s dB = 197.50 + 100 - 147.56$$

$$L_s = 149.94 \text{ dB}$$

Then, using the formula for receiving power -

$$W_r dB = W_t dB + G_t dB + G_r\,dB - L_s\,dB$$

$$W_r\,dB = 13.01 + 30 + 30 - 149.94$$

$$W_t = -\,76.93 \text{ dB}$$

In watt, $W_r dB = 10 \log 10\ W_r$ (Watts)

$$-\,76.93 = 10 \log_{10} W_r$$

$W_r = 0.0202\ \mu W.$ **Ans.**

Example 33. Calculation the power density reacting the moon's surface from a 10 mw pulse transmitter located on the earth. The antenna gain is 65 dB. Take the distance between the moon and the earth as 4,00,000 km. Assume that transmission medium is loss less?

[Raj. Univ. 2004, 2000]

Solution: As $P_o = \dfrac{W_T\,G_T}{4\pi d^2}$

Given $W_T = 10$ Mw $= 10\times10^6$ *watts*

$G_T = 65$ dB

or $(G_T)\,dB = 10\log_{10} G_T$

$65 = 10 \log_{10} G_T$

G_T = anti log 6.5

$= 3.126 \times 10^6$

d = 4.00,000 km

= 400 x 10^6 m

$$P_o = \frac{10\times 10^6 \times 3.162 \times 10^6}{4\times \pi \times \left(400 \times 10^6\right)^2}$$

$$= \frac{3.162 \times 10^{13}}{2.0096 \times 10^{18}}$$

$$= 1.5734\times 10^{-5}$$

$$= 15.734\ \mu W$$ **Ans.**

Example 34. Assume a 12-km communication link at 2.1 GHz carrier frequency. Calculate the radius of the first Fresnel zone at the midpoint between trans-receivers. **[Raj. Univ. 1997, 2003]**

Solution: According to question following parameters are given

(i) Distance between trans-receivers, D = 12 km

(ii) Carrier frequency, f = 2.1 GHz

(iii) Distance between Transmitter and mild point (d_1) = 6 km

Now for calculation of first Fresnel zone, using the formula

$$R = \sqrt{\frac{Cd_1\left(D-d_1\right)}{fD}}$$

$$R = \sqrt{\frac{3 \times 10^8 \times 6\times 10^3 \times (12 - 6)\times 10^3}{2.1 \times 10^9 \times 20 \times 10^3}}$$

R = 16.03 meter **Ans.**

Example 35. The receiver located at 10 km away from transmitter. If the carrier frequency of microwave link is 28 GHz. Calculate the first, second and third Fresnel zone radius at the point which is located 4 km away from transmitter. **[Raj. Univ. 1993, 2004]**

Solution: According to question following parameters are given

(i) Distance between trans-receivers, D = 10 km

(ii) Distance of point from transmitter, d_1 = 4 km

(iii) Carrier frequency, f = 28 GHz

For calculation of first, second and third Fresnel zone radius, we use the formula

$$R_n = \sqrt{\frac{ncd_l\,(D - d_l)}{fD}}$$

Put the value of n = 1, 2 and 3

(a) For first Fresnel zone radius

$$R_1 = \sqrt{\frac{3 \times 10^8 \times 4 \times 10^3 \times (10 - 4) \times 10^3}{28 \times 10^9 \times 10 \times 10^3}}$$

R_1 = 5.070 meter **Ans.**

(b) For second Fresnel zone radius

$$R_2 = \sqrt{\frac{2 \times 3 \times 10^8 \times 4 \times 10^3 \times 6 \times 10^3}{28 \times 10^9 \times 10 \times 10^3}}$$

R_2 = 7.17 meter **Ans.**

(c) For third Fresnel zone radius

$$R_3 = \sqrt{\frac{3 \times 3 \times 10^8 \times 4 \times 10^3 \times 6 \times 10^3}{28 \times 10^9 \times 10 \times 10^3}}$$

R_3 = 8.781 meter **Ans.**

THEORITICAL QUESTIONS

Q1. Analyze surface wave propagation as a function of numerical distance and polarization. Explain Brewster angle?

Q2. Explain microwave communication through Troposcatter propagation. What are the major conditions of its operation?

Q3. Explain with the help of a neat sketch the Troposcatter mode of wave propagation. What are its advantages?

Q4. Discuss the salient features of sky wave propagation? Bring out the various problems associated with this mode of propagation. How are these problems overcome?

Q5. Draw a sketch showing tropospheric scatter and explain its basic principles?

Q6. Describe ground wave propagation. What is the angle of tilt? How does it affect the field strength at a distance from the transmitter?

Q7. Bring out the important difference between ground wave propagation, space wave propagation, high lighting the application areas of each of them?

Q8. Describe propagation characteristics of radio waves of different frequencies in relation to problems of practical radio communication?

Q9. Discuss briefly the various kind of wave propagation?

Q10. Derive fundamental equation for free space propagation?

Q11. What you mean by plane waves and explain the polarization states?

Q12. Calculate the phase difference due to path length differences in UHF mobile communication.

Q13. If h_f and h_r are the heights in meters of transmitting and receiving antennas above the ground, show that maximum separation Between them for line of sight transmission is $D_{max} = 3.57 \left[\sqrt{h_t} + \sqrt{h_r}\right]$ km. Derive the relation and calculate the factor by which D_{max} gets modified when the refraction of radio waves due to decrease in refractive index of air with height for a standard atmosphere is considered. Given earth's radius = 6370 km and gradient of refractive index of air = -0.04×10^{-6} per meter.

SHORT QUESTIONS

1. What is propagation?
2. How is a wave defined as it applies to wave propagation?
3. What is wave motion?
4. What are some examples of wave motion?
5. What type of wave motion is represented by the motion of water?
6. What are some examples of transverse waves?
7. What example of a longitudinal wave was given in the text?
8. What are the three requirements for a wave to be propagated?
9. What is a cycle?
10. What is wavelength (λ)?
11. What is the law of reflection?
12. When a wave is reflected from a surface, energy is transferred. When is the transfer of energy greatest?
13. When is the transfer of energy minimum?
14. A refracted wave occurs when a wave passes from one medium into another medium. What determines the angle of refraction?
15. The apparent change in frequency or pitch because of motion is explained by what effect?

16. What term describes sounds capable of being heard by the human ear?
17. Are all sounds audible to the human ear? Why?
18. Sound waves transmitted from a source are sometimes weak when they reach the detector.

What instrument is needed to boost the weak signal?

19. What are the three basic requirements for sound?
20. What are the two general groups of sound?
21. What are the three basic characteristics of sound?
22. What is the normal audible range of the human ear?
23. What is intensity as it pertains to sound?
24. What characteristic of sound enables a person to distinguish one musical instrument from another, if they are all playing the same note?

ANSWERS

1. Induction field and radiation field.
2. Induction field.
3. Radiation field.
4. Fundamental frequency.
5. Harmonic frequency or harmonics.
6. 30 meters.
7. 5 megahertz.
8. Vertically polarized.
9. Direction of wave propagation.
10. Shifting in the phase relationships of the wave.
11. Troposphere, stratosphere, and ionosphere.
12. Stratosphere.
13. Whether the component of the wave is traveling along the surface or over the surface of the earth.
14. Sea water
15. Radio horizon is about one-third farther. .
16. (a) electrical properties of the terrain
    (b) Frequency
    (c) Polarization of the antenna
17. High energy ultraviolet light waves from the sun.
18 D, E, F1, and F2 layers.

19. D layer is 30 to 55 miles, E layer 55 to 90 miles, and F layers are 90 to 240 miles.
20. Thickness of ionized layer.
21. Critical frequency.
22. (a) density of ionization of the layer

    (b) Frequency

    (c) Angle at which it enters the layer
23. A zone of silence between the ground wave and sky wave where there is no reception.
24. Where ionization density is greatest.

UNIT 5
RADIO WAVE PROPAGATIONS-II

CONTENTS

RADIO WAVE PROPAGATION - II

5.1 IONOSPHERIC ABNORMALITIES

Ionosphere is highly dependent up sum and hence its conditions vary continuously.

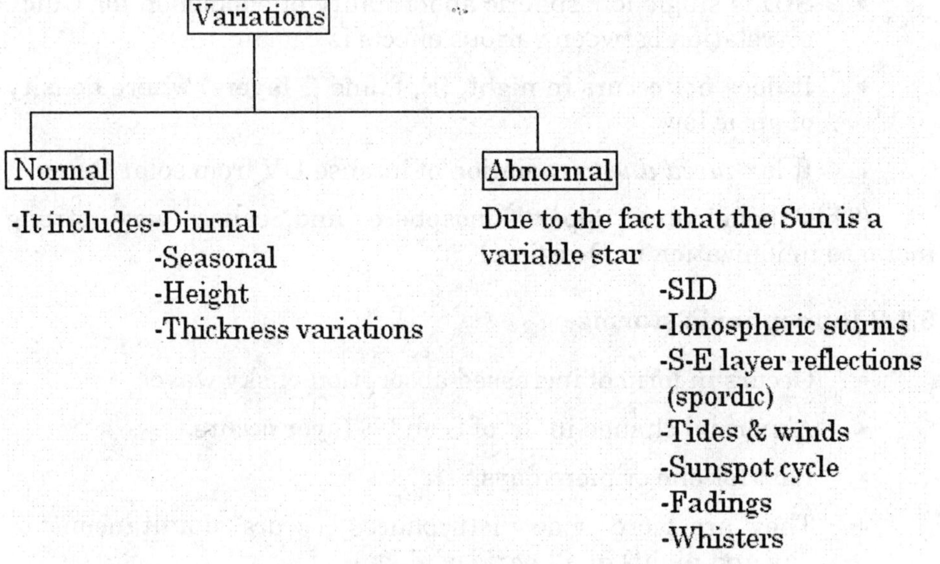

∴ Here we are going to discuss the latter topic

5.1.1 SID (Sudden Ionospheric Disturbances)(Mongel - Dellinger Effect/Dellinger Fadeouts)

- SID is due to sudden appearance of solar flares i.e. bright spots on solar disc which are due to gigantic emission of H_2 from sun
- Sun flares are sudden unpredictable.

- X-Rays radiations following solar flares increases tremendously the ionization density right down to D-layer. As a result of intense, rise in ionization density of D-region causes the increases in absorption of short wave signals and increased reflection of atmospheric noise as well.

∴ LUF *increases > MUF* ∴ **resulting in complete blackout of all high frequency sky wave communication via ionosphere**.

- **The phenomenon of complete black out is known as SID**
- **During SID, VLF propagation improves**
- SID rarely occurs during sun spot minimum and its `f' of occurrence increases rapidly as sun spot cycle advances.
- **The intensity of disturbances tends to be maximum in the region where sun's radiation is perpendicular i.e. greater at noon, at equator.**
- SID is single ionospheric abnormality phenomenon for which correlation between various effects is certain.
- It does not occurs in night, (E, f, and f_2 layers) where density of air is low
- It is *caused due to* radiation of intense UV from solar flares

(This penetrates upper Ionosphere and cause very strong increase in ionization in D-region.)

5.1.2 Ionospheric storms

- Occurs in form of increased absorption of sky waves
- Abnormal change in `fc' of f_2 and E-layer occurs.
- Last for one or more days
- They are word wide disturbances (occurs simultaneous in day and night) in all parts of globe)
- They are disturbances in ionosphere. These storms are correlated with rapid and excessive fluctuations associated with magnetic storms. This occurs in earth's magnetic field.
- During storms ionosphere is turbulent. It losses it's normal stratification. The result is RWP becomes erratic and ∴ necessary to decreases working frequency of communication.

- This abnormality f_c (*decreases*) *and* ∴ *virtual height* of f_2 increases, and ∴D-layer ionization increases and so the absorption.
- Effects of storm on short wave radio transmission is to narrow the range of frequencies of communication over a given circuit.
- *Caused by* particle emissions from sun $\begin{cases} \alpha - rays \\ \beta - rays \end{cases}$
- *Reoccurrence*: 27.3 days interval (period of skin's rotation)
- Its effect is more near Polar Regions.
- **Most easily observed indication of solar variation is size and number of sun spot, which can be seen on sun discs. The number of sun spots changes over a cycle from min. to maximum and again to minimum is called as sun spot cycles and it has mean period of 11.1 years.**

5.1.3 Sun spot cycles

- Although the light variations are small, but solar O/P of UV, coronae, flares, particles radiation. The sun spot may vary 50 fold over that period.
- The extension of solar disturbances is measured by method of sunspot counting.
- Prediction of ionospheric characteristics is basically 11 years sun spot cycles which can be made in advances ∴ frequency selection can accordingly be made for communication.

5.1.4 Tides and winds

- Tides-Atmosphere also experiences tidal pulls to sun and moon as in case of oceans. In its case, solar tidal effects get more amplified than in oceans. **The reasons behind is the resonance phenomenon as free period of oscillation of atmosphere coincides approx. With tidal period of 12 hrs.**
- **Effect is further complicated by thermal heating of atmosphere by radiation of sun for 24 hrs.**
- Winds tides are also detectable though feeble.
- Ionized particles, of upper atmosphere. During exciting these tidal motions, is affected by Magnetic field of earth as moving

ion or e^- constitute current and so affected by mag. field. Thus ions and e^- in upper atmosphere follow earth's mag. field rather than moving horizontally.

- F_2 layer has greatest speed of tidal motions because it is situated highest having lowest particle density so the effect of tides is experienced most by f_2 layer.
- Due to this effect is causes small peak of max. Ionization density is in region of about mid night.
- Winds-(motion of ionized particles) in ionosphere caused by tides.

5.1.5 Fading

- Fluctuations in the received strength at receiver or random variation in received signal is known as fading.
- Also defined as - undesirable variations in intensity or loudness of waves received at receiver.
- It is caused by changes in heights and density of ionization in different layers of ionosphere.
- The reason behind is interference between 2 waves of different path lengths.

The signal received at any point is vector sum of all waves received. The result is as the signal waves leaves the Tx at same time but arrive at Rx following diff. paths. It is more common in high `f'.

- 10 dB to 20 dB - Severe fading.
- Total fading in SID or sun spot cycle.

Various types of Fading

(a) Selective fading

- Prevalent in high freq.

- Fading is frequency selective.

- It is severe for AM (Amplitude modulated) signals.

It can be reduced by exalted carrier Reception and also SSB systems.

(b) Interference fading

Most serious:

1. Reflected from E layer.

2. Reflected from F layer
3. Ground wave interfering at point M producing louder weak signals depending welting in phase or out phase.

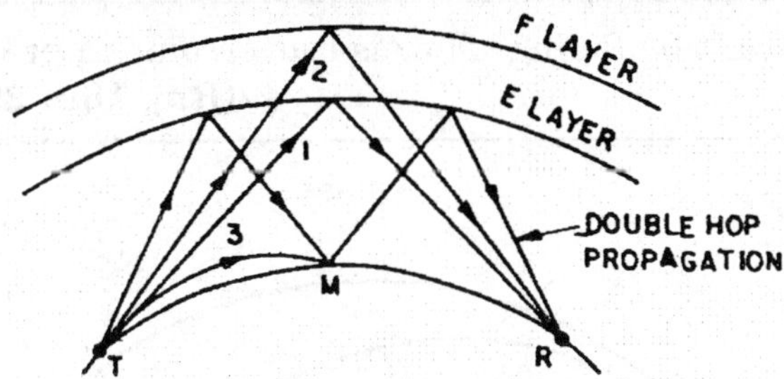

Figure 5.1

It is produced between upper or lower rays of sky wave, (2) between sky waves reaching the Rx by different number of Hops or different paths and (3) even between a ground wave and sky wave. (Particularly at lower end of HF - band)

It also occurs because of variations of height (or ionic density) in ionospheric layer if a single sky wave frequency is in use.

As we can also atmosphere have continuous changes and as the path length of each wave is subjected to continuous. Small changes ∴ relative phase of all waves of Rx vary randomly. ∴ Amplitude of resultant varies continually. This effect is called interference fading.

It can be minimized by space diversity of freq diversity reception.

(c) Absorption fading

Occurs due to variations of signal strength with different amount of absorption of wave absorbed by Tx n medium

(d) Polarization fading

This type of fading occurs due to the change in polarization of down coming sky waves. This is continuously changing.

This is caused by superimposition of ordinary and extra ordinary waves (with random any and Phase) which are oppositely polarized.

Polarization w.r.t. antenna is variable giving rise to changes of amplitude in Rx and produce polarization fading.

(e) Skip fading

This type of fading occurs at distance near skip distance. Any changes in height or density of ionized layer may move the receiving point in out of skip zone.

(1) To minimize any fading used (AGC or AVC) in Rx.

(2) To minimize other way is to employ diversity reception system.

> **Q.** What is meant by fading? How is it overcome in practical radio system? **(Raj. Uni. 2002, 2005)**

5.1.6 Whistlers

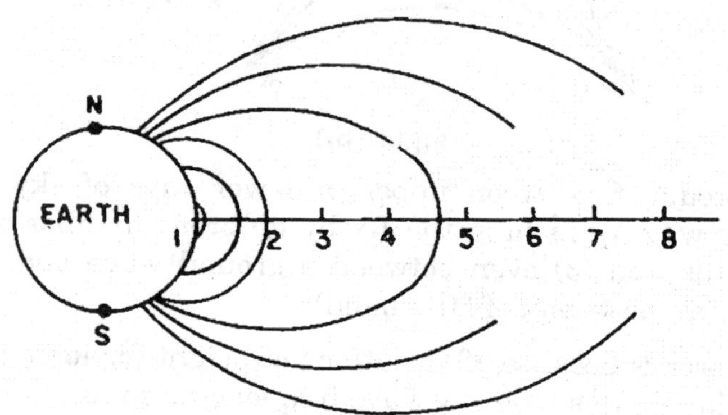

Figure 5.2

- Naturally occurring transient descending electromagnetic disturbances are called whistlers.
- These atmospheric whistlers are related with EM pulses of audio frequency radiation propagation along lines of earth's magnetic field between conjugate points in northern and southern hemisphere.
- Lightening discharges generate the pulses which may ``bounce" back and forth several times before disappearing between the 2 hemisphere.
- Long whistlers
- Short whistler and noise whistlers.
- Characteristics of whistlers are found by, pulse trajectory at height of 2/10 earth radii.
- It is powerful means in study of composition of upper atmosphere.

5.1.7 Luxemburg effect or Ionospheric cross Modulations

While listening to distant medium frequency broad casting station, sometimes it happens that there is a continuous background of some more powerful station, usually of lower frequency.

Interference is most marked when

(1) Two transmitting stations were closely on same great circle bearing from Rx.

(2) The interfering station is geographically between the Rx and wanted Tx stations.

Possibility of production of cross-modulation products by the reason of interaction of two strong radio-waves passing through ionosphere, simultaneously. Under favorable conditions magnitude of cross - modulation is sufficient to produce appreciable interference, as has been observed to occur between two modulating radio frequencies signals inside ionosphere.

Cross modulation is found in ionosphere and the modulation of unwanted signal is superimposed on wanted signal and hence the annoyance or interference occurs.

Q. Describe the structure of ionosphere and part played by it in long distance transmission of radio signal in HF band. **(R.T.U. 2009)**

5.1.8 Statics or Atmospheric Noise or Atmospherics

The receives contains background of various dicks, bangs, rumbles, rattles and crashes etc. and disappear when antenna is disconnected from receiver. Noises are of 2 kinds

(1) Statistics or Atmospheric Noise

(2) Manmade Noise

(i) (a) Noise picked up by antenna

(b) Due to natural electrical disturbances = thunder storms.

(c) Frequency increases Noise decrases ∴ found basically in low frequency signal.

(d) Seasonal changes

(ii) In frequency range (9-21 MHz) - Though No atmospheric or manmade noise are present still it is found that some noise is picked up in antenna. This noise is generated from radio stars which emit radio waves and from certain region of interstellar space. This is generally limiting noise.

(iii) Above 30 MHz – No energy is reflected from ionosphere and hence distant atmospheric has no effect and atmospheric due to local thunder storms may be heard.

- Airplane passing through snow or rain, it becomes electrically changed to upper level potential w.r.t. surrounding space. That a (corona/bluish) discharge - breaks out at some sharp point on plane.
- This discharge reception and is called perception static and is serious at short wave and lower frequencies.
- Can be reduced by insulated wires or shielded loop and equipping plane with discharger.

(iv) Manmade noise: by electrical devices.

- Strong in cities, industrial areas, near ignition system of internal combustion engines.
- Ignition system diathermy machines, power line discharges, sparking brushes all causes interference.
- Are of Hiss type or Impulse type generally
- Minimized by using line filters.

5.1.9 Polar Lap Absorption (SPE)

- Disturbances which occur throughout polar region in high geomagnetic latitudes, these are called polar cap absorption event or solar proton events (SPE)
- They are caused by high energy solar protons and are guided by earth's magnetic field towards polar region.
- They are accompanied by increase in D-region ionization resulting in strong absorption of HF & VHF radio waves.
- Black out may exist for up to a day but SPE itself can last up to week or more.
- SPE is preceded by major flare & occurs most frequently at solar sun spot maximum.

Q. Describe the characteristics of different ionized regions.

(Raj. Univ. 2006)

Q. How is ionosphere formed? What are the various layers of the ionosphere? Which layers disappear at night and why?

(Raj. Uni. 2002, 2008)

5.2 IONOSPHERIC ABSORPTION

- It also limits radio transmission over large distance.
- When radio waves pass through ionized layer it cause e^- to vibrate. This vibrating electron gives some of its energy by the reason of colliding with neighboring molecules and of its energy by colliding with neighboring molecules and ions. This energy is used up in heating air and hence wasted. This increases with distance of travelling.
- Absorption is greater in day time (D and E^- region)
- It is less than or equal to 1.4 MHz absorption greater ∴ MUF are selected.

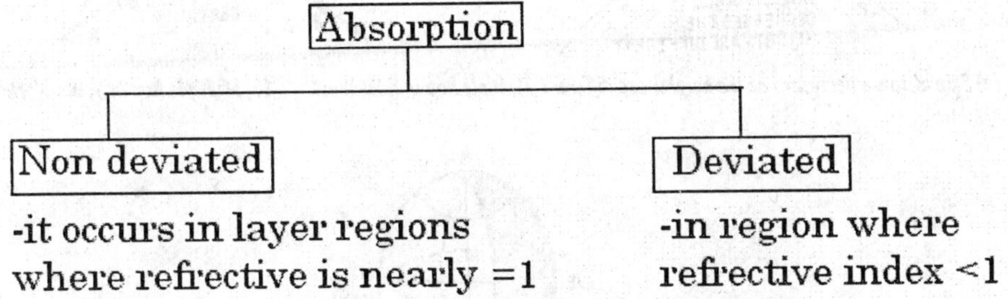

- Non-deviated is maximum in lower most D-region in daytime where collision frequency is highest.
- Absorption *increases with frequency decreases*
- If magnetic field of earth is not countable then for non deviating absorption $\mu \approx 1$ the absorption coefficient.

$$K = \frac{Vf_c^2}{2\,C\,f_c^2}$$

$$\text{Or} \qquad k \;\alpha\; \frac{1}{f^2}\;\alpha V$$

$$V \to \textit{collision frequency}$$

- This shows *V inreases f decreases ∴ absorption inreases*

∴ During daytime when D-layer is present it becomes necessary to use a high transmission frequency.

> **Q.** Prove the refractive index of ionosphere is less than 1 where that an unionized medium is 1. **(Raj. Univ. 2006)**

5.3 MULTI HOP PROPAGATIONS

The transmission path is limited by skip distance and curvature of earth. The longest single hop propagation is obtained when transmitted ray is tangential at earth surface.

Maximum distance covered (practically) = 2000 km - E - Region (in single hop)

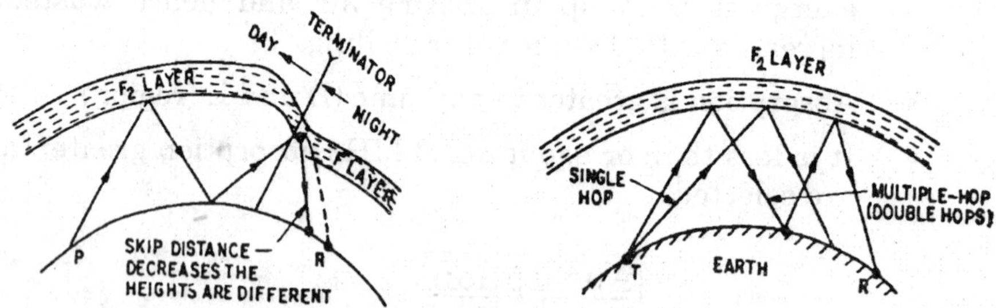

(a) Skip distance decreases as the heights are different. Multiple hop in East-West. *(b) Multiple hop in South-North.*

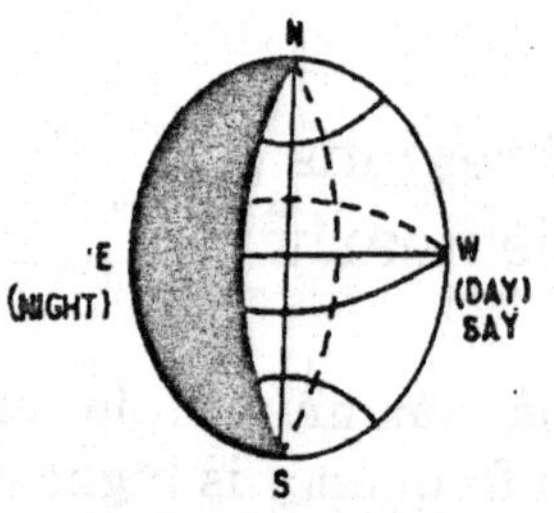

(c) Illustration of day and night.

Figure 5.3

∵ Semi circumference of earth is over = 20,000 km, multi hop propagation path are occurring. It is found that long distance short wave communication generally involves 2-4 Tx paths & each contribute appreciable energy to Rx.

Q. Explain with suitable diagram the multiple hop transmission. **[Raj. Univ. 2004, 2002]**
Q. Explain Define short questions: i. Multi Hop propagation. **(R.T.U. 2009)**

5.4 SPACE WAVE PROPAGATION

- Space wave/ line of sight propagation is useful at high frequencies i.e. V.H.F., U.H.F. and microwaves. (Few Km)

- It is limited Tx-line of sight distance.
- Curvature of earth
- **Line of sight distance is the distance between the Tx & Rx in which if a direct ray passes from Tx to Rx without being intercepted by bulge in earth's surface, while considering variation of ref. index (m) of earth's atmosphere with height, the transmitting antenna must `see' at least top of Rx antenna.**
- Propagation occurs in troposphere region. (16 km above earth's surface)

5.4.1 Range of space wave propagation (or line of sight distance) (LOS)

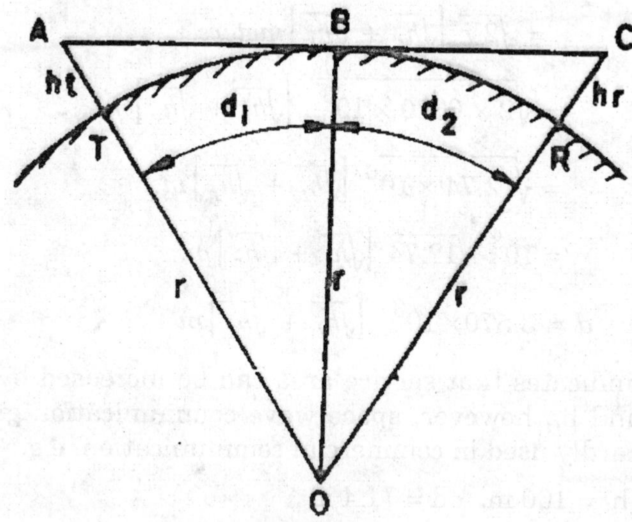

Figure 5.4

It is possible only or slightly beyond LOS distance, distance is determined by heights of Tx & Rx antenna.

$$d = d1 + d_2 \qquad (1)$$

Where d = distance between Tx & Rx

h_t = height of transmitter above ground

h_r = height of receiver above ground

r = Radius of earth (= 6370 km)

$\therefore$ *from* ΔABO *and* ΔCBO

$$d_1 = \sqrt{(h_t + r)^2 - r^2} \qquad d_2 = \sqrt{(h_r + r)^2 - r^2}$$

$= \sqrt{h_t^2 + r^2 + 2h_t r - r^2}$ $\qquad$ $= \sqrt{h_r^2 + r^2 + 2h_r . r - r^2}$

$= \sqrt{h_t^2 + 2 h_t . r}$ $\qquad$ $= \sqrt{h_r^2 + 2 h_r . r}$

$as\ h_t^2 << 2h_t . r$ $\qquad$ $again\ as = h_r^2 << 2h_r . r$

$$\therefore d_1 \approx \sqrt{2r\, h_t} \tag{2}$$

$$\therefore d_2 \approx \sqrt{2\, h_r . r} \tag{3}$$

Now on putting values of d_1 and d_2 from - (2) - (3) equation in (1) equation

We have $= d = \left(\sqrt{2r . h_t} + \sqrt{2 r . h_r}\right)$ *meters*

$$= \sqrt{2\, r}\left[\sqrt{h_t} + \sqrt{h_r}\right] meters.$$

$$= \sqrt{2 \times 6370 \times 10^3}\ \left[\sqrt{h_t} + \sqrt{h_r}\right] m$$

$$= \sqrt{12.74 \times 10^6}\left[\sqrt{h_t} + \sqrt{h_r}\right] m.$$

$$= 10^3\ \sqrt{12.74}\left[\sqrt{h_t} + \sqrt{h_r}\right] m.$$

$$d = 3.570 \times 10^3\ \left[\sqrt{h_t} + \sqrt{h_r}\right] m \tag{4}$$

Eq.(4) indicates that service area can be increased by increasing heights h_t and h_r, however, space wave communication greater than 100 km. is hardly used in commercial communication. e.g.:- h_t = 100 m

h_r = 100 m. d = 71.4 m

Q. If h_t and h_r are the heights in meters of transmitting and receiving antennas above the ground. Show the maximum separation between them for line of sight transmission is

$$D_{\max} = 3.57\left[\sqrt{h_t} + \sqrt{h_r}\right] km,$$

Derive the relation calculate the factor by which D_{max} gets modified when the refractions of radio waves due to decrease in refraction index of air with height for a standard atmosphere is considered. Given earth's radius = 6370 km and gradient of refractive index of air = - 0.04 x 10^{-6} per meter.

(Raj. Uni. 2002, 2004)

5.4.2 Effective Earth's Radius

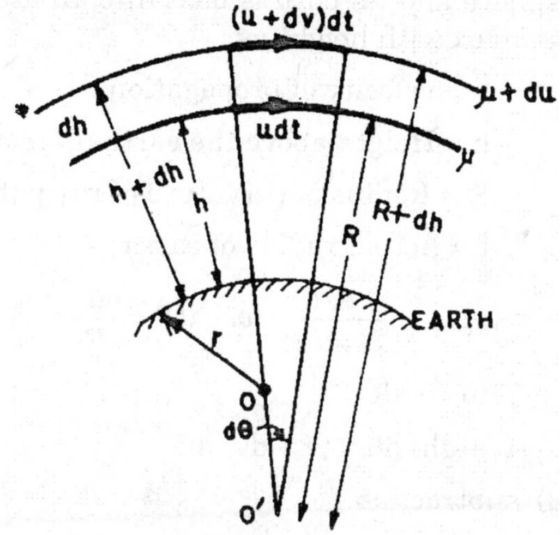

Figure 5.5

Radio wave traveling horizontally in the earth's atmosphere follows a path which has a slight downward curvature. The reason is the refraction of the wave in atmosphere. This curvature of path tends to overcome partially loss of signal due to curvature of earth & permits direct ray to reach point slightly beyond horizon as found by straight line or line of sight path.

Assuming-

- Effective radius of > Actual Radius curvature of earth
- Straight line path (without refraction) in atmosphere.
- **As the dielectric constant i.e. reference index of atmosphere varies with height above ground and so refraction of radio wave takes place.**
- Dielectric constant near earth > 1
- Dielectric constant at more height < 1 where air density = 0
- **This decrease in reference index with height causes refraction of radio wave, it results in the bending of radio wave towards region of higher dielectric constant or reference index i.e. towards earth.**
- *Relation between* = Radius of curvature of ray path in troposphere and changes of refractive index with height by assuming curvature of earth.

- Consider a radio wave which is travelling nearly horizontally in troposphere and its path is bent into an arc by variation of reference index with height as

Let v = velocity of propagation

h = Height above the earth (of travelling ray)

R = Radius of curvature of ray path

r = Actual radius of earth.

$$Angle = \frac{Arc}{Radius} \quad or \quad d\theta = \frac{vdt}{R}$$

$$Rd\theta = v\,dt \qquad (1)$$

Similarly $(R + dh)\,d\theta = (v + dv)\,dt$ (2)

or (1) and (2) subtraction.

$$= (R + dh - R)\,d\theta = (v + dv - v)\,dt$$

$$= dh\,.\,d\theta = dv\,.\,dt$$

$$\boxed{or \quad \frac{d\theta}{dt} = \frac{dv}{dh}} \qquad (3)$$

But

$$\boxed{v = \frac{c}{\sqrt{k_r}} = \frac{C}{\sqrt{\mu}}} \qquad (4)$$

Here, k_r = dielectric constant (relative)

c = velocity of light

μ = Ref. index at height (h)

∴ Change in ref. index with height is obtained by differentiating equation (4) w.r.t. `h'

$$\frac{dv}{dh} = \frac{-C}{v^2}\frac{du}{dh}$$

$$= -\left(\frac{C}{\mu}\right)\left(\frac{1}{\mu}\right)\frac{d\mu}{dh}$$

$$= -\frac{v}{\mu}\frac{d\mu}{dh}$$

$$\boxed{\therefore \frac{dv}{dh} = -\frac{vd\mu}{dh}} \qquad (5)$$

$$\because \mu \approx 1$$

From eq. (1) $R = \frac{vdt}{d\theta} = \frac{v}{d\theta / dt}$

From eq. (3) and (5)

$$R = \frac{v}{\left(\frac{d\theta}{dt}\right)} = \frac{v}{\left(\frac{dv}{dh}\right)} = \frac{v}{-v\frac{d\mu}{dh}}$$

$$\therefore R = -\frac{dh}{d\mu} \qquad (6)$$

This shows that radius of curvature of wave path is a function of rate of change of dielectric constant or ref. index with height, varies from hour to hour, day to day and season to season.

Q. Prove that the effective earth's radius, R is expressed by

$R = -\frac{dh}{d\mu}$ *where* μ = Re*fractive index* **(Raj. Univ. 2006)**

But, an average value, 4 times the radius of earth is used for calculation purposes.

In real world working with propagation problems, it is usually convenient to assume ray paths as straight lines instead of being curved, as they in real are, and to compensate for the curvature by using a large value for effective radius of earth in Figure 5.6.

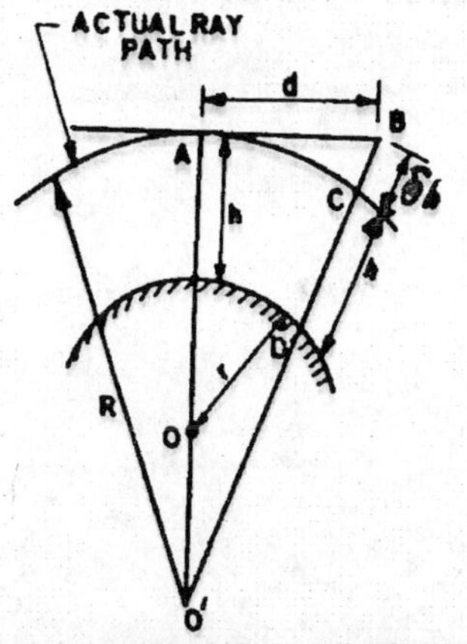

(a) Equivalent ray path over earth with actual earth's radius (r).
d = horizontal distance.

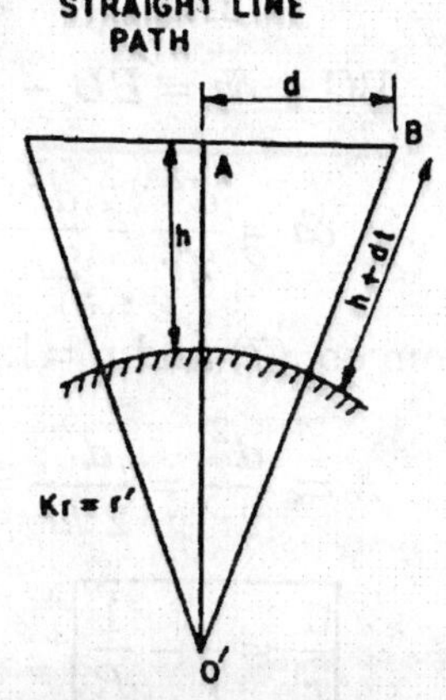

(b) Equivalent ray path over the earth with effective earth's radius.
(r' = kr).

Figure 5.6

For straight line path in II figure to be equivalent to (I) it becomes necessary that dh (change in height) be the same in 2 cases for same horizontal distance `d'.

Now from figure, Right angle $\Delta O'AB$ (II)

$$d^2 + (r' + h)^2 = [r' + (h + \delta h)]^2$$

$$\Rightarrow d^2 + r'^2 + 2hr' + h^2 = r'^2 + (h + \delta h)^2 + 2r'(h + \delta h)$$

$$\Rightarrow d^2 + r'^2 + 2hr' + h^2 = r'^2 + h^2 + (\delta h)^2 + 2(h)(\delta h) + 2r'[h] + [\delta h]$$

$$\Rightarrow d^2 = (\delta h)^2 + 2(h)(\delta h) + 2r'[h + \delta h - h]$$

$$\Rightarrow d^2 = (\delta h)^2 + 2(h)(\delta h) + 2r'(\delta h)$$

$$\Rightarrow d^2 = 2(\delta h)[r' + h] + (\delta h)^2$$

$$\Rightarrow d^2 \cong 2(\delta h)(r' + h) \qquad \because \delta h^2 << 2\,\delta h\,(r' + h)$$

$$\therefore \delta h = \frac{d^2}{r' + h} \qquad \because r' >> h$$

$$\boxed{\therefore \delta h \approx \frac{d^2}{2r'}} \qquad (7)$$

Similarly from figure (I)

$$DC = \frac{d^2}{2R} \quad \text{and} \quad BD = \frac{d^2}{2r} \quad \text{since } r' >> h.$$

$$BC = \delta h = BD - DC$$

$$\therefore \delta h = \frac{d^2}{2r} - \frac{d^2}{2R}$$

From eq. (7) and r' = kr

$$\therefore \Rightarrow \frac{d^2}{2r'} = \frac{d^2}{2kr} = \frac{d^2}{2r} - \frac{d^2}{2R}$$

$$\boxed{\frac{1}{r'} = \frac{1}{r} - \frac{1}{R}} \qquad (8)$$

Or

$$\boxed{\frac{1}{kr} = \frac{1}{r} - \frac{1}{R}} \qquad (9)$$

Where r′ = kr = effective radius of earth which is nothing but the ratio of effective earth's radius to the actual earth's radius & is denoted by a factor k

$$\boxed{\Rightarrow k = \frac{r'}{r}} \tag{10}$$

From eq. (9) $\Rightarrow \frac{1}{r'} = \frac{R - r}{rR}$

or $r' = \frac{rR}{R - r}$

So $$\boxed{r' = \frac{r}{r - \frac{r}{R}}} \tag{11}$$

Now as

$$k = \frac{r'}{r} = \frac{1}{1 - \frac{r}{R}}$$

or

$$\boxed{k = \frac{1}{1 + r\frac{d\mu}{dh}}} \tag{13}$$

> **Q.** Show that the ratio of the effective earth's radius to the actual earth's radius :
>
> $K = \frac{1}{1 + \frac{d\mu}{dh}}$ *where r =radius of actual earth*
>
> **(Raj. Uni. 2006)**

From eq. (6)

From Eq. (12), if radius of curvature `R' of ray path is equal to 4 times actual earth's radius, then effective radius of earth is 4/3 times the real radius of earth. For value of $\frac{d\mu}{dh}$ corresponding to `standard' atmosphere is taken as 0.040 x 10^{-6} per meter & radius of earth r = 6370 km = 6.37 x 10^6 m. if substituted in eq. (13) we get

$$k = \frac{1}{1 - 6.37 \times 10^6 \times 0.040 \times 10^{-6}}$$

$$= \frac{1}{1 - 6.37 \times 0.04} = \frac{1}{1 - 0.2548} \approx \frac{1}{1 - 0.25}$$

$$= \frac{1}{1 - \frac{1}{4}} = \frac{4}{4-1} = \frac{4}{3}$$

$$\boxed{\therefore k = \frac{4}{3} = \frac{r'}{r}} \quad (14)$$

$$\boxed{\therefore r' = \frac{4}{3} r} \quad (15)$$

Hence for standard atmospheric refraction effective earth's radius is 4/3 times of the actual earth's radius. Further eq. $d = \sqrt{2r'}\left[\sqrt{h_r} + \sqrt{h_t}\right]m$. is modified by replacing r by r' & LOS distance

$$\therefore d = \sqrt{2r'}\left[\sqrt{h_t} + \sqrt{h_r}\right]$$

$$= \sqrt{2 \times \frac{4}{3} r}\left[\sqrt{h_t} + \sqrt{h_r}\right]$$

$$= \sqrt{\frac{8 \times 6.37 \times 10^6}{3}}\left[\sqrt{h_t} + \sqrt{h_r}\right]$$

$$= 10^3 \sqrt{8 \times 2.12}\left[\sqrt{h_t} + \sqrt{h_r}\right]$$

$$= 10^3 \sqrt{16.96}\left[\sqrt{h_t} + \sqrt{h_r}\right] \quad \because 1\ km = 10^3\ m$$

$$\boxed{\therefore d = 4.12\left[\sqrt{h_t} + \sqrt{h_r}\right] km} \quad (16)$$

Where, d = LOS distance (km)

h_t, h_r = heights of Tx & Rx antennas (m)

∴ Eq. (16) is eq. for calculating radio horizon or LOS distance when h_t and h_r are given in m i.e. in metric system. However, sometimes the height of Tx and Rx antennas are also given in fact and LOS distance or radio horizon distance is required to be in miles. This can be calculated.

$$d = \sqrt{2r'}\left[\sqrt{h_t} + \sqrt{h_r}\right]$$

r and d = in statute miles.

h_f & h_r = in fact

Where r = Radius of earth = 3960 miles

1 statute mile = 5280 feet

Then $$r' = \frac{4}{3}r = \frac{4}{3} \times 3960$$

Putting these values in the given equation

$$d = \sqrt{2r'}\left[\sqrt{hr} + \sqrt{h_t}\right]$$

$$d = \sqrt{2 \times \frac{4}{3} \times 3960}\left[\sqrt{\frac{h_t}{5280}} + \sqrt{\frac{h_r}{5280}}\right] \textit{ miles}$$

$$= \sqrt{2 \times \frac{4}{3} \times \frac{3960}{5280}}\left[\sqrt{h_t} + \sqrt{h_r}\right] \textit{miles}$$

$$= \sqrt{2 \times \frac{4}{3} \times \frac{99}{132}}\left[\sqrt{h_t} + \sqrt{h_r}\right] \textit{miles}$$

$$= \sqrt{2 \times \frac{4}{3} \times \frac{9}{12}}\left[\sqrt{h_t} + \sqrt{h_r}\right] \textit{ miles}$$

$$= \sqrt{2 \times \frac{4}{3} \times \frac{3}{4}}\left[\sqrt{h_t} + \sqrt{h_r}\right] \textit{miles}$$

$$\boxed{\therefore\ d = \sqrt{2}\left[\sqrt{h_t} + \sqrt{h_r}\right] = 1.4142\left[\sqrt{h_t} + \sqrt{h_r}\right] \textit{miles}} \quad (17)$$

5.4.3 Effect of Earth's curvature on Tropospheric Propagation

On troposphere propagation following 2 - effects are introduced by curvature of earth

(i) Difference in path lengths between direct & ground reflected waves is reduced according to the point of reflection on the ground is raised. As a result, it tends to reduce the signal strength at receiving point.

(ii) Further, since the reflection at the ground takes place at a spherical point rather than a flat point & so reflected ray becomes divergent. It results in weaker at receiving point. This tends **to increase field strength of total space wave at the receiving point.**

Both these effects are of opposite nature & so neutralize each other. Hence it is assumed that the plane earth in above derivation is justified, inspite their existence considerable curvature in the earth.

5.4.4 Field strength of space or Tropospheric wave

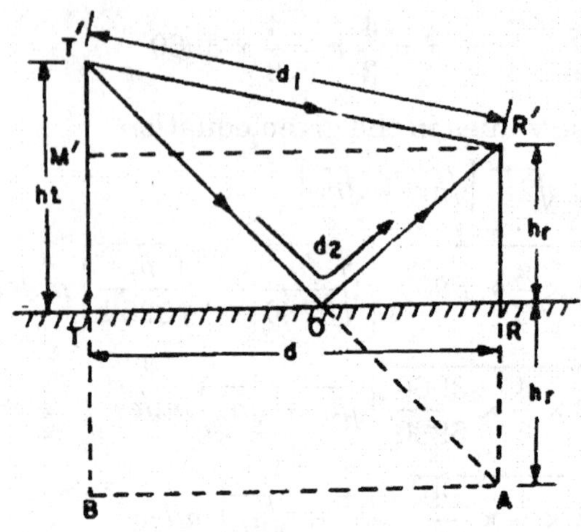

Figure 5.7

If the curvature of earth is neglected, there propagation is shown in figure. In which energy can be received in 2 ways

(i) direct path (OT' R')

(ii) Indirect path (after Reflection from ground) (T' OR')

Field strength = (i) + (ii)

∴ *angle of incidence at ground is also small.*

h_t, h_r = height of antenna (T_x, R_x)

d = distance between them

d_1 = direct ray's path

d_2 = Indirect ray's path

E_o = Field strength at R' due to direct rays

By assumption of earth to be perfect, magnitude of field strength at R' would be equal to E_o as the distance travelled by both rays are approx. equal. However, if reflection coefficient at the ground is k, then the magnitude of ground reflected wave is kE_o. The 2 rays combine at receiving point according to vector. Now from figure $\Delta T'R'M' \Rightarrow$

$$(h_t - h_r)^2 \neq d^2 = d_1^2 \tag{18}$$

And similarly from right angle $\Delta T'' AB$,

$$(h_t + h_r) + d^2 = d_2^2 \tag{19}$$

From equation (18) $\because RR' = RA$

$$d_1 = \left[d^2 + (h_t - h_r)^2\right]^{1/2}$$

$$= d\left[1 + \left(\frac{h_t - h_r}{d}\right)^2\right]^{1/2}$$

$$= d\left[1 + \frac{1}{2}\left[\frac{(h_t - h_r)}{d}\right]^2 + \ldots\right]$$

$$d_1 = d + \frac{(h_t - h_r)^2}{2d} \qquad (20)$$

And similarly

$$d_2 = d + \frac{(h_t + h_r)^2}{2d} \qquad (21)$$

Path difference between direct and indirect rays

P. d = $d_2 - d1$

$$= d + \frac{(h_t + h_r)^2}{2d} - d - \frac{(h_t - h_r)^2}{2d}$$

$$= \frac{h_t^2 + h_r^2 + 2(h_r)(h_t) - h_t^2 - h_r^2 + 2(h_r)(h_t)}{2d}$$

$$= \frac{4(h_t)(h_r)}{2d} = \frac{2(h_t)(h_r)}{d} \qquad (22)$$

$$\therefore \textit{Path difference} = \frac{2h_t h_r}{d} \qquad (22)$$

It is known from the optics that

Phase difference (α) $= \dfrac{2\pi}{\lambda}$ (Path difference) from eq. (22)

$$= \frac{2\pi}{\lambda}\left(\frac{2h_t h_r}{d}\right)$$

$$\therefore \alpha = \frac{4\pi(h_t)(h_r)}{\lambda d} \text{ Radians} \qquad (23)$$

It is phase difference caused due to path difference Along with it there is another phase difference of 180° due to reflection from the ground (i.e. B = 180°) Hence total phase difference (θ) is given by

$\theta = \alpha + \beta$ phase difference due to reflection at ground = 180°

Where phase difference due to path difference

∴ Resultant field strength = at point R

$$\boxed{E_R = E_o\left(1 + ke^{-j\theta}\right)}$$

$$E_R = E_o\left[1 + k\left(\cos\theta - j\sin\theta\right)\right]$$

$$\therefore |E_R| = E_o\sqrt{(1 + k\cos\theta)^2 - (jk\sin\theta)^2}$$

$$= E_o\sqrt{1 + k^2\cos^2\theta + 2k\cos\theta - j\,k^2\sin^2\theta} \qquad \because d^2 = -1$$

$$= E_o\sqrt{1 + K^2\cos^2\theta + 2k\cos\theta + k^2\sin^2\theta}$$

$$= E_o\sqrt{1 + k^2\left(\cos^2\theta + \sin^2\theta\right) + 2k\cos\theta}$$

as $\sin^2\theta + \cos^2\theta = 1$

$$= E_o\sqrt{1 + K^2 + 2K\cos\theta}$$

if earth = perfect = k = 1.

B = 180° or π radians.

$$\therefore |E_R| = E_o\sqrt{1 + 1^2 + 2(1)\cos\theta}$$

$$= E_o\sqrt{2 + 2\cos\theta} \qquad \left[\because \cos\theta = 2\,Cos^2\frac{\theta}{2} - 1\right]$$

$$|E_R| = E_o\sqrt{2 + 4\cos^2\frac{\theta}{2} - 2}$$

$$= E_o\left(2\cos\frac{\theta}{2}\right)$$

$$= 2E_o\left(\cos\frac{\theta}{2}\right)$$

$$|E_R| = 2E_o\,Cos\left(\frac{\alpha + \pi}{2}\right) \qquad as\ \theta = \alpha + B$$

$$= \alpha + T_1$$

$$as\ Cos\left(\frac{\pi}{2} + \frac{\alpha}{2}\right) = \sin\frac{\alpha}{2}$$

$$= 2\ E_o\ Cos\left(\frac{\pi}{2}+\frac{\alpha}{2}\right)$$

$$=2\ E_o\ \sin\left(\frac{\alpha}{2}\right)$$

$$|E_R| = 2\ E_o\ \sin\left(\frac{4\pi\ h_t\ .\ h_r}{2d\ \lambda}\right) \quad \textit{from eq. (23)} \tag{24}$$

As d > > h_t or h_r

$$\therefore \sin\frac{4\pi\ h_t\ h_r}{2d\lambda} \approx \frac{4\ \pi\ h_t\ h_r}{2\ d\ \lambda}$$

$$\therefore |E_R| = 2\ E_o\left(\frac{4\pi\ (h_t)(h_r)}{2d\ \lambda}\right)$$

$$|E_R| = E_o\left(\frac{4\pi\ h_t\ .\ h_r}{d\lambda}\right) \tag{25}$$

Where h_t, h_r, d (m)

If E_f = Electric field strength of direct ray i.e. free space field strength at unit distance

Then $$E_o = \frac{E_f}{d} \tag{26}$$

But $$E_f = 7\sqrt{p}\ V/m \tag{27}$$

P = effective power radiated (watts)

$$\therefore E_o = \frac{7\sqrt{p}}{d} \tag{28}$$

Putting this value of E_o from eq. (28) in (25) we have

$$|E_R| = \frac{7\sqrt{p}}{d}\cdot\frac{4\pi\ h_t\ .\ h_r}{d\ .\ \lambda}$$

$$= \frac{7\times4\times3.14\sqrt{P}\ .\ h_t\ .\ h_r}{d^2\ \lambda}$$

$$= \frac{87.92\ \sqrt{P}\ (h_t)\ (h_r)}{d^2\ \lambda}$$

or $$|E_R| \approx \frac{88\sqrt{P}\,(h_t)(h_r)}{d^2\,\lambda} \quad V/m \qquad (29)$$

Hence for a propagation of space wave, we see that

$E_R \; \alpha \; \sqrt{P}$ *(Effective power radiation)*

$\alpha \; h_t$ *(height of T_x antenna)*

$\alpha \; h_r$ *(height of R_x antenna)*

$\alpha \; \frac{1}{\lambda}$ *(inverse relation with λ)*

$\alpha \; \frac{1}{d^2}$ (inverse square with square of distance from T_x antenna)

Eq. (29) is used to calculate field strength at R_x. point, if all above quantities are given.

Q. Explain tropospheric scatter propagation and applications.

(R.T.U. 2009)

5.5 DUCT PROPAGATION (Super Refraction)

At VHF, UHF and microwaves, the waves are neither reflected by ionosphere nor propagated along earth's surface, but transmission does occur much beyond LOS due to refraction of such high freq. waves (μ-waves especially) in troposphere (16 km above).

Region above it is ``stratopause" where constant temp. = –50°C (minus)

Inside troposphere atmosphere has dielectric constant slightly higher than unity at earth's surface where density is most dense and this $\downarrow$ *to* 1 at great heights where the air density approaches zero.

Dielectric of dry air > 1.

(Presence of water vapour increases it further)

∴ Dielectric constant depends upon air condition (weather)

Dielectric constant is assumed to decrease uniformly with height to valve of = 1 at height where air density = 0.

⇒ at normal/ standard atmosphere (virtual)

The air is generally turbulent with different layers. These layers are having different temperature & water vapour contents.

This condition has other phenomenon than the normal ones (scattering, refraction & reflection) it has super refraction **or duct propagation.**

Two boundary surfaces between layers of **air form a duct or wave guide (leaky) which guides EM wave** between its walls

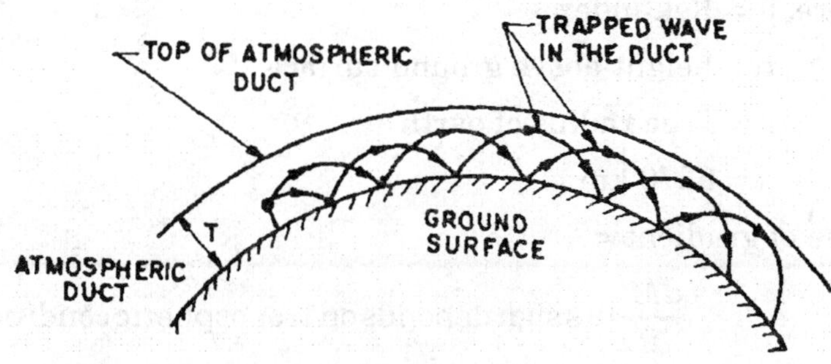

Figure 5.8

When the frequency is sufficiently high, the region where the variation of dielectric constant or reference index is usually high (ref. index decreases with height), actually traps **the energy and causes it to travel along the earth surface as happens in a waveguide.**

This happens near the ground ($(\leq 50m)$ **of troposphere).**

The higher frequencies or μ-waves are thus continuously refracted in duct and reflected by the ground that's why they propagate around curvature for beyond LOS, (even upto 1000 km)

This special refraction of EM waves is called super refraction and the process is called as DUCT PROPAGATION.

* **The main requirement for the formation of duct is "*temperature inversion*"** i.e. in inversion layer the temperature increases with height rather usual decreases of temperature at the rate of 6.5°C km in standard atmosphere.

In order to give necessary curvature practical ref. index μ at a height 'h' must be replaced by modified ref. index given by

$$\boxed{N = \mu + \frac{h}{r}}$$

Although N = 1, yet its real value is important & hence when dealing with numerical values, it is convenient to introduce excess *modified index of ref. modulus (M)*, related to N as

$$N - 1 = \mu - 1 + \frac{h}{r}$$

$$(N-1) \times 10^6 = \left(\mu - 1 + \frac{h}{r}\right) \times 10^6$$

or $$M = (N-1) \times 10^6 = \left(\mu - 1 + \frac{h}{r}\right) \times 10^6$$

Where, μ = Ref. index

h = height above ground surface

r = True radius of earth

= 6370 km

Value of gradient =

$$\frac{dM}{dt}\text{(its sign depends on tropospheric condition)}$$

= M against h.

Duct = Tube = Channel = Waveguide (leaky)

(I) From above figure it is clear that *duct is formed only when value of gradient dM/dh is negative i.e. m ↓ with h ↑*. The height for which this process of decreasing M continues forms the inversion layer.

(1) (d_1C_1); (d_2C_2); (d_3 C_3) = inversion layer.

(2) (d_1 a_1); (d_2 a_2); (d_3 a_3) = respective heights.

(3) Width of duct = determined by dropping projection vertically downward which cut at a_1, b_2 & b_3.

(4) The horizontals drawn at these points a_1, b_2 and b_3 give lower ends of duct

(5) Horizontals at d_1, d_2 and d_3 = top end of (6) duct width = (d_1 a_1); (d_2 b_2) and (d_3 b_3)

When h-m curve has negative slope, ray enters duct with sufficiently small angles are tilted until they become horizontal.

These rays are trapped between upper and lower walls of duct and are oscillating between ground and upper wall of duct in case of gnd. based duct and between 2 walls in atmosphere in case of an elevated duct. This process is called - super refraction of Duct propagation.

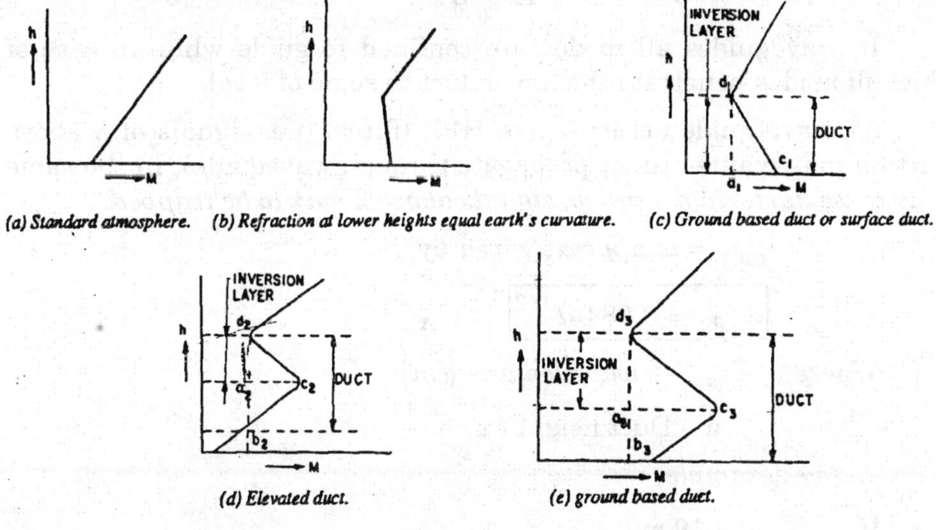

(a) Standard atmosphere. (b) Refraction at lower heights equal earth's curvature. (c) Ground based duct or surface duct.

(d) Elevated duct. (e) ground based duct.

Figure 5.9

Note : Only those waves are trapped which are entering with small angles w.r.t. horizontal but if this angle exceeds a critical angle (α_c) it is pen entering duct as in case of ray 3 and 4 while ray 1 and 2 have been trapped as their angle of entrance was less than critical α_c.

Duct propagation depends on =

(1) Whether Transmitter is inside duct/above/below the duct.

(Favorable→ When Tx is inside duct

→ in this condition all the waves do not get trapped)

→ Ducts are formed mostly over water; same of ducts are formed on land also but they are less frequently used.

→ Ducts are at coastal areas (elevation of 1000-5000 feet)

(Width=1000 feet)

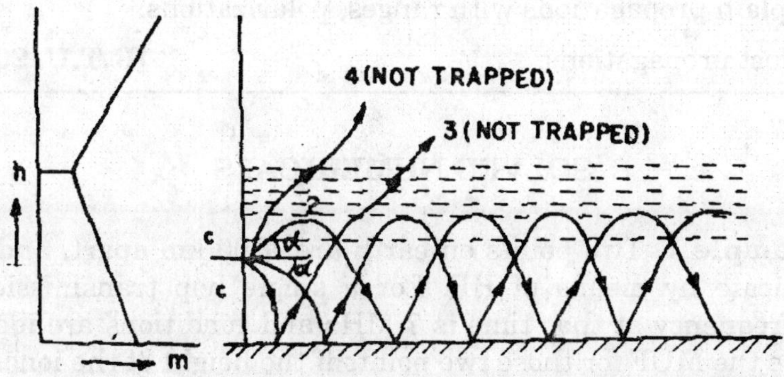

Figure 5.10

Physical picture of wave trapping is similar to waveguide→

In waveguides all modes are confined to guide while in case of duct all modes penetrate the top of duct to some of level.

As wave guide action→ like HPF (filter) (i.e. signals of λ above certain max. value cannot propagate through waveguide). In the same way *these ducts do not allow, signals above λ max to be trapped.*

$$\lambda_{max} = \textit{is approx. given by}$$

$$\boxed{\lambda_{max} = 0.084\, d^{3/2}}$$

Where, $\lambda_{max} \rightarrow$ *wave lengths* (*cm*)

d – Duct height (m)

∴ *say* (*example*) : –

If d = 10 m

$$\text{Then } \lambda_{max} = 0.084\,(10)^{3/2}$$

$$= 0.084 \times 10\sqrt{10}$$

$$= 0.084 \times 3.162$$

$$= 2.656$$

$$\approx 2.6 \text{ cm}$$

⇒ Duct phenomenon is important at. →UHF

→ Microwave freq

Q. Write short note on Duct propagation. **[Raj. Univ. 2008, 2006]**   **Q.** Write short notes duct propagation, use of Rhombic Antenna, and effect of ground conductivity? **[Raj. Univ. 2003]**
Q. Explain propagations with ranges; polarizations: i. Duct propagations. **[R.T.U. 2009]**

SOLVED NUMERICALS

Example 1. Two points on earth are 1500 km apart, and are to communicate by means of HF. For a single hop transmission, the critical frequency at that time is 7 MHz and conditions are idealized. Calculate the MUF for those two points if the height of the ionospheric layer is 300 km?

Solution: $f_{muf} = f_c \sqrt{1 + \left(\frac{D}{2h}\right)^2} = 7 \sqrt{1 + \left(\frac{1500}{2 \times 300}\right)^2}\ MHz$

$= 7\sqrt{1 + (2.5)^2} = 7\sqrt{1 + 6.25} = 7\sqrt{7.25}$

$= 7 \times 2.693 = 18.851\ MHz$ **Ans.**

Example 2. The observed critical frequencies of E and F layers at Gauhati at particular times are 2.5 MHz and 8.4 MHz respectively. Calculate the maximum electron concentrations of the layer?

Solution: We know

$$f_c = 9\sqrt{N_{max}} \text{ or } N_{max} = \frac{f_c^2}{81}$$

Where, f_c is in Hz an N_{max} is in M^{-3}. If f_c is in, KHz, N_{max} is in cubic centimeters.

(i) For E layers,

$f_c = 2.5$ MHz

$$\therefore N_{max} = \frac{(2.5 \times 10^6)^2}{81} = \frac{6.25 \times 10^{12}}{81}$$

$$= 0.07716 \times 10^{12}\ m^{-3}$$

Or $N_{max} = 0.7716 \times 10^{11}$ per cubic meters. Ans

(ii) For F layer

$f_c = 8.4\ MHz$

$$\therefore N_{max} = \left(\frac{8.4 \times 10^6}{81}\right)^2 = \frac{70.56 \times 10^{12}}{81} = 0.8711 \times 10^{12}\ m^{-3}$$

Or $N_{max} = 87.11 \times 10^{11}$ *per cubic meters* **Ans.**

Example 3. A High frequency radio link has to be established between two points on the earth 2000 km away. If the reflection region of the ionosphere is at a height of 200 km, and has a critical frequency of 6 MHz. Calculate the MUF for the given path.

Solution: $f_m = f_c \sqrt{1 + \left(\frac{D}{2h}\right)^2}$

$$f_m = 6\sqrt{1+\left(\frac{2000}{2\times 200}\right)^2} = 6\sqrt{1+(5)^2}$$

$$= 6\sqrt{26} = 6 \times 5.099$$

$$= 30.594\ MHz \approx 30.6\ MHz$$ **Ans.**

Example 4. Communication is to be established between two stations 1500 km apart. Derive the necessary formula and calculate the maximum frequency you may choose for communication using the ionosphere as reflector if the height and the plasma frequency of the ionosphere at the point of reflection are respectively 250 km and 12 MHz. Assume the ionosphere to be thin and the earth to be flat.

Solution: As above

D = 1500 km; h = 250 km

f_c = 12 MHz and f_{muf} = ?

$$f_{MUF} = f_c\sqrt{1+\left(\frac{D}{2n}\right)^2}$$

Hence, $f_{muf} = 12\sqrt{1+\left(\frac{1500}{2\times 250}\right)^2}\ MHz = 12\sqrt{1+(3)^2}$

f_{MUF} = 37.944 MHz

Example 5. If the critical frequency of an ionized layer is 2 MHz, find the electron density of the layer?

Solution: The critical frequency is defined as $f_c = 9\sqrt{N_{max}}$

Or $$N_{max} = \frac{f_c^2}{81} = \frac{(2\times 10^6)^2}{81} = \frac{4\times 10^{12}}{81}$$

$$N_{max} = 4.93\times 10^{10}\ electrons/m^3$$ **Ans.**

Example 6. A sky wave is incident on D layer at an angle of 45°, find the angle of refraction of the frequency of the transmitted wave in 40 MHz. The electron density in D layer is N_{max} = 380 electrons/cm³.

[Raj. Univ. 2002]

Solution: N_{max} = 380 electrons / cm³ = $380\times 10^6\ electrons/m^3$

Now $$n = \sqrt{\varepsilon_r} = \left(1 - \frac{81\, N_{max}}{f_c^2}\right)^{1/2}$$

So $$n = \sqrt{\varepsilon_r} = \left(1 - \frac{81 \times 380 \times 10^6}{16 \times 10^{14}}\right)^{1/2} \approx 1$$

So $n = 1$

Now using Snell's law $\sin \theta_i = n \sin q_r$ where n for free space = 1

Or $\sin \theta_r = \sin \theta_i$

$= \sin 45°$

The angle of refraction $\theta_r = 45°$ **Ans.**

Example 7. At what frequency L a wave must propagate for the D region to have an index of refraction 0.6. Given N_{max} = 300 electron/ cubic centimeter for D region? **(Raj. Univ. 1997, 2004)**

Solution: Since $$n = \sqrt{1 - \frac{81\, N_{max}}{f^2}}$$

$$\Rightarrow (0.6)^2 = 1 - \frac{81\, N_{max}}{f^2} \Rightarrow 1 - 0.36 = 0.64$$

$$f^2 = \frac{81 \times 300}{0.64} = 194\ KHz$$

f = 194 KHz **Ans.**

Example 8. What is the critical frequency for reflection at vertical incidence if the maximum value of electron density 2.25 x 10^7 cm^{-3}? **(Raj. Univ. 1994, 2001)**

Solution: We know, critical frequency can be defined as

$$f_c = 9\sqrt{N_m} \qquad Now\ N = 2.25 \times 10^7\ cm^{-3}$$

$$N_m = 22.5\ m^{-3} \qquad = 2.25 \times 10^7 \times 10^{-6}\ m^{-3}$$

So $$f_c = 9\sqrt{22.5}\ MHz \quad = 22.5\ m^{-3}$$

$$f_c = 9 \times 4.74\ MHz$$

$= 42.69$ MHz

$f_c = 42.69$ MHz **Ans.**

Example 9. Calculate the transmission path distance for an ionospheric transmission that utilizes a layer of height 250 km. The angle of elevation of the antenna beam is 30°. The earth's radius can be taken as 6400 km? **(Raj. Univ. 1993, 2000)**

Solution: Diameter of earth for curved surface

$$D = \frac{2h}{\tan\beta} = \frac{2 \times 250}{\tan 30^\circ} = \frac{500}{0.577} = 866\,km$$

R = 6400 km

If earth is assumed to be curved

$$D = 2R\left[(90-\beta) - \sin^{-1}\left(\frac{R\cos\beta}{R+h}\right)\right]$$

$$= 2R\left[\left(\frac{\pi}{2}-\beta\right) - \sin^{-1}\left(\frac{R\cos\beta}{R+h}\right)\right]$$

$$= 2R\left[(90^\circ - 30^\circ) - \sin^{-1}\left(\frac{6400 \times \cos 30^\circ}{6400+250}\right)\right]$$

= 792.67 km. **Ans**

Example 10. Calculate the maximum range obtainable in a single hop transmission utilizing E layer for which earth's radius is assumed to be 6400 km and height of layer is 150 km?

(Raj. Univ. 1997)

Solution: The maximum range in a single transmission is given by

$$d = 2R.\theta = 2R\cos^{-1}\frac{R}{R+h_m}$$

$$= 2 \times 6400 \times \cos^{-1}\left(\frac{6400}{6400+150}\right)$$

$$= 12800 \times \cos^{-1}(0.9465)$$

$$= 12800 \times 12.3^\circ \times \frac{\pi}{180} = 1280 \times \frac{38.62}{18}$$

$$= 1280 \times 2.14 = 2739.20\,km$$

$d = 2739.20\,km$ **Ans.**

Example 11. F_2 layer of the ionosphere has an electron density of 0.81×10^{12} per m^3 at a height of 350 km from the earth's surface.

Find the critical frequency of this layer. Also find the maximum usable between two stations 1500 km apart. Neglect earth's curvature? **(Raj. Univ. 2004)**

Solution: The critical frequency is defined as

$$f_c = 9\sqrt{N_{max}} \qquad (Given)\ Where\ N_{max} = 0.81 \times 10^{12}\ /m^3$$

$$\text{Therefore, } f_c = 9\sqrt{0.81 \times 10^{12}}$$

$$= 9 \times 0.9\ MHz$$

$$= 8.1\ MHz$$

$$f_c = 8.1\ MHz$$

The maximum usable frequency will be

$$f_{MUF} = f_c\sqrt{1+\left(\frac{D}{2h}\right)^2} \qquad where\ D = 1500\,km$$

$$h = 350\,km$$

So

$$f_m = 8.1 \times 10^6\sqrt{1+\left(\frac{1500}{2 \times 350}\right)^2}$$

$$f_m = 8.1 \times 10^6\ \sqrt{5.59}$$

$$= 8.1 \times 2.36\ MHz$$

f_{muf} = 19.1 MHz **Ans.**

Example 12. F_2 layer of the ionosphere at a height of 275 km from the earth's surface has an electron density of 10^{12} per m^3. Find the critical frequency for this layer. Also find the maximum usable frequency between two stations 1500 km apart. Neglect earth's curvature? **(Raj. Univ. 2002)**

Solution: $f_c = 9\sqrt{N_{max}}$ Given $N_{max} = 10^{12}\ per\ m^3$

$$f_c = 9\sqrt{10^{12}} = 9\ MHz$$

Now

$$f_{MUF} = f_c\sqrt{1+\left(\frac{D}{2h}\right)^2} \qquad D = 1500\,km \quad h = 275\ km$$

So $$f_{MUF} = 9 \times 10^6 \times \sqrt{1 + \left(\frac{1500}{2 \times 275}\right)^2}$$

$$F_{muf} = 9 \times 10^6 \times 2.90 \; Hz = 26.1 \; MHz$$ **Ans.**

Example 13. The electron concentration at a height of 250 km is $10^{12}/m^3$. What is the maximum angle of incidence ψ_t that can be used at a frequency of 15 MHz? What is the horizontal skip distance if the maximum angle ψ_t is used? **(Raj. Univ. 2000, 2001)**

Solution: The critical frequency is given by

$$f_c = 9\sqrt{N_{\max}}$$

$$f_c = 9\sqrt{10^{12}} = 9 \; MHz$$

The maximum angle of incidence can be calculated as

$$\sin \psi_t \geq \sqrt{1 - \frac{81 \, N_m}{f^2}}$$

$$\geq \sqrt{1 - \frac{81 \times 10^{12}}{(15 \times 10^6)^2}}$$

$$\geq \sqrt{1 - \frac{81}{225}}$$

$$\sin \psi_t \geq \sqrt{1 - 0.36}$$

$$\sin \psi_t \geq 0.8$$

$$\psi_t \geq 53^\circ$$

Assuming flat earth, skip distance is given by

$$\frac{D^2}{4h^2} = \left(\frac{f_{muf}}{f_c}\right)^2 - 1$$

or $$D = 2h\sqrt{\left(\frac{f_{muf}}{f_c}\right)^2 - 1}$$

$$= 2 \times 250 \sqrt{\left(\frac{15}{9}\right)^2 - 1}$$

$$= 500\sqrt{2.1481}$$

$$D = 732.81\ Km$$ **Ans.**

Example 14. Calculate the maximum usable frequency for an electromagnetic wave incident at an angle of 45° to be propagated through ionospheric layer having electron density of 4 x 10^4 electron/m^3. **(Raj. Univ. 1999)**

Solution: Since

$$\sin^2 t = 1 - \frac{f_c^2}{f_{MUF}^2}$$

Now $$f_c = 9\sqrt{N_{\max}}$$

$$= 9 \times \sqrt{4 \times 10^4}\ Hz$$

$$= 9 \times 2 \times 10^2\ Hz$$

$$= 1800\ Hz$$

or $$\sin^2 45° = 1 - \frac{(1800)^2}{f_{MUF}^2}$$

or $$\frac{(1800)^2}{f_{MUF}} = 1 - \sin^2 45° = 0.5 \qquad \left[\sin 45° = \frac{1}{\sqrt{2}}\right]$$

Therefore $$f_{MUF}^2 = 2 \times 1800 \times 1800$$

or $$f_{MUF} = 2545.58\ Hz$$ **Ans.**

Example 15. Calculate the value of the factor by which the horizon distance of a transmitter will be modified if the gradient of refractive index of air near the ground is –0.5 x 10^{-6} m^1. The radius of the earth is given to be 6370 km. **(Raj. Univ. 1996)**

Solution: Since

$$k = \frac{1}{1 + r\dfrac{dx}{dh}}$$ *where x is refractive index*

Given $$r = 6370\ km \quad \frac{dx}{dh} = -\ 0.05 \times 10^{-6}$$

Substituting

$$k = \frac{1}{1 - 0.05 \times 10^{-6} \times 6.37 \times 10^{6}}$$

$$= \frac{1}{1 - 0.3185} = \frac{100}{68} = \frac{25}{17}$$

$$r^1 = kr = \frac{25}{17} r$$

Therefore the horizon distance of the transmitter can be modified by replacing r by r^1

i.e. $r' = kr = \frac{25}{17} r$

(Where is the radius of earth)

Example 16. What is the critical frequency for reflection at vertical incidence if the maximum value of electron density is $1.24 \times 10^6 \ cm^{-3}$?

Solution: $f_c = 9\sqrt{N_m}$

$$= 9\sqrt{1.24} \ MHz = 9 \times 1.114 \ MHz$$

$$f_c = 10.026 \, MHz$$ **Ans.**

Example 17. Radio communication is desired two places separated by a distance of 1000 km. On the earths surface via an ionospheric layer whose height is 400 km and critical frequency is 11 MHz.

Solution: $f_o = f_c \sqrt{1 + \left(\frac{D}{2h}\right)^2}$

f_c = 11 MHz

h = 400 km.

D = 1000 km

f_o = ?

$$f_o = 11 \times 10^6 \sqrt{1 + \frac{(1000 \times 10^3)^2}{(2 \times 400 \times 10^3)^2}}$$

$$= 11 \times 10^6 \sqrt{1 + \left(\frac{5}{4}\right)^2}$$

$$= 11 \times 10^6 \sqrt{1 + \frac{25}{16}}$$

$$= 11 \times \sqrt{\frac{41}{16}} \; MHz$$

$$f_o = 11 \times \sqrt{2.5625} = 17.6 \; MHz$$ **Ans.**

Example 18. Find out the value of refractive index by substituting the constants in equation.

$$\mu = \sqrt{1 - \frac{Me^2}{m\omega^2 K_o}}$$

Solution: $m = 9.107 \times 10^{-31} \; kg; \; e = 1.602 \times 10^{19} \; c;$

$$K_o = 8.854 \times 10^{-12} \; f/m; \; w = 2 \; \pi \; f$$

$$\frac{Ne^2}{mw^2 \, k_i} = \frac{N \, . \, (1.602)^2 \times 10^{-38}}{4 \, \pi^2 \, f^2 \times 9.107 \times 10^{-31} \times 8.854 \times 10^{-12}}$$

$$= \frac{N \, . \, (1.602)^2 \times 10^5}{4 \, \pi^2 \, f^2 \times 9.107 \times 8.854} = \frac{81 \, N}{f^2}$$

2 log 1.602 = 2 × 0.2049 =0.4098

5 log10 = 5.000 = 5.4098

And,

$\log \pi^2 = 0.9944$

log 4 = 0.6021

log 9.107 = 0.9563

log 8.854 = 0.9741 = 3.5029

$$\frac{Ne^2}{mw^2 \, K_o} = A' \log (5.4098 - 3.5029) = A' \log 1.9069$$

$$= 80.71 \approx 81$$

$$\mu = \sqrt{1 - \frac{81 \, N}{f^2}}$$ **Ans.**

Example 19. What do you understand by the term critical frequency? Calculate the critical frequencies for the F_1, F_2 and E layers for which the maximum ionic densities are 2.3 x 10^6, 3.5 x 10^6 and 1.7 x 10^6 electrons per c.c. respectively.

Solution: Since

$$N_{max} = 2.3 \times 10^6 m^{-3} = 2.3 \times 10^6 \times 10^{-6} = 2.3$$

$$f_c = 9\sqrt{N_{max}} = 9\sqrt{2.3} = 9 \times 1.517 = 13.653\ MHz$$

Similarly,

$$f_c = 9\sqrt{3.5 \times 10^6 \times 10^{-6}} = 9\sqrt{3.5} = 9 \times 1.871 = 16.839\ MHz$$

$$and\ \ f_c = 9\sqrt{1.7 \times 10^6 \times 10^{-6}} = 9\sqrt{1.7} = 9 \times 1.304 = 11.736\ MHz$$

Thus, $f_c = 13.653\ MHz, 16.839\ MHz\ and\ 11.736\ MHz$

$For\ F_1, F_2\ and\ E-layers$ **Ans.**

Example 20. Determine the change in the electron density of E-layer when the critical frequency changes from 4 MHz to 1 MHz between Mid-day and sun-set.

Solution: We know,

$$f_c = 9\sqrt{N_{max}}$$

$$f_{c1} = 9\sqrt{N_{m1}}\ ;\ f_{c2} = 9\sqrt{N_{m2}}$$

$$\Rightarrow f_{c1} - f_{c2} = 9\left[\sqrt{N_{m1}} - \sqrt{N_{m2}}\right]$$

$$\Rightarrow 4 \times 10^6 - 1 \times 10^6 = 9\left[\sqrt{N_{m1}} - \sqrt{N_{m2}}\right]$$

$$\text{or} \Rightarrow f_{c1} - f_{c2} = 9\left[\sqrt{N_{m1}} - \sqrt{N_{m2}}\right] = 3 \times 10^6$$

Then

$$\Rightarrow f_{c1} + f_{c2} = 9\left[\sqrt{N_{m1}} + \sqrt{N_{m2}}\right] = 1 \times 10^6 + 4 \times 10^6 = 5 \times 10^6$$

Now, adding above equations, we have

$$18\sqrt{N_{m1}} = (3 + 5) \times 10^6 = 8 \times 10^6$$

By subtracting, we have

$$-18\sqrt{N_{m2}} = -2 \times 10^6\ or\ 18\sqrt{N_{m2}} = 2 \times 10^6$$

$$(18)^2 N_{m2} = 4 \times 10^{12}\ or\ 324\sqrt{N_{m2}} = 4 \times 10^{12} \qquad \text{(i)}$$

Also squaring $\sqrt{N_{m1}}$ eq. we get

$$(18)^2 \times N_{m1} = \left(8 \times 10^6\right)^2 \; or \; 324\, N_{m1} = 64 \times 10^{12} \qquad \text{(ii)}$$

Hence, from equation (i) and (ii)

$$324\,(N_{m1} - N_{m2}) = 64 \times 10^{12} - 4 \times 10^{12} = 60 \times 10^{12}$$

$$N_{m1} - N_{m2} = \frac{60}{324} \times 10^{12} \; \frac{1}{54} \times 10^{12} \; electrons/m^3$$

$$= 0.0185 \times 10^{12} = 1.85 \times 10^{10}\, m^{-3}$$ **Ans.**

Example 21. Explain MUF and give any method of calculating a high frequency radio link has to be established between two points at a distance of 2500 km on earth's surface considering the ionospheric height to be 200 km and its critical frequency 5 MHz, calculate the MUF for the given path.

Solution: $$f_{muf} = f_c \sqrt{1 + \left(\frac{D}{2h}\right)^2}$$

Given that D = 2500 km

f_m = ?

f_c = 5 MHz

h = 200 km

$$f_{MUF} = 5\sqrt{1 + \left(\frac{2500}{2 \times 200}\right)^2}$$

$$= 5\sqrt{1 + \frac{625}{16}}$$

$$= 5\sqrt{\frac{641}{16}}$$

$$= \frac{5}{4} \times 10 \times 2.532$$

f_{MUF} = 31.6450 MHz **Ans.**

Example 22. The electron concentration at a height of 300 km is 10^{11} / m^3. What is the maximum angle of incidence ψ_t that can be used at a frequency of 10 MHz? What is the horizontal skip distance if the maximum angle ψ_t is used? **(Raj. Univ. 2000)**

Solution:

$$n = \sqrt{1 - \frac{81N}{f^2}}$$

For maximum angle of incidence = 90°

$$n = \sin \psi_t$$

$$\sin \psi_t = \sqrt{1 - \frac{81 \times 10^{11}}{(10 \times 10^6)^2}}$$

$$= \sqrt{1 - 0.081}$$

$$\sin \psi_t = 0.9586$$

$$\psi_t = \sin^{-1} (0.9586)$$

$$\psi_t = 73.46$$

Skip distance

$$D_{skip} = 2h \sqrt{\left(\frac{f_{MUF}}{f_c}\right)^2 - 1}$$

$$\left(\frac{f_{MUF}}{f_c}\right)^2 = \frac{1}{1 - n^2}$$

$$= \frac{1}{1 - (0.9586)^2}$$

$$= 12.346$$

$$D_{skip} = 2h \sqrt{\left(\frac{f_{MUF}}{f_c}\right)^2 - 1}$$

$$= 2 \times 300 \sqrt{12.346 - 1} \qquad = 600 \sqrt{11.346}\ km$$

$$D_{skip} = 2021.03\ km$$ **Ans.**

Example 23. Assume that reflection takes place at a height of 400 km and that the maximum density in the ionosphere corresponds to a 0.9 refractive index at 10 MHz. What will be the range (assume earth is flat) in which the MUF is 10 MHz?

Solution: We know that

$$\mu = \sqrt{1 - \frac{81\,N}{f^2}}$$

$$0.9 = \sqrt{1 - \frac{81\,N_{max}}{(10 \times 10^6)^2}}$$

$$N_{max} = \frac{(1 - 0.81)\,(10 \times 10^6)^2}{81}$$

$$= \frac{0.19 \times 10^{14}}{81}$$

therefore, $f_c = 9\sqrt{N_{max}}$

$$= 9\sqrt{\frac{19 \times 10^{14}}{81}} = 43.588 \times 10^5\ HZ$$

$$= 4.3588 \times 10^6\ HZ$$

$$D_{SKIP} = 2h\sqrt{\left(\frac{f_{MUF}}{f_c}\right)^2 - 1}$$

$$= 2 \times 400\sqrt{\left(\frac{10 \times 10^6}{4.3588 \times 10^6}\right)^2 - 1}$$

$$= 800\sqrt{\left(\frac{10}{4.3588}\right)^2 - 1} = 800\sqrt{(2.2942)^2 - 1}$$

$$= 800\sqrt{4.2633} = 800 \times 2.0647 = 1651.76\ km$$ **Ans.**

Example 24. The reflection takes place at a height of 350 km and the maximum density in the ionosphere corresponds to a 0.75 refractive index at 10 MHz. What will be the range (assume earth is flat) for which the MUF is 12 MHz?

Solution: $\mu = \sqrt{1 - \frac{81N}{f^2}}$

$$0.75 = \sqrt{1 - \frac{81\,N}{(10 \times 10^6)^2}}$$

$$N_{max} = \frac{\left[1-(.75)^2\right]\left[\left(10\times10^6\right)^2\right]}{81} = 5.40\times10^{11}$$

$$f_c = 9\sqrt{N_{max}} = 6.614\times10^6\ Hz$$

$$D_{SKIP} = 2h\sqrt{\left(\frac{f_{MUF}}{f_c}\right)^2 - 1}$$

$$= 2\times350\sqrt{\left(\frac{12\times10^6}{6.614\times10^6}\right)^2 - 1}$$

$$= 631.64 \text{ km}$$ **Ans.**

Example 25. A sky wave is incident on D layer at an angle of 30°. Find the angle of refraction of the frequency of the transmitted wave is 50 MHz.

$$\left[N_{max} = 400\, e/cm^3 \text{ for } D \text{ layer}\right]$$

Solution: $$\sqrt{\varepsilon_r} = \left(1 - \frac{81\,N_{max}}{f_c^2\ (KHz)}\right)^{1/2}$$

$$= \sqrt{1 - \frac{81\times400}{5\times10^4}}$$

$$\varepsilon_r = 1$$

As refractive index $n = \sqrt{\varepsilon_r}$

$$n = 1$$

According to snell's law

$$n \sin\theta_r = \sin\theta_i$$

$$[\because n = 1]\ \sin\theta_r = \sin\theta_i$$

$$\sin\theta_r = \sin 30^\circ$$

Angle of refraction qr = 30° **Ans.**

Example 26. What is the critical angle of propagation for D layer if the transmitter and receiver are separated by 500 km? [Height of D layer = 70 km]

Solution: $\theta_c = \sin^{-1}\left[\dfrac{h}{\sqrt{h^2 + \dfrac{d^2}{4}}}\right]$

$$= \sin^{-1} \frac{70}{\sqrt{4900 + \dfrac{(500)^2}{4}}}$$

$$\theta_c = 15.64°$$ **Ans.**

It can also be calculated from

$$\theta_c = \tan^{-1}\left(\frac{2n}{d}\right)$$

$$\theta_c = \tan^{-1} \frac{2 \times 70}{500}$$

$$\theta_c = 15.64°$$ **Ans.**

Example 27. Find the skip distance assuming earth to be flat for ionospheric propagate at frequency 10 MHz and critical frequency is 4.35 MHz and height of ionosphere length is 400 km.

Solution: $D_{skip} = 2h\sqrt{\left(\dfrac{f_{MUF}}{f_c}\right)^2 - 1}$

$$= 2 \times 400 \sqrt{\left(\frac{10 \times 10^6}{4.35 \times 10^6}\right)^2 - 1}$$

$$= 800\sqrt{\frac{100}{18.92} - 1}$$

$$= \sqrt{4.285}$$

$$D_{skip} = 2.07\ km$$ **Ans.**

Example 28. Calculate the transmission path distance for an ionospheric transmission that utilize a layer height 200 km and angle of electron of the antenna beam is 20° for a flat earth assumption.

Solution: $D = \dfrac{2h}{\tan \beta}$

According to question following parameters are given:

$$h = 200 \text{ km}$$

$$\beta = 20^\circ$$

$$D = \frac{2 \times 200}{\tan 20^\circ} = \frac{400}{0.3640}$$

$$D = 1099\ km$$ **Ans.**

Example 29. Assume that reflection takes place at a height of 400 km and that the maximum density ionosphere corresponds to a 0.9 refractive index at 10 MHz. What will be the range assume that earth) for which the MUF is 10 MHz.

Or

In the ionospheric propagation, consider that the reflection takes place at a height of 400 km and that the maximum density in the ionospheric corresponds to a refractive index of 0.9 at a frequency of 10 MHz. Determine the ground range for which this frequency is the MUF. Take the earth's curvature into consideration.

Solution: We know that

$$\mu = \sqrt{1 - \frac{81\,N}{f^2}}$$

$$D_{skip} = 2h\sqrt{\left(\frac{f_{MUF}}{f_c}\right)^2 - 1}$$

Given,

$h = 400$ km; $\mu = 0.9$; $f_{muf} = 10$ MHz;

$f = 10$ MHz and $D_{skip} = D_{range} = ?$

Putting these values in above equation we get -

$$0.9 = \sqrt{1 - \frac{81 N_{max}}{f^2}} \text{ or } 0.81$$

$$= 1 - \frac{81 N_{max}}{f^2}$$

$$\Rightarrow \frac{81\,N_{max}}{f^2} = 1 - 0.81 = 0.19$$

$$N_{max} = \frac{0.19 \times f^2}{81} = \frac{0.19 \times (10 \times 10^6)^2}{81}$$

$$= \frac{0.19 \times 10^{14}}{81}$$

$$= 0.0023456 \times 10^{14} = 23.456 \times 10^{10}\ m^{-3}$$

$$f_c = 9\sqrt{N_{max}} = 9\sqrt{23.456 \times 10^{10}} = 9 \times 4.8431 \times 10^5\ MHz$$

$$= 43.588 \times 10^5\ Hz = 4.3588 \times 10^6\ Hz$$

Case I: When earth is flat

$$D_{skip} = 2 \times 400\sqrt{\left(\frac{f_{muf}}{f_c}\right)^2 - 1}$$

$$D_{skip} = 800\sqrt{\left(\frac{10 \times 10^6}{4.3588 \times 10^6}\right)^2 - 1} = 800\sqrt{\left(\frac{10}{4.3588}\right)^2 - 1}$$

$$= 800\sqrt{(2.2942)^2 - 1} = 800\sqrt{5.2633 - 1}$$

$$= 800\sqrt{4.2633} = 800 \times 2.0647$$

$$D_{skip} = 1651.76\ km$$ **Ans.**

Case II: When earth's curvature is taken into account, then R=6370 km = Radius of earth h = Height of reflecting layer from earth

$$\frac{f_{muf}}{f_c} = \sqrt{\frac{D^2}{4\left(h + \frac{D^2}{8R}\right)^2} + 1}$$

or

$$\frac{D^2}{4\left(h + \frac{D^2}{8R}\right)} = \left(\frac{f_{muf}}{f_c}\right)^2 - 1$$

$$D = 2\left(h + \frac{D^2}{8R}\right)\sqrt{\left(\frac{f_{MUF}}{f_c}\right)^2 - 1}$$

Now putting the values, we have

$$= 2\left(400 + \frac{1651.76 \times 1651.76}{8 \times 6370}\right)\sqrt{4.2633}$$

$$= \left(800 + \frac{272831.1}{25480}\right) \times 2.0647$$

$$= (800 + 10.707656) \times 2.0647$$

$$= 80.707656 \times 2.0647$$

$$= 1673.868 \; km \cong 1673.86 \; km.$$ **Ans.**

THEORITICAL QUESTIONS

Q.1 Prove that the highest frequency which can be reflected by particular layer at vertical incidence is given by $9\sqrt{N_{max}}$; where N_{max} = maximum electron density.

(Raj. Univ. 2001, 2006, 2008)

Q. 2. Define MUF show that

$$(f_{muf})_{max} = f_c \frac{\left[\frac{D^2}{4} + \left(h + \frac{D^2}{8R}\right)^2\right]^{1/2}}{\left[4 + \frac{D^2}{8R}\right]}$$

Where D = distance between transmitter and receiver

f_c = critical frequency

h = height of layer **(Raj. Uni. 2001, 2003, 2006, 2008)**

Q. 3. Define virtual height and skip distance. **(Raj. Univ. 2005, 2008)**

Q.4. Show that the ionosphere behaves as a medium of refractive index

$$\mu = \sqrt{1 - \frac{81N}{f^2}}$$

Where N is electron density and f is the frequency.

(Raj. Uni. 2005)

Q.5 Define the expressions for MUF frequency

$f_{MUF} = f_c \sec i$ **(Raj. Uni. 2005, 2003)**

Q.6 Explain the formation of ionosphere. What are various layers of the ionosphere? What are various layers of the ionosphere? With the help of neat diagram show their respective electron density and heights above the ground. **(Raj. Univ. 2002, 2003, 2004)**

Q.7. How do the earth's magnetic field and solar active affect line of sight and ionospheric propagation. **(Raj. Uni. 2003)**

Q.8. Explain the following terms used for sky wave propagation.

(1) MUF (2) Optimum Frequency

(3) LUHF (4) Dellinger effect **(Raj. Uni. 2001)**

Q.9 Briefly explain the following.

(i) Effect of frequency on free space loss.

(ii) Effect of height of antennas of radio horizon.

Q.10 Discuss the effect of earth's magnetic field on ionospheric wave propagation. **(Raj. Uni. 1999)**

Q.11. Write short notes on -

(i) Method for the detection of winds in the ionosphere.

(ii) Measurement of ionospheric absorption

(iii) Minimization of fading of radio signals.

(iv) Ionospheric characteristics.

Q.12. Describe how the ionospheric layers D, E, F_1, F_2 are formed and how they affect the propagation of radio waves?

Q.13. Derive an expression for the refractive index of the ionosphere in terms of electron number density and frequency.

Q.14. What is meant by Critical frequency? How can it be measured? What information does it convey regarding the ionosphere?

Numerical questions

Q1. Find the maximum range for Tropospheric propagation for which the transmitting antenna is having height 25 m and the receiving antenna of height 16 m for effective point to point communication. **(Ans. d = 37.08 km)**

Q.2 Calculate the value of radio horizon for a TV antenna of height 166 km. If the signal is to be received at a distance of 66 km. Find the height of the receiving antenna. **(Ans. d = 9.858 km).**

Q.3. Calculate the maximum electron density of the ionosphere layer corresponding to refractive index at 10 MHz. **(Ans. N_{max} = 18.96 x 10^{10} electron/m^3)**

Q.4. If electron density in D region is 400 electron / cm^3 then calculate the frequency at which the wave can propagate in D region. [μ = 0.5] **(Ans. f_c = 207.82 KHz)**

Q.5. Calculate maximum usable frequency of transmission between two stations 500 km apart, if electron density of the reflecting layer is 10^{12} electron per cubic meter at height of 240 km [f_c = 9 MHz, f_{MUF} = 13 MHz]

Q.6 A high frequency radio link has to be established between two points at a distance of 2500 km on earth's surface. Considering the ionospheric height to be 20 km. and its critical frequency 5 MHz, Calculate the MUF for the given path.

$$\text{Hint: } \left[f_{MUF} = f_c \sqrt{1 + \left(\frac{D}{2h}\right)^2} = 31.650\ MHz \right] \qquad \textbf{Ans.}$$

SHORT QUESTIONS

1. How do density and temperature affect the velocity of sound?
2. What term is used to mean the science of sound?
3. A sound wave that is reflected back toward the source is known as what type of sound?
4. What is the term for multiple reflections of sound waves?
5. A cavity that vibrates at its natural frequency produces a louder sound than at other frequencies. What term is used to describe this phenomenon?
6. What do we call a disturbance that distracts or distorts the quality of sound?
7. What are three means of producing light?
8. What is the smallest unit of radiant energy?
9. What unit is used to measure the different wavelengths of light?
10. What are the three primary colors of light?
11. What are the three secondary colors of light?
12. White light falls upon a dull, rough, dark-brown object. Will the light primarily be reflected, diffused, or absorbed by the object?
13. What color will be emitted by a dull, rough, black object when white light falls upon it?
14. A substance that transmits light but through which an object cannot be seen clearly is known as what kind of substance?
15. At what speed does light travel?

16. A light wave enters a sheet of glass at a perfect right angle to the surface. Is the majority of the wave reflected, refracted, transmitted, or absorbed?
17. When light strikes a piece of white paper, the light is reflected in all directions. What do we call this scattering of light?
18. What three examples of electromagnetic energy are mentioned in the text?
19. What is the main difference between the bulk of the electromagnetic spectrum and the visible spectrum?
20. What are the two components (fields) that make up the electromagnetic wave?
21. What do we call a conductor (or set of conductors) that radiates electromagnetic energy into space?
22. What do we call the field that is created between two rods when a voltage is applied to them?
23. When current flows through a conductor, a field is created around the conductor. What do we call this field?
24. An induction field is created around a conductor when current flows through it. What do we call the field that detaches itself from the conductor and travels through space?

ANSWERS

1. A term used to describe the multiple patterns a radio wave may follow.
2. Selective fading.
3. Natural and man-made interference.
4. Natural.
5. Man-made.
6. (a) filtering and shielding of the transmitter
   (b) Limiting bandwidth
   (c) Cutting the antenna to the correct frequency
7. (a) physical separation of the antenna
   (b) Limiting bandwidth of the antenna
   (c) Use of directional antennas
8. Regular and irregular variations.
9. Regular variations can be predicted but irregular variations are unpredictable.
10. Daily, seasonal, 11-year, and 27-day variations.

11. Sporadic E, sudden disturbances, and ionospheric storms.
12. MUF is maximum usable frequency. LUF is lowest usable frequency. FOT is commonly known as optimum working frequency.
13. MUF is highest around noon. Ultraviolet light waves from the sun are most intense.
14. When LUF is too low it is absorbed and is too weak for reception.
15. Signal-to-noise ratio is low and the probability of multipath propagation is greater.
16. Frequent signal fading and dropouts.
17. FOT is the most practical operating frequency that can be relied on to avoid problems of multipath, absorption, and noise.
18. They can cause attenuation by scattering.
19. It can cause attenuation by absorption.
20. It is a condition where layers of warm air are formed above layers of cool air.
21. It can cause VHF and UHF transmission to be propagated far beyond normal line-of-sight distances.
22. Troposphere.
23. VHF and above.
24. Near the mid-point between the transmitting and receiving antennas, just above the radio horizon.

UNIT 6
ADVANCE ANTENNA APPLICATIONS

CONTENTS

ADVANCE ANTENNA APPLICATIONS

6.1 ANTENNAS FOR BIOMEDICAL APPLICATIONS

The antenna used in many biological applications operates under different conditions than do its more traditional free-space, far-field counterparts. Near fields and mutual interaction with the body dominate; also, the antenna radiates in a lossy environment rather than free space. Many antennas, from microstrip antennas to phased arrays, operating at various frequencies, have been developed to couple electromagnetic energy in or out of the body. Main medical applications can be classified into two groups: therapeutic and informational. Examples of therapeutic applications are hyperthermia for cancer therapy, enhancement of bone and would healing, nerve simulation, neural prosthesis, microwave angioplasty, treatment of prostatic hyperplasia, and cardiac ablation. Examples of applications of information are tumor detection using microwave radiometry, imaging using microwave tomography, measurement of lung water content, and dosimeter.

Therapeutic applications are further classified as invasive and noninvasive. Both applications require different types of antennas and different restrictions on their design. In the noninvasive applications (i.e., not penetrating the body), antennas are used to generate an electromagnetic field to heat some tissue. Antennas like as helical-coils, ring capacitors, dielectrically loaded waveguides and microstrip radiators are attractive because of their compactness. Phased arrays are also used to provide focusing and increase the depth of penetration. The designer has to choose the right frequency, antenna size, and spot size that the beam has to cover in the body. The depth of penetration - as the medium of propagation is lossy; it is determined by the total power applied or available to the antenna.

Invasive applications require some kind of implantation in the tissue. Many single antennas and phased or non-phased arrays have been used extensively for treating certain tumors. A coaxial cable with an extended center conductor is a typical implanted antenna. This type of antenna has also been used in arteries to soften arterial plaque and enlarge the artery. Antennas have also been used to stimulate certain nerves in the human body. As the technology advances in the areas of materials and in the design of more compact antennas, many antenna applications will be found in the areas of biology and medicine.

6.1.1. Introduction

Rapid researches in technology have been accompanied by an increasing application of technology to health care. More and more interaction of scientists and engineers with medical people has resulted in significant developments in improved health care using highly sophisticated equipment. Correspondingly, the use of antennas of one type or another in the practice of medicine has increased. Most of the medical applications of antennas involve coupling electromagnetic energy into the human body or into other biological systems, such as animals used in experimental measurements.

Primarily, these applications can be divided into mainly two categories: therapy and diagnostics. In therapy, antennas have been used mainly for producing hyperthermia for treating cancer. This therapy consists of heating tumors, alone or in combination with either X-ray irradiation or chemotherapy. In diagnostics, electromagnetic energy is coupled into and out of the body to monitor various physiological parameters. Topographic photography of the body by electromagnetic measurements has been investigated, but not yet developed to the point of practical implementation.

The problems encountered in designing antennas for medical applications are somewhat different from those involved in the design of antennas for other applications. For example, in coupling electromagnetic energy into biological systems the near fields are frequently more important than the far fields, in dissimilarity to other applications. Near field problems are especially important in the design of hyperthermia applicators. Since biological systems usually have high dielectric constant the antennas for coupling into them must be designed differently from those radiating into free space. It is also very significant that antennas used for coupling electromagnetic energy into the human body be designed to produce minimum leakage radiation outside the body. Leakage radiation can sometimes be hazardous to medical personnel, and it can produce artifacts in measurements.

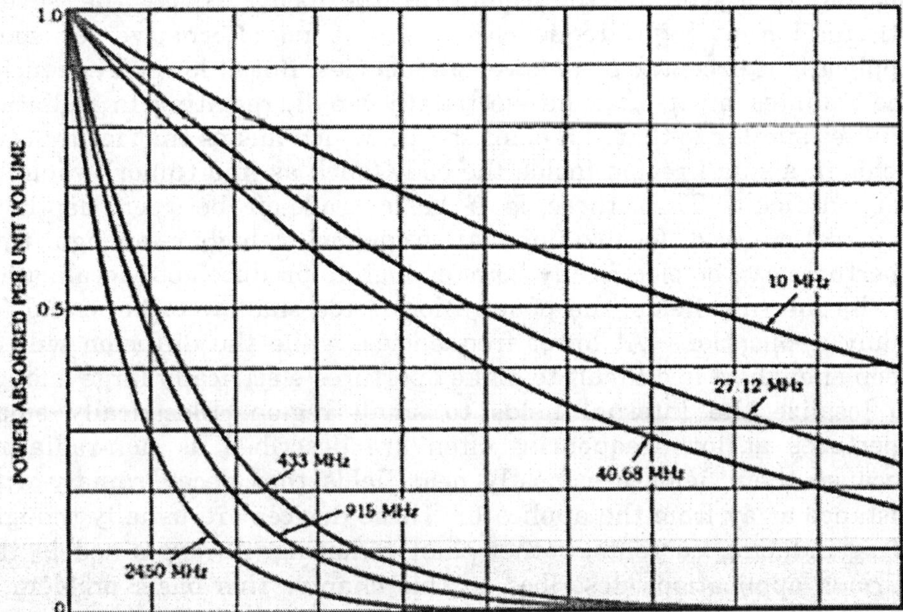

Figure 6.1 Relative absorbed power per unit volume versus depth for a plane wave propagating into a dielectric half-space having the properties of tissue.

Furthermore, since biological systems are usually quite lossy, there is a basic trade-off between depth of penetration and localization of the fields inside the body. In many applications, localization of the internal fields is important. In hyperthermia, for example, it is often desirable to concentrate the internal fields at the tumor, thereby heating it without appreciably affecting the surrounding normal tissue. Concentrating the fields, however, amounts to having narrow-beam radiation, which requires an electrically large aperture. To limit apertures to physical sizes convenient for use with patients and still get narrow-beam radiation, higher-frequency operation, in the high-megahertz or low-gigahertz range, is necessary. At these high frequencies, however, the penetration of the electromagnetic radiation is too shallow to be useful. Some imminent into this problem can be gained from the data presented in Fig. 6.1, which shows the plane-wave penetration depth versus frequency in a dielectric half-space with a permittivity equal to the average permittivity of the human body. The power magnitude is normalized to the value at the surface. The profile of electromagnetic field penetration into the body is, of course, a strong function of aperture size. Since the data in Fig. 6.1 is for an incident plane wave, it corresponds to that of a very large aperture. The curves show that even for a very large aperture the penetration at high frequencies is too shallow to be useful for any application that requires transmission of a signal through the body or heating of tissue in the center of the body. Given the severe attenuation at high frequencies, there is no alternative for some applications except to go to lower frequencies. But at lower frequencies the radiates must often be electrically small, resulting in radiation profiles similar to that of a point source, which means that focused the fields in a small region inside the body (such as in a tumor) would be very difficult. Thus there is a basic trade-off between depth of penetration and localization. At frequencies higher enough that apertures can be electrically large enough to produce good localization of the internal fields, the penetration is too shallow to be useful for many applications. At lower frequencies, while the diffusion will be deep enough, it is difficult to make apertures electrically large enough to localize the internal fields to small region. Electrically small apertures at low frequencies often are described as non-radiating because their fields are mostly near fields that decay rapidly with distance away from the applicator. These devices are usually thought of as inducing currents, rather than radiating. As explained in the various applications described in this chapter this basic problem is encountered over and over again. In producing hyperthermia, for example, the shallow penetration at higher frequencies forces the designer to use low frequencies to get deeper heating; otherwise the surface overheats, even with surface cooling, before sufficient heating occurs deeper inside the body. At lower frequencies, though, the radiation produced by the electrically small apertures is often dominated by near fields, which, because they die away rapidly with distance from the applicator, also cause the surface to overheat. The main problem in hyperthermia applications, therefore, is how to obtain deep heating without overheating the surface.

Another factor that is significant in coupling electromagnetic energy into biological systems is the reflection at the interfaces between different kinds of tissues. This reflection depends not only on the specific types of tissues at the interface, but also on the frequency, the polarization, and the angle of incidence of the propagating waves. Fig. 6.2 illustrates the variation of the magnitude of the reflection coefficient as a function of frequency for various tissue interfaces.

6.1.2. Waveguide- and Radiation-Type Antennas

One class of usually bused antennas for coupling electromagnetic energy into the human body consists of open-ended waveguides of one kind or another. Such radiators have been used primarily for inducing hyperthermia for cancer therapy, which requires effective heating of muscle tissue without excessive heating of the fat layer that overlies the muscle. Waveguide radiators generally give good coupling of the electromagnetic energy into the tissue, but they tend to be bulky and awkward. Since the waveguides must be physically very large at low frequencies, they have been used mostly at frequencies of 915 MHz and above, although a ridged waveguide loaded with water have been used at frequencies as low as 27 MHz for inducing hyperthermia. Other radiation-type applications that are used in electromagnetic hyperthermia applications comprise various forms of reflector antennas and the annular phased array (APA) applicator.

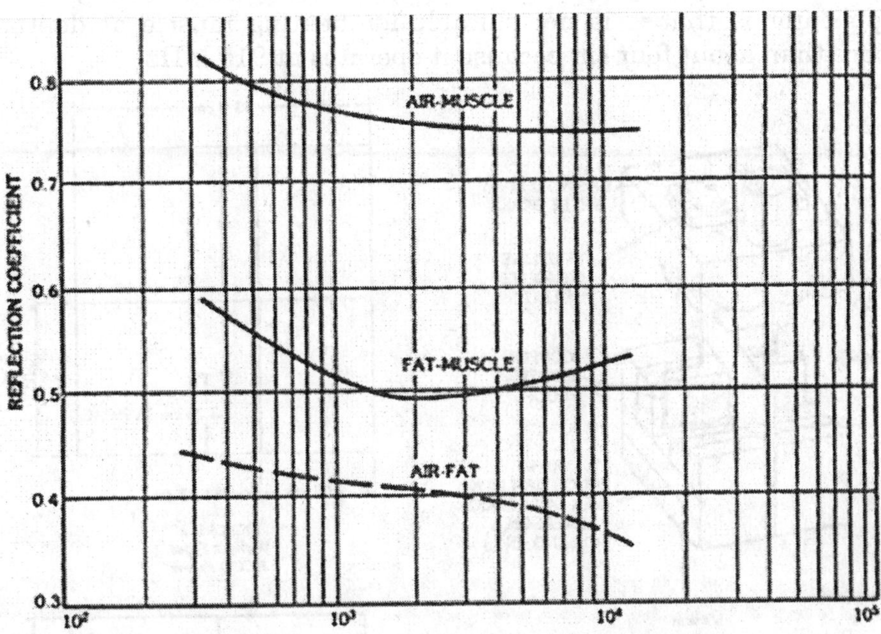

Figure 6.2 Plane-wave reflection coefficients for normal incident at the interfaces between different tissue layers

(1) Direct-Contact Waveguide Applicators

Perhaps the simplest effective waveguide applicator is an open-ended dielectric loaded waveguide. Loading the waveguide with a lossless dielectric of permittivity approximately is equal to that of tissue provides a good impedance match and ensures high-quality transmission into the tissue. The dielectric loading also reduces the size of the waveguide. However, the effective heating of the tissue depends strongly on the size of the aperture and the field distribution in it. In an extensive analysis of the internal fields produced by a TE_{10} - mode rectangular aperture in a two-layer semi-infinite tissue model, Guy calculated internal fields as a function of the size of the aperture and the thickness of the fat layer. He showed that of the two frequencies, 915 MHz and 2450 MHz, that are authorized for medical applications in the higher-frequency ranges (in the United States), 915 MHz produces a higher ratio of muscle-to-fat heating, which is pleasing for therapeutic applications. At 915 MHz, the optimum aperture size was found to be 13-cm aperture applicator that includes provisions for cooling the skin was designed and tested. Details are shown in Fig. 6.3 measured heating patterns correlated well with calculated values.

This applicator offers the benefit of providing optimal heating of a fat-muscle configuration at 915 MHz. The disadvantage of the applicator is that it is not suitable for heating muscle at depths of more than about four cm because it operates at 915 MHz.

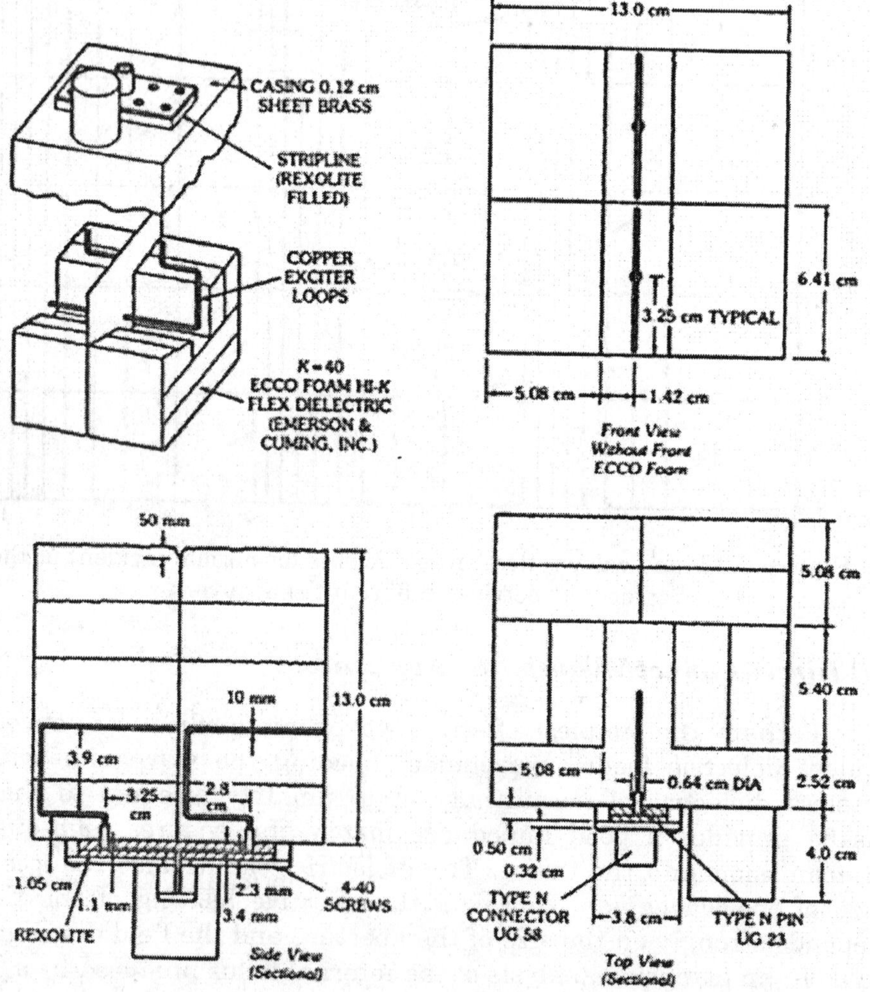

Figure 6. 3 Details of the air-cooled applicator designed by Guy and others (After Guy *et al.* [5] 1978 IEEE)

(2)TEM Waveguide Antenna

Van Koughnett and Wyslouzil suggested the use of an in homogeneously loaded dielectric waveguide to simulate TEM-mode wave propagation for biological studies. Cheung, Dao and Robinson used open-ended TEM waveguide as an applicator for hyperthermia to take advantage of the uniformity of the fields across the aperture of the waveguide. Usually it is desirable to have a power absorption pattern that is uniform in the cross section of the region to be heated.

6.1.3 Microstrip Antennas and Applicators

Antennas consisting of radiating structures on microstrip transmission line are attractive applicators for pairing electromagnetic energy into biological systems because they can be very flat, compact, and mechanically convenient. These kinds of applicators are also relatively easy to design and fabricate. However, they tend to be narrow-band, produce some leakage radiation, and would be physically too large to be useful at lower frequencies where the penetration is best. They have been originate to be useful for heating tumors near the body surface, typically using frequencies of 433, 915 or 2450 MHz and they have been used for diagnostics.

Since microstrip antennas are treated extensively elsewhere in this book, the discussion in this chapter is limited to their medical applications. The primary difference between designing microstrip antennas for medical applications and other applications is that for medical applications the antenna must usually be intended to radiate into a lossy medium of relatively high permittivity, whereas in other applications the antennas often is designed to radiate into free space.

(1) Microstrip Path Antennas

A microstrip path antenna consists of a conducting patch on the side opposite the ground plane of a microstrip transmission line, as shown in Fig. 6.4.

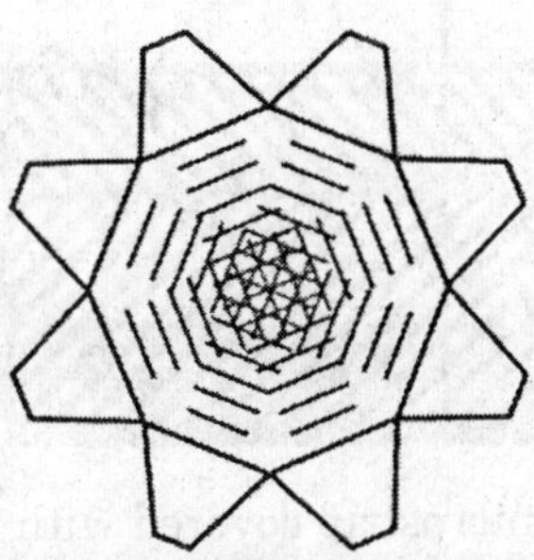

Figure 6.4

(2) Microstrip Loop Radiators

Bahl and colleagues designed and tested microstrip transmission-line loop radiators in the pattern shown in Fig. 6.5. The ring conductor, which is the radiating element, is fed on one side by a coaxial line protruding through the substrate. A pin shorting the ring to the ground plane through the substrate is located 180° from the feed point. In live out, it was found that best results for heating biological tissue were obtained when a water bolus was placed between the loop radiator and the biological body. Although the relationship between the size of the ring and the resonant frequency is rather complex, Bahl and colleagues developed a simple empirical formula that applies when the ring is placed on a thick, lossy, dielectric medium. For this case the mean radius R of the ring is given in centimeters by $R \simeq \dfrac{6}{f\sqrt{|\varepsilon_r|}}$

To obtain a high-quality impedance match between the transmission line and the antenna, their dimensions for an applicator at 2.45 GHz are given away in Fig. 6.8. In this case the dimensions of the slot are approximately 1λ by 2λ when the applicator is in contact with muscle tissue($\varepsilon_r \cong 48 - j14$), where λ is the wavelength in muscle. The usual design procedures for a slot antenna radiating into a dielectric medium can be used to determine dimensions of the slot. The radiator in Fig. 6.8 produced a heating pattern in simulated muscle material similar to that of an open-ended waveguide aperture. The heating was relatively uniform over an area of about 1.5 cm by 1 cm. The leakage radiation was about 35 dB below the transmitted power.

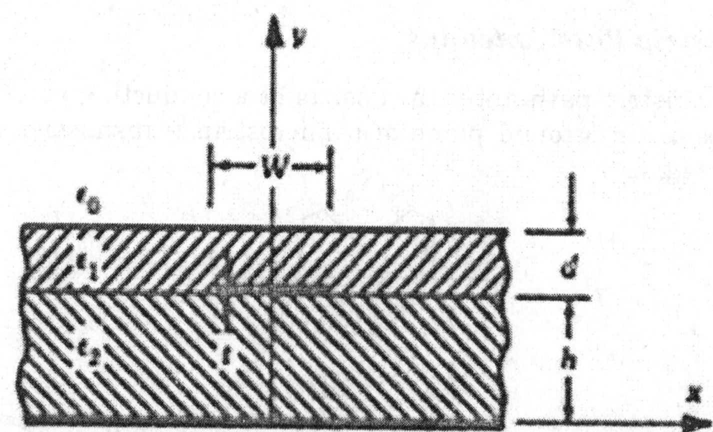

Figure 6.5 Geometry of a microstrip covered with dielectric as used by Bahl and Stuchly to calculate the effective dielectric constant.

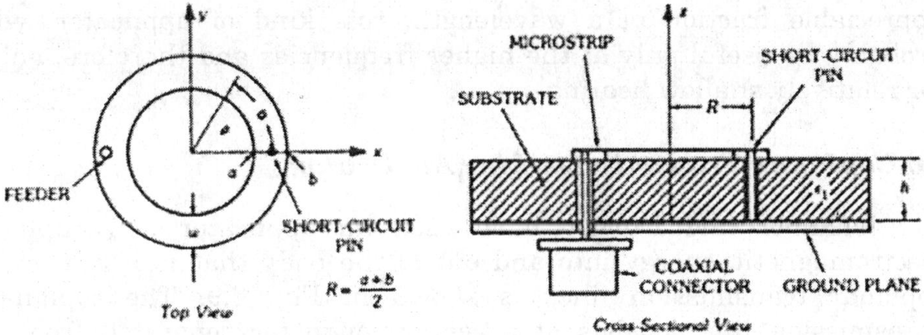

Figure 6.6 The microstrip loop radiator designed by Bahl and others

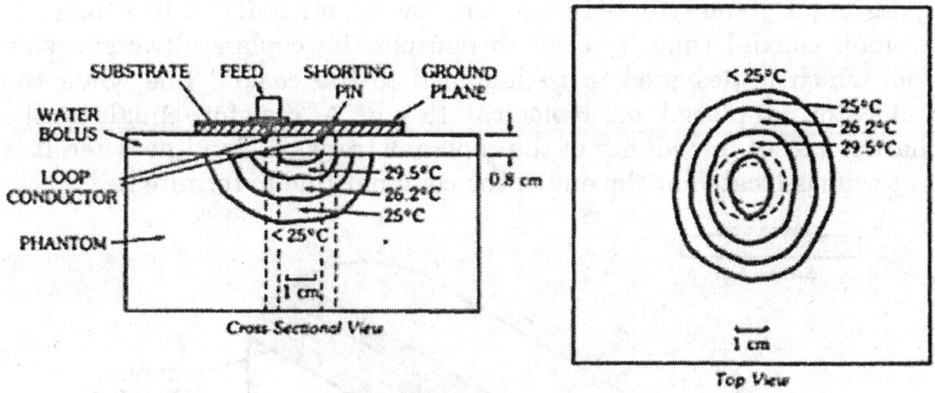

Figure 6.7

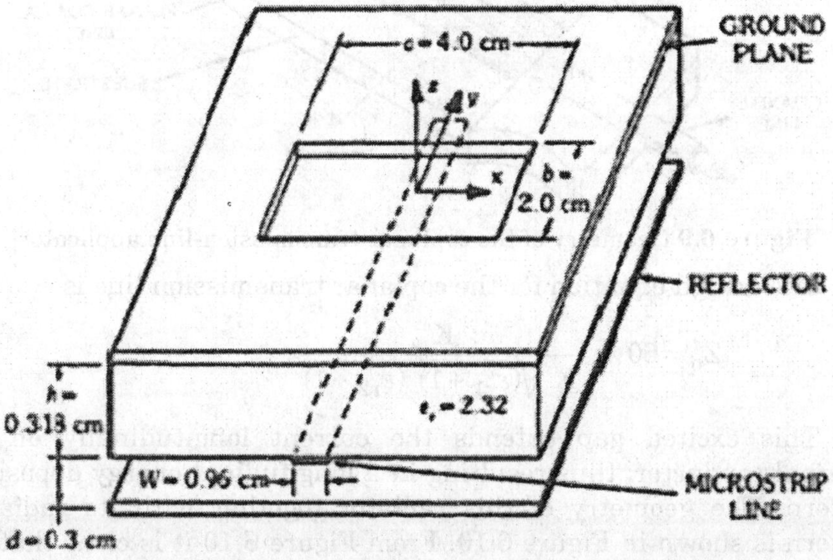

Figure 6.8 Geometry of microstrip slot radiator

The microstrip slot radiator has the reward of being flat, compact, and lightweight, with good impedance matching and a relatively uniform heating pattern. Since the slot size must be an appreciable fraction of a wavelength, this kind of applicator will probably be useful only at the higher frequencies and therefore, only for relatively shallow heating.

(3) Coplanar Transmission-Line Applicator

Iskander and Durney have used an applicator for coupling electromagnetic energy into and out of the body that is based on a coplanar transmission line, as shown in Fig. 6.9. The coplanar transmission line consists of a center conductor separated from a coplanar ground plane by an air gap, with both the center conductor and ground plane situated on a very low permittivity, thin substrate. A small coaxial cable is used to nourish the coplanar transmission line, which is designed to be matched to the coaxial line when the applicator is placed on biological tissue. A resistor equal to the characteristics impedance of the coplanar transmission line when it is on tissue is located at the end of the coplanar line to termite it.

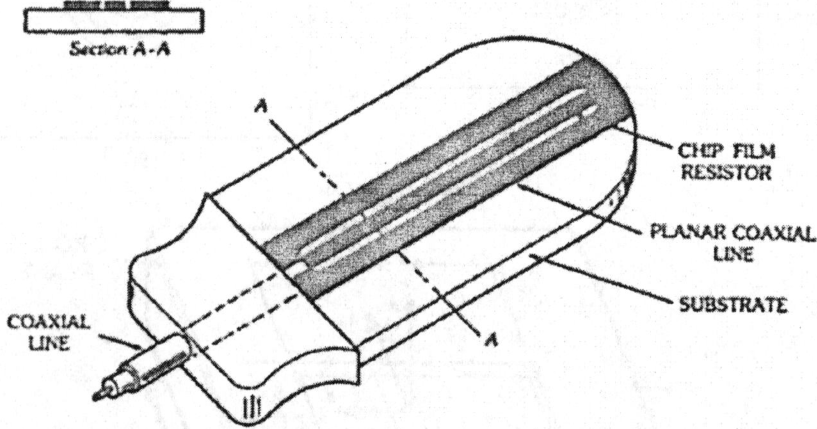

Figure 6.9 Geometry of the coplanar transmission-line applicator

The design equation for the coplanar transmission line is

$$Z_0 = 60\,\pi \frac{K''}{K\sqrt{(\varepsilon_{r1}+1)\,(\varepsilon_{r2}+1)}}$$

This excited gap extends the current longitudinally on the external conductor, thus resulting in a longitudinal energy deposition pattern. The geometry of this radiator together with its radiation pattern is shown in Figure 6.10. From Figure 6.10 it is clear that the radiation pattern displaying two principal peaks with an interference utmost in between is very roughly that of a collinear array of two half-wave dipoles.

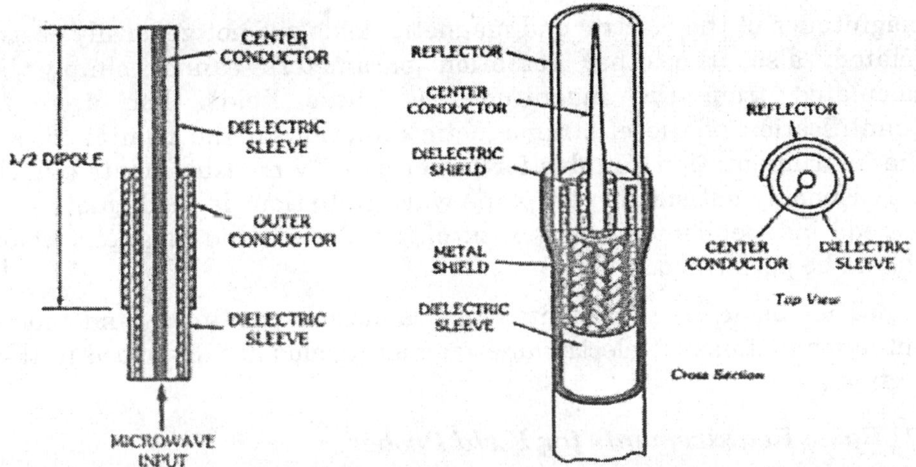

Figure 6.10 Dipole antenna and applicator with reflector (a) Longitudinal section of dipole antenna (b) Coaxial applicator with directive reflector.

6.1.4 Antennas for Monitoring Rf Radiation

In any situation that could involve exposure of people to significant amounts of electromagnetic radiation, monitoring the electromagnetic radiation levels is important. In some cases a device that measures radiation in a given area would be satisfactory. In other cases it would be attractive to have a device that could be worn by a person to monitor the radiation to which that person is exposed. Area radiation monitors are commercially available, but a satisfactory personnel dosimeter is not, almost certainly because of the many difficulties encountered in the design and construction of such a device.

An important aspect of either kind of radiation monitor is what parameter of the electromagnetic fields should be measured. It is usually known that for a simple plane wave any one of a number of field parameters constitutes a valid index and the director of propagation. These parameters include the magnitudes of the electric field E, of the magnetic field H, and the time-average power density. Provided that the source does not show super gain, plane-wave approximations are appropriate in the practical sense when the distance from the source is large compared to $\mathbf{2D^2}/\lambda$, where, λ is the wavelength in the medium and D is the largest dimension of the source. Therefore, for quantifying this far-field plane-wave-type radiation, it is sufficient to calculate the magnitude of either the electric or magnetic field. Other parameters of interest can easily be computed from these magnitudes without approximation for given properties of the propagation medium. In the near field, however, magnitudes of the electric and magnetic fields are not generally easily related; also, the other radiation parameters cannot simply be calculated from the magnitudes of these fields. For accurate quantification of the electromagnetic radiation in the near field, so, the instrumentation suitable for power density measurements (which is reasonably satisfactory in plane-wave radiation) is inadequate and instead independent measurement of the electric and magnetic fields should be performed.

The basic requirements for radiation monitors and some antennas to measure electric and magnetic fields are described in this section.

(1) Basic Requirements for Field Probes

Desirable features in the design of an rf and microwave radiation monitor include the following:

1. The measuring probes should upset the electromagnetic fields minimally.
2. The probe response should be isotropic and therefore independent of the field polarization.
3. It should have a flat frequency response over a broad frequency band. This characteristic is important in making field measurement without precise knowledge of the frequency.
4. It should exhibit linear dynamic response with the field or power quantity being measured.
5. It should be appropriate for operation in the near- or far-field regions. This simply requires independent measurement of the electric- and magnetic field components in three orthogonal directions.
6. The field probes (E- and H-field sensors) as well as the connecting leads should rather be made of high-resistance material (carbon-loaded Teflon) to minimize their interference with the fields being measured.

In addition to these basic requirements the probe should also have fast response and be small enough to make spatially determined measurements in small regions comparable in size to organs of the body.

(2) Detection Methods

In designing a appropriate antenna system for detecting the E and H fields of RF radiation, it is essential to include in the probe a means for detecting the induced currents. The type of detectors used can generally be segregated into two basic types:

1. Thermocouple type devices that measure the electromagnetic power by sensing alter in temperature between the hot and cold junctions of a thermocouple device. Examples of this thermocouple detecting device include the electromagnetic radiation monitor developed by the Narda Microwave Corporation. In this device the thermocouple materials of the sensor are antimony and bismuth deposited on a substrate of plastic or mica. The sensitivity of the sensor is controlled by difference of the substrate and the dimensions of the hot junction of the thermocouple.
2. Diode detectors, which build a current or voltage related to the fields of the electromagnetic radiation.

The usefulness of thermocouple detection devices is limited by their sensitivity to changes in ambient temperature and other changes in all forms of heat. These devices also have slow response time and may even burn out if overloaded. Their main advantage is linearity of response with intensity of electromagnetic power. Diode detectors are more sensitive than thermocouple detectors, but the diode response is nonlinear with electromagnetic power. For example, if the signal level is such that the diode is operating in the square-law region, the induced current will be proportional to the square of the E-field intensity.

(3) E- and H-Field Probes

As indicated in a previous section, a measurement of power density is not sufficient to characterize electromagnetic radiation in the near-field region. The question that is often raised, `however, is whether or not we need to calculate both the electric- and magnetic-field intensities. Since these fields are indeed related to each other by Maxwell's equations, one can generally measure either one of the fields, say, the electric field, and calculate the other, if necessary, from it. Although the above statements are surely correct, practical problems in relating these fields make it easier to measure them both. For instance, it is known that the ratio of E to H in the near field is not the same as intrinsic wave impedance in free space and that H and E are not generally perpendicular to each other. In other words, no straightforward relation exists between these near E and H fields, and full utilization of Maxwell's equations is required.

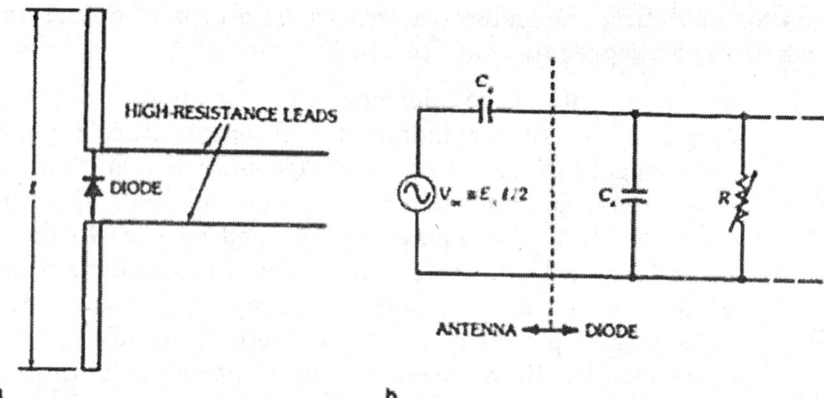

Figure 6.11 Diagram of the E-field probe and its equivalent circuit (a) The E-field probe (b) Approximate equivalent circuit

6.1.5 Antenna Design for Microwave Hepatic Ablation using an Axisymmetric Electromagnetic Model

(1) Introduction

Liver cancer is a significant worldwide public health issue. The disease has a death rate of 100% at 5 years in untreated cases and results in the deaths of more than one million people each year worldwide. Although liver cancer can be treated successfully by surgical resection of the malignant tissue, approximately 90% of patients with the disease are not entitled for the procedure due to factors such as insufficient hepatic reserve and the close proximity of tumors to blood vessels. One promising alternative for these patients is hepatic microwave ablation (MWA), an experimental procedure in which an antenna is inserted percutaneously or during surgery to induce cell necrosis through the heating of deep-seated tumors. Unlike additional alternatives to resection such as RF ablation and cry ablation, MWA systems are able to produce large lesions in the presence of blood perfusion and are not restricted by tissue charring. MWA also has favorable one, two, and three year survival rates of 96%, 83%, and 73%. The many apparent advantages of microwave ablation have driven researchers to develop innovative antennas to effectively treat deep-seated, non-respectable hepatic tumors. These designs have focused mainly on thin, coaxial- based interstitial antennas, which are minimally invasive and capable of delivering a large amount of electromagnetic power. These antennas can usually be classified as one of three types (dipole, slot, or monopole) based on their physical features and radiative properties to assist in antenna design for MWA, many researchers have employed the use of mathematical models rooted in computational electromagnetic (CEM), a discipline that employs numerical methods to explain propagation of electromagnetic waves. These methods center around the formulation of discrete solutions to the fundamental electromagnetic equations collectively referred to as Maxwell's equations. Three principal techniques exist within CEM. The first of these, the finite-difference time-domain (FDTD) method, is based on the Yee algorithm and uses finite difference approximations of the time and space derivatives of Maxwell's curl equations to generate a discrete three-dimensional representation of the electric and magnetic fields. This method has been widely used to numerically assess the electro-magnetic radiation patterns of antennas in tissue, although long computation times are generally required. Another commonly used CEM technique is method of moments (MoM), in which approximate numerical solutions to integral equations are formulated in the frequency domain to determine an unknown current distribution for an antenna. This distribution can then be subsequently extended to yield the antenna's radiation pattern. An alternative to the two techniques above is the finite element method (FEM), which has been extensively used in simulations of cardiac and hepatic radio-frequency (RF) ablation. FEM models can provide users with quick, precise solutions to multiple systems of differential equations and as such, are well suited to heat transfer problems like ablation. In this study, we have adapted an existing axisymmetric electromagnetic FEM model that was implemented using FEMLAB™ 3.0 to demonstrate proper techniques for the design of an antenna for hepatic MWA.

(2) Materials and methods

Antenna design

Several antenna designs have been shown to be effective for the treatment of deep-seated hepatic tumors. Freshly, Saito *et al* presented a design for a double slot antenna that is capable of achieving a higher degree of SAR localization than a standard single slot antenna and has been used in initial clinical trials. Another popular design, the cap-choke antenna, was originally designed for cardiac MWA applications but has been also found through independent investigation to be an effective applicator for hepatic MWA. In this plan, the choke acts as a balun (which isolates the antenna and feed line to provide a impartial output) to prevent current backflow along the axial length of the antenna, resulting in high SAR localization near the distal tip of the antenna. Fig. 6.2 shows the general schematic for a double slot choked antenna. Antenna geometry parameters, the slot spacing, choke offset, choke length, etc, were chosen based on the effective wavelength in bovine liver tissue at 2.45 GHz, which was calculated using where *c* is the speed of light in free space (m/s), *f* is the operating frequency of the microwave generator (2.45 GHz), and εr = 44.4 is the relative permittivity of bovine liver tissue at the operating frequency; this yielded 18.4 mm for the effective wavelength. However, because the catheter and its thickness also affect the optimal geometry and performance of the antenna, equation 7 only provides a very basic approximation for the design. Fig. 6.2 shows that slot spacing, choke offset, and choke length correspond to 0.25λeff, 0.5λeff, and λ of respectively, which were chosen to achieve localized power deposition near the distal tip of the antenna. Tip length and catheter thickness were adjusted to achieve resonance.

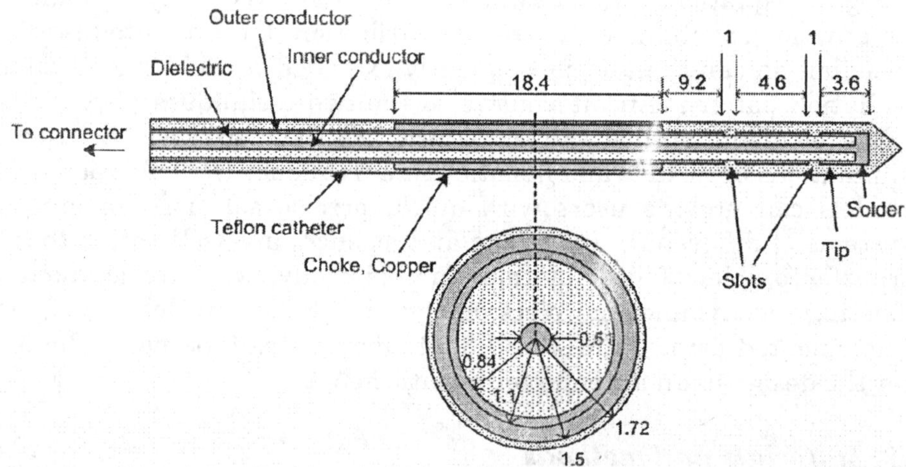

Figure 6.12 Axial and radial schematics of a coaxial-based double slot choked antenna designed for hepatic microwave ablation (MWA). All units are in mm.

6.2 IR ANTENNAS-INTRODUCTION

This study of IR antennas and frequency-selective surfaces is through collaboration with Dr. Glenn Boremanand the IR Systems Lab at University of Central Florida.

Basically at any wavelength, an antenna allows a sub-wave-length-sized sensor to bring together incident EM radiation. Where incident radiation induces IR-frequency current waves to flow in the arms of the antenna, and puts an IR-frequency sinusoidal voltage across the sensor element. Various sensors can be used are: bolo meters (Figure 6.13), tunnel junctions (fast rectifiers), Schottky diodes, etc. The Sensor signal typically read out with "dc" connection.

And here spectral response, determined by antenna dimensions rather than by sensor-material characteristics, is visible to IR to mmW demonstrated. Antenna design determines polarization response.

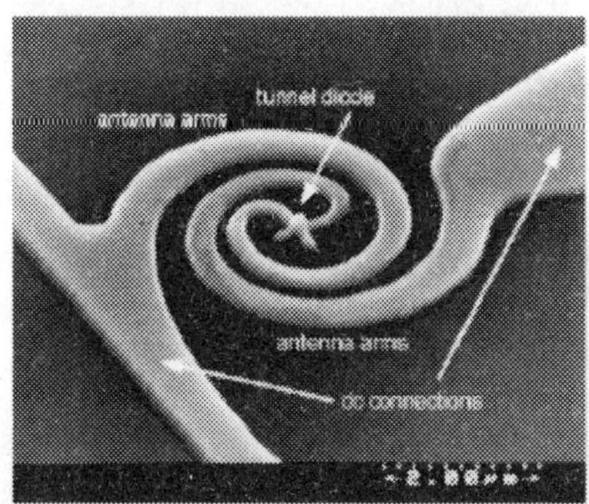

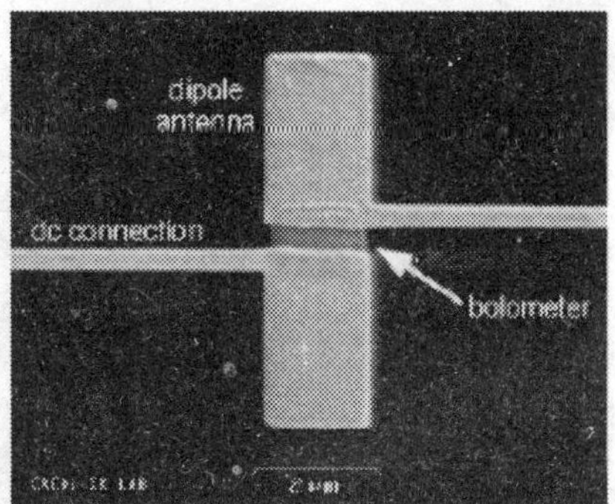

Fig 6.13

Control of pixel dimensions –it have to provide freedom to set pixel pitch in focal-plane array. Here the decrease in bolometer mass gives fast sensor response -higher frame rates. Overall dimensions of the antenna determine the primary resonant frequency.

Typically a half-wave resonance for a linear antenna is like a dipole or bow-tie. Figure 6.14 One circumference equals a full wave for a circular antenna like a spiral or log periodic. It is the "guided "wavelength on the antenna that is significant, intermediate between the free-space and dielectric wavelengths.

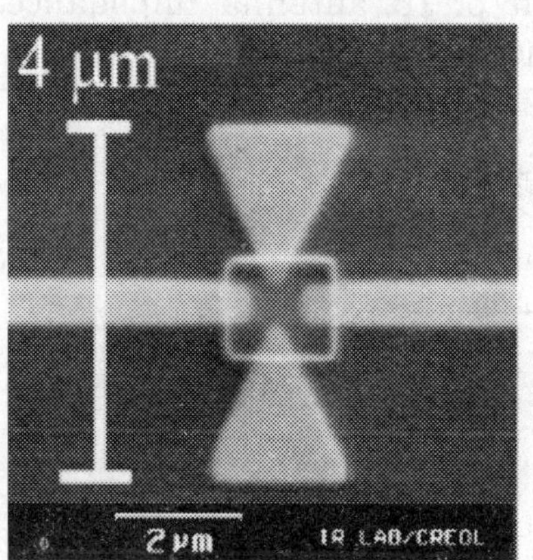

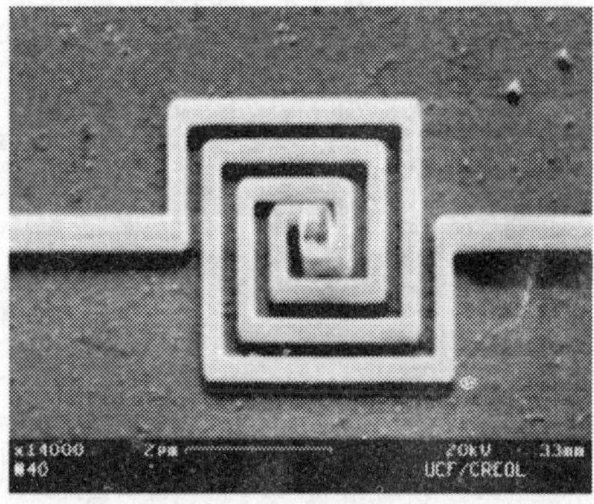

Fig 6.14

6.2.1 Materials Characterization

Firstly the correct material information is needed, to get accurate agreement between design and measured performance data. So they used variable-angle IR ellipsometry at UCF IR Systems Lab to calculate real and imaginary parts of permittivity from 1-30 microns.

It is important for thin (200-nm) metallic films and plastics. They measured films in as-deposited state using witness samples.

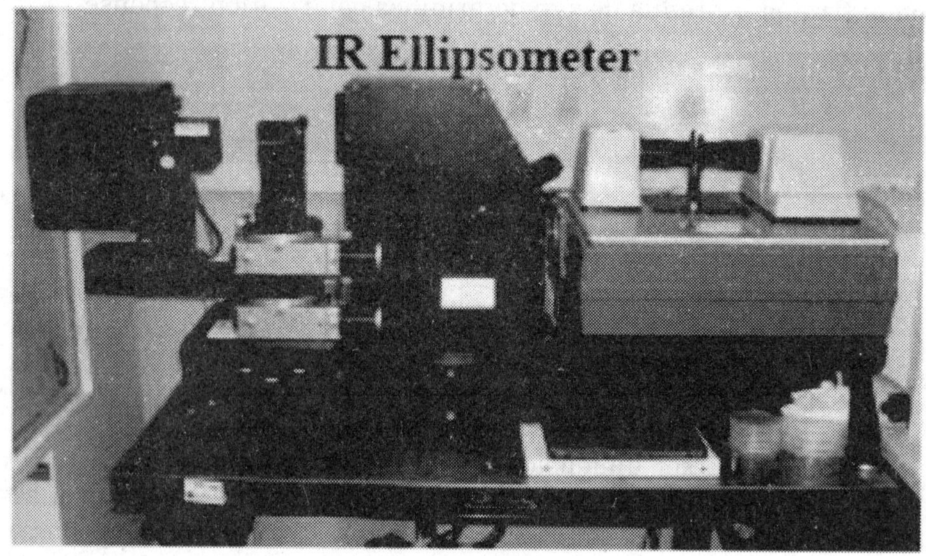

Figure 6.15

6.2.2 Impedance Matching

Next one is impedance matching, which is important. Antenna and load should be impedance matched for maximum power transfer. There should be no direct measurement of IR antenna impedance. Example of deviation of input impedance is based on Des Champs predictions for multi-arm self-complimentary antennas Figure 6.16.

Figure 6.16

- Parasitic third arm
- Improved matching
- Broadband
- Real impedance

See the Figures 6.17 and 6.18

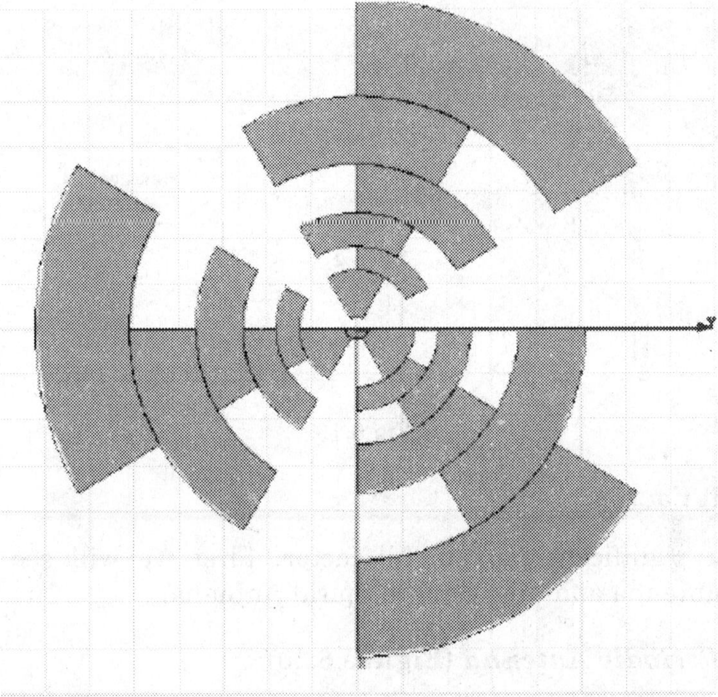

Fig 6.17

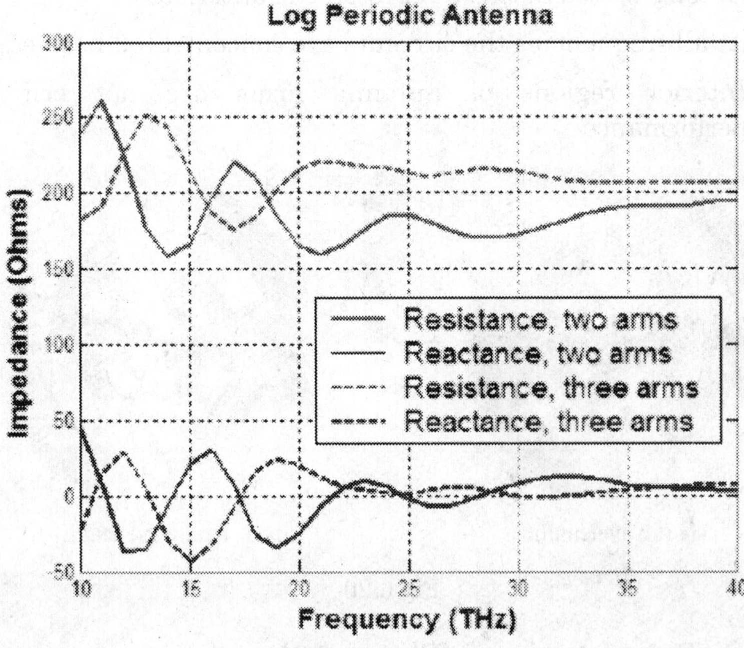

Fig. 6.18

Further increases in impedance are seen for 4-terminal log-periodic design, 2 parasitic arms Figure 6.19.

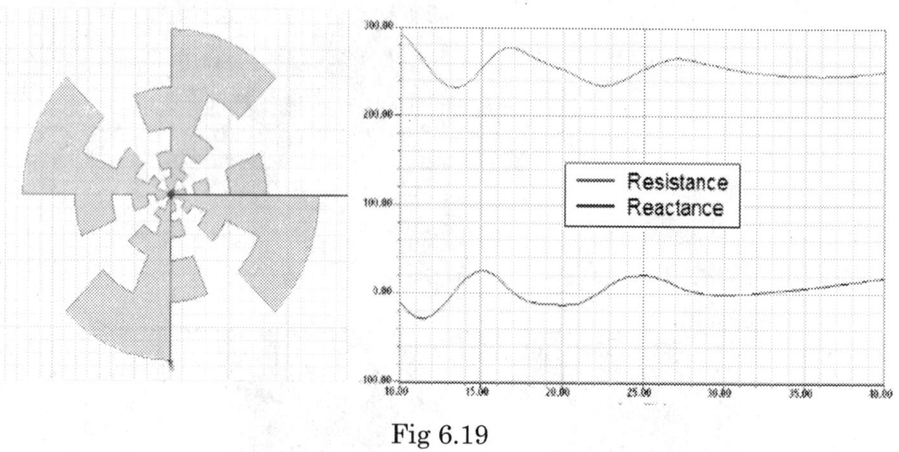

Fig 6.19

6.2.3. Fill Factor

Next significant part is fill factor. First we will see the log periodic antenna and then Square Spiral Antenna.

(a) Log Periodic Antenna **(Figure 6.20)**

- At short wavelength currents occupy the centre portion.
- At long wavelength currents fill the antenna.
- At all frequencies the currents are concentrated near edges.
- Interior regions of antenna arms are not critical to performance

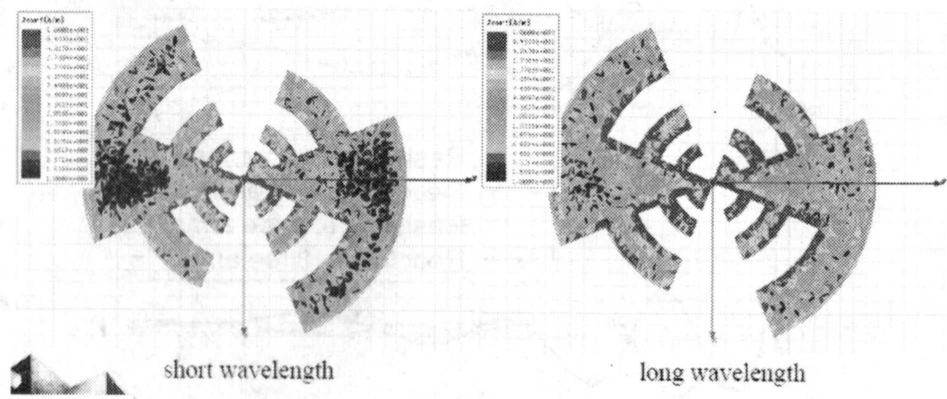

Fig 6.20

(b) Square Spiral Antenna: **(Figure 6.21)**

- As line width increases, overall dimension increases, but active area remains fixed.

- By dropping the number of turns, both active and physical areas are kept constant.
- Impedance response is maintained.
- Fill Factor is maintained Figure 6.22.

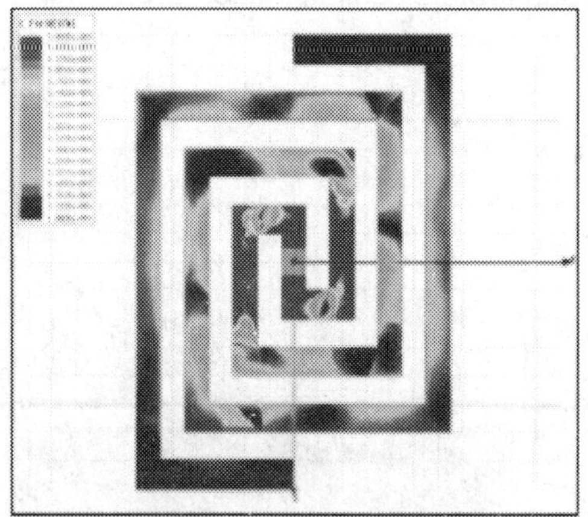

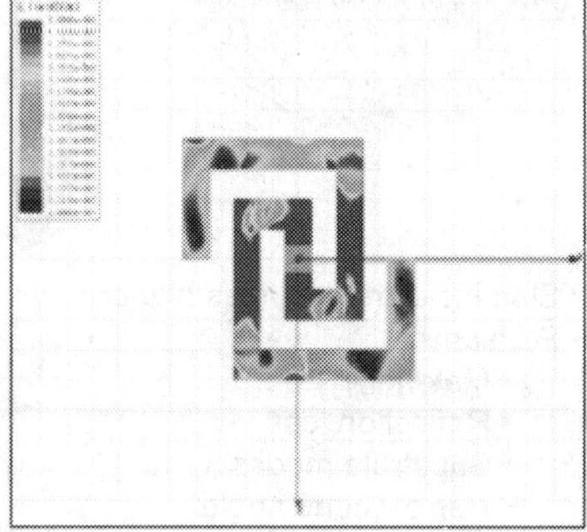

Fig 6.21

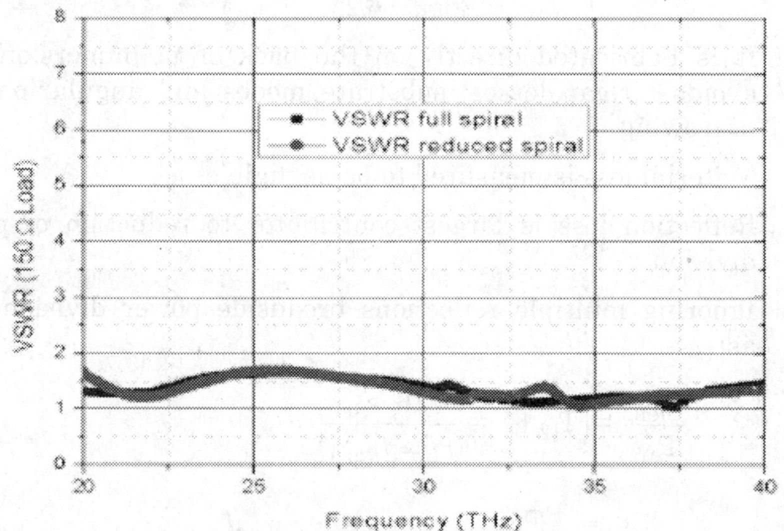

Fig. 6.22

The main applications of IR antenna are:

6.2.4 Immersion Lens

An infrared light condensing apparatus that focuses an infrared light of several tens microns in wavelength at a micro fine area of submicron or less and also a near-field from a micro fine area of submicron or less, and permits a scanning image to be obtained.

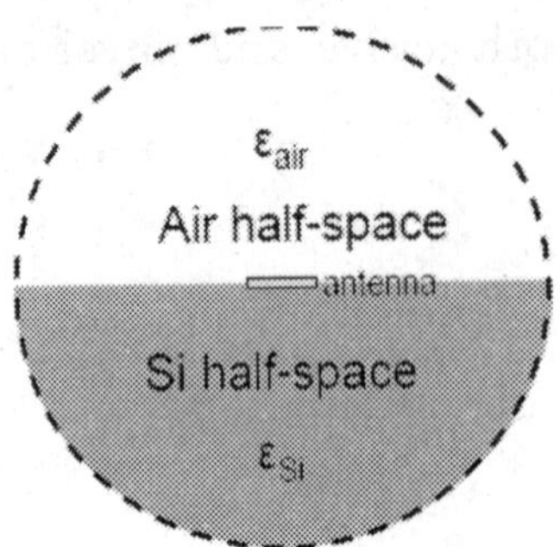

• Radiated power $\sim\varepsilon^{3/2}$.

• $P_{Si}/P_{air} = (\varepsilon_{Si}/\varepsilon_{air})^{3/2} \cong 40$.

•Substrate side illumination preferred.

• Slab dielectric provides higher ε.
• Problems:
    • Material loss.
    • Reflection loss.
    • Substrate modes.
    • Narrow cone angle.

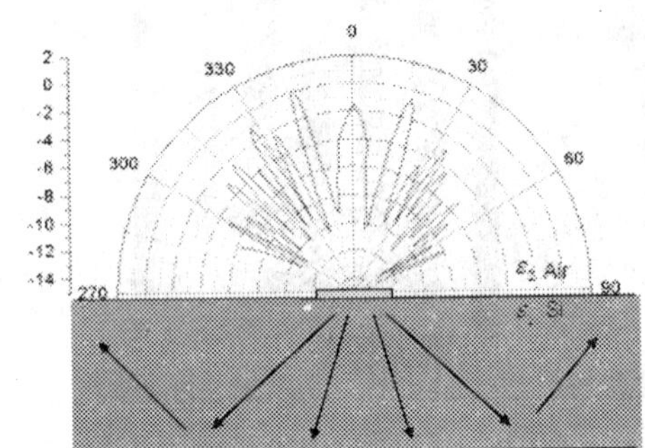

Figure 6.23

- It is Fabricated directly on the back of Si immersion lens avoids air-gap losses, substrate modes and angular-pattern narrowing.
- Material loss is measured to be negligible.
- Reflection loss is largest contributor to reduction of power division.
- Ignoring multiple reflections broadside power division goes as:

$$\frac{P_{lens}}{P_{air}} = |T_{12}|^4 = \frac{16\ \varepsilon_{Si}^2}{(1+\sqrt{\varepsilon_{Si}})^4}$$

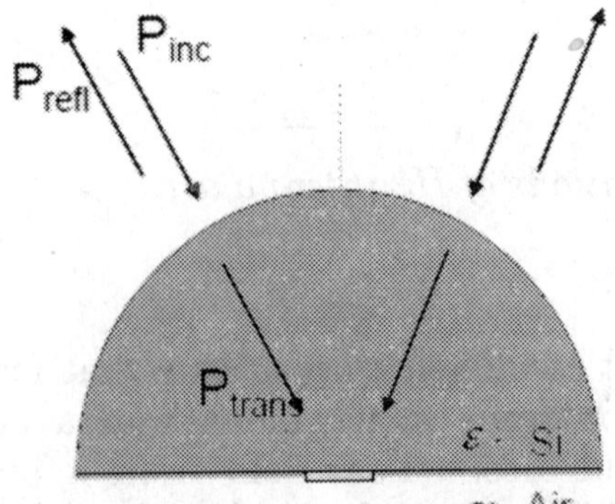

Hemispherical Immersion Lens Illumination

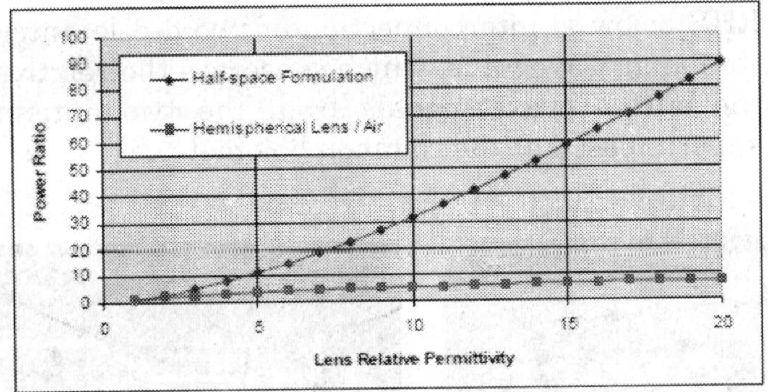

For Si lens in air: $\dfrac{S_{Si}}{S_{air}} \approx 5.8$

Figure 6.24

- HFSS receiving model with incident-field PML predicts normalized pattern.
- Power dissipated in load is computed as received signal.
- Analytical formulation predicts relative power for lens-side against air-side.

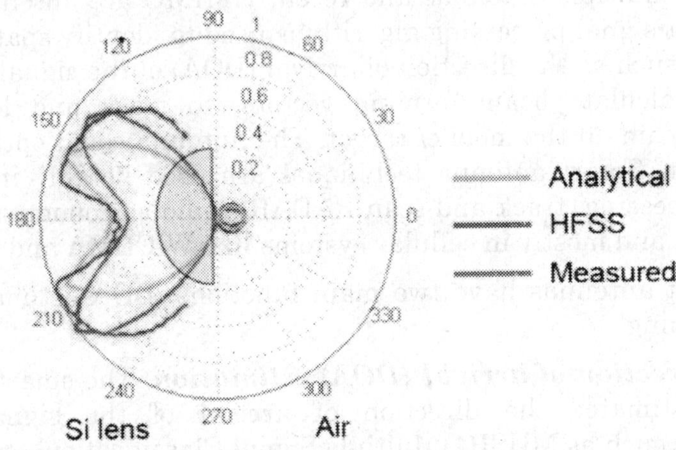

Figure 6.25

6.2.5 Phased Array

- Distinguish between phase-incoherent and phase-coherent interconnections.
- LHS below is a series interconnection of bolometers, each with their own IR antenna -results in an extended spatial response but not a phased array.

- RHS below is interconnection of two dipole antennas with co-planar waveguide, which conserves the relative phase of the current-wave signals from the two antennas when superimposed at the common bolometer.

See the Figure 6.26

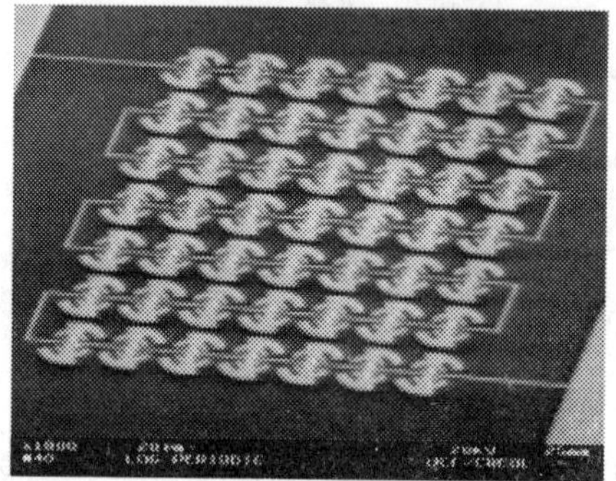

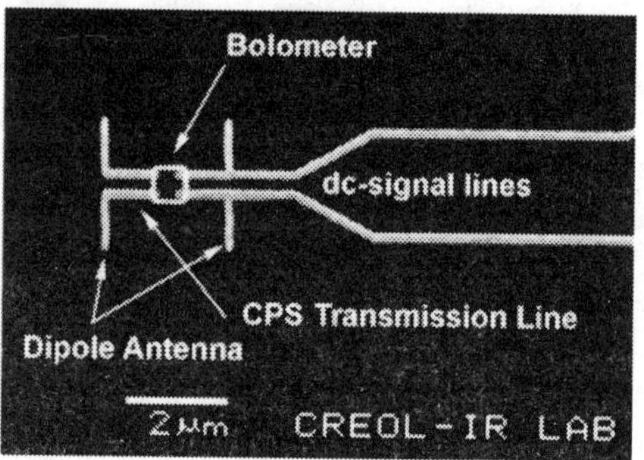

Figure 6.26

6.3 SMART ANTENNAS

Smart antenna is the name which also known as adaptive array antennas, multiple antennas and recently MIMO are antenna arrays with smart signal processing algorithms used to identify spatial signal signature such as the direction of arrival (DOA) of the signal, and it is used to calculate beam forming vectors, to track and locate the antenna beam on the mobile/ target. The antenna could optionally be any sensor. Smart antenna techniques are used notably in acoustic signal processing, track and scan RADAR, radio astronomy and radio telescopes, and mostly in cellular systems like W-CDMA and UMTS.

Smart antennas have two main functions: DOA estimation and Beam forming.

1. Direction of arrival (DOA) estimation:-The smart antenna system estimates the direction of arrival of the signal, using techniques such as MUSIC (Multiple Signal Classification), estimation of signal parameters via rotational invariance techniques (ESPRIT) algorithms, Matrix Pencil method or one of their derivatives. They involve finding a spatial spectrum of the antenna/sensor array. They also involve in calculating the DOA from the peaks of this spectrum. These calculations are computationally intensive.

Matrix Pencil is very efficient in case of real time systems, and under the correlated sources.

2. Beam forming:-Beam forming is used to create the radiation pattern of the antenna array by adding constructively the phases of the signals in the direction of the targets/mobiles desired, and nulling the pattern of the targets /mobiles that are undesired/interfering targets. It can be done with a simple FIR tapped delay line filter. The weights of the FIR filter may also be changed adaptively, and used to provide optimal beam forming, in the sense that it reduces the MMSE between the desired and actual beam pattern formed. Typical algorithms are the steepest descent, and LMS algorithms.

6.3.1 Smart-Antenna Analogy

The functionality of many engineering systems is readily understood when it is related to our human body system. Therefore, we will take a view into how a smart-antenna system works. First let us imagine two persons carrying on a chat inside a dark room [refer to Figure 6.27(a)]

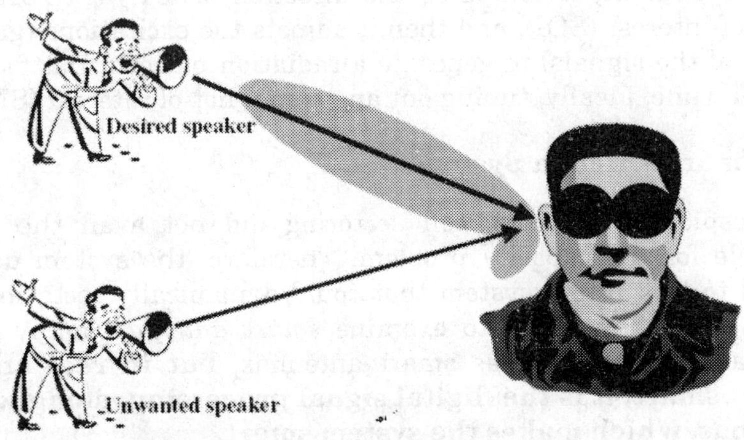

(a) Human analogy [3]

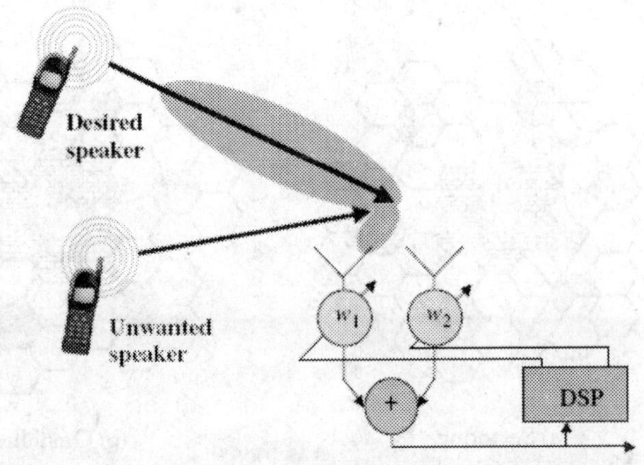

(b) Electrical equivalent

Figure 6.27

The listener between the two persons is capable of recognizing the location of the speaker as he moves in the room because the voice of the speaker arrives at each audio sensor, the ear, at a different time. The human signal processor (the brain) analyses the direction of the speaker from the time differences or delays of the voice received by the two ears. After that, the brain adds the strength of the signals from each ear so as to concentrate on the sound of the computed direction. Furthermore, if additional speakers join in the conversation, the brain can tune out unwanted interferers and concentrate on one conversation at a time. And same as, the listener can react back to the direction of the desired speaker by orienting the transmitter (mouth) toward the speaker Electrical smart-antenna systems work the similar way using two antennas in its place of the two ears and a digital signal processor instead of a brain [refer to Figure 6.27(b)] so, after the digital signal processor measures the time delays from each antenna element, it computes the direction of arrival (DOA) of the signal-of-interest (SOI), and then it adjusts the excitations (gains and phases of the signals) to generate a radiation pattern that focuses on the SOI while, ideally, tuning out any signal-not-of interest (SNOI).

6.3.2 Smart-Antenna Systems

Despite its benefits, cell sectoring did not avail the solution desirable for the capacity problem. Therefore, the system designers started to look into a system that could dynamically sectaries a cell. That's why, they began to examine *smart antennas* many refer to smart-antenna systems as smart antennas, **but in real antennas are not smart; it is the digital signal processing, along with the antennas, which makes the system smart.**

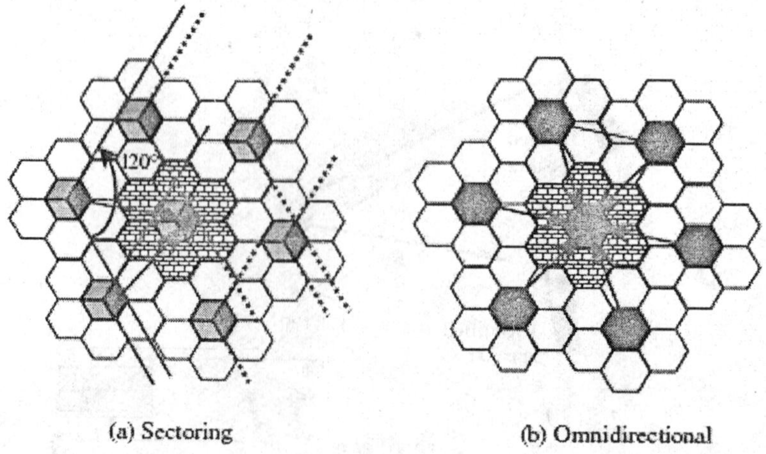

Figure 6.28

Although it seems that smart-antenna system is a new technology, the basic theory of smart (adaptive) antennas is not new. Even, it has been applied in defense-related systems since World War II. Until recent years, with the appearance of powerful low-cost digital signal processors (DSPs), general purpose processors (and Application-Specific Integrated Circuits—ASICs), as well as innovative signal-processing algorithms, smart-antenna systems have become practical for commercial use. Smart-antenna systems are basically an expansion of *cell sectoring* in which the sector coverage is composed of multiple beams. This is achieved by the utilization of antenna arrays, and the number of beams in the sector (e.g., 120°) is a function of the array geometry. Because smart antennas can focus their radiation pattern toward the desired users while rejecting surplus interferences, they can make available greater coverage area for each base station. Moreover, because smart antennas have a greater rejection interference, and therefore lower bit error rate (BER), they can provide a substantial capacity improvement. **These systems can generally be classified as either *Switched-Beam* or *Adaptive Array*.**

Types of smart antennas

smart antennas

switched beam smart antennas   adaptive array smart antennas

There are mainly two types of smart antennas include switched beam smart antennas and adaptive array smart antennas. Switched beam systems have a lot of available fixed beam patterns. Only a decision is made as to which beam to access, at any given point in time, based on the requirements of the system. Adaptive arrays allow the antenna to steer the beam to any direction of interest while simultaneously nulling interfering signal. Beam direction can be estimated using the so-called direction-of-arrival (DOA) estimation methods.

In 2008, the United States NTIA started a major effort to assist consumers in the purchase of digital television converter boxes. Through this effort, many people have been come to know the concept of smart antennas for the first time. In the context of consumer electronics, a "smart antenna" is one that conforms to the EIA/CEA-909 Standard Interface.

Extension of smart antennas: Smart antenna systems are also a defining characteristic of MIMO systems, such as the IEEE 802.11n standard. Conventionally, a smart antenna is a unit of a wireless communication system and performs spatial signal processing with multiple antennas. Multiple antennas can be used at either the transmitter or receiver. The technology has been extended to use the multiple antennas at both the transmitter and receiver recently; this type of system is called a multiple-input multiple-output (MIMO) system. As extended Smart Antenna technology, MIMO supports spatial information processing, in the sense that conventional research on Smart Antennas has focused on how to provide a beam forming advantage by the use of spatial signal processing in wireless channels. Spatial information processing includes spatial information coding such as spatial multiplexing and Diversity Coding, as well as beam forming.

(1) Switched-Beam Systems

A *switched-beam system* is a system that can choose from one of many predefined patterns in order to improve the received signal (see Figure 2), and it is obviously an expansion of cell sectoring as each sector is subdivided into smaller sectors. As the mobile unit is moving through out the cell, the switched-beam system detects the signal strength, chooses the appropriate predefined beam pattern, and continually switches the beams as necessary. The overall aim of the switched-beam system is to increase the gain according to the location of the user. However, since the beams are fixed, the intended user may not be in the middle of any given main beam. If here is an object that is creating interference near the center of the active beam, it may be enhanced more than the desired user.

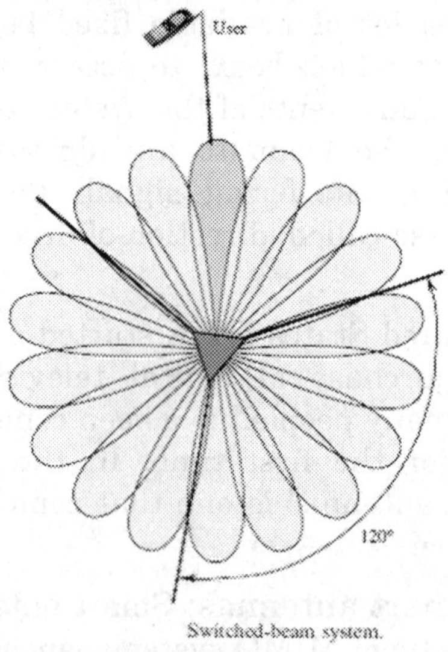

Switched-beam system.

Figure 6.29

(2) Adaptive Array Systems

Adaptive array systems provide higher degrees of freedom since it have the ability to adapt in real time the radiation pattern to the RF signal environment. In other way, they can direct the main beam toward the pilot signal or SOI while suppressing the antenna pattern in the direction of the interferers or SNOIs. In simple language, adaptive array systems can customize an appropriate radiation pattern for each individual user. This is far superior to the presentation of a switched-beam system, as shown in Figure 4.

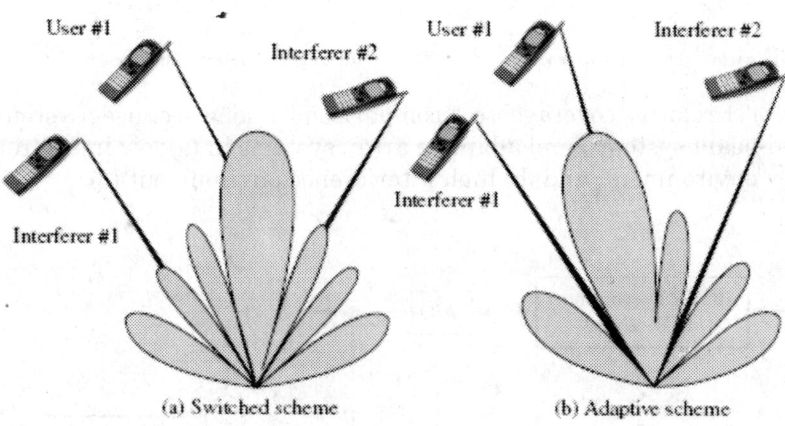

Figure 6.30 Comparison of (a) switched-beam scheme, and (b) adaptive array scheme.

This figure shows that not only the switched-beam system may not able to place the preferred signal at the maximum of the main lobe, but also it exhibits the inability to fully reject the interferers. Because of the ability to control the in general radiation pattern in a greater coverage area for each cell site, as illustrated in Figure 6.31, adaptive array systems greatly increase capacity.

Figure 6.31 shows a comparison, in terms of relative coverage area, of conservative sectaries switched-beam and adaptive arrays. In the presence of a low-level interference, both types of smart antennas provide significant gains over the conventional sectored systems. However, when a high-level interference is present, the interference rejection capability of the adaptive systems provides significantly more coverage than either the conventional or switched-beam system. Adaptive array systems can locate and track signals (users and interferers) and energetically adjust the antenna pattern to enhance reception while minimizing interference using signal-processing algorithms. A functional block diagram of such a system is shown in Figure 6.31.

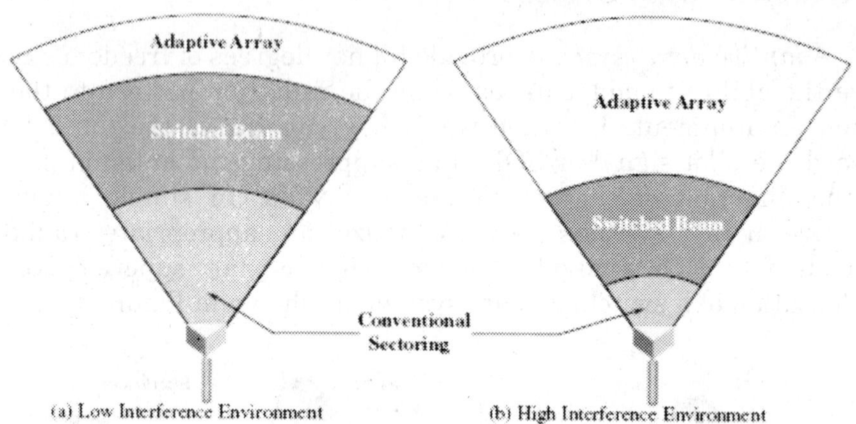

Figure 6.31 relative coverage area comparisons among sectaries systems, switched-beam systems, and adaptive array systems in (a) low interference environment, and (b) high interference environment (3).

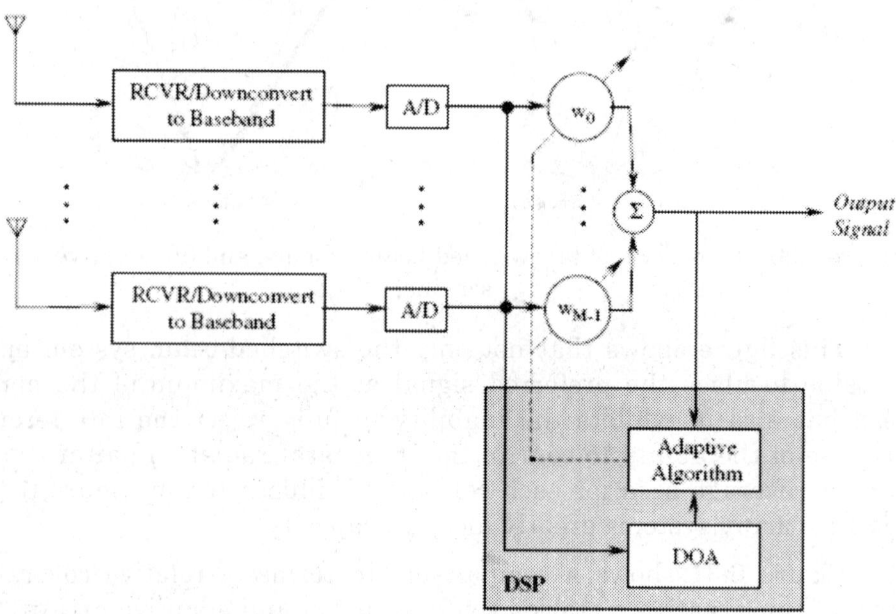

Figure 6.32. Functional block diagram of an adaptive array system.

This figure shows that after the system down converts the received signals to base band and digitizes them, it locates the SOI using the direction-of-arrival (DOA) algorithm, and it continuously tracks the SOI and SNOIs by dynamically altering the weights (amplitudes and phases of the signals). Basically, the DOA computes the direction of arrival of all signals by computing the time delays between the antenna elements, and afterward the adaptive algorithm, using a cost function, computes the appropriate weights that result in an optimum radiation pattern. The details of how the time delays and the weights are computed are discussed later in this chapter. Because adaptive arrays are generally more digital-processing concentrated and require a complete RF portion of the transceiver behind each antenna element, they tend to be more costly than switched-beam systems.

(3) Spatial Division Multiple Access (SDMA)

The final goal in the development of cellular radio systems is SDMA. SDMA is among the most-sophisticated use of smart-antenna technology; its advanced spatial-processing capability enables it to locate many users, creating different beam for each user, as shown in Figure 6.33.

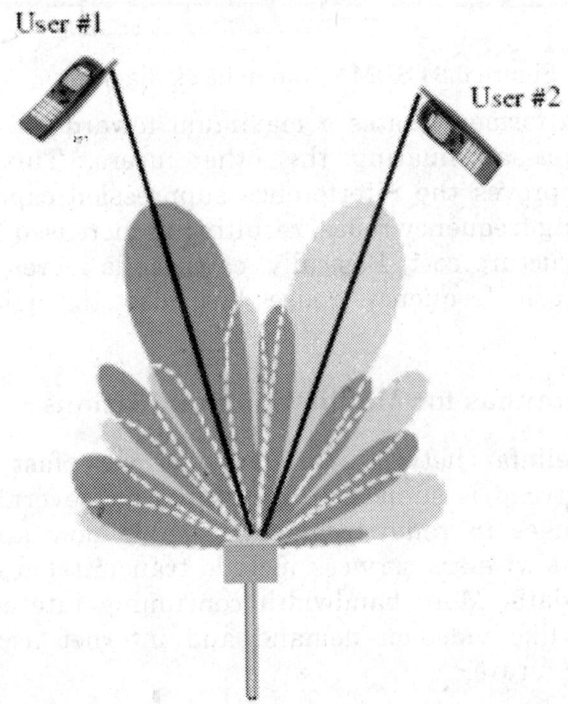

Figure 6.33 SDMA multi-beam systems

This means that more than one user can be owed to the same physical communication channel in the same cell simultaneously, with only an angle separation. This is gifted by having N parallel beam formers at the base station operating independently, where each beam former has its own adaptive beam forming algorithm to control its own set of weights and its own direction-of-arrival algorithm (DOA) to determine the time delay of each user's signal (Figure 6.34).

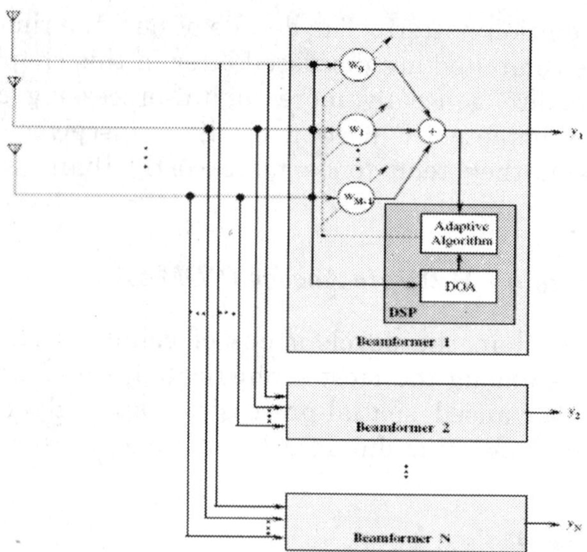

Figure 6.34 SDMA system block diagram.

Each beam former creates a maximum toward its desired user while nulling or attenuating the other users. This technology dramatically improves the interference suppression capability while greatly increasing frequency reuse, resulting in increased capacity and reduced infrastructure cost. Basically, capacity is increased not only through inter cell frequency reuse but also through intra cell frequency reuse

6.3.3 Smart Antennas for Mobile Communications

Wireless cellular networks are growing very fast around the world and this trend is supposed to continue for several years. The impressing changes in radio technology enable new and improved services. Current wireless services include transmission of voice, fax and low-speed data. More bandwidth consuming interactive multi-media services like video-on demand and internet access will be supported in the future.

Wireless networks should provide these services in a wide range of environments, spanning dense urban, suburban, and rural areas. Varying mobility requirements must also be addressed. Wireless local loop networks serviced subscribers. Micro-cellular networks serve pedestrians or slow moving users, and macro-cellular networks serve high speed vehicle-borne users. Several competing standards have been developed for terrestrial networks. AMPS (advanced mobile phone system) is an example of first-generation frequency division multiple access analog cellular system. Second generation principles include GSM (Global System for Mobile) and IS-136, using Time division multiple access (TDMA) and IS-95 using code division multiple access (CDMA). IMT-2000 is proposed to be the third generation standard and will use either a wide-band CDMA or TDMA technology.

Amplified services and lower costs have resulted in an increased air time usage and number of subscribers. Since the radio (spectral) resources are limited, system capacity is a primary challenge for current wireless network designers. Other major challenges include: (1) An unfriendly transmission medium, due to the occurrence of multipath, noise, interference and time-variations. (2) The limited battery life of the user's hand-held terminal. (3) Efficient radio resource management to offer high quality of service. Current wireless modems use signal processing in the time measurement alone through advanced coding, modulation and equalization techniques. The primary goal of smart antennas in wireless communications is to integrate and exploit efficiently the extra dimension offered by multiple antennas at the transceiver in order to enhance the over-Space-time processing offers various leverages: the first is array gain. Multiple antennas capture more signal energy, which can be combined to improve the signal-to-noise ratio (SNR). Next, spatial diversity obtained from multiple antennas can be used to combat channel fading. Finally, space-time processing can help mitigate inter-symbol interference (ISI) and co-channel interference (CCI). These leverages can be traded for improvements in

- Coverage: Square miles/base station.
- Quality: Bit error rate (BER)/MOS/Outage.
- Capacity: Erlangs/Hz/base station.
- Data rates: Bit per sec./Hz/base station.

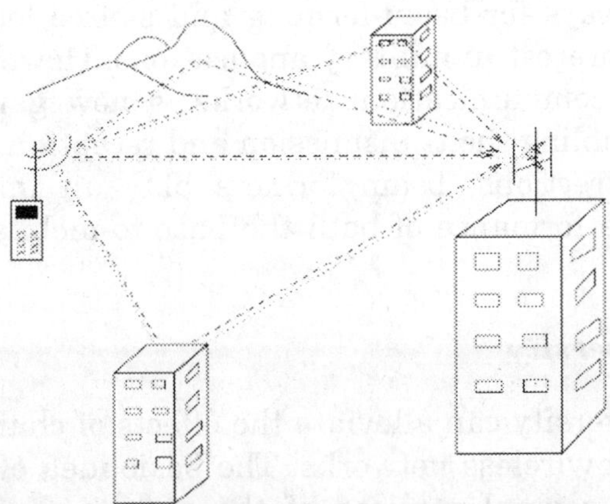

Figure 6.35 The user signal experiences multi path propagation and is Impinging on a two-element array on a building roof-top.

6.3.4 Early forms of spatial processing

(1) Adaptive antennas

The use of adaptive antennas dates back to the 50's with their applications to radar and ant jam problems. The primary goal of adaptive antennas is the automatic generation of beams (beam forming) that track a desired signal and possibly reject (or\null") intrusive sources through linear combining of the signals captured by the different antennas. An early contribution in the field of beam forming was made in 1956 by Altman and Sichak who proposed a combining device based on a phase-locked loop. This work was later refined in order to incorporate the adjustment of antenna signals both in phase and gain, allow improving the performance of the receiver in the presence of strong jammers. Howells proposed the side-lobe canceller for adaptive nulling. Optimal combining schemes were also introduced in order to minimize different criteria at the beam former output. These include the minimum mean-square error (MMSE) criterion, as in the LMS algorithm proposed by Widrow, the signal-to-interference and Noise ratio (SINR) criterion proposed by Applebaum, and the minimum variance beam former distortion-less response (MVDR) beam former proposed by Capon. Further Advances in the field were made by Frost, Griffiths and Jim among several others.

Besides beam forming, another application of antenna arrays is direction -of-arrival (DOA) judgment for source or target localization purposes. The leading DOA estimation methods are the MUSIC and ESPRIT algorithms. Note that in many of the beam forming techniques (for instance in Capon's method), the estimation of the source direction is an essential step. DOA estimation is still an area of active research.

Antenna arrays for beam-forming and source localization are of course of great interest in military applications. However, their use in civilian cellular communication networks is now gaining increasing attention. By enabling the transmission and reception of signal energy from selected directions, beam formers play an important role in improving the performance of both the base-to-mobile and mobile-to-base links

(2) Antenna diversity

Antenna diversity can alleviate the effects of channel fading, and is used widely in wireless networks. The basic idea of space diversity is as follows: if several replicas of the same information carrying signal are received over multiple branches with comparable strengths and which exhibit independent fading, then there is a high probability that at least one (or more) branch will not be in a become lighter at any given instant of time. When a receiver is equipped with two or more antennas that are sufficiently separated (typically several wavelengths) they offer useful diversity branches. Diversity branches tend to fade independently, therefore, a proper selection or combining of the branches increases link reliability. Without diversity, the protection against deep channel fades requires higher transmits power to make sure the link margins. Therefore, diversity at the base can be traded for reduced power consumption and longer battery life at the user terminal. Also, lower transmit power decreases the amount of co-channel interference and increases the system capacity. Independent fading across antennas is possible when radio waves impinge on the antenna array with sufficient angle spread. Paths coming from different arriving directions will add differently (constructive or destructive manner) at each antenna. This requires the presence of significant scatterers in the propagation medium, such as in urban environment and hilly terrain. Diversity also helps to combat large-scale fading effects caused by shadowing from large obstacles, e.g. buildings or terrain features. However, antennas located in the same base station experience the same shadowing. Instead, antennas from different base stations can be combined to offer a protection against large-scale fading (*macro* diversity). Note that antenna diversity can be complemented by other forms of diversity. Polarization, time, frequency and path diversity are some examples. These are particularly useful when physical constraints prevent the use of multiple antennas (for instance at the hand-held terminal). Combining the different diversity branches is an important issue. The main options used in current systems are briefly described below. In all cases, independent branch fading and equal mean branch powers are assumed. However, in non ideal situations, branch correlation and unequal powers will result in a loss of diversity gain. A correlation coefficient as high as 0:7 (between instantaneous branch envelope levels) are considered acceptable.

(2.1) Selection diversity

Selection diversity is one of the simplest forms of diversity combining. Given several branches with unreliable carrier-to noise ratios (*C=N*), selection diversity consists in choosing the branch having the highest instantaneous *C=N*. The performance improvement from selection diversity is evaluated as follows: Let us suppose that *M* branches experience independent fading but have the same mean *C=N*, denoted by. Let us now denote by □*s* the mean *C=N* of the selected branch. Then it can be shown that: Then it can be shown that:

$$\tau_s = \tau \sum_{j=1}^{M} \frac{1}{j}$$

For instance, selection over two branches increases the mean C=N by a factor of 1:5. Note that selection diversity requires a receiver behind each antenna. Switching diversity is a variant of selection diversity. In this method, a chosen branch is held until it falls below a threshold T, at which point the receiver switches to another branch, regardless of its level. The threshold can be fixed or adaptive. This strategy performs almost as well as the selection method described above, it also reduces the system cost, since only one receiver is required.

(2.2) Maximum ratio combining

Maximum ratio combining (MRC) is an optimal combining approach to combat fading. The signals from M branches are first co-phased to bring mutual coherence and then summed after weighting. The weights are chosen to be proportional to the respective signals level to maximize the combined C=N. It can be shown that the gain from MRC is directly proportional to the number of branches, i.e.:

$$\Gamma_s = M\Gamma$$

(2.3) Equal gain combining

Although optimal, MRC is expensive to implement. Also MRC requires an accurate tracking of the complex fading, which is difficult to achieve in practice. A simpler alternative is agreed by equal gain combining which consists in summing the co-phased signals using unit weights. The performance of equal gain combining is found to be very close to that of MRC. The SNR of the combined signals using equal gain is only 1 dB below the SNR provided by MRC.

6.3.5. Emerging application of space-time processing

While the use of beam forming and space diversity proves useful in radio communications applications, an inherent limitation of these concepts lies in the fact that the aforementioned techniques exploit signal combining in the space dimension only. Directional beam forming, in particular, heavily relies on the use of the spatial signatures of the incoming signals but does not consider their temporal structure. The techniques which combine the signals in *both time and space* can bring new leverages, and their importance in the area of mobile communications is now recognized. The main reason for using space-time processing is that it can exploit the rich temporal structure of digital communication signals. In addition, multipath propagation environments introduce signal delay spread, making techniques that exploit the total space-time structure more natural. The typical structure of a space-time processing device consists in using a bank of linear filters, each of them being located behind a branch, followed by a summing network. The received space-time signals can also be processed using non-linear schemes, e.g. in maximum-likelihood sequence detection. The space-time receivers can be optimized to make best use of array and diversity gains, and to minimize 1. Inter-symbol-interference, induced by the delay spread in the propagation channel. ISI can be suppressed by selecting a space-time filter that equalizes the channel or by using a maximum-likelihood sequence detector. 2. Co-channel-interference, coming from neighboring cells operating at the same frequency. CCI is suppressed by using a space-time filter that is orthogonal to the interferer channel. The key point is that CCI that cannot be rejected by space-only filtering may be handled more effectively using space-time filtering. Smart antennas can also be used at the transmitter to maximize array gain and/or diversity, and to mitigate ISI and CCI. In the transmit case, however, the efficiency of space-time processing schemes is usually limited by the lack of precise channel information. The major effects induced by radio propagation in a cellular environment are pictured in Figure 2. The leverages offered by space-time processing for receive and transmit are summarized in Table 1.

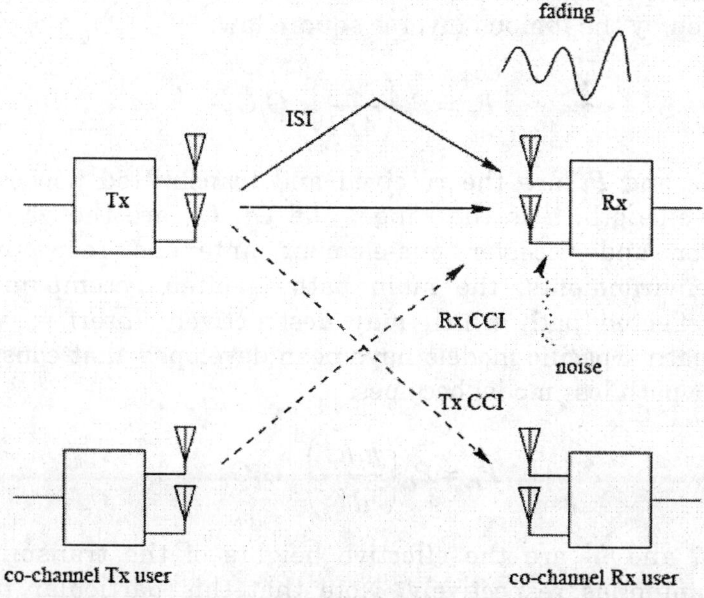

Figure 6.36 Smart antennas help mitigate the effects of cellular radio propagation

In the following sections, we describe channel models and algorithms used in space-time meting out. Both simple and advanced solutions are presented and trade-offs highlighted.

Finally, we describe current applications of smart antennas.

6.3.6 Channel models

Channel models capture radio propagation effects and are useful for simulation studies and performance prediction. Channel models also help in motivating suitable signal processing algorithms. The effects of radio propagation on the transmitted signal can be generally categorized into two main classes, fading and spreading. Fading refers to the propagation losses experienced by the radio signal (on both the forward and reverse links).

One type of fading called selective fading causes the received signal level to vary around the average level in some regions of space, frequency, or time. Channel spreading refers to the dispersion of the information-carrying signal energy in space, and on the time or frequency axis. Selective fading and spreading are dual descriptions.

(1) Channel fading

(1.1) Mean path loss

The mean path loss describes the attenuation of a radio signal in a free space propagation situation, due to isotropic power spreading, and is given by the famous inverse square law:

$$P_r = P_t \left(\frac{\lambda}{4\pi d} \right)^2 G_t G_r$$

Where, *Pr* and *Pt* are the received and transmitted powers, is the radio wave-length, *d* is the range, and *Gt*, *Gr* are the gains of the transmitter and receiver one-element antennas respectively. In cellular environments, the main path is often accompanied by a surface reflected path which may destructively interfere with the primary path. Specific models have been developed that consider this effect. The path loss model becomes:

$$P_r = P_t \left(\frac{h_t h_r}{d^2} \right)^2 G_t G_r$$

Where, h_t and h_r are the effective heights of the transmitter and receiver antennas respectively. Note that this particular path loss model follows an inverse fourth power law. In fact, depending on the environment, the path loss exponent may vary from 2:5 to 5.

(1.2) Slow fading

Slow fading is caused by long-term shadowing effects of buildings or natural features in the land. It can also be described as the local mean of a fast fading signal (see below). The statistical distribution of the local mean has been studied experimentally and was shown to be influenced by the antenna height, the operating frequency, and the type of environment. It is therefore difficult to predict. However, it has been experimental decision that when all the above mentioned parameters are fixed, then the received signal fluctuation approaches a normal distribution when plotted in a logarithmic scale (i.e. in dB's). Such a distribution is called log-normal. A typical value for the standard deviation of shadowing distribution is 8 dB.

(1.3) Fast fading

The multipath propagation of the radio signal causes path signals to add up with random phases, constructively or destructively, at the receiver. These phases are determined by the path length and the carrier frequency, and can vary tremendously rapidly along with the receiver location. This gives rise to the so-called fast-fading phenomenon that describes the both rapid and large fluctuations of the received signal level in space. If we assume that a large number of scattered wave fronts with random amplitudes and angles of arrival arrive at the receiver with phases consistently distributed in [0; 2f) then the in-phase and quartered phase components of the vertical electrical field *Ez* can be shown to be Gaussian processes. In turn, the envelope of the signal can be well approximated by a Rayleigh process. If there is a direct path present, then it will no longer be a Rayleigh distribution but becomes a Rician distributed instead.

(2) Channel spreading

Propagation to or from a mobile user, in a multipath channel, causes the received signal energy to spread in the frequency, time and space dimensions (Figure 3, and also table 2 for typical values). The characteristics of the spreading (that is to say, the particular dimension(s) in which the signal is spread) affects the design of the receiver.

(2.1) Doppler spread

When the mobile is in motion, the radio signal at the receiver experiences a shift in the frequency domain (also called Doppler shift), the amplitude of which depends on the path direction of arrival. In the presence of nearby scatterers with multiple directions, a pure tone is *spread* over a finite spectral bandwidth. In this case, the Doppler power spectrum is defined as the Fourier transform of the time auto-correlation of the received signal and the Doppler spread is the support of the Doppler power spectrum. Assuming scatterers uniformly distributed in angle, the Doppler power spectrum is given by the so-called classical spectrum:

$$S(f) = \frac{3\sigma^2}{2\pi f_m}\left[1 - \left\{\frac{f - f_c}{f_m}\right\}^2\right]^{-1/2}, \quad f_c - f_m < f < f_c + f_m$$

Where, *fm* = *v=f* is the maximum Doppler shift, *v* is the mobile velocity, *fc* is the carrier frequency and f2 is the signal variance. When there is a dominant source of energy coming from a particular direction (as in line-of-sight situations), the appearance for the spectrum needs to be corrected, accordingly to the Doppler shift of the dominant path *fD*:

$$S(f) + B\delta\,(f - f_D)$$

Where, *B* denotes the ratio of direct to scattered path energy.

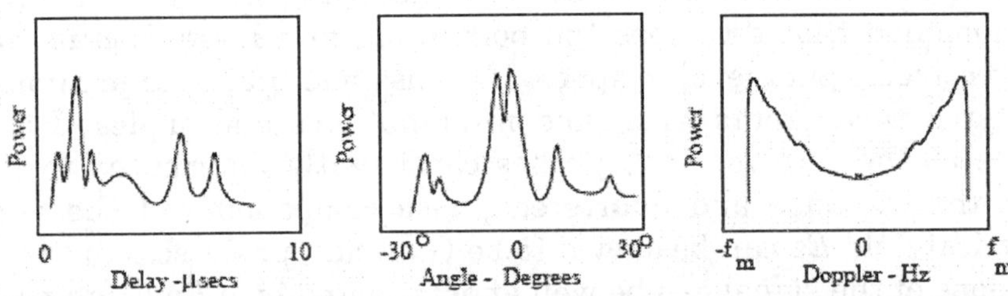

Figure 6.37 The radio channel induces spreading in several dimensions. These spreads strongly affect the design of the space-time receiver.

The Doppler spread causes the channel characteristics to change rapidly in time, giving rise to the so-called *time selectivity*. The coherence time, during which the fading channel can be measured as constant, in inversely proportional to the Doppler spread. A typical value of the Doppler spread in a macro-cell environment is about 200 Hz at 65 Mph, in the 1900 MHz band. A large Doppler spread makes good channel tracking an essential feature of the receiver design.

2.2. Delay spread

Multipath propagation is often characterized by several versions of the transmitted signal arriving at the receiver with dissimilar attenuation factors and delays. The spreading in the time domain is called *delay spread* and is responsible for the selectivity of the channel in the frequency domain (different spectral components of the signal carry different powers). The coherence bandwidth, which is the maximum range of frequencies over which the channel response can be viewed as constant, is inversely proportional to the delay spread. Significant delay spread may cause strong inter-symbol interference which makes necessary the use of a channel equalizer.

2.3. Angle spread

Angle spread at the receiver refers to the spread of directions of arrival of the incoming paths. Likewise, angle spread at the transmitter refers to the spread of departure angle of the multipath. As mentioned earlier, a large angle spread will cause the paths to add up in a random manner at the receiver as a function of the location of the receive antenna, hence it will be source of *space* selective fading. The range of space for which the fading remains constant is called the coherence distance and is inversely related to the angle spread. As a result, two antennas spaced by more than the coherence distance tend to experience uncorrelated fading. When the angle spread is large, which is usually the case in dense urban environments; a significant gain can be obtained from space diversity. Note that this usually conflicts with the possibility of using directional beam forming, which typically requires well defined and dominant signal directions, i.e. a low angle spread.

(3) Multipath propagation

3.1. Macro-cells

A macro-cell is characterized by a large cell radius (up to a few tens of kilometers) and a base station located above the roof top. In macro-cell environments, the signal energy received at the base station comes from three main scattering sources: scatterers local to the mobile, remote dominant scatterers, and scatterers local to the base (Figure 4 for an illustration). The following description refers to the reverse link but applies to the onward link as well. The scatterers local to the mobile are those located a few tens of meters from the hand-held terminal. When the terminal is in motion, these scatterers give rise to a Doppler spread which causes time selective fading. Because of the small dispersion radius, the paths that emerge from the vicinity of the mobile and reach the base station show a small delay spread and a small angle spread. Of the paths emerging from the local-to-mobile scatterers, some reach remote dominant scatterers, like hills or high rise buildings, before ultimately traveling to the base station. These paths will typically reach the base with medium to large angle and delay spreads (Depending of course on the number and locations of these remote scatterers). Once these multiple wave fronts reach the vicinity of the base station, they usually are further scattered by local structures such as buildings or other structures that are close to the base. These scatterers local to the base can cause large angle spread, therefore, they can cause severe space-selective fading.

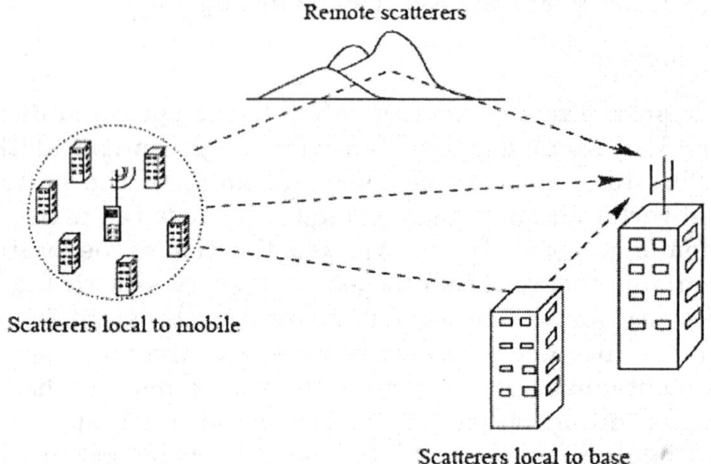

Figure 6.38 Each type of scatterers introduces specific channel spreading characteristics.

3.2. Micro-cells

Micro-cells are characterized by highly dense built-in areas, and by the user's terminal and base being comparatively close (a few hundred meters). The base antenna has a low elevation and is typically below the roof top, causing significant scattering in the vicinity of the base. Micro-cell situations make the propagation difficult to analyze and the macro-cell model described earlier does not necessarily hold anymore. Very high angle spreads along with small delay spreads are likely to occur in this situation. The Doppler spread can be as high as in macro-cells, although the mobility of the user is predictable to be limited, due to the presence of mobile scatterers.

(4) Parametric channel model

A complete and accurate understanding of propagation effects in the radio channel requires a detailed description of the physical environment. The *specular model*, to be presented below, only provides a simplified report of the physical reality. However, it is useful as it describes the main channel effects and it provides the means for a simple and efficient mathematical treatment. In this model, the multiple elementary paths are grouped according to a (typically low) number of L main path clusters, each of which contains paths that have roughly the same mean angle and delay. Since the paths in these clusters originate from different scatterers, the clusters typically have near independent fading. Based on this model, the continuous time channel response from a single transmit antenna to the ith antenna of the receiver can be written as:

$$f_i\,(t) = \sum_{l=1}^{L} a_i\,(\theta_t)\; a_l^R\,(t)\; \delta\,(t-\tau_l) \qquad 1$$

(*t*), *fl*, and *fl* are respectively the fading (including mean path loss, slow and fast fading), angle, and delay of the lth receive path cluster. Note that this model also includes the response of the ith antenna to a path from direction *fl*, denoted by $a_i(fl)$. In the following we make use of the specular model to represent the structure of the signals in space and time. Note that in the situation where the path cluster assumption is not acceptable, other channel models, called *diffuse* channel models are more appropriate.

6.3.7 Smart Antennas' Benefits

The primary reason for the rising interest in smart-antenna systems is the *capacity* increase. In densely populated areas, mobile systems are usually interference-limited, meaning that the interference from other users is the main source of noise in the system. This means that the signal-to-interference ratio (SIR) is much smaller than the signal-to-noise ratio (SNR). In wide-ranging, smart antennas will, by simultaneously increasing the useful received signal level and lowering the interference level, increase the SIR.

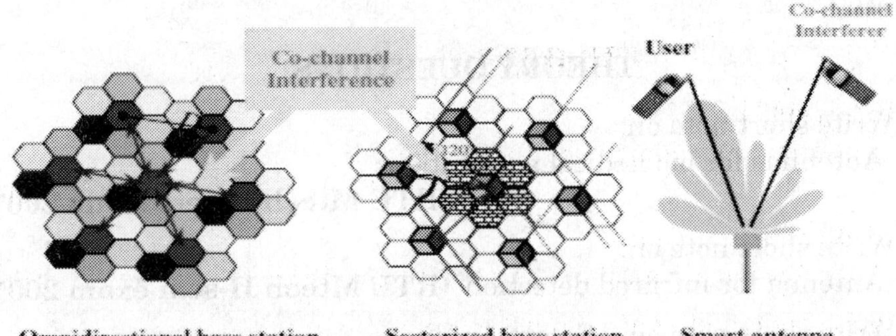

Figure 6.39 Co-channel interference comparisons of the different systems

Another benefit that smart-antenna systems provide is *range* increase. Because smart antennas are more directional than omnidirectional and sectorized antennas, a range increase potential is avail-able. In other words, smart antennas are able to focus their energy toward the intended users, as an alternative of directing it in other unnecessary directions (wasting) like omnidirectional antennas do. This means that base stations can be placed further apart, leading to a more cost-efficient development. Therefore, in rural and sparsely populated areas, where radio coverage rather than capacity is more important, smart-antenna systems are also well suited.

Another added advantage of smart-antenna systems is *security*. Ina society that becomes more dependent on conducting business and transmitting personal information, security is an important issue. Smart antennas make it more difficult to tap a connection because the intruder must be located in the same direction as the user as seen from the base station to successfully tap a connection. Finally, because of the spatial detection nature of smart-antenna systems, they can be used to locate humans in emergencies or for any other location-specific service.

6.3.8 Smart Antennas' Drawbacks

While smart antennas provide many benefits, they do suffer from certain drawbacks. For example, their transceivers are much more complex than customary base station transceivers. The antenna needs separate transceiver chains for each array antenna element and accurate real-time calibration for each of them. Moreover, the antenna beam forming is computationally intensive, which means that smart-antenna base stations must be equipped with very powerful digital signal processors. This tends to boost the system costs in the short term, but since the benefits outweigh the costs, it will be less expensive in the long run. For a smart antenna to have pattern-adaptive capabilities and reasonable gain, an array of antenna elements is essential.

THEORY QUESTIONS

Q.1 Write short note on:
Antennas for biomedical applications
(RTU Mtech II-sem exam 2007)

Q.2 Write short note on:-
Antenna for infrared detection **(RTU Mtech II-sem exam 2007)**

Q.3 Write short note on:- Smart Antenna

Q.4 Explain the smart antennas for mobile communication.

Q.5 What are the early forms of spatial processing? Explain.

Q.6 Define a smart antenna and what is useful analogy for adaptive smart antenna. **(RTU Mtech II-sem exam 2007)**

Q.7 Write short note on:

i. Smart Antenna **(R.T.U. Mtech II SEM 2010)**

UNIT 7

IE3D – ANTENNA DESIGNING SOFTWARE

CONTENTS

IE3D – ANTENNA DESIGNING SOFTWARE

7A. BASIC TECHNIQUES

IE3D is a general purpose EM simulator. It is used to design high frequency circuits and antennas. Without loss of generality, we will call a circuit or an antenna as a structure. On IE3D, a structure is described as a set of polygons, and a polygon is described as a set of vertices. This chapter will explain the basic editing skill by going through, step by step, the simulation on a microstrip bend structure.

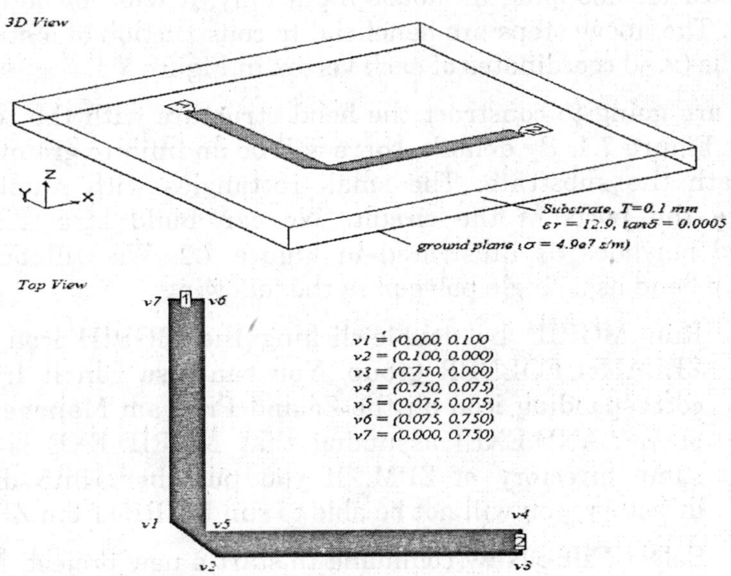

Figure 7.1 A simple chamfered microstrip bend structure.

There are many ways to create a structure on IE3D (see flow chart in Figure 7.5). You can import it from other standard formats such as GDSII, DXF, ACIS and GERBER using MGRID and ADIX. You can also convert it directly from GDSII, Cadence Virtuoso and Cadence Allegro using AGIF. You can convert it directly from Microwave Office using EM Socket. You can create them from a set of library elements on MGRID or IE3DLIBRARY. You can even draw it. Drawing mode is the default mode on MGRID. When you manage to draw a structure with the many advanced geometry modeling commands on MGRID, you will be able to model and design very complicated circuits. All other ways on MGRID, IE3DLIBRARY and AGIF are based upon the geometry modeling capability underneath MGRID. If you can manage MGRID in doing geometry modeling, you will be able to manage the different ways in other applications of IE3D. They are built on top the same geometry modeling engine.

Section 7.1 Defining Basic Parameters

We start from MGRID. The parameters of the microstrip bend are shown in Figure 7.1. Before you draw the structure, please make notes to your structure. First you should introduce the x and y-coordinate system. The x and y-coordinates of each vertex should be marked. Distances between vertices should be calculated if it is necessary. Try to break a structure into several pieces when a complicated structure is modeled. Try to find a grid size such that your structure can be best fitted into the uniform grid. The uniform grid is used for snapping in mouse input only; it will not be used for meshing. The above steps are beneficial to construction of a structure. We list the (x, y) coordinates of each vertex in Figure 7.1.

We are going to construct the bend structure with the top view shown in Figure 7.1. By default, there will be an infinite ground plane underneath the substrate. The small rectangles with numbers on them are the ports of the circuit. We can build structure as 3 connected polygons as illustrated in Figure 7.2. We will build the microstrip bend as a single polygon in the following.

Step 1 Run MGRID by double-clicking the MGRID icon in the ZELAND FOLDER group. You can also run it from the corresponding icon on the Zeland Program Manager (ZPM or ZELAND.EXE) assuming that MGRID.EXE is in the same directory of ZPM. If you put them into different directory, you will not be able to run MGRID from ZPM.

Step 2 Select **File->New** command to start a new project. MGRID will prompt you for the *Basic Parameters* dialog. You need to enter the basic parameters first.

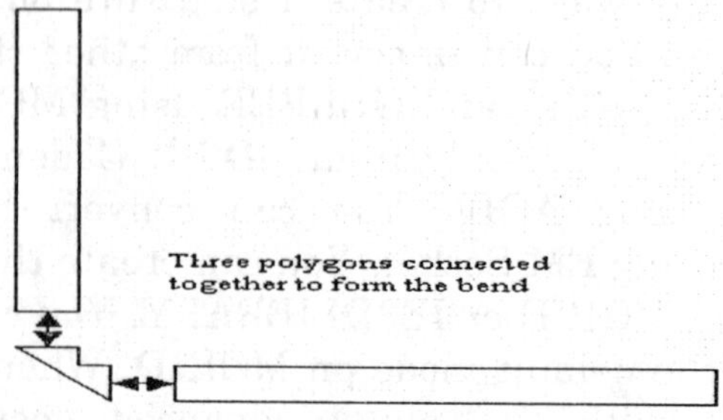

Figure 7.2. The chamfered bend broken down into three polygons.

The Basic Parameters include 6 groups of parameters: (1) Comment: The comment for the structure; (2) Length: the Length Unit and Minimum Length for the structure; (3) Layouts and Grids: the parameters for the uniform grid system for the layout editing; (4) Enclosure: The walls for the enclosures used in the structure; (5) Meshing Parameters: the parameters control the meshing of the geometry; (6). Substrate Layers: The parameters of substrates and infinite ground planes; (7) Metallic Strip Types: The parameters for different types of metallic strips used in the structure; (8) Dielectric Types: The dielectric types for 3D dielectrics and finite size substrates.

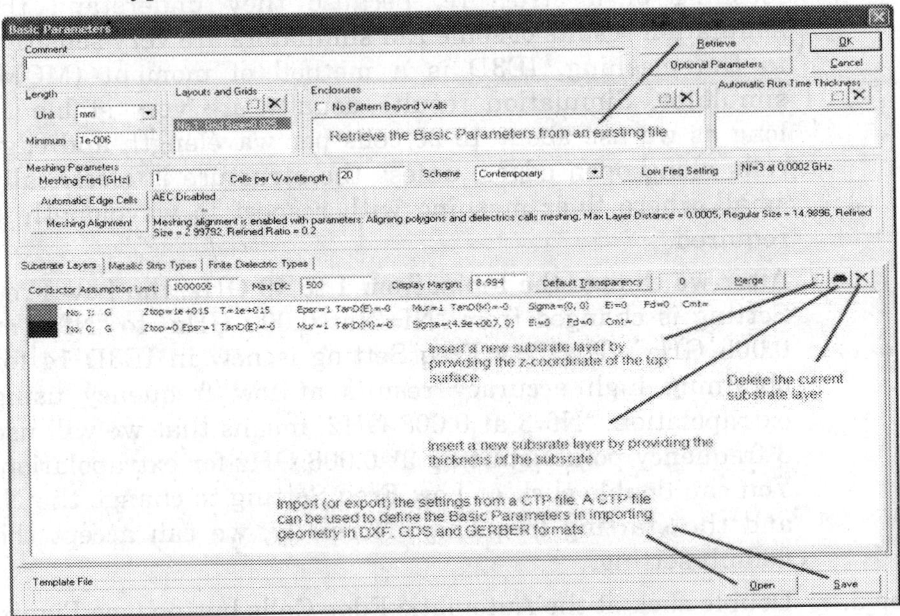

Figure 7.3 The Basic Parameters dialog.

For this structure, we will use "mm" for the length unit. We can accept the default Minimum Length which controls the accuracy of vertex location in mouse input. Please note the buttons above some list boxes in Figure 7.3. The meanings of the buttons are shown in Figure7.4.

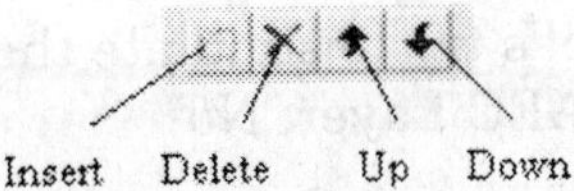

Figure 7.4 The meaning of the buttons for the list boxes in Figure 7.3.

Step 3 You can define multiple "Layout and Grid" items. There is a default one in the list box. Please double click at the only item in the Layouts and Grids list box (Figure 7.3). MGRID prompts you for the Edit Grid Size. The default Grid Size = 0.025 mm. It is what we want.

Select OK to accept the default value for the default grid size. You are back to the Basic Parameters dialog.

Step 4 Change the Meshing Frequency (Fmax) from 1 to 40 GHz because we want to simulate the structure up to 40 GHz. Remain the Cells per Wavelength (N cell) as 20.

The higher the Meshing Frequency is, the finer mesh a structure will be discretized to. Finer meshing is preferred for higher accuracy, but you pay for the substantial increase in the simulation time. Many users are concerned about the meshing of a structure, because they understand the simulation results of some EM simulators are very sensitive to the meshing. IE3D is a method of moment (MOM) simulator. Simulation results usually are very stable as long as we use about 15-20 cells per wavelength and edge cells along open edges, unless the structure is electrically small where finer meshing with respect to wavelength is required.

After we change the Fmax from 1 to 40 GHz, the Low Freq Setting is changed from "Nf=3 at 0.002 GHz" to "Nf=3 at 0.008 GHz". The Low Freq Setting is new in IE3D 14 for obtaining high accuracy results at low frequency using extrapolation. "Nf=3 at 0.008 GHz" means that we will use 3 frequency points starting at 0.008 GHz for extrapolation. You can double click at Low Freq Setting to change the Nf and the starting frequency. Normally, we can accept the default setting.

Step 5 Double click at the Automatic Edge Cells button (see Figure 7.3). The Automatic Meshing Parameters comes up 7. There are many parameters in the dialog: Fmax and Ncell are in this dialog too.

Automatic Edge Cells (AEC) improves simulation accuracy significantly. It can yield expert results for novice users. However, IE3D may still be able to yield accurate results for many simple structures without strong coupling. For the 1st example, let's try to simulate the structure without AEC. Please select AEC Layer: No"

You can select Contemporary meshing scheme. Contemporary scheme is a newer scheme which normally may yield similar accuracy with fewer unknowns. Classical scheme is an old scheme and it is normally more reliable.

Section 7.2 Defining Substrate Parameters

Substrate parameters are also part of Basic Parameters. They include the number of dielectric substrate layers, the z-coordinate of the top surface of each substrate layer, the substrate permittivity, permeability and conductivity. Complex permittivity, permeability and conductivity are accepted in IE3D. The No.0 substrate layer is assumed to be the ground plane by default. The infinite ground plane is defined as a substrate with a very high conductivity. The z-coordinate of the top surface of No.0 substrate layer is always equal to 0 and cannot be modified. Other parameters can be modified according to the actual cases. For example, you can define the conductivity of the No.0 layer as 0 in order to replace the default ground plane with air.

For multiple layer substrates, you just need to define the top surface z-coordinate of each layer and define the corresponding permittivity, permeability and conductivity. You can define as many substrate layers as you like. You have to define at least one substrate layer in the upper half space. It means there should be at least 2 substrate layers (including the No.0 layer) in a structure. When the upper half space is filled with a single type of dielectrics, you can define a substrate layer with the top surface z-coordinate of top surface as a very large number (such as 1.0e+15 mm).

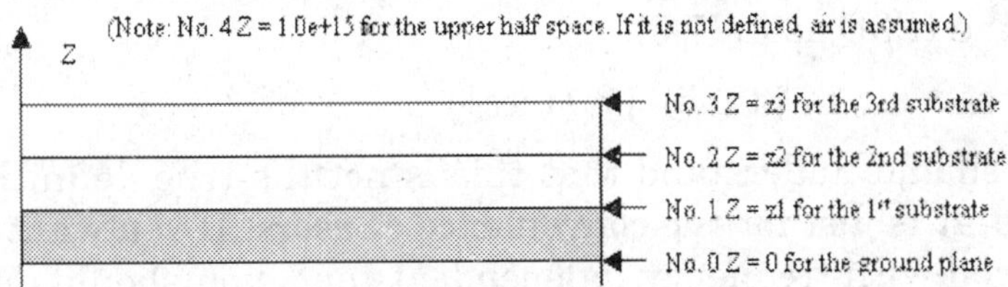

Figure 7.5 The substrate stack up and definition on MGRID.

By default, MGRID creates 2 substrate layers for you. The No.0 layer is a good conductor with conductivity = 4.9e+7 s/m, which is gold. The No.1 layer is air and its top surface Z-coordinate, Ztop = 1.0e+15 mm, meaning that the whole upper space is filled with air.

For the structure we are going to simulate, we have three dielectric layers (including the ground plane). The No.0 layer is the ground plane. The No.2 layer is the upper half space air, which can be omitted. The No.1 layer is the dielectric slab with the following parameters:

Top Surface Z-Coordinate, Ztop = 0.1 mm

Real Part of Permittivity, Re(EPSr) = 12.9

Loss Tangent of EPSr = 0.0005

Real Part of Permeability, Re(MUr) = 1.0

Loss Tangent of MUr = 0.0

Real Part of Conductivity = 0.0 s/m

Imaginary Part of Conductivity = 0.0 s/m

Dielectric permittivity is also called dielectric constant. There are 2 ways to define substrate loss in IE3D: (1) defining the imaginary part (or loss tangent) of the permittivity and (2) defining the conductivity. Theoretically, they are interchangeable, and they are defined as:

$$\varepsilon_{rc} = \varepsilon_r - j\,\sigma / (\omega\varepsilon_0) = \varepsilon_r (1 - j\tan\delta) \tag{7-1}$$

where ε_r is the real dielectric constant; σ is the conductivity; ω is the angular frequency; ε_0 is the complex dielectric constant in free space (8.86×10^{-12} F/m); $\tan\delta$ is the loss tangent. In the other word, we have,

$$\tan\delta = -\mathrm{Im}(\varepsilon_{rc}) / \mathrm{Re}(\varepsilon_{rc}) = \sigma / (\omega\varepsilon_0\varepsilon_r) \tag{7-2}$$

The Loss Tangent is always a non-negative number and the imaginary part of permittivity is always a non-positive number. In practical applications, users normally use either $\tan\delta$ or σ as frequency independent values. However, we can see from (7-2) that, when $\tan\delta$ is frequency independent, σ will be frequency dependent, or vice-versa. In order to resolve this inconsistency, we define the $\tan\delta$ and σ in the IE3D as in (7-3):

$$\varepsilon_{rc} = \varepsilon_r (1 - j\tan\delta) - j\,\sigma / (\omega\varepsilon_0) \tag{7-3}$$

You should understand that this is not the true definition of the $\tan\delta$ andσ. It is just for the convenience of users. If you want to define the material with frequency independent $\tan\delta$, you should define $\sigma = 0$ for the dielectrics. If you want to define the material with frequency independentσ, should define $\tan\delta = 0$ for the substrate. However, if you define non-zero values for both $\tan\delta$ andσ, both will be used to calculate the ε_{rc} value in (7-3).

You may notice that a parameter called Conductor Assumption Limit (CAL) is displayed in the dialog. In IE3D, highly conductive substrates are treated differently from normal substrates while they are defined in the same section. The CAL is used to identify which substrate will be treated as high conductivity. Each substrate will be displayed with a CFactor. If the CFactor of a substrate exceeds the CAL, the substrate will be considered as high conductivity substrate or ground plane. Another parameter is called Max DK. The Max DK is used to identify which substrates will be used to estimate the waveguide wavelength for the meshing. If the CFactor of a substrate exceeds Max DK, the substrate will not be used for estimation for the meshing density. Other Parameters are Substrate Display Margin and Default Transparency. They are used to control how we display the substrates in the 3D view window. They will not affect the meshing and the simulation. Starting from IE3D 14, Substrate Display Margin is in the true length dimension, while it is the ratio with respect to the wavelength in the older versions.

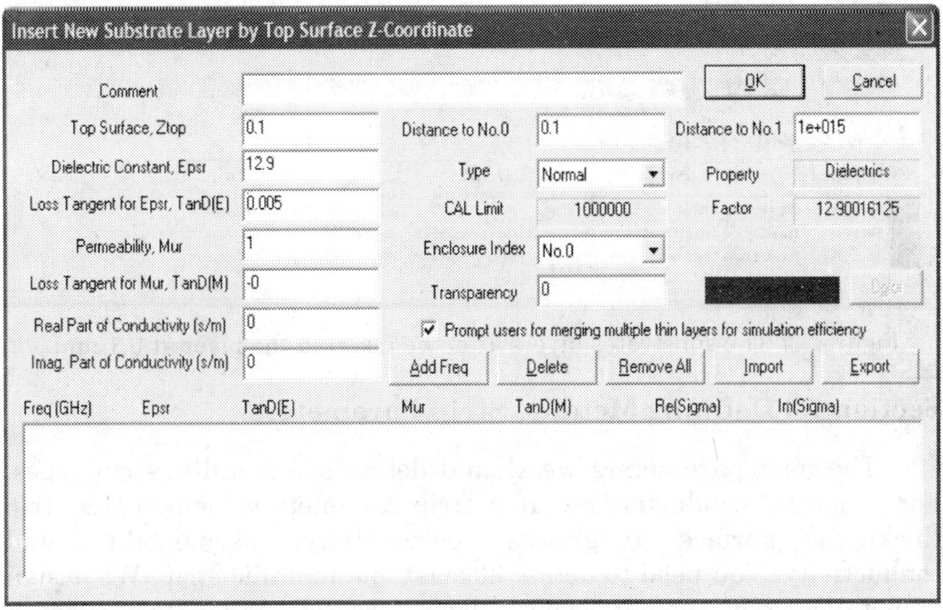

Figure 7.6 The Edit Substrate dialog with the No.1 substrate parameters entered.

Step 6 Select the "Insert" button (see Figure 7.3 and Figure 7.4) above the Substrate Layers list box. MGRID will prompt you for a new substrate layer (see Figure 7.6).

Step 7 Make sure it is Normal type. Enter the parameter of the No. 1 substrate into the dialog (see Figure 7.6). We can specify an **Enclosure Index** to each substrate. This feature allows us to define multiple enclosures separated by ground planes. We can define an enclosure between each pair of ground plane. Typical applications are cavity-backed antennas. Multiple enclosures are discussed in Appendix H. Also, you can select HTS II type for the substrate. We will explain the HTS II type when we discuss the metallic strip parameters in the next.

Please select OK to enter the new substrate into the dialog. We will have the Substrate Layers listed in Figure 7.7.

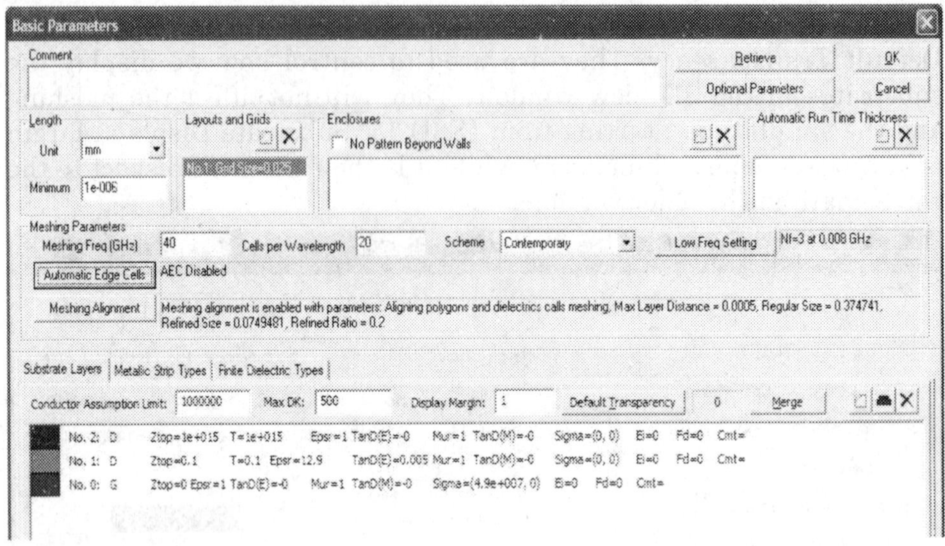

Figure 7.7 The substrates layers after we inserted the layer at 0.1 mm.

Section 7.3 Defining Metallic Strip Parameters

The next parameters we should define are metallic strip types. For a normal conductor, metallic strip parameters include the strip thickness, surface roughness, permittivity, permeability and conductivity. You need to define at least one metallic strip. We assign a metallic strip type to each polygon created. When you enter a polygon during geometry editing, the polygon is normally assumed to be of the 1st printed strip type. To modify the metallic strip type of some polygons, you select the polygons you want to change the metallic strip type. Then, you select **Edit->Object Properties** command. Then, select **Group Change Metal Type** button in the dialog to change it.

You can also define HTS I, HTS II and thin film resistor for the metallic strip types. The HTS I and HTS II do not mean they are different types of HTS material. They just mean we use two different formulas for the HTS modeling.

To define a metallic strip type, please make sure you select the Metallic Strip Types tab and select "Insert" button above the list box, the layout editor will prompt you to define the parameters.

For this circuit, we have one metallic strip type. The parameters of the No.1 metallic strip type are listed below. The parameters are the same as the default No.1 metallic strip type in the list box.

Strip thickness = 0.002 mm

Surface Roughness = 0.0

Real part of permittivity = 1.0

Imaginary part of permittivity = 0.0

Real part of permeability = 1.0

Imaginary part of permeability = 0.0

Real part of conductivity = 4.9e+7 s/m

Imaginary part of cónductivity = 0.0 s/m

Step 8 Double click at the No.1 type in the Metallic Strip Types list box of the Basic Parameters dialog. The Edit Metallic Type dialog comes up and let you edit the parameters (see Figure 7.8).

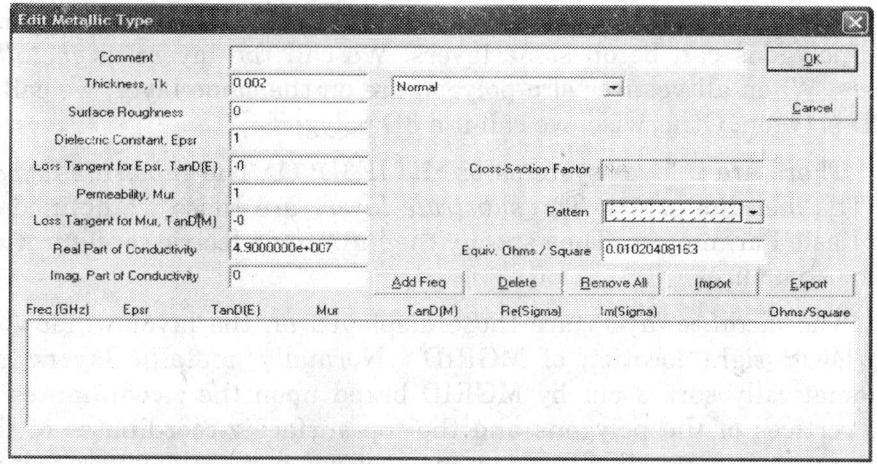

Figure 7.8 The Edit Metallic Type dialog with the No.1 metallic strip type parameters.

You are allowed to define frequency dependent metallic strip types. 7

Also, MGRID lists the Equiv. Ohms/Square value for the provided conductivity value. You should understand that the Equiv. Ohms/Square value is the value at DC. It will change at high frequency. In fact, it should also have imaginary part at high frequency.

We do not need to change the parameters because they are what we want. Select OK and MGRID will show all the Basic Parameters on the dialog.

Step 9 Select OK to close the Basic Parameters dialog. MGRID will be ready for us to create the polygons for the structure.

Another group of parameters in Basic Parameters are the Dielectric Types. They are for the modeling of finite dielectrics. For this example, it does not involve finite dielectrics and we do not need to worry about it.

The layout and grid of the MGRID is completely shown now. The small red dots are the grids. The red lines represent the x and y-coordinate system. For the open boundary formulation, the IE3D assumes the boundary of the substrate and ground plane (if defined) extend to infinity. For closed boundary formation, the IE3D allows a user to define the boundary of the enclosure. There are still some other parameters in the Parameters menu. We do not need to change the Optional Parameters every time we construct a structure.

Section 7.4 Editing Polygons on Layers

What we need to do is to build a structure as a set of polygons. The polygons can be on some layers. We call the layers as *metallic layers*. When all vertices of a polygon are on the same layer, we call it a 2D polygon. Otherwise, we call it a 3D polygon.

There are 2 layer systems on the IE3D: (1) The *substrate layers*; (2) The *metallic layers*. The *substrate layers* are those we defined in the Basic Parameters. They specify the infinite dielectric configuration of the structure.

The *metallic layers* are those displayed on the layer window at the lower right location of MGRID. Normally, metallic layers are automatically sorted out by MGRID based upon the z-coordinates of the vertices of the polygons and the top surface z-coordinates of the substrates. It is possible that you may see some metallic layers you do not expect to have 2D polygons on them. This may mean that some vertices of 3D polygons are on the layer or some substrate's top surface z-coordinate is on the layer. When you are building strong 3D structures, you may have many different layers on the layer windows even there are no 2D polygons on them. If you want MGRID not to display those 3D layers, you can un-check the *List All Layers* (see Figure 7.9) on the *layer window* to hide them.

Also, it is possible that a metallic layer does not define some metallic polygons on it. When the layer is on the interface of high conductivity (or metallic) substrate, the substrate is basically a ground plane and the metallic layer defines a layer of slots on the ground plane. When such case happens, you will see a horizontal line across the color strip representing the metallic layer. The Figure 7.9b is an example. Z = 0 is the ground plane and the metallic layer on z = 0 has a horizontal line across it, meaning that any 2D polygon on this layer defines a slot on the ground plane.

There are many ways to enter vertices in MGRID. The simplest way is to use the mouse to draw them. Mouse input can be easy and accurate when there exists a good minimum length in the circuit. Good minimum length means that you can fit the circuit into a uniform grid with the minimum length as the grid size. Unlike some other simulators, the layout grid on the IE3D has nothing to do with the meshing in a simulation. Smaller layout grid does not slow down an IE3D simulation because we do not need to fit a structure into the uniform grid. We will use mouse to construct the chamfered bend in the following.

Please pay attention to the layer window on the lower right corner. Depending upon whether the List All Layers checkbox is checked or unchecked, we will get either Figure7.9a or Figure7.9b.

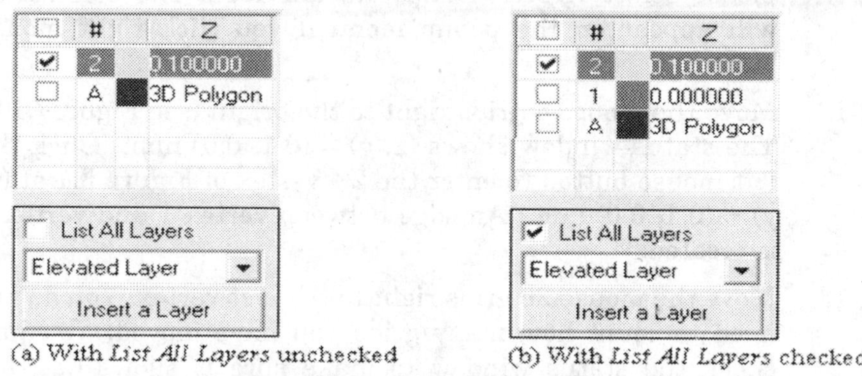

(a) With *List All Layers* unchecked (b) With *List All Layers* checked

Figure 7.9 The layer window (a) with *List All Layers* unchecked; (b) with *List All Layers* checked.

Step 10 Please click at the color strip representing No.2: Z = 0.1 layer in the Layer Window to shift the 2D input focus on the No.2 layer at z = 0.1 mm.

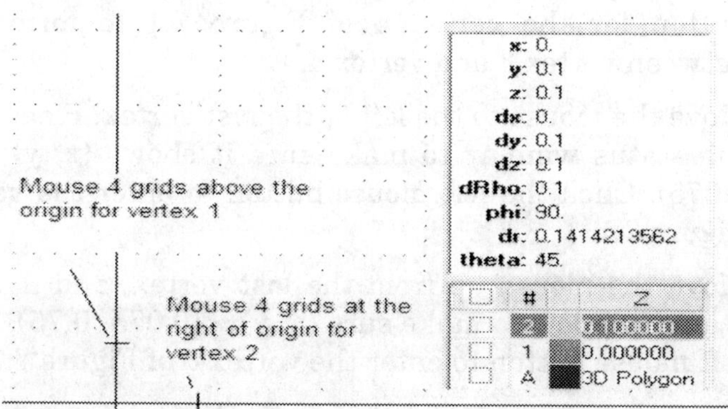

Figure 7.10 The mouse location and the status window.

Step 11 Move the mouse 4 grids over the origin. You will see x=0 and y=0.1 on the status window at the upper right corner (see Figure 7.10). Click the left mouse button. The vertex 1 of Figure 7.1 is entered at (x, y) = (0.0, 0.1) mm. It is displayed as a small dot on the window. A line linking the vertex to the mouse cursor is also displayed on the window.

We need not to locate the mouse cursor exactly on the grid point at (x, y) = (0.0, 0.1) mm as long as the cursor is closest to the grid point. MGRID will automatically snap the vertex to the grid. When you move the mouse, you can count how many grids the mouse cursor is from a reference point (such as the origin), and keep checking the coordinates displayed in the status window. If a vertex is entered at an unexpected grid point, you can drop the vertex by selecting **Input->Drop Last Vertex** To drop all entered vertices, please select **Input->Drop All Vertices**. The commands will appear on the popup menu if you click right mouse button.

Step 12 Move the mouse 4 grids right to the origin (see Figure7.10). The status window shows (x, y) = (0.1, 0.0) mm. Click the left mouse button to enter the 2nd vertex of Figure 3.1 at (x, y) = (0.1, 0.0) mm. An edge between vertex 1 and vertex 2 is created.

Step 13 Move the mouse 26 grids right to the last vertex. You do not need to count how many grids you have moved. You just check the status window to make sure it shows (x, y) = (0.75, 0.0), and click the left mouse button. The vertex 3 of Figure 7.1 is entered and an edge between vertex 2 and vertex 3 is formed.

Step 14 Move the mouse 3 grids up from the last vertex. The status window shows (x, y)=(0.75, 0.075). Click the left mouse button for the vertex 4 of Figure 7.1 to form an edge between vertex 3 and vertex 4.

Step 15 Move the mouse to the left of the last vertex. Please monitor the status window to make sure it shows (x, y) = (0.075, 0.075). Click the left mouse button to enter the vertex 5 of Figure 7.1.

Step 16 Move the mouse up from the last vertex, and monitor the status window to make sure (x, y) = (0.075, 0.75). Click the left mouse button to enter the vertex 6 of Figure 7.1.

Step 17 Move the mouse 3 grids left to the last vertex, and monitor the status window for (x, y) = (0.0, 0.75). Click the left mouse button to enter the vertex 7 of Figure 7.1.

Vertexes 1 thru 7 are sequentially connected, but they still have not formed the polygon. To form a polygon, we need to connect vertices 1 and 7. We can enter the last vertex at the same location as vertex 1 to close it. However, there are other ways.

Step 18 Select **Input->Form Polygon** command. The polygon is formed and colored. The color of the polygon is the same as the color strip labeled with z=0.1 in the layer window. It means that the polygon is on the layer with a vertical coordinate at z=0.1mm. The final result is shown in Figure 7.11.

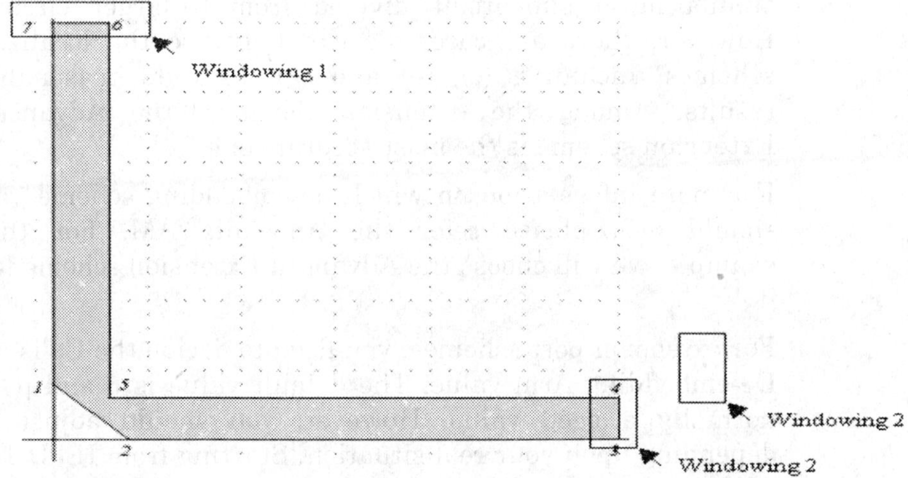

Figure 7.11 The entered polygon of the bend (vertex labels are no on actual MGRID window).

Section 7.5 Defining Ports and the Advanced Extension Scheme

Construction of the bend is finished at Step 18. Now we need to define the ports. The simulation engine will not run without any port (or plane wave excitation) on a structure. A port is defined on polygon edges. There are multiple ways to define a port. The Port->Define Port command allows you to select an edge by clicking on it. The Port->Port for Edge Group command allows you to select a group of edges by windowing them. The Port for Edge Group command is more convenient even though there are cases the Define Port is a better choice. No matter which command you use, the end result should be the same if you select the same edges.

Step 19 Select **Port->Port for Edge Group** command. MGRID will prompt for the de-embedding scheme to be used (see Figure 7.12).

Every port is tied with a de-embedding scheme on IE3D. Different port may need different de-embedding scheme. One de-embedding scheme can only solve some kind of problems. To make the IE3D more flexible, we implemented total 7 de-embedding schemes: (a) **Advanced Extension**; (b) **Extension for MMIC**; (c) **Localized for MMIC**; (d) **Extension for Waves**; (e) **Vertical Localized**; (f) **50-Ohms for Waves**; (g) **Horizontal Localized**.

The extension de-embedding schemes, including (a), (b), (d) and (f), normally are the most accurate ones when we are simulating a sub-circuit divided from a larger circuit. However, there are cases we can only use the localized schemes, including (c), (e) and (g), to yield reasonable results. Among the extension schemes, the Advanced Extension scheme is the most accurate one.

For more information on which de-embedding scheme you should use, please refer the Appendix AM. For this example, we will choose the Advanced Extension scheme for it.

For extension port schemes, you need to decide the Cells on De-embedding Arm value. The default value is 5 and it is normally a good value. However, you should adjust it depending upon your real situation. Starting from IE3D 11, we have implemented the Auto Adjustment which makes it even easier for you to decide. The parameters for the General Extension Port Parameters section in Figure 7.12 are good values. In IE3D 14, we have implemented accurate extraction of s-parameters with shift of reference planes for coupled ports and differential ports.

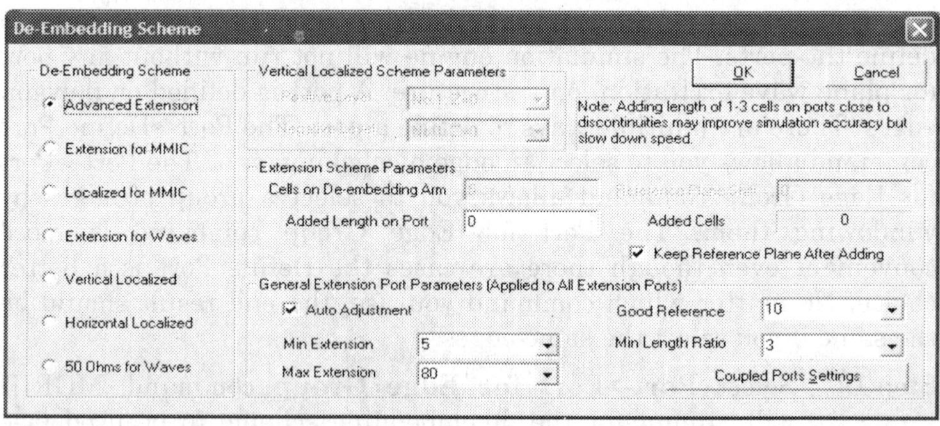

Figure 7.12 The De-Embedding Scheme dialog.

Step 20 Select **Advanced Extension** scheme. Check Auto Adjustment. Select OK to continue.

MGRID will be waiting for you to select the edges for defining a port. Please pay attention to the Status Window and the Layer Window. The Status Window is no longer display the mouse coordinates. It is displaying a message "Port for Edge Group: Window edge group on a vertical plane..." The Status Window basically tells what mode you are in and give you a hint what you should do.

On the Layer Window, you may notice that there is a check box for each layer. Depending upon the Default Selection Focus setting in Parameters->Optional Parameters dialog, you may have different default states for the check boxes on each layer.

Please make sure the check box for No.2: Z = 0.1 is checked (see Figure 7.9). The reason we want the Z = 0.1 layer to be checked is the following: When we window some edges to define the edges of the polygons as a port, only the edges on the checked layer will be selected. If some vertex of an edge is on a layer not checked, the edge will not be selected even it is being windowed.

If you are an old IE3D user, you may notice that the polygon where mouse is pointing it while you are moving the mouse is highlighted. This is a new feature on IE3D 14. You are able to access the highlighted object (polygon, port or text) easily on IE3D 14 or later.

Step 21 Move the mouse to somewhere on the upper left corner of vertex 7 of Figure 7.12. Press down the left button. Drag the mouse to the lower right corner of vertex 6 while keeping the left button pressed. Release the left button at the lower right corner of vertex 6.

The above action is called windowing. You just defined the windowing 1 in Figure 7.12. The port 1 is defined on the edge formed by vertex 6 and 7 (see Figure 7.13a). A small rectangle with "1" is displayed on the edge by vertices 6 and 7. You will see 2 gray color lines extending from the strip. The gray color lines represent the port extension. The length of the gray color lines defines how long a fictitious section of transmission line will be attached on the port location in an actual simulation. You should be careful not to let other polygons to come across this port extension. Otherwise, you may get unreasonable results.

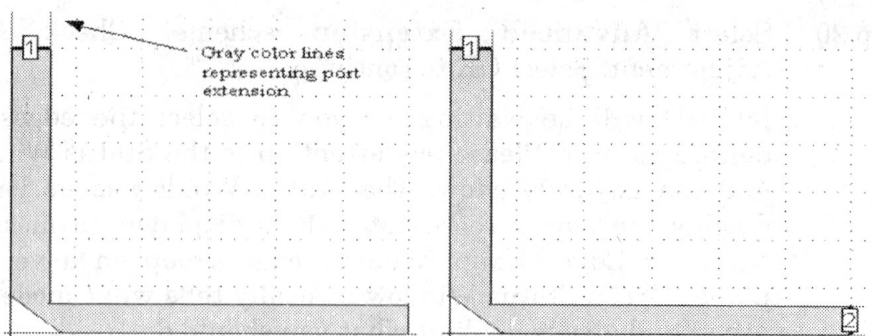

(a) The structure with port 1 defined (b) The structure with both ports defined.

Figure 7.13a The structure with ports defined.

Step 22 MGRID is still in **Port->Port for Edge Group** mode. Please move the mouse to upper left corner of vertex 4 and window the vertex 4 and vertex 3 (see Figure7.12). After you release the left mouse button, the port 2 is defined. The structure will look like what is shown in Figure 7.13b.

Step 23 MGRID is still in the **Port->Port for Edge Group** mode. Select **Port->Exit Port** command. MGRID will resume to the drawing mode. You can also select the button on the toolbar for Port for Edge Group mode to release it from the mode.

Step 24 Select **File->save** command, and enter "**.\ie3d\practice\c_bend.geo**" for the geometry file name.

You can always intermediate result anytime. It is always recommended to save changes frequently when constructing large structure in order to avoid accidental loss of data.

Section 7.6 Meshing Structure on MGRID for Visualization

We have defined the polygons representing the strip and we have defined the ports of the structure. It is ready for simulation. We can select **Process->Simulate** command to simulate the structure now. However, you may be curious about how it looks like after a structure is meshed. We will demonstrate how to mesh the structure. The meshing on MGRID is just for display only, and its result is not used in the IE3D engine. The meshing is always performed on the IE3D simulation engine when you start a simulation.

Step 25 Select **Processes->Display Meshing**. MGRID will prompt you for the Automatic Meshing Parameters. Meshing Parameters are defined in the Basic Parameters dialog. However, you can change them when you perform a meshing or a simulation setup.

Statistics of Meshed Structure

Total Cells	25
Rectangles	22
Triangles	3
Sides	73
Volumes	0
Volumes Surfaces	0
Total Unknowns	26
2D Cells	25
3D Cells	0
Minimum Cell Size	0.0237171
Maximum Line Size	0.142147
Maximum Cell Size	0.16072
FASTA Factor	N/A

☐ Convert Meshed Cells into Polygons
☐ Find Nodes of Shorted Edge
☑ Update Meshing Parameters of Structure

Comment on RAM Requiremnts:

The structure may require about 0.01 to 0.01 MB RAM to run using SMSa solver.

Continue

Figure 7.14 The meshing results in tabular format.

Step 26 Please choose Contemporary meshing scheme. It is a new scheme implemented in IE3D 11. It is supposed to create less cells and more regular edge cells for higher efficiency and higher accuracy.

Remember to select AEC Layer = No to disable the Automatic Edge Cells, which we will discuss later. Select **OK** to accept the default values. A pop-up window comes up to indicate the meshing process. After the process, MGRID will prompt you with the dialog "Statistics on Meshed Structure (see Figure 7.14). Only 25 cells are created in the meshing process with Maximum Line Size = 0.142 mm and Maximum Cell Size = 0.1607 mm. Maximum Line Size is the maximum size of any adjacent vertices of all the cells. Maximum Cell Size is the maximum distance between any two vertices of a cell for all the cells. You may see parameters such as "Volumes" and "Volume Surfaces". Those are for the finite dielectrics. There are no finite dielectrics defined in this structure.

The meshed cell density is determined by the product of the Fmax and Ncell, as well as the AEC. If you need to change the cell density, you can change either Fmax or Ncell for the same result as long as the product of Fmax and Ncell is the same. If you choose a too big Fmax or Ncell such that the structure becomes too big and the estimated minimum cell number exceeds the Warning Limit, MGRID will issue a warning on it to prevent the case of endless meshing due to large number of cells created. The Warning Limit can be changed in **Parameters-> Optional Parameters** dialog.

There are additional information and options available on the IE3D 14. It indicates "The structure may require about

0.01 to 0.01 MB RAM to run using SMSa solver. This is a tiny structure for IE3D because the major data size is the matrix and it is about 0.01 MB. The SMSa solver is the default matrix solver for IE3D. It is very efficient and accurate and it makes use of multi-CPUs for multi-core systems.

On the dialog, you also see an option "Update Meshing Parameters of the Structure". For the current structure or any un-simulated structure, the default for this option is checked. If you select CONTINUE, MGRID will update the meshing parameters with the ones defined in the Meshing command. If you uncheck it and select OK, MGRID will display you the meshing results only. It will not change the meshing parameters of the structure. Starting from IE3D 12, the simulation results or the s-parameters are saved into the current geometry file. Changing of the geometry including the meshing parameters will cause removal of the simulation results. On IE3D 12, users can't perform meshing again after a structure is saved. Otherwise, the results will be lost in the geometry. If you save the geometry file after meshing, the simulation results will be lost forever. If you change the geometry and you don't want to lose the results, you should not save the changed geometry. It is inconvenient, even it is not impossible, to check the meshing of a geometry after a simulation. On IE3D 14, we make it easier. You can still perform a meshing with different meshing parameters. If there are results in the current geometry, the default for the option "Update Meshing Parameters of the Structure" will be un-checked. After you select CONTINUE, the meshing para-meters of the structure will not be updated if you leave the option unchecked.

Step 27 Select **Continue** button in the **Statistics of Meshed Structure** dialog. MGRID will show you the meshed structure in a separate 3D Meshing View window. IE3D 14 no longer displays the meshing on the Polygon Editor window. The menu system of MGRID will be changed if you select the 3D Meshing View window. IE3D 14 adopts a new 3D view with many powerful commands and features (in Manipulate, Options and Windows menu). For example, you can switch between Perspective View and Orthographic View easily by selecting the corresponding menu items in the Options dialog. You see the infinite substrates are displayed as some frames on the 3D view window. The size of the frames is defined as the Display Margin in the

Substrates section of Basic Parameters or the menu item Parameters->Substrate Display Parameters or Options->Structure Display Parameters->Substrate Display Parameters. On IE3D 14, the ports are also displayed on the 3D view window. The handling of 3D view may be different too. For example, to zoom on the 3D view, you can just slide the mouse wheel. To pan the 3D view, you press down SHIFT while you move the mouse. In the earlier versions, you hold down the right mouse button while you move the mouse. Clicking right mouse button is reserved for bringing up a popup menu for other commands on IE3D 14. To zoom a portion of the structure, you press down CTRL while you window the portion.

As you can see, rectangular cells are used in the regular region while triangular cells are used in the irregular region. The efficiency of rectangular cells and the flexibility of triangular cells are smartly combined in IE3D.

To get a hard copy of the meshed geometry, you can select **File->Print** command while the 3D Meshing View is the active window. To get a snap shot of the top view, please select File->Save To Bitmap Files command. MGRID will allow you to save the picture into a bitmap file. You can also select ALT+ PrtScrn buttons to capture the MGRID window or the 3D Meshing View window into the clipboard. Then, you can paste it onto the MS-Paint program in the Accessory group or other graphic application programs.

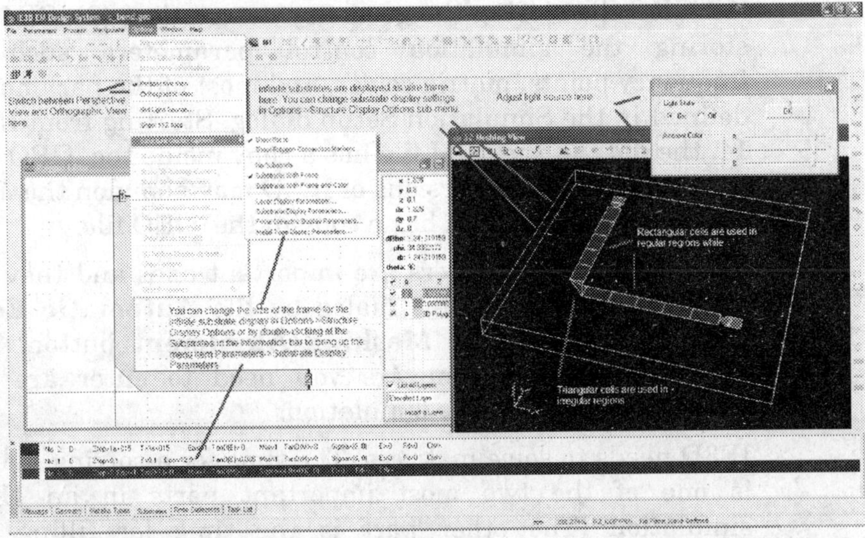

Figure 7.15 The meshed geometry displayed on 3D Meshing View and the different options.

Section 7.7 Electromagnetic Simulations and Adaptive Intelli-Fit (AIF)

We have just demonstrated how to view the meshed structure on MGRID. As it is mentioned earlier, meshing on MGRID is not necessary for a simulation. It is just for display only. We can perform an EM simulation without the meshing process on MGRID. We will show how to setup a simulation in this section.

Step 28 Select the Polygon Editor window and you will see the menu system switch back to the one for editing. Select **Process->Simulate** command. MGRID prompts you for the Simulation Setup dialog (see Figure 7.16). This is a shared dialog for Simulation Setup and Optimization Setup. Some controls for EM tuning and EM optimization may be disabled because MGRID detects no optimization variables are defined. The Simulation Setup dialog allows you to define and change Meshing Parameters, Accuracy Control Parameters and the simulation frequency points.

There is a Capture button for you to capture the frequency points from a simulated result saved in the .sp (.spt or .s2p) files. There is another button called "Retrieve". It allows you to retrieve most of the data for the Simulation Setup dialog from some previous created simulation input file with extension .SIM. The data retrieved including the frequency points and optimization objectives (for optimization), etc. The .SIM file is created after a simulation setup process. There are two files involved in an IE3D simulation: (1) .GEO file storing the geometry description; (2) .SIM file storing the simulation control parameters such as frequency points, matrix solver, and most of the parameters defined in the Simulation Setup dialog. Starting from IE3D 14, the data in the .SIM file has a copy inside the .GEO file. It is possible future versions of IE3D may abandon the .SIM file because all its data has a copy in the .GEO file.

The Meshing Parameters are important ones and they are also accessible from this dialog by the **Automatic Edge Cells** button and the **Meshing Alignment** button. The most important parameters you need to enter are the frequency points for the simulation.

IE3D needs to solve matrices. The time for a solving matrix is one of the two most important parts in an IE3D simulation (The other part is the time for filling the matrix). We have implemented multiple matrix solvers for different purposes. In IE3D 11, the default matrix solver is the Adv. Symmetric Matrix Solver (SMSi). It does not support multi-thread and multi-CPUs. We have implemented the Adaptive Symmetric Matrix Solver (SMSa). SMSa can utilize multiple threads and multiple-CPUs. It can be more than 4 times faster than SMSi for large structures. However, SMSa may take about twice as much memory as SMSi. An adaptive scheme is implemented into SMSa. In case no enough RAM is available, SMSa automatically switches to the SMSi scheme. The default matrix solver is defined in the **Parameters->Optional Parameters** command.

In most cases, you are interested in the frequency responses of the s- (y-, z-) parameters. If you are simulating the structure with many frequency points, you should always enable the Adaptive Intelli-Fit (AIF). The AIF is an extremely efficient and robust way to get fast and accurate frequency response. AIF can adaptively select some of the defined simulation frequency points, simulate them and extrapolate at other frequency points from the solved frequency points. It will automatically detect convergence. Normally, it can save 90% of the simulation time with guaranteed accuracy at the extrapolated frequency points. You can feel safe to use it without worrying about whether the extrapolated data are accurate.

In case you need to save the current distribution for visualization and/or the radiation pattern, you can check the **Current Distribution File (.Cur)** and **Radiation Pattern File (.Pat)**. It will save the current distribution and/or radiation pattern at the simulated frequency points. When AIF is enabled, you will not know which frequency points will be simulated (except the Start, Mid and End points). In case you want to make sure some specific frequency points will be simulated with AIF enabled, you can check the check boxes before the specific frequency points in the list box of Frequency Parameters.

In IE3D 12, you have the option "Save S-Parameters into Fast EM Data". This option allows IE3D to save the s-parameters into the geometry file (.geo) in addition to the s-parameter file (.sp). You can do integrated s-parameter visualization and processing on MGRID 14. Automatic lumped element extraction is also implemented into IE3D 12. By default, "No Lumped Model Output" is selected. If you select it, IE3D will output the frequency dependent lumped element model for you automatically.

Traditionally, IE3D s-parameter results are saved into a Touchstone compatible file (.sp), and they are displayed and processed on MODUA, a bundled circuit simulator. On IE3D 14, you are able to visualize and process s-parameters on MGRID. Most of the s-parameter processing functionalities of MODUA are integrated into MGRID, while the circuit simulation part is not. Circuit simulation is better implemented on MDSPICE, another product of Zeland. MDSPICE will be eventually integrated with IE3D to cover the full functionalities.

If you want to display the s- (y-, z-) parameters immediately after simulation without selection, you can select the Define Graphs button on Simulation Setup dialog (Figure 7.16) to define some graphs. Then, after simulation, the results will be automatically display. If you select Invoke MODUA in Post Processing, MODUA will also be invoked to display the s-parameters saved into the .sp file.

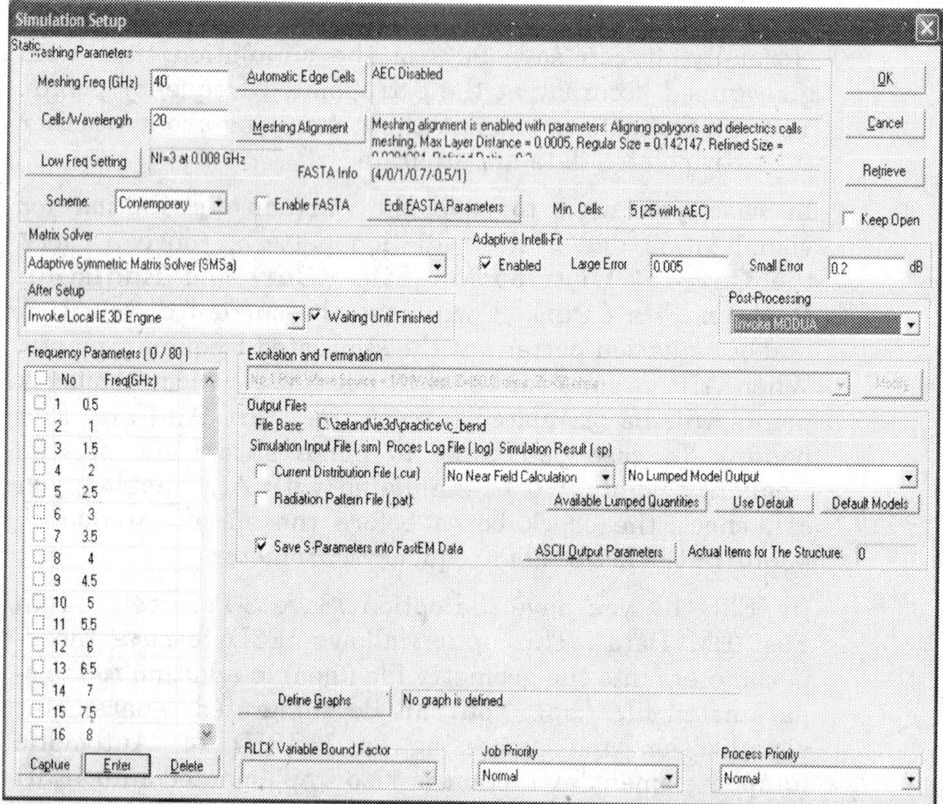

Figure 7.16 The Simulation Setup dialog with frequency points defined.

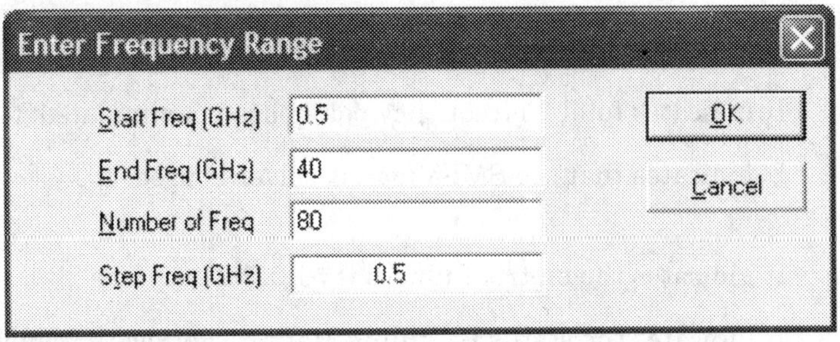

Figure 7.17 The Enter Frequency Range dialog.

Step 29 Please select the Enter button in the **Frequency Parameters** section. MGRID will prompt you for the **Enter Frequency Range** dialog (Figure 7.17). Please enter Start Freq = 0.5 GHz, End Freq = 40 GHz and Number of Freq = 80. MGRID will automatically calculate the Step Freq = 0.5 GHz. Select OK to continue. You will get the exact frequency points in the **Frequency Parameters** list box. You are allowed to enter multiple frequency ranges. MGRID will automatically sort the frequency points for you.

Step 30 Please select Invoke Local IE3D Engine in the After Setup combo box and check **Wait until Finished**. Starting from IE3D 12.1, you have more choices on how you want MGRID to perform an IE3D simulation. The different choices are documented in Table 7.1.

Step 31 Select OK to continue. MGRID will invoke the IE3D engine in the background to perform the simulation while the progress is monitored on the IE3D dialog.

IE3D dialog will display the intermediate simulation information (Figure 7.18). If you want to see whether the IE3DOS.EXE is running, you can activate Task Manager of Windows and find the IE3DOS.EXE from the Processes tab. You can see how much memory and how much CPU resource it is using.

Table 7.1 The explanation on some of the intermediate parameters in the IE3D dialog in Figure 7.18.

Item	Description
Information	This item describes that the structure has 2-ports and it is meshed into 25 cells and 26 unknowns. Minimum 0 MB RAM is required for the SMSa matrix solver.
Total Freq Points	It indicates total 80 frequency points are required (Not all of them will be simulated if AIF is enabled).

Finished	It indicates 5 frequency points are finished.
Remained	It indicates total 75 frequency points are not simulated.
Matrix Solver	It indicates that the SMSa matrix solver is used.
Fullness	It indicates the matrix fullness is 51.92%.
Status	It indicates the status is "Filling Matrix". For every frequency point, it will go through "Filling Matrix" and "Solving Matrix" processes. IE3D dialog will display the progress of "Filling Matrix". It cannot display the progress of "Solving Matrix".
Iteration	This is not useful for the SMSa matrix solver. It is only meaningful for iterative matrix solvers.
Comment	It displays "LVres=0.046 (0.005) at 33 GHz and SVres = 10.2 (0.2) at 36 GHz". It is the AIF convergence information. It means that the current maximum large value error for AIF is 0.046 at 33 GHz and the convergence criteria are 0.005. The current maximum small value error is 10.2 at 36 GHz and the convergence criteria are 0.2. The criteria "0.005" and "0.2" are the "Large Error" and "Small Error" defined in the AIF section of the Simulation Setup dialog (see Figure 3.16). The default values are fully tested and you do not need to change them. The LVres and SVres values are for AIF convergence only. They are not related to the accuracy of the simulation.

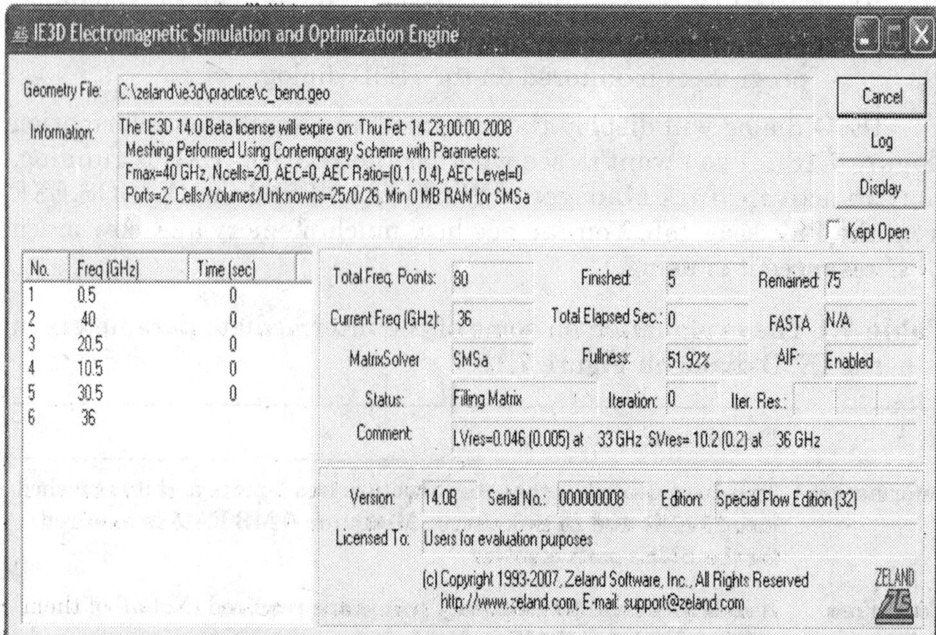

Figure 7.18 The IE3D dialogs displaying the intermediate simulation data.

For this particular example, the whole IE3D engine may take less than 1 second. You may not be able to see how the IE3D dialog looks like and see the IE3DOS.EXE information from the Task Manager of Windows before it finishes.

A snapshot of the IE3D dialog in Figure 7.18 is captured in a very short time. As you can see, 5 frequency points are done while it takes less than 1 second. There are many parameters displayed in the dialog. Their meanings are explained in Table 7.1.

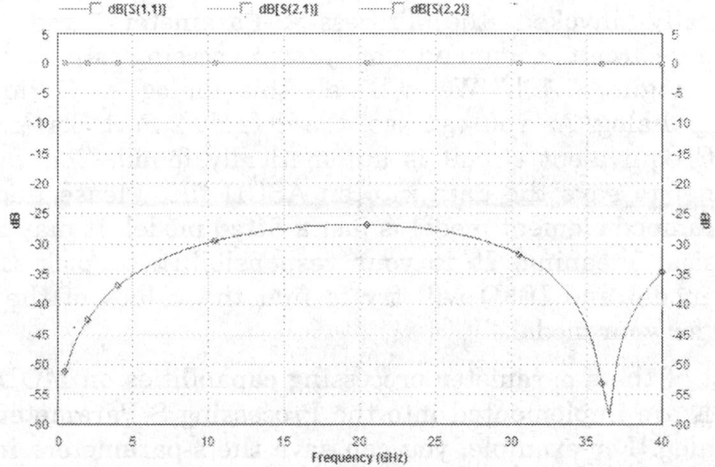

(a) S-parameters displayed on MODUA.

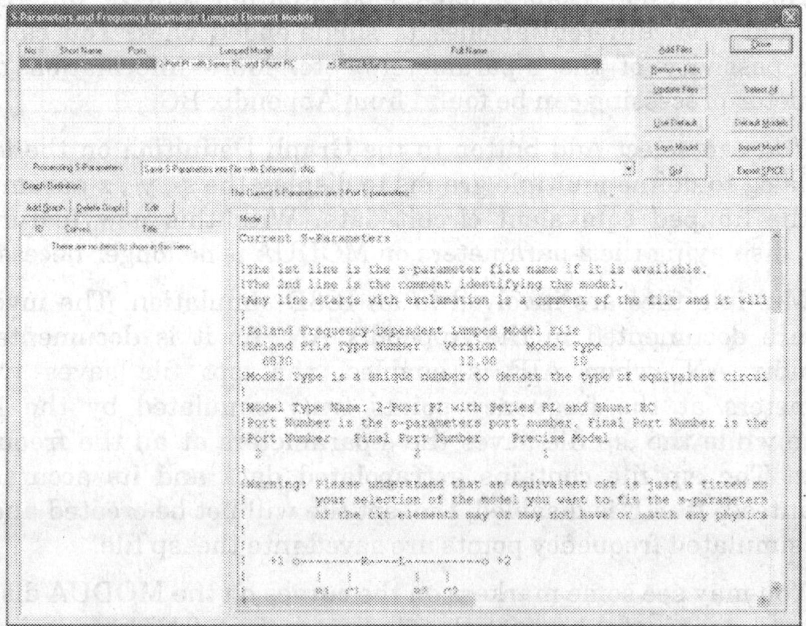

(b) S-parameters displayed on MGRID.

Figure 7.19 The simulated s-parameters of the microstrip bend structure.

After the simulation, IE3D will save the s-parameters into .\ie3d\practice\output\c_bend.sp & .\ie3d\practice\output\ c_bend. spt files. Because you selected "Invoke MODUA" in Post-Processing in the Simulation Setup dialog, it will invoke the MODUA to display the s-parameters (Figure 7.19a). Starting from IE3D 14, it is not necessary to invoke MODUA for processing s-parameters. You can do it on MGRID or IE3D EM Design System internally. You have checked Save S-Parameters into Fast EM Data in the Simulation Setup dialog, the s-parameters is also saved into the .geo file. It automatically invoked the Process->S-Parameters and Lumped Equivalent Circuit command for post-processing s-parameter as shown in Figure 7.19b. We will call this dialog as S-Parameters Processing dialog. As you can see, the default 2-Port Series RL and Shunt RC equivalent circuit is automatically found. You can select Save Model to save the data into an ASCII file. Please understand that the lumped element model is just a fitted model. It may not have any physical meaning. It is your responsibility to pick the right physical model and IE3D will try to find the values of the lumped elements for your model.

Most of the s-parameter processing capabilities on MODUA and MDSPICE are implemented into the Processing S-Parameter section in the dialog. For example, you can save the s-parameters into .sNp format with the default file extension for Touchstone format. You can perform conversion from single-ended s-parameters to differential ones, and from differential ones to single-ended ones. You can also check passivity of the s-parameters, etc. More information on s-parameter processing can be found from Appendix BG.

You can select Add button in the Graph Definition on the left of the dialog to define multiple graphs to display the s-, y-, z-parameters and the lumped equivalent circuit data. With this new feature on IE3D, displaying the s-parameters on MODUA is no longer necessary.

Multiple files are involved in an IE3D simulation. The involved files are documented in the Appendix AN. As it is documented in Appendix AN, when AIF is enabled, the .spt file saves the s-parameters at the frequency points truly simulated by the IE3D engine while the .sp file saves the s-parameters at all the frequency points. The .sp file contains extrapolated data and its accuracy is guaranteed. If AIF is disabled, the .spt file will not be created and the truly simulated frequency points are saved into the .sp file.

You may see some markers on the curves on the MODUA display. They are meaningful because they indicate which frequency point has true EM simulation from IE3D. Other frequency points are accurately obtained from the simulated frequency points using some extremely robust interpolation scheme with AIF. For this particular example, the EM simulation is actually performed at 7 frequency points (0.5, 2, 4, 10.5, 20.5, 30.5, 36 and 40 GHz) and the results are in the .spt file. All the other frequency points are interpolated data with guaranteed accuracy saved in the .sp file and the .geo file.

You can use MODUA to display S, Y and Z-parameters in different forms (Cartesian graphs, data lists, and Smith Chart). Please read Appendix AO on using the MODUA for different display. Basically, you can select Control->Define Display Data, Control->Define Display Graph and Control->Define Display Smith Chart to display S, Y and Z-parameters in different form after you have simulated a circuit or displayed an s-parameter on MODUA. For example, while we have MODUA displaying the dB[S(1,1)] , dB[S(2,1) and dB[S(2,2)], please select Control->Define Display Graph command on MODUA. Please select "dB and Phase of S-Parameters" in "Display Parameters" and select OK. Uncheck "dB[S(2,2)]". Check "Ang[S (1,1)]" and "Ang[S(2,1)]". Make sure only 4 items ("dB[S(1,1)]", "dB[S(2,1)]", "Ang[S(1,1)]" and "Ang[S(2,1)]" are selected and elect OK. MODUA will display the frequency responses of the 4 items as shown in Figure 7.20.

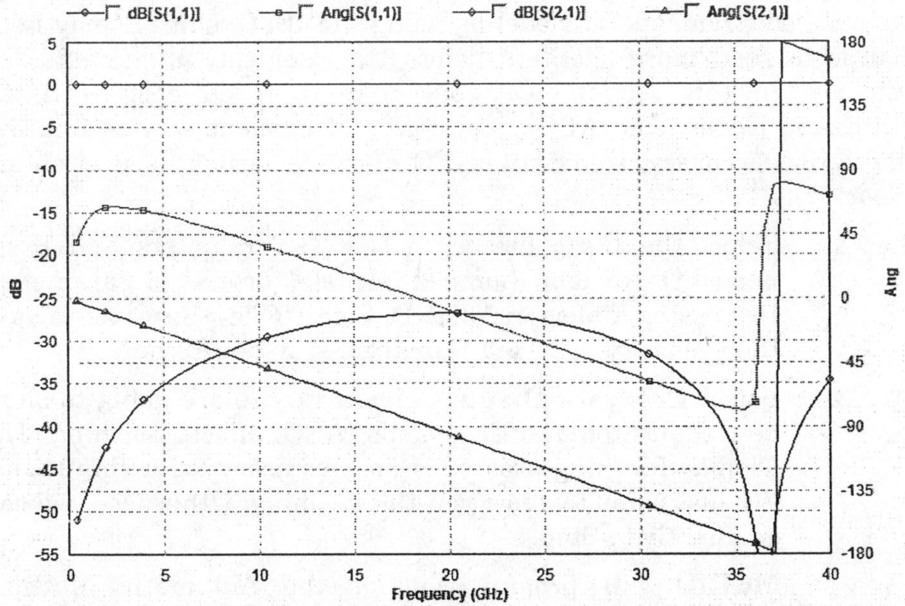

Figure 7.20 The dB and Ang of different s-parameter items displayed on MODUA.

You can do much display and analysis on the s-parameters on MODUA. For example, you can define markers and find the exact value of an item at a specified frequency.

Starting from IE3D 14, it may not be necessary to display and processing the s-parameters on MODUA. You may be able to do a better job on MGRID.

Section 7.8 High Accuracy Simulation Using Automatic Edge Cells (AEC)

Are the results displayed in Figure 7.20 accurate? The answer can be yes or no depending upon your requirements for accuracy.

We have claimed that the AIF scheme yields "accurate" result over a wide bandwidth by just simulating a few frequency points. However, when we say "accurate", we do not mean that we have proved the result is accurate compared to the measured data or the converged data. When we say the AIF provides "accurate results", we just mean that the result of AIF will perfectly match the true simulation results under the same condition (same meshing and same accuracy control parameters). Whether a simulation is accurate or not, we need to do many validations by comparison to measured data or converged data. In the past 15 years, we have been validating the IE3D by comparing its results to measured data or by performing convergence studies. We have come up with different rules for accurate electromagnetic modeling using the IE3D. since it may need a user to read more chapters before the user can understand the schemes. In this section, we would like to touch base with the Automatic Edge Cells (AEC), a simple and efficient way to improve simulation accuracy. Generally, AEC allows a novice user to obtain expert results.

Step 1 Select the Close button in the "S-Parameters and Frequency Dependent Lumped Models" or the S-Parameters Processing dialog on MGRID. Select **File->Save as** to save the geometry as "**.\ie3d\practice\ c_bend3.geo**".

Step 2 Select **Process->Display Meshing**. We are going to show how the Automatic Edge Cells (AEC) affects meshing. The Display Meshing process is not necessary for a simulation. We just want to show you the meshing. Otherwise, we can go directly to Step 4.

MGRID will prompt you for the Automatic Meshing Parameters.

Step 3 Select **AEC Layer** = 1. Make sure the **AEC Ratio = 0.1**. MGRID will automatically calculate the AEC Width = 0.0142147 mm (Figure 7.22). Again, starting from IE3D 11, we allow you to enter the AEC Ratio instead of the AEC Width. In this way, the AEC Width will be in normally range and it will avoid accidentally problem caused by a user forgetting to define the correct AEC Width value after the Fmax and Ncell are changed.

Select OK to continue. MGRID will prompt you the Statistics on Meshed Structure. It indicates that total 81 cells and 134 unknowns are created. Select CONTINUE button. MGRID will display the meshed structure (Figure 7.21b). You can see some narrow cells are created along the edges. You can repeat the Process->Display Meshing multiple times with different AEC Layer selections; you will get different meshing results. When you choose different AEC Layer value, please choose the Multi-Layer Ratio = 0.4. Normally, you should not change this value.

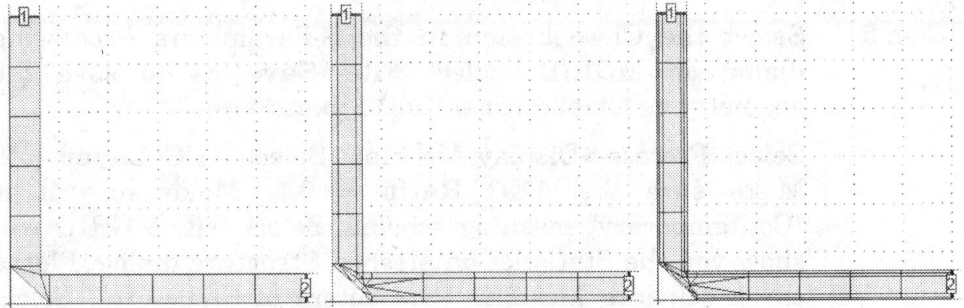

(a) AEC Layer = No (b) AEC Layer=1, AEC Ratio=0.1 (c) AEC Layer = 2, AEC Ratio=0.1

Figure 7.21 The meshed structure with different AEC Layer values.

When you get the Statistics on Meshed Structure, please pay attention to the option "Update Meshing Parameters of Structure". The default for this time is un-checked. Please remember that the default was checked when we do it the 1st time. Why is it un-checked this time? The difference is that there was no s-parameters in the .geo the 1st time we do the meshing after we created the geometry. This time, we have simulated the structure and the s-parameters are saved in the geometry. If we change the parameters of the structure (including the shapes and the meshing parameters), the results will be lost. However, we can still do meshing to check how the meshed structure looks like with other meshing parameters. By default, we will not update the meshing parameters. Please leave it un-checked.

Step 4 Select **Process->Simulate**. If you check the Automatic Edge Cells section, you will see AEC is not enabled because we did not update the meshing parameters in the last step by checking the "Update Meshing Parameters of Structure".

Please select Automatic Edge Cells button to bring out the Meshing Parameters dialog. Please select AEC Layer = 1. Select OK to go back to Simulation Setup dialog. Select OK again. MGRID will warn you that s-parameters results are available and re-simulation to cause removal of the existing s-parameters results. Please select YES to continue because we don't need the old results. It will take 1-2 seconds to finish the simulation. MODUA is invoked to display the results and the "S-Parameters and Frequency..." dialog on MGRID is also invoked. Comparing the new results with the old one and you will see the dB[S(1,1)] and Ang[S(1,1)] are different while the S(2,1) items are almost the same. |S(1,1)| is a small quantity. Its values are more affected by simulation accuracy.

Step 5 Select the Close button in the S-Parameters Processing dialog on MGRID. Select **File->Save as** to save the geometry as **".\ie3d\practice\c_bend5.geo"**.

Select Process->Display Meshing. Select **AEC Layer** = 2. Make sure the **AEC Ratio = 0.1**. Make sure it is "Contemporary" meshing scheme. Select OK. MGRID will show you the Statistics on Meshed Structure dialog. Please check "Update Meshing Parameters of Structure". Select CONTI-NUE button. This time, MGRID warns you that updating the meshing parameters will cause removal of the s-parameters. Select YES to continue. We will get what is shown in Figure 7.21c. As you can see, the strip is divided into 5 cells with 2 narrow edge cells on each side. Total 139 cells and 247 unknowns are created.

Step 6 Select **Process->Simulate**. The Simulation Setup dialog comes up and you will see the AEC is enabled with **AEC Layer** = 2 because we have updated the meshing parameters in the last step. Please select OK to continue. MGRID will invoke the IE3D engine internally to do the simulation. It takes 2-3 seconds to finish it. The results will be saved in: and .\ie3d\practice\output\c_bend5.sp.

Section 7.9 S-Parameter Visualization and Processing

Traditionally, s-parameters visualization and process are on the MODUA. You also perform circuit simulation on a circuit consisting of s-parameter black boxes and lumped elements such R, L, C and mutual L on MODUA. On IE3D 14, we have implemented more s-parameter visualization and processing capabilities into MGRID. MODUA is no longer the default visualization tool even though we still need to perform circuit simulations and mixed EM and circuit simulations on MODUA.

We will discuss how we can use MGRID for s-parameter visualization and processing. With s-parameters processing integrated into MGRID, we normally don't need MODUA for s-parameter processing. By default, we suggest users to select "No Display" for Post-Processing. In case, you want to display the s-parameters immediately after simulation, you should select Add Graphs in Simulation Setup to pre-define some graphs before simulation starts. Certainly, we can always define graphs after simulations.

Step 1 Run MGRID and open file: ".\ie3d\practice\c_bend.geo".

It is a simulated structure while the results are automatically saved into the .geo file. Please note that any change to the structure and meshing parameters will cause removal of the s-parameters in the data. When you do some potential change, MGRID will warn you anyway. In case you change it and the s-parameter data is lost, you will not be able to access the data from MGRID. However, it is still available in the file if you do not save the change into the file.

Step 2 Select Process->S-Parameters and Lumped Equivalent Circuit command or Window->S-Parameters Display->Define Plots (Note: We provide the commands with the same functionalities for your convenience). The "S-parameters and Frequency Dependent Lumped Element Models" or S-Parameters Processing dialog comes out (see Figures 7.22 or 7.19b).

Please pay attention to the message at the middle right of the dialog "The Current S-Parameters are 2-port s-parameters". It is for the selected item or the only No.0 item in the list box. You are allowed to define the graphs in Simulation Setup dialog or in Window->S-Parameters Display->Define Plots even though the s-parameters may not be available. In such a case, MGRID will indicate to you that the s-parameters are fake ones for defining the graphs only in this message.

Step 3 Please double click at the Short Name section of the No.0 item in the list box to change its Short Name from "" to "AEC Layers = 0" (see Figure 7.22).

Step 4 Select "Add Files" button on the right. Please select both c_bend3.sp and c_bend5.sp in the directory: \ie3d\practice\output by holding down the Ctrl key. The two files will be added into the list box in the dialog. MGRID will automatically detect the strings "c_bend3" and "c_bend5" for their short names, respectively. Please double click at the Short Name section of the two items to change them to "AEC Layers = 1" and "AEC Layers = 2".

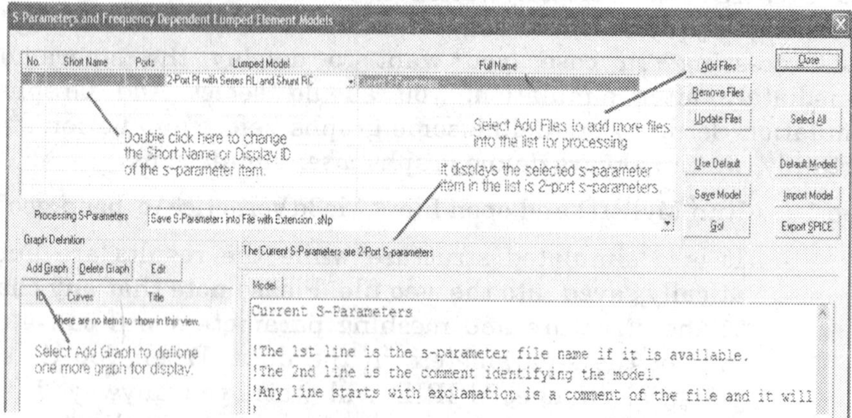

Figure 7.22 The S-Parameters Processing dialog.

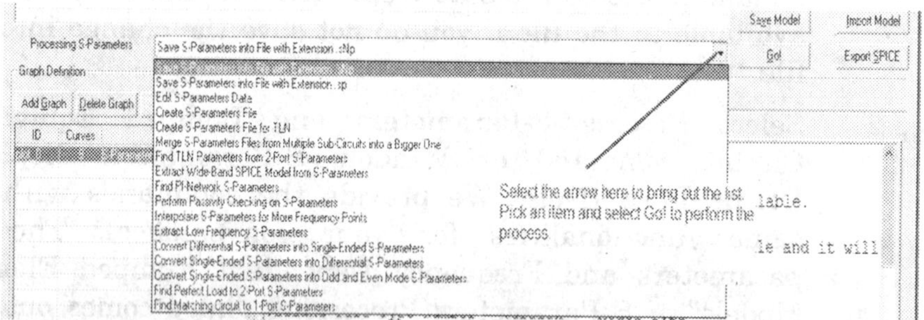

Figure 7.23 The S-Parameters Processing dialog for s-parameter processing.

Step 5 Select the No.0 S-parameters item in the top list. Select the downward button on the right for the combo box in Processing S-Parameters (see Figures 7.22 and 7.23). It will bring out a list of possible processing features for the selected s-parameters or others.

Step 6 Select "Perform Passivity Checking on S-Parameters" from the list. Select "Go!" button. MGRID will perform the passivity checking for you and it will create a dialog for the report (see Figure 7.24). It tells you whether the s-parameters data is passive at a specific frequency in the list. When the "Minimum Value" is non-negative, the s-parameters are passive at the frequency. For this particular example, MGRID detect the circuit passive for all the 80 frequency points. You can select Save Report to save the data into an ASCII file. In case, there are some non-passive points, you have the option to enforce passivity at the points. MGRID will try to adjust the s-parameters to enforce passivity. Certainly, it will change the s-parameters and the accuracy of the final results can't be guaranteed.

For a structure without active devices, it certainly should be passive. However, it is impossible to always guarantee an EM simulation result will always be passive even though non-passive results may happen occasionally. There are multiple reasons: (1) there are numerical errors involved. (2) There might be positive feedback loop between the source and your open structure and the positive feedback loop may cause non-passive s-parameters. There are more chances for such a case to happen especially for extension ports if the ports are not defined properly so that the port extensions, which are supposed to be not part of the circuit, are coupled with the structure. In fact, it is impossible to de-couple the port extensions and the structure. We just need to avoid the coupling is to be significant so that it will affect our results.

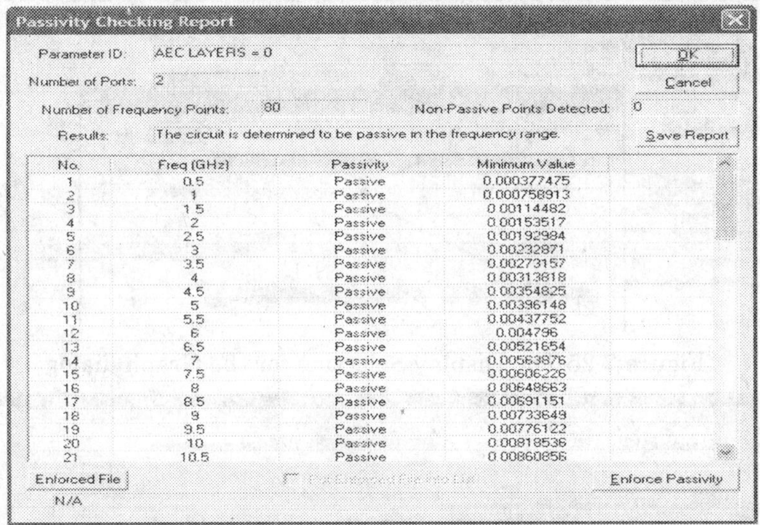

No.	Freq (GHz)	Passivity	Minimum Value
1	0.5	Passive	0.000377475
2	1	Passive	0.000758913
3	1.5	Passive	0.00114482
4	2	Passive	0.00153517
5	2.5	Passive	0.00192984
6	3	Passive	0.00232871
7	3.5	Passive	0.00273157
8	4	Passive	0.00313818
9	4.5	Passive	0.00354825
10	5	Passive	0.00396148
11	5.5	Passive	0.00437752
12	6	Passive	0.004796
13	6.5	Passive	0.00521654
14	7	Passive	0.00563876
15	7.5	Passive	0.00606226
16	8	Passive	0.00648663
17	8.5	Passive	0.00691151
18	9	Passive	0.00733649
19	9.5	Passive	0.00776122
20	10	Passive	0.00818536
21	10.5	Passive	0.00860856

Figure 7.24 The passivity checking results.

Step 7 Please select OK to close the passivity checking result dialog.

There are many s-parameters processing features in the list and you can explore them. There will be more implemented in the future.

Step 8 Select Add Graph on the left. MGRID will prompt you for the Graph Type. Please select S-Parameters from the list.

MGRID will set the default title of the graph as "S-Parameters Display" and you can change it to whatever you like. Select OK and MGRID will prompt you for the display selection (see Figure 7.25). Please click at the No.0 item in the list box "Root=Current S-Parameters". Check the items "dB[S (1, 1)]", "dB[S (2, 1)]" and "Ang[S (2, 1)]". The repeat the selections for the "c_bend3.sp" and "c_bend4.sp". Select OK to go back to the S-Parameters Processing dialog. The defined graph "plot_0" is listed in the list box. The preview window for "plot_0" is created as a tab page on the right. Click at the tab will show you the results. The s-parameters from the three files are compared (see Figure 7.26).

You can define more graphs if you like. Each additional graph will correspond to a tab page on the right. You certainly can check the data in the tab pages even though they are small. They are actually for your preview only and they don't contain the tables for the user defined markers for measurement.

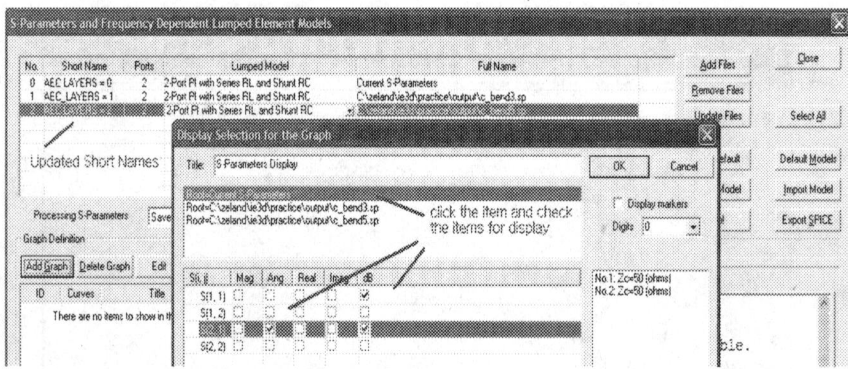

Figure 7.25 The Display Selection for the Graph dialog.

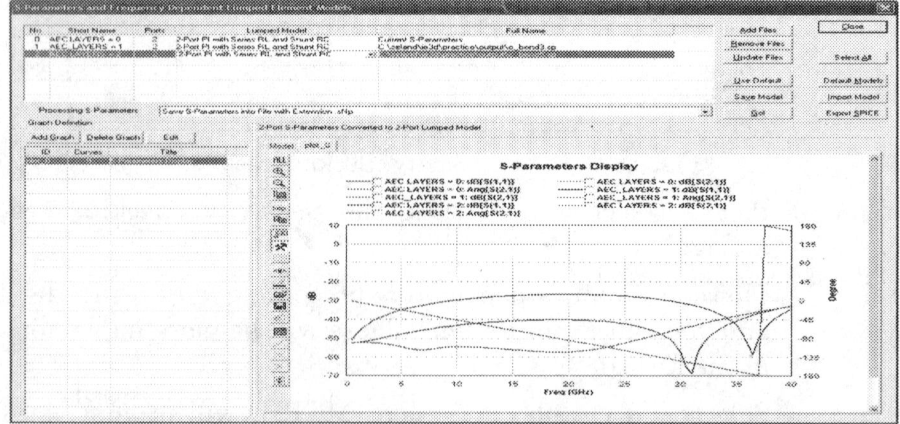

Figure 7.26 The defined s-parameters plot is previewed in the dialog.

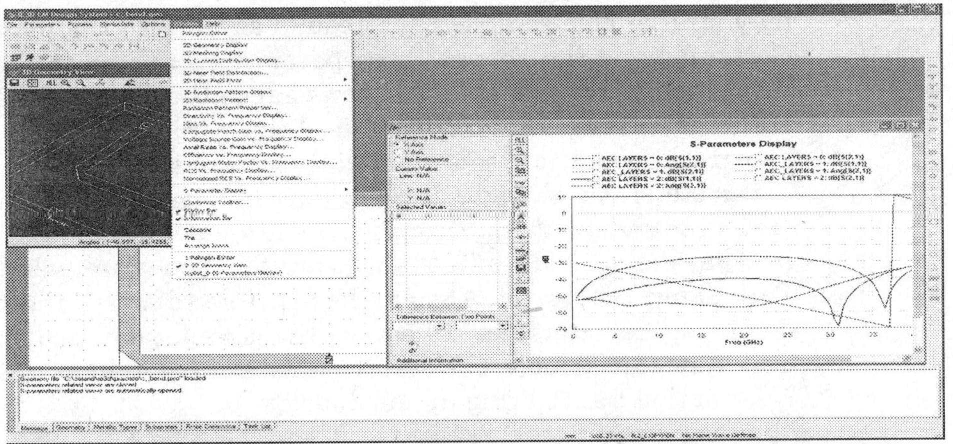

Figure 7.27 The defined plot is displayed on the MGRID main workspace.

Step 9 Select Continue button in the S-Parameters Processing dialog. The s-parameter plot is displayed in the MGRID workspace.

Old IE3D users may realize the big difference between IE3D 14 and older versions. The older MGRID is a SDI (Single-Document Interface). It can only show one window, and the user can open the 3D view. However, the 3D view is always on top of the drawing window and you will feel the 3D view window is in the way. We invoke MODUA to process the s-parameters, PATTERNVIEW to process radiation patterns, and another instance of MGRID to display the current and near field distributions. All the different windows from different application programs (MGRID, MODUA and PATTERNVIW) are overlapping each other and it causes confusion to users.

Starting from IE3D 14, we have integrated all the functionalities into MGRID. You can do all the different post-processing in the single interface MGRID or IE3D EM Design System. You have a big list of functionality in the Window menu of MGRID (Figure 7.27). For this particular case, we have not saved the current distribution data (.cur file), the radiation pattern (.mpa and .pat files) and the near fie- ld (.nff). We can't process them right now. If you have enabled them, you will be able to access the data from the menu.

Again, the simulation results are saved either in the .geo or other related files. You can't modify the parameters of current structure that may affect the simulation results. Otherwise, the results will be removed even though you may still be able to access the results saved into the files. Certainly, you can change those parameters related to the display only. For example, you can change the colors of the polygons layers, the colors and display margin of the substrate layers, etc. For changing the substrate display parameters, you should not go to Parameters->Basic Parameters to change them because there are many parameters in the dialog will affect the results. You can select Parameters->Substrate Display Parameters to access those display parameters not affecting the simulation results. You can also double click at the Substrates tab in the Information Bar to bring up the dialog.

For comparison, the major parameters for the 3-cases are listed in Table 7.2 and a summary is listed in Table 7.3. The simple example here demonstrates how easy you can obtain reliable accurate and reliable results from IE3D efficiently.

Table 7.2 The various simulation cases with different AEC Layer settings.

File Name	C_BEND.GEO	C_BEND3.GEO	C_BEND5.GEO
AEC Layer	0 (or No)	1	2
AEC Ratio	N/A	0.1	0.1
Multi-Layer Ratio	N/A	N/A	0.4
Cells	25	83	142
Unknowns	26	137	251
Zc at 10 GHz	52.429-j0.163	50.831-j0.193	50.320-j0.206
Total Time	1	2	4

Table 7.3 Summary on Simulation of the Simple Microstrip Bend.

Summary	Explanation
Single cell in the cross-sectional direction may still yield good results for simple and isolated structures	When the structure is not strongly coupled, edge condition may not affect the s-parameters very significantly. As you can see from Figure 7.27, the dB[S(2,1)] and Ang[S(2,1)] are still quite reasonable using just single cell in the cross-section. The most difference is in the dB[S(1,1)] while the dB[S(1,1)] is below -27 dB. It mainly comes from the difference in predicting the Zc value (about 4% as shown in Table 7.2). How to find the Zc of a TLN is discussed in Appendix AQ. For such a simple structure, there is no doubt we should use AEC. However, for some large structures, the number of unknown will become too big

	with AEC. We may have to disable AEC or use edge cells wisely.
AEC Layer = 1 normally can yield very accurate results. For most applications, we can use AEC Layer = 1 to obtain precise results. For normal applications, there is no need to use AEC Layer larger than 1.	From Figure 7.27, you may still see 10 dB difference between AEC Layer = 1 and AEC Layer = 2. However, the dB[S(1,1)] is below -40 dB and it is very small. For transmission line structure close to matching, S(1,1) is the most sensitive parameter. The sensitivity also comes from the accuracy in modeling the Zc of the TLN. As you can see from Table 7.2, the difference between the Zc values predicted by AEC Layer = 1 and AEC Layer = 2 is less than 1%. It is very small. However, the difference it brings to dB[S(1,1)] is 10 dB with the base at -40 dB.
Fine-tuning AEC Ratio value may improve the accuracy.	Using AEC will certainly improve the accuracy significantly. The AEC Width is also very critical. AEC Width should be about 5-20% of the cell width. The default value of AEC Ratio = 0.1, or the AEC Width is about 10% of the standard cell size. IE3D is using non-uniform meshing. It is impossible to control AEC Width to be exact 10% or it may need to adjust for some structures. For example, if there is a narrow gap between two traces, we may need to adjust AEC Ratio to make the edge cell width not significantly larger than the gap width in order to get high accuracy results.
Normally, more and smaller edge cells will make the simulator predict more accurate loss effect.	More edge cells will capture the edge effects better. One of the major edge effects is the non-uniform current distribution in the cross-section. More and smaller edge cells can capture such a change better.

Section 7.10 Simple Ways to Build a Chamfered Bend and Other Path Structures

If a structure contains many bends and other geometry, it will be tedious to construct it as one or more polygons with many vertices. You can import complicated structures from other CAD tools using GDSII, DXF, GERBER and ACIS. Numerous powerful commands are imple-mented into MGRID to help you build and clean structures. We will show you some simple ways to build the bend. Let's analyze it first.

This is a bend with the centerline defined by three vertices: (0.0375, 0.75), (0.0375, 0.0375) and (0.75, 0.0375). The three vertices and the width determine the basic shape of the bend except the corner. We can use the Adv Edit->Build Path command to build the bend easily. Please note that the three vertices are not on the grid points. We cannot use mouse input to enter the 3 points. We can use keyboard input for them.

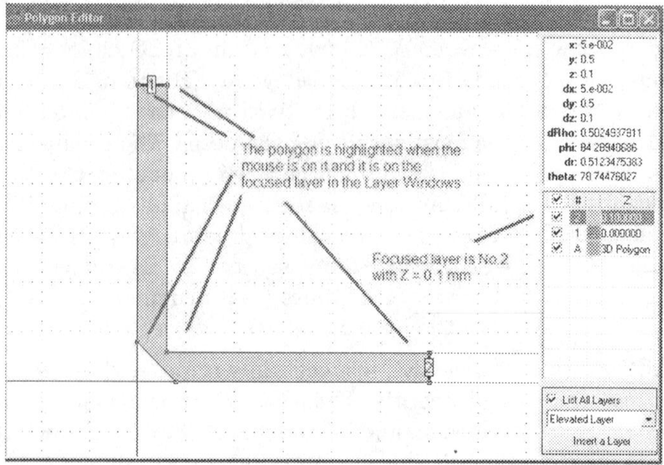

Figure 7.28 The object or polygon is highlighted when mouse is pointing at it.

Step 1 Run MGRID. Open the file "**.\ie3d\practice\c_bend.geo**". Click at the No.2 layer with Z = 0.1 mm in the Layer Window to focus the input on the layer. Move the mouse to point at the bend. The vertices of the bend are displayed with some markers indicating the bend polygon is highlighted.

We are going to delete the bend and rebuild it using another way.

Step 2 While the bend polygon is highlighted, click the right mouse button and you will see three useful commands in the pop-up menu:

Highlighted Object Properties

Delete Highlighted Object

Select Highlighted Object

Allowing highlighted objects is a new feature on MGRID. The object can be a polygon, a text object or a port. For our case, we just want to delete the polygon.

Step 3 Please select Delete Highlighted Object from the pop-up menu. MGRID will warn you that the s-parameters results will be lost. Select Yes to continue. The bend polygon will be removed. In case the s-parameter display graph was still opened, the graph will be automatically closed.

There are many ways to delete the polygon in IE3D 14 while the above way is the simplest. You can select Edit-

>Select All->Polygons to select the polygon. The selected polygon will turn black color when it is selected. The status window will display "Select Polygon Mode..." Then, you can select Edit->Delete command to delete the selected polygon.

Step 4 Select **Input->Key in Absolute Location** command. MGRID will prompt you to enter the X and Y-coordinates of the vertex. Enter X-Coordinate = 0.0375. Type the **Tab** key to shift the focus to the Y-Coordinate edit control. Enter the Y-Coordinate = 0.75. Then, hit **Enter** key (or select OK) to create the 1st vertex at (x, y) = (0.0375, 0.75).

Step 5 Type **Shift+A**. The **Shift+A** means that you press down the **Shift** key and hold it. Then, you type the **A** key. Finally, you release the hold **Shift** key. It is equivalent to selecting **Input->Key In Absolute Location** command. You will be prompted for the location of the next vertex. Enter (X, Y) =(0.0375, 0.0375) at the two fields and hit **Enter** key. The 2nd vertex is entered at (x, y) = (0.0375, 0.0375).

The **Shift+A** is the accelerator for the menu item **Input->Key in Absolute Location** command. It is listed in the right hand side of the menu item. You can access the menu item using the accelerator.

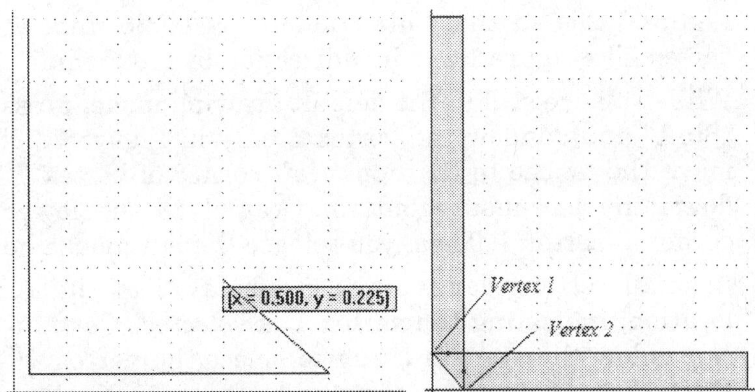

(a) The entered 3 vertices (b) The bend built using Build Path command

Figure 7.29 The chamfered bend created using the Build Path in Adv Edit menu.

Step 6 Type **Shift+R**. You will be prompted for the offset of the next vertex to the last vertex. The next vertex should be located at (x, y) = (0.75, 0.0375). The last vertex is at (x, y) = (0.0375, 0.0375). The offset should be (dx, dy) = (0.7125, 0).

The relative location between the 2nd and 3rd vertices is not straightforward. We use this command just because we want to demonstrate how we can use the Input->Key In Relative Location command. The **Shift+R** is the accelerator for it.

Step 7 Enter X-offset = 0.7125 and the Y-offset = 0.0. Hit **Enter** key. The 3rd vertex is entered at (x, y) = (0.75, 0.0375). We will see the 3 entered vertices forming a right angle in Figure 7.24a.

The above steps demonstrate how we can enter vertices using keyboard input (**Input->Key in Relative Location** and **Key in Absolute Location** commands).

Step 8 Select **Adv. Edit->Build Path** command. Enter the Path Width = 0.075 and select the Path Style as Smooth Corner (the default). Hit **Enter**. A chamfered bend is automatically created on MGRID as three polygons (Figure 7.29b).

The bend is at the same location as the one created in Sections 4 and 5. However, the cut at the corner is smaller. This bend is automatically created. You may not have the control over some of the small detail. The small difference in the corner may not have significant effect to the performance of the structure. You may leave it as it is. However, if you want to make it exactly like the bend we constructed in Section 4, you need to move the vertex 1 in Figure 7.29a in the y-direction in by 0.025 mm, and the vertex 2 in Figure 1.29b in x-direction by 0.025 mm.

Step 9 While you are still in the default drawing mode, press down "Shift" and window the vertex 1 only in Figure 7.29b. You move the mouse to the upper left corner of vertex 1. Press down the left mouse button. Drag it to the lower right corner of vertex 1. Then, you release the left mouse button.

A small rectangular marker is displayed at the vertex 1 location, indicating the vertex 1 is selected. Certainly, you can follow the standard way to select the vertex 1: Select **Edit->Select Vertices** command. Then, the window vertex 1 only in Figure 7.29b.

Step 10 Press down "Shift". Press down the left mouse button and move. MGRID automatically gets into the **Edit->Move Objects** mode. This is a short cut to select **Edit->Move Objects** command. It only works when MGRID is in selection mode.

Move the mouse upward somewhere and click the left button. The "Move Object Offset to Original" dialog comes up and display the offset values (Figure 7.30).

MGRID is a very precise layout editor. The coordinates of each vertex of the polygons are described precisely using double precision floating point values. When we use the move and copy commands, we need to preserve the precision. Therefore, we always prompt you to confirm the co-ordinates on the move and copy commands. You can always enter precise values which mouse entry cannot capture.

There are in fact 3 snapping modes for the move and copy commands: (1) Snap to Vertex, (2) Snap to Edge, (3) Snap to Vertex. The default snapping mode is "Snap to Vertex". To change the snapping mode, you can type F4 while you are moving the vertex. If you use the **Edit->Move Objects** command, you don't need to keep the left mouse button pressed. You can select **Edit->Change Snapping** command (corresponds to F4) to change the snapping mode. In the **Edit->Move Objects** mode, you can also change the **Moving Reference Vertex** by clicking the right mouse button or use the **Input->Set Moving Reference Vertex**, or **Shift Moving Reference by Vertex** or Shift **Moving Reference by Polygon** command. The Moving Reference Vertex is the vertex following the cursor in the move and copy command.

For our current moving, there isn't any other vertex near the vertex 1. It will not snap to any vertex. We do not need to worry about it because we know the offset values as: dx = 0 and dy = 0.025 mm.

Step 11 Enter the **X-offset** = 0 and the **Y-offset** = 0.025. Hit **Enter** to continue. The vertex 1 in Figure 7.29b is moved in the y-direction by 0.025 mm. The vertex 1 is de-selected after the moving and the result is shown in Figure 7.31a. Vertex 1 is moved to the right location right the vertex 2 still needs to be moved.

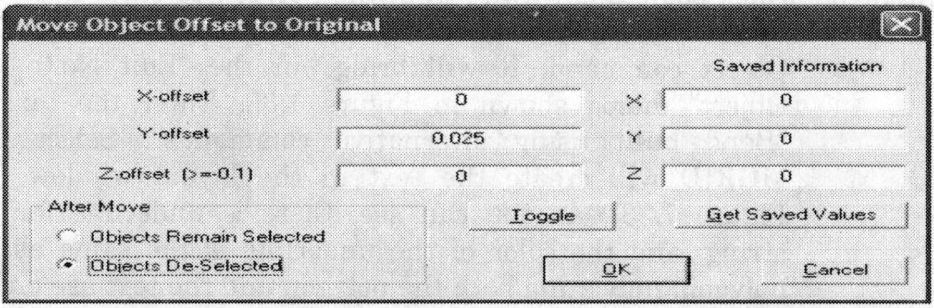

Figure 7.30 The Move Object Offset to Original dialog.

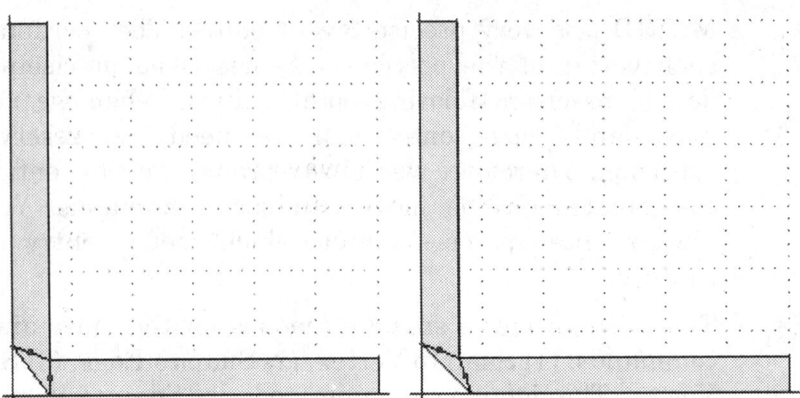

(a) The shape after moving vertex 1 (b) The shape after moving vertex 2.

Figure 7.31 The shape of the bend after the vertices are moved.

Step 12 Press down "Shift" and window the vertex 2 only in Figure 7.29b to select it. Press down "Shift" and press down the left mouse button and move the mouse to move the vertex 2. Click somewhere at the right. The "Move Object Offset to Original" dialog will come up. Change the **X-offset** = 0.025 and the **Y-offset** = 0. Hit Enter.

The vertex 2 is moved with the offset as (dx, dy) = (0.025, 0). MGRID resumes to the geometry Drawing mode automatically because the changed default for **After Move** is **Objects De-Selected**.

The bend after Step 12 will be exactly as the bend we want (see Figure 7.31b). We built the bend as a single polygon in Section 4 while we built it as 3 connected polygons. The results of both structures will be identical if the meshing is the same. In case different meshing is used, the difference between the results should be very small.

We just demonstrate a new way to create the bend. We would like to define some comment in the file to denote it.

Step 13 Click some point away from the polygon to enter a vertex. Type "t". It is equivalent to selecting Input->Define Text Object command. It will bring out the "Edit No.0 Text Object" dialog shown in Figure 1.32. Enter the text as "Bend built using alternative commands". Select OK. MGRID will create the text in the layout window (see Figure 7.33). As you can see, there is underline for the string, and the color of the underline is the same as the polygon, indicating both the polygon and the text are on the same layer. A text string also has layer and location.

Allowing text objects is the new feature of IE3D 14. Text objects are comments only. They will not affect the simulation results.

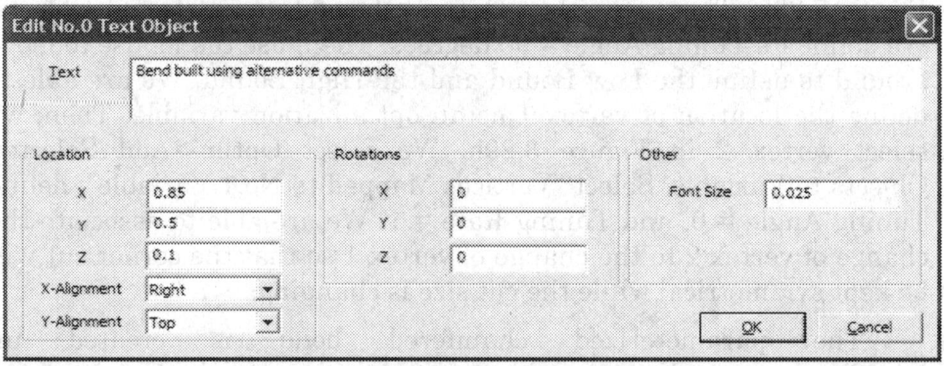

Edit No.0 Text Object

Text: Bend built using alternative commands

Location — X: 0.85; Y: 0.5; Z: 0.1; X-Alignment: Right; Y-Alignment: Top

Rotations — X: 0; Y: 0; Z: 0

Other — Font Size: 0.025

OK Cancel

Figure 7.32 The Edit Text Object dialog.

Step 14 Define the port 1 and port 2 on the structure. We will not repeat the steps because they are discussed earlier. Save the geometry file as: .\ie3d\practice\c_bend2.geo. If you simulate it, you should get the same or close results as the one built in c_bend.geo.

There is still another simpler way to build the chamfered bend. It is using the **Entity->Chamfered Bend** command documented in the Appendix I. You can try it out if you are interested in it.

When we build the structure as 3 polygons we need the 3 polygons to be connected "electrically". We have built the 3 polygons using automatic way. The "electrical connections" among the 3 polygons are guaranteed. The guaranteed electrical connections are indicated by the red dots at the centers of the common edges (see Figures 7.26). In case we need to build a structure with many polygons with some of them connected and some of them not connected, how could we guarantee the "electrical connection" between them? Polygon connection or "electrical connection" is a very important concept in the IE3D.

Section 7.11 Parameterization of Structures

IE3D is not only good for analysis. It is also good for tuning and optimization. To tune or optimize a structure, we will need to parameterize it first. For this simple bend structure, we know we can optimize the cut of the corner for best performance. The question is how we can optimize it.

Tuning and optimization start from parameterization of the structure. The 1st step is to parameterize the corner cut of the bend.

We are able to parameterize the cut on MGRID by controlling the location of the vertices in the same way as we discussed to change the shape of the cut by moving the vertices. We select the vertex 1 in Figure 7.29b. Then, we select Optim->Variable for Selected Objects. We define its Tuning Angle = 90 degrees. Then, use the mouse to move around to define the Low Bound and the High Bound. We are able to define the location of vertex 1 as an optimization variable. Then, we select vertex 2 in Figure 3.29b. We select Optim->Add Selected Objects to Variable. Select "Vertices Mapped to No.1 Variable", define Tuning Angle = 0, and Tuning Rate = 1. We are able to associate the change of vertex 2 to the change of vertex 1 so that the corner cut will be kept symmetrical while the cut size is changing.

The parameterized chamfered bend is created in: .\ie3d\samples\c_bend2_tuning.geo. You can open it. Then, please select Optim->Geometry Tuning. Select the Variables Slides tab on the right. Drag the slider representing the only variable in the geometry, you will be able to see how the variable is controlling the cut size of the corner and the structure is kept symmetry while it is changing.

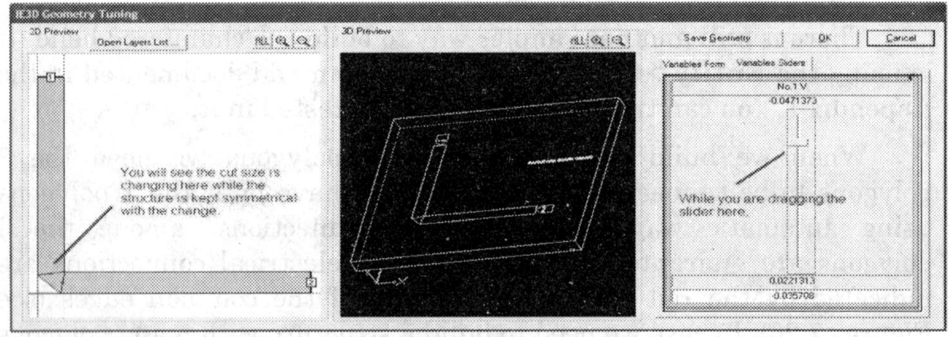

Figure 7.33 Geometry tuning for parameterized structure.

As MGRID is able to parameterize structures by controlling the vertices of the polygons, IE3DLibrary is much easer and more powerful in parameterize structures. On IE3DLibrary, a structure is created as a group of parameterized objects. An object is not a polygon, but a higher level abstraction such as a T-junction, a straight-strip or a circular patch. An object is described as a set of dimensional parameters such as length and width etc. Each dimensional parameter can be an equation and the equation can be a function of tuning variables. We can easily parameterize the chamfered bend on IE3DLibrary. Saved in .\ie3d\practice\c_bend2_tuning2.ie3 file is an example of the IE3DLibrary. For this structure, the cut size is parameterized as a tuning variable while the line length and the line width are parameterized as global variables. You can easily tune the cut size and even the line length and the line width easily if you hook up the global variables with some more tuning variables. Open the file from IE3DLibrary, and then select Edit->Geometry Tuning. You will see something similar to what is shown in Figure 7.33.

After a structure is parameterized, we can do a preparation for real-time full-wave EM tuning and optimization on the Fast EM Design Kit, which is an integrated part of IE3D. We can also perform IE3D full-wave EM optimization on a parameterized structure.

7B. MODELING OF WIRE ANTENNAS AND OTHER RF ANTENNAS

IE3D is not only good for patch antennas but also good for wire antennas and other RF antennas. In this chapter, we will discuss the modeling of wire antennas and other RF antennas.

Section 7.12 Discussion on Modeling Wire Antennas

Traditionally, a wire is modeled with separated source point and observation point or field test point (see Figure 7.34a). In order to avoid the singularity in numeral calculation, we assume the electric current is along the axis of the wire and the location where we test the field is on the surface of the wire. It is a good approximation when the wire is very thin compared to the length and wavelength. Normally, it can yield accurate real part of the input impedance for a typical half-wavelength dipole antenna. We call the model the thin wire approximation. The thin wire approximation certainly introduces some error, because we know the current is mainly flowing on the surface of the wire.

On the IE3D, we always model the surface current on metallic structures and we also test the field on the surface of metallic structures. Depending upon the accuracy requirement, we can model a thin wire as a metallic strip or a ribbon (see Figure 7.34b), a rectangular tube (see Figure 7.34c) or a tube with more sides.

The metallic strip model in Figure 7.34b and the tube model in Figure 7.34c assume both the current source and the observation point to be on the surface. They are closer to the reality and they can yield much more accurate results than the thin wire model. Normally, the tube model should take more time to simulate.

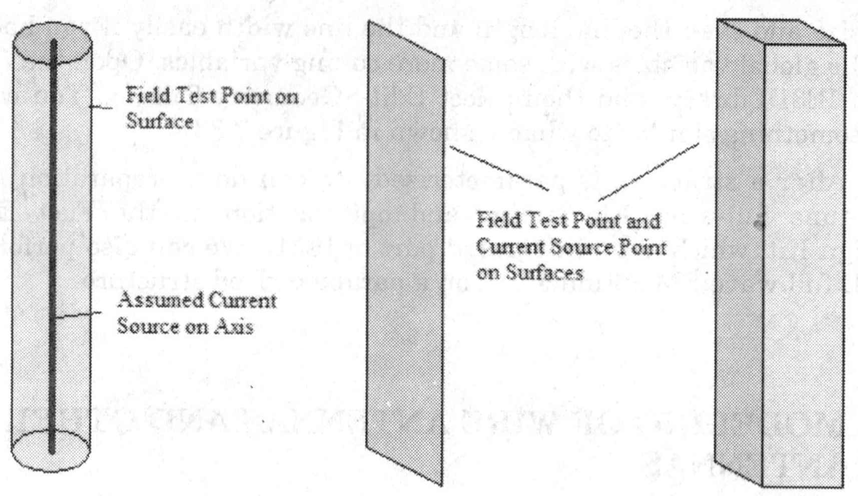

(a) Thin wire model (b) Ribbon model on IE3D (c) Tube model on IE3D

Figure 7.34 Different models of wire antennas.

Section 7.13 Discussion on Dipole Antennas

We start from a simple dipole antenna. The total length of the dipole is 15 mm. The radius of the dipole is 0.075 mm (0.5% of the length). We expect the resonant frequency to be slightly lower than 10 GHz. There are a few different ways to model such a dipole antenna depending upon which de-embedding scheme we use. We will try to use the different model on it. The 1st model we want to use is the differential **Advanced Extension** scheme. The file is saved in c:\ie3d\samples\dipole6.geo. We will re-build it for practice:

Step 1 Run MGRID. Open c:\ie3d\samples\dipole6.geo. Please select Basic Parameters to check the dielectric configuration.

No. 0 Z top = 0 mm, $\varepsilon_r = 1$, $\mu_r = 1$, $\sigma = 0$

No. 1 Z top = 1.0e+15 mm, $\varepsilon_r = 1$, $\mu_r = 1$, $\sigma = 0$

It is free space everywhere. Select OK to continue. MGRID is in the default drawing mode.

Step 2 Please select all the polygons on the window and delete them. We are going to rebuild the structure.

Select **Input->Create and Edit Vertices** command. Select the **Import** button. Select **OK** when prompted. Select **c:\ie3d\samples\dipole.txt**. MGRID will inform you that 6 vertices are imported and whether you want to give it some offset values. Select OK to continue. The 6 vertices are imported. The 6 vertices define the centerline of the dipole we are going to build. Select OK and MGRID resumes to drawing mode with the 6-vertices entered.

Step 3 Select **Adv Edit->Build Wire Path** command. Enter the **Number of Segments for Circle** = 6 and the **Radius** = 0.075. We will get the dialog shown in Figure 7.35. Select **OK** to accept the other default settings. MGRID will build a wire path with a small bend at the center (see Figure 7.36a). Select **View->Whole Circuit** command display the whole structure.

Wire Parameters

Total Vertices: 6 Wire Length: 31.6

No.1 (X,Y,Z) = (0, 0.5, 0.5)
No.2 (X,Y,Z) = (0, 7.85, 0.5)
No.3 (X,Y,Z) = (0.4, 7.85, 0.5)
No.4 (X,Y,Z) = (0.4, 8.15, 0.5)

Leave Start Point Open

Leave End Point Open

Number of Segments for Circle 6

Terminal Separation 0.0375

Radius 0.075

Division Option Circumscribed Polygons

OK Cancel

Figure 7.35 The Build Wire Path dialog.

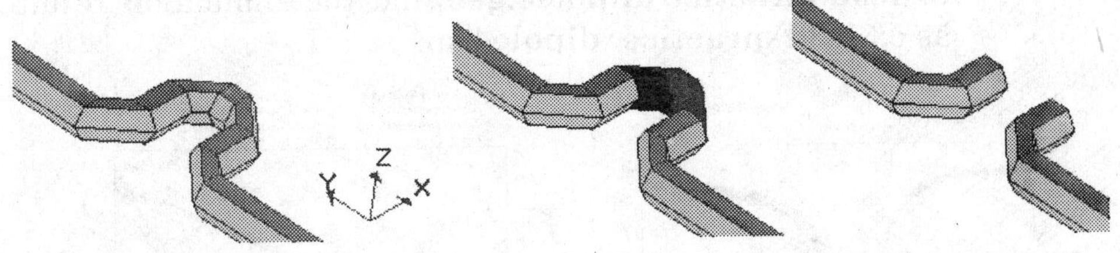

(a) Before the bends are deleted (b) The bends selected (c) After the bends are deleted.

Figure 7.36

Step 4 Press down "Shift" button and window the two right-angle bends at the center to select the polygons (see Figure 7.36b). Select **Edit->Delete** command to delete the polygons on the bends. We will get the picture in Figure 7.36c.

Step 5 Select **Port->Port For Edge Group** command. Select **Advanced Extension** scheme. Please make sure all the layers are checked on the Layer Window. Please window one of the terminals to define the port 1 on it. Please click the right mouse button to bring up the pop-up window. Select **Define Negative Port**. Then, window the edges of the other terminal to define the port –1 on it. The port 1 automatically becomes port +1. Select **Port->Exit Port** command to resume the drawing mode. The final result at the feed location is shown in Figure 7.37.

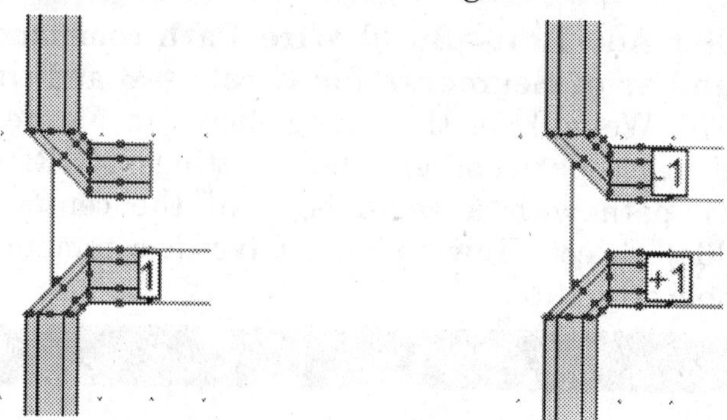

(a) Port 1 is defined. (b) Port 1 changed to Port +1 after Port -1 defined.

Figure 7.37 The bends with ports defined.

Step 6 Save the structure as **c:\ie3d\practice\dipole6.geo**. Simulate the structure from 5 to 15 GHz with 101 frequency points. It takes couple seconds to finish it.

Step 7 Select all the polygons of the structure. Delete them. Repeat Step 2 to Step 5. In Step 3, you should enter the **Number of Segments for Circle** as 4. You will build the dipole with a rectangular tube. Save the structure as **c:\ie3d\practice\dipole4.geo** and the simulation result as **c:\ie3d\practice\dipole4.sp**.

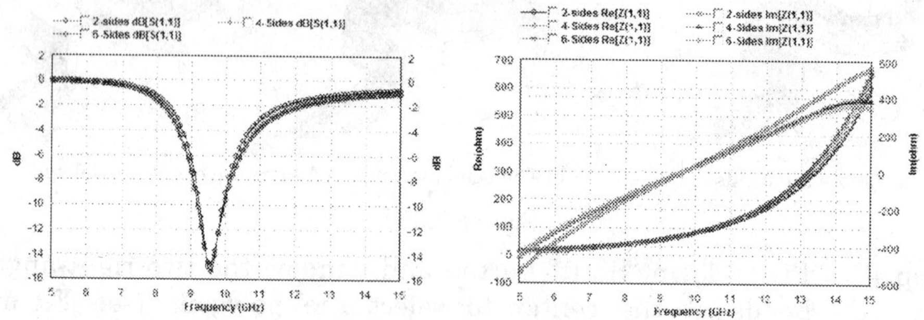

Figure 7.38. The s-parameters input impedance of the dipole for 3 different models.

Step 8 Select all the polygons of the structure. Delete them. Repeat Step 2 to Step 5. In Step 3, you should enter the **Number of Segments for Circle** as 2. You will build the dipole with a flat strip. Save the structure as **c:\ie3d\practice\dipole2.geo** and the simulation result as **c:\ie3d\practice\dipole2.sp**. Comparisons among the 3 cases are shown in Figure 7.38 and Table 7.4. Please note that the

Re[Z(1,1)] is using the left vertical scale and the Im[Z(1,1)] is using the right vertical scale and they are different. The s-parameters look quite the same for the 3-models. There is some difference in the Im[Z(1,1)] of the resonance.

Table 7.4 The comparison among the 3 models.

Wire Approximation	1-Side (or 2-Sides)	4-Sides	6-Sides
Total Time (seconds)	0	2	6
Resonant (Im[Zin] = 0)	≈9.617 GHz	≈9.629 GHz	≈9.628 GHz
Re[Zin] at Resonance	70.82 Ω	72.89 Ω	73.37 Ω
Directivity	2.157 dB	2.162	2.163
Efficiency	95.6%	95.8%	95.7%

As we can see, the differences among the 3 models are quite small. The flat strip model is reasonably accurate around the resonance for the thin wire antennas. However, when the radius of the wire becomes bigger or when it is getting farther from the 1st resonance, we will need the more accurate tube models.

Section 7.14 Modeling Dipole Antennas using Localized Ports.

We used the **Extension for MMIC** scheme for modeling of the above examples. In fact, we can do equally well or even better using the **Vertical Localized** and **Horizontal Localized** ports to model a vertical dipole.

Step 1 Open file: **c:\ie3d\practice\dipole6.geo**. Select all polygons and delete them.

Step 2 Select **Input->Create and Edit Vertices** command. Select the **Import** button. Select **OK** when prompted. Select **c:\ie3d\samples\dipolev.txt**. MGRID will inform you that 3 vertices are imported and whether you want to give it some offset values. Select OK to continue. The 3 vertices are imported and listed. The 3 vertices define the centerline of the dipole we are going to build. The three vertices are at: (X, Y, Z) = (0, 0.5, 0.5), (0, 0.5, 7.95) and (0, 0.5, 15.5). We purposely make the 3 vertices non-symmetrically. The distance from the 1st and 2nd vertices is 7.45 mm and the distance from the 2nd and 3rd vertices is 7.55 mm. The reason for it is that we will use the Adv Edit->Build Wire Path command to build the dipole with a vertical localized port on it automatically. The vertical localized port will be defined from Z = 7.95 to Z = 8.05. Then, the two arms will be of the same length.

Step 3 Select OK and MGRID resumes to drawing mode with the 3-vertices entered.

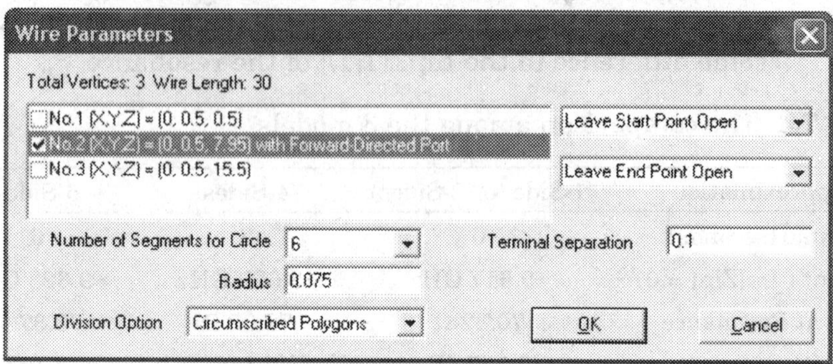

Figure 7.39 The dialog for building the wire path with a vertical localized port.

Step 4 Select **Adv Edit->Build Wire Path** command. Enter the **Number of Segments for Circle** = 6 and the **Radius** = 0.075. Please check the No. 2 vertex in the list (see Figure 11.6). It defines MGRID to build a vertical localized port at the 2nd vertex. The Terminal Separation will be automatically set to 0.0375. Change the Terminal Separation to 0.1. We will get the dialog shown in Figure 7.39. Basically, we are trying to build a wire path as 3 segments. From the No. 1 vertex at (0, 0.5, 0.5) to the No. 2 vertex at (0, 0.5, 7.95) is the 1st segment. It is the 1st arm of the dipole. From the No.2 vertex at (0, 0.5, 7.95) to (0, 0.5, 7.95+0.1)=(0, 0.5, 8.05) as the 2nd segment. The 2nd segment will be defined as a vertical localized port automatically. The 3rd segment is from (0, 0.5, 8.05) to (0, 0.5, 15.5) and it is the 2nd arm of the dipole.

Step 5 Select OK to continue. MGRID will create a vertical dipole with a vertical localized port exactly at the center. Save the geometry file as: c:\ie3d\practice\diplole6v.geo. Simulate it from 5 to 15 GHz with 101 points.

Step 6 Select all polygons and delete them.

Step 7 Select **Input->Create and Edit Vertices** command. Select the **Import** button. Select **OK** when prompted. Select **c:\ie3d\samples\dipoleh.txt**. MGRID will inform you that 3 vertices are imported and whether you want to give it some offset values. Select OK to continue. The 3 vertices are imported and listed. The 3 vertices define the centerline of the dipole we are going to build. The three vertices are at: (X, Y, Z) = (0, 0.5, 0.5), (0, 7.95, 0.5) and (0, 15.5, 0.5). The vertices in dipoleh.txt are basically the same as those in dipolev.txt file with the Y and Z-coordinates swapped. Using the dipolev.txt, we built the vertical dipole. Using the dipoleh.txt file, we will build a horizontal localized port.

Step 4 Select **Adv Edit->Build Wire Path** in **Adv Edit** menu. Enter the **Number of Segments for Circle** = 6 and the **Radius** = 0.075. Check the No.2 vertex in the list. The Terminal Separation will be automatically set to 0.0375. Change the Terminal Separation to 0.1. We will get the dialog similar to the one shown in Figure 7.39. Only the vertices are different. Select OK to continue. MGRID will build a horizontal dipole with a horizontal localized port at the center. Save the file as c:\ie3d\practice\dipole6h.geo. Simulate it from 5 to 15 GHz with 101 points. The comparison among the 3 schemes, (1) Advanced Extension (Dipole6), (2) Vertical Localized (Dipole6v), (3) Horizontal Localized (Dipole6h), is shown in Figure 11.7. The Dipole6v and Dipole6h are almost identical. The Dipole6 is slightly off. The difference does not indicate which scheme is better in accuracy. All of them are accurate. However, the slight difference in the feed proximity is changing the properties of the antenna slightly. If your dipole is, in fact, a monopole over a large ground plane and fed by a coaxial, the vertical localized and the horizontal localized ports are closer to reality. If your dipole is fed by a parallel line, the Advanced Extension port is the better choice.

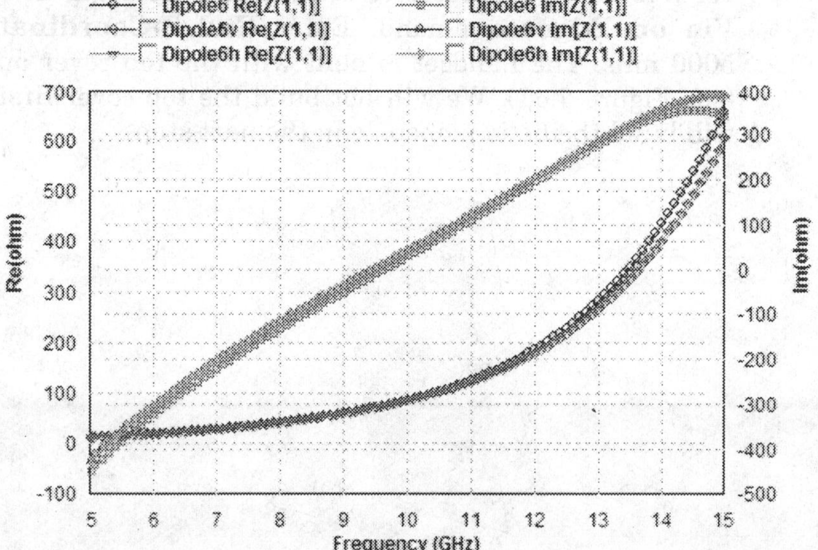

Figure 7.40 The comparison among the 3 different port schemes.

Section 7.15 Modeling of a Cylindrical Helix on a Handset

There is great interest in the modeling of a cellular handset. In this section, we will try to build a simple handset model. Assuming the handset's dimensions are 5 inches by 2 inches by 0.8 inches. The length of the helix portion is 600 mils. The radius of the helix is 200 mils. There are total 4 turns for the helix. The wire radius is 30 mils. The length of the wire before the helix is 150 mils. The length of the wire after the helix is 3.5 inches. We will neglect the plastic portion.

Step 1 Delete all the polygons on the window. Select Basic Parameters in Param menu. Change the Length Unit to mil. Define the **Metallic Strip Parameters** as:

No. 1 Thickness = 4 mils, $\varepsilon_r = 1$, $\mu_r = 1$, $\sigma = 4.9e7$ s/m

Step 2 Define the **Meshing Parameters** as: **Fmax** = 1.2 GHz and **Ncell** = 15. Change the Grid Size for the layout as 100 mils. Select OK to continue.

Step 3 We are going to build the bottom of the handset first. Select Insert a Layer on the Layer Window. Enter Z = 0. Click the left button at x = 0 and y = 0 to enter the first vertex at (0, 0, 0). Type **Shift+R** and enter dx = 2000 mils and dy = 0 to enter the second vertex at (2000, 0, 0). Type **Shift+R** and enter dx = 0 and dy = 800 mils to enter the third vertex at (0, 800, 0). Type **Shift+F** to form the rectangle for the bottom of the handset.

Step 4 Select **Edit->Select Vertices** command. Window the 4 vertices of the rectangle to select them. Select **Edit->Add Via on Edges** command. Enter **End Z-Coordinate** as 5000 mils. The handset is built with the top cover opened (see Figure 7.41). We will not build the top cover first. We will build the helix monopole in the next step.

Figure 7.41 The handset is built with the top cover removed.

Step 5 Select **Entity->Cylindrical Helix** command. MGRID will prompt you to enter the parameters for the cylindrical helix. Enter the following parameters:

Wire Parameters:

Number of Segments for Circle = 2

Radius = 30 mils

Cylindrical Helix Parameters:

Segments each Turn = 12

Total Segments = 48

Helix Length = 600 mils

Radius = 200 mils

Start Vertical Wire Length = 150 mils

End Vertical Wire Length = 3500 mils

Start Angle = 0

Center X-Coordinate = 400 mils

Center Y-Coordinate = 400 mils

Center Z-Coordinate = 5010 mils

Select **OK** to continue. The cylindrical helix will be built. The feed proximity is shown in Figure 11.9. The bottom end of the helix is at Z = 5010 mils. It is 10-milis above the handset top cover, which is not built.

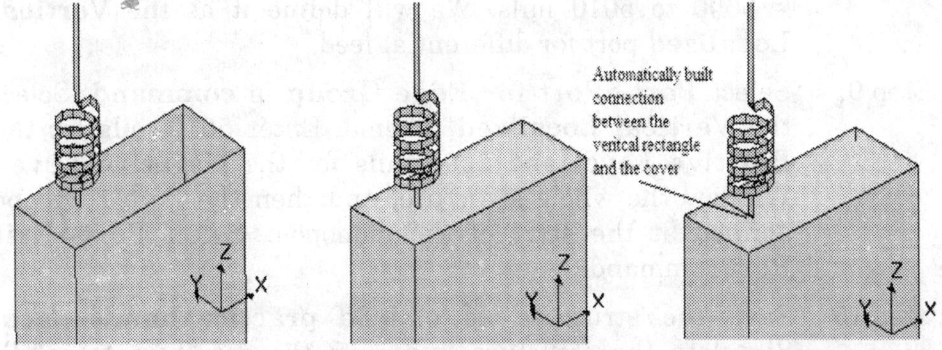

(a) The helix is built 10-mils above the handset (b) The cover is built (c) The connection is built

Figure 7.42 The helix is built 10-milis above the handset without cover.

Step 6 Select **Edit->Select Vertices** command. Try to find out the layer with z = 5010 milis on the layer window. Un-check all layers. Then, check the Z = 5010 layer to focus the selection on it.

Step 7 Window the whole structure. The 2 vertices at the start of the monopole at z = 5010 mils are selected.

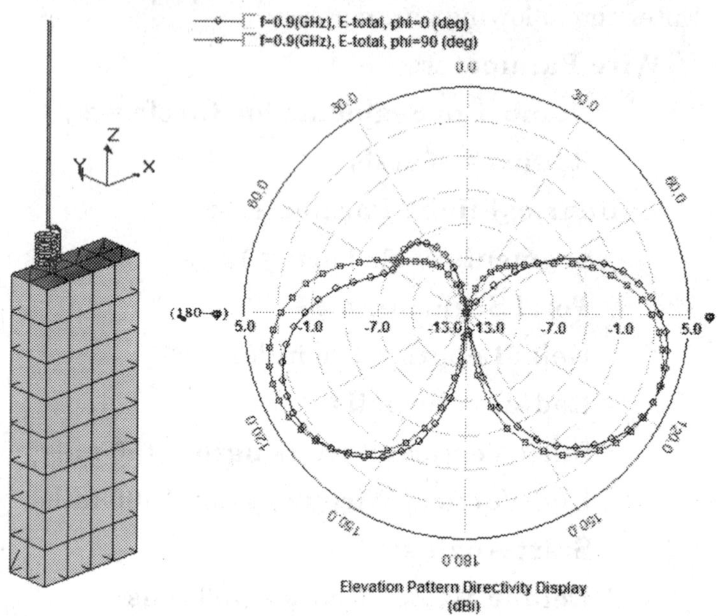

Figure 7.43 The meshed handset and the radiation pattern.

Step 8 Select **Edit->Add via on Edges** command. Enter **End Z-Coordinate** as 5000 mils. A vertical rectangle should be built from Z = 5010 to Z = 5000. It will automatically create the connection at the top cover.

The vertical rectangle is at the start of the monopole from z = 5000 to 5010 mils. We will define it as the **Vertical Localized** port for differential feed.

Step 9 Select **Port->Port for Edge Group** in **command**. Select the **Vertical Localized** scheme. Enter 5010 mils for the **Positive Level** and 5000 mils for the **Negative Level**. Window the whole structure, and then the Port 1 will be defined at the start of the monopole. Select **Port->Exit Port** command.

Step 15 Save the structure as **c:\ie3d\practice\handset.geo**. Simulate the structure and save the result in **c:\ie3d\practice\handset.sp**. It takes a fraction of a minute to simulate it. The meshed structure and the radiation pattern at 0.9 GHz are shown in Figure 7.43. Again, we would like you to understand that this is not a design. The dimensions are arbitrarily chosen and they are not optimized. We just demonstrate the concepts and approaches to model complicated structures.

Section 7.16 Modeling of Inverted F Antennas

Inverted F antennas or similar antennas are frequently used in RF applications due to their compact sizes. In this section, we will demonstrate how you can model an inverted F antenna. Shown in Figure 11.11 is an inverted F antenna in **c:\ie3d\samples\invf.geo**. The s-parameter result is saved into **c:\ie3d\samples\output\invf.sp**. Basically, it is a horizontal rectangle of length 2520 mils and width 900 mils. The horizontal rectangle is 700 mils above the infinite ground plane. On the left end, there is a via connecting the horizontal rectangle to the ground plane. The via connected to the port is of width 60 mils. The via for the port is 990 mils from the via on the edge and it is 180 mils from the closest edge,

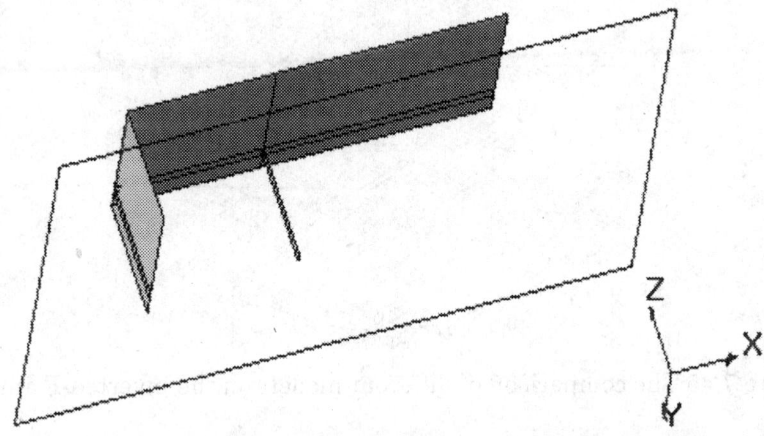

Figure 7.44 An inverted F antenna on an infinite ground (shown as a wire frame).

In practical application, 60-mil thick aluminum is normally used to make the antennas. The thickness of 60-mils is quite thick and we should model it with thickness in order to get high accuracy result. Also, the ground plane for the antenna may be small and we need to model the finite ground plane. The inverted F antenna with finite ground plane is saved in **c:\ie3d\practice\invf1.geo**. The finite ground plane is 500-mil larger than the antenna on each side. The finite ground plane and finite thick metallic strip model is saved in **c:\ie3d\practice\invf2.geo**.

The comparison of the 3 models is shown in Figure 7.45. The invf.geo is the "Infinite GND" model. The invf1.geo is the "Finite GND" model. The invf2.geo is the "Finite TK" model. The "Infinite GND model is quite different from the other two models. Inverted-F antenna is not a low profile antenna. The ground plane size does have significant effect to the performance of the antenna.

The locus of the "Finite GND" model is similar to the one for "Finite TK" model. However, the resonant frequency shifts a few percents. Certainly, 60-mil thick strips are quite significant. We cannot neglect it. We will not show how to build the models here. The examples serve to demonstrate the importance of the ground plane and finite thickness for non-low-profile antennas.

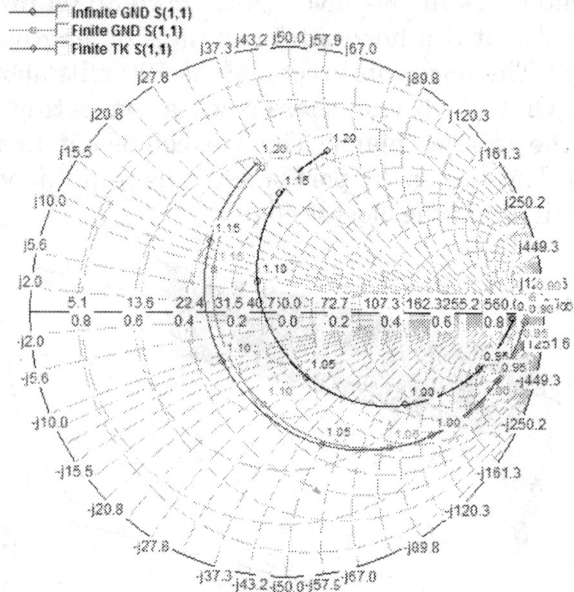

Figure 7.45 The comparison of different models on the inverted-F antenna.

Section 7.17 Simulations of RFID Antennas

RFID has been a hot topic in the last few years. IE3D is excellent for RFID applications due to its flexibility and high accuracy. There are normally two frequency bands for RFID applications: (1) Spiral-like antennas working at 13.56 MHz; (2) Half-wavelength long antennas working at about 868 MHz.

The design of RFID antennas should be similar to other types of antennas except there are special requirements for the input impedance for an RFID antenna. Normally, the antenna is connected to a RFID chip. The chip's internal impedance is of a complex value at the target frequency. To receive maximum efficiency, we should try to design an antenna such that the impedance of the antenna is a conjugate match to the chip internal impedance.

IE3D has excellent built-in optimizers. You are allowed to define some dimensions of an antenna as optimization variables and let IE3D to optimize them for specified goals. For RFID applications, you can specify the optimization goals as: Re[Z(1,1)] = Ri and Im[Z(1,1)] = - Xi, where Zi = Ri + j Xi is the internal impedance of the RFID chip at the target frequency.

On IE3D, we have defined two efficiency values for an antenna: (1) Radiation Efficiency; (2) Antenna Efficiency. The definitions can be found in Appendix AV. We also have defined Antenna Gain based upon the definition of Antenna Efficiency. The Antenna Efficiency and Antenna Gain are more defined for constant wave sources.

For RFID applications, we have defined Conjugate Match Factor for judging how good an RFID antenna is. Please read the Appendix BI for more discussion on RFID designs.

Section 7.18 Simulation of Ultra-Wide Band (UWB) Antennas

IE3D is an excellent tool for designing various antennas. We have implemented advanced features for UWB applications. In this section, we will not discuss how we can achieve wide band. We will just demonstrate how we can use IE3D and MDSPICE to perform simulation in both frequency and time domain.

Saved in .\ie3d\samples\oval_dipole.geo is a printed dipole antenna with arms of elliptical shapes. We would like to investigate how the waveform looks like if we have excite one Tx antenna and receive the signal from an RX antenna. For the oval_dipole.geo, we have defined one port and a plane wave excitation for it.

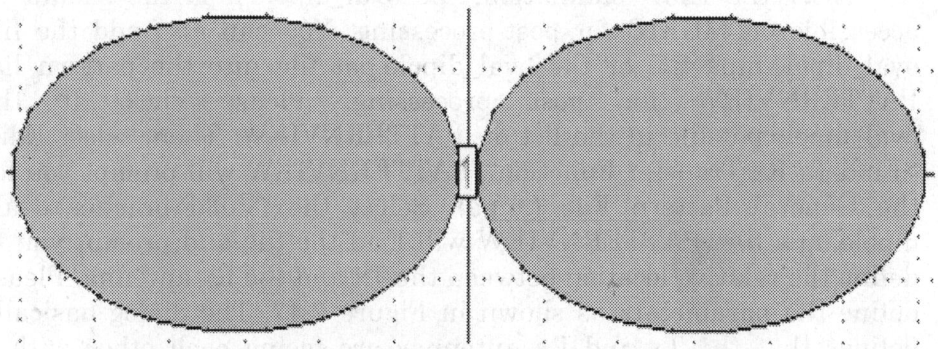

Figure 7.46 A printed oval shape dipole antenna.

Please open the file. Save it into .\ie3d\practice\oval_dipole.geo (see Figure 7.46). Simulate it at 0.1, 0.2, 0.5, 1.0, 1.5, 2.0 ... 14.5 and 15 GHz. Please enable pattern calculation with the default setting (only the port is excited while the plane waves do not present). It will take a few minutes to finish because pattern calculation will take some time. Actually, we are not just interested in the radiation pattern with specified excitation. What is of interests to us is the General Pattern File (.mpa) created in the simulation. As it is discussed in Appendix AJ, we can find the 2-port s-parameters from the .mpa file of an antenna with 1-port and plane wave excitations defined simultaneously.

Assume we would like to find the 2-port s-parameters of the two antennas with a separation of 3 feet. We can do it in the following.

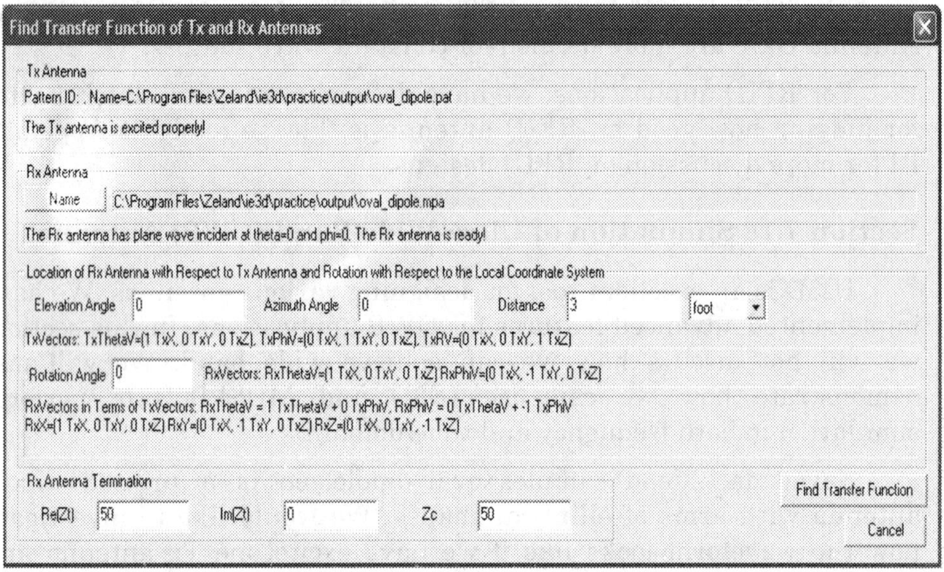

Find Transfer Function of Tx and Rx Antennas

Tx Antenna

Pattern ID: , Name=C:\Program Files\Zeland\ie3d\practice\output\oval_dipole.pat

The Tx antenna is excited properly!

Rx Antenna

Name C:\Program Files\Zeland\ie3d\practice\output\oval_dipole.mpa

The Rx antenna has plane wave incident at theta=0 and phi=0. The Rx antenna is ready!

Location of Rx Antenna with Respect to Tx Antenna and Rotation with Respect to the Local Coordinate System

Elevation Angle 0 Azimuth Angle 0 Distance 3 foot

TxVectors: TxThetaV=(1 TxX, 0 TxY, 0 TxZ), TxPhiV=(0 TxX, 1 TxY, 0 TxZ), TxRV=(0 TxX, 0 TxY, 1 TxZ)

Rotation Angle 0 RxVectors: RxThetaV=(1 TxX, 0 TxY, 0 TxZ) RxPhiV=(0 TxX, -1 TxY, 0 TxZ)

RxVectors in Terms of TxVectors: RxThetaV = 1 TxThetaV + 0 TxPhiV, RxPhiV = 0 TxThetaV + -1 TxPhiV

RxX=(1 TxX, 0 TxY, 0 TxZ) RxY=(0 TxX, -1 TxY, 0 TxZ) RxZ=(0 TxX, 0 TxY, -1 TxZ)

Rx Antenna Termination

Re(Zt) 50 Im(Zt) 0 Zc 50

Find Transfer Function

Cancel

Figure 7.47 The definition of the Tx and Rx antenna's relative location.

After the IE3D simulation, the oval_dipole.mpa file should be accessible on MGRID for post-processing. You can also add the file oval_dipole.mpa file or the oval_dipole.pat file into the pattern list PATTERNVIEW for post processing. Please click at the oval_dipole.pat file in the list on PATTERNVIEW. Then, select Edit->Find Tx Rx Transfer Function. PATTERNVIEW will prompt you for the General Pattern File (.mpa). Select the .\ie3d\practice\oval_dipole.mpa file. PATTERNVIEW will load the file and prompt you to define the relative location between the Tx and the Rx antenna. Please define the parameters as shown in Figure 7.47. The dialog basically defines that the Tx and Rx antennas are facing each other with a distance of 3 feet. Please select Find Transfer Function button. PATTERNVIEW will prompt you for the file name. Please select OK to accept the default setting of TX_RX_0.TXT file name. PATTERNVIEW will prompt you for the Save Transfer Function Data dialog. You have the option to find different transfer functions and the 2-port s-parameter file. Please select 2-Port S-Parameters for Tx Rx. PATTERNVIEW will automatically update the file name s: TX_RX_0.SP. Select OK. PATTERNVIEW will save the s-parameters into TX_RX_0.SP and open it on Notepad.

Figure 7.48 The Save Transfer Function Data dialog.

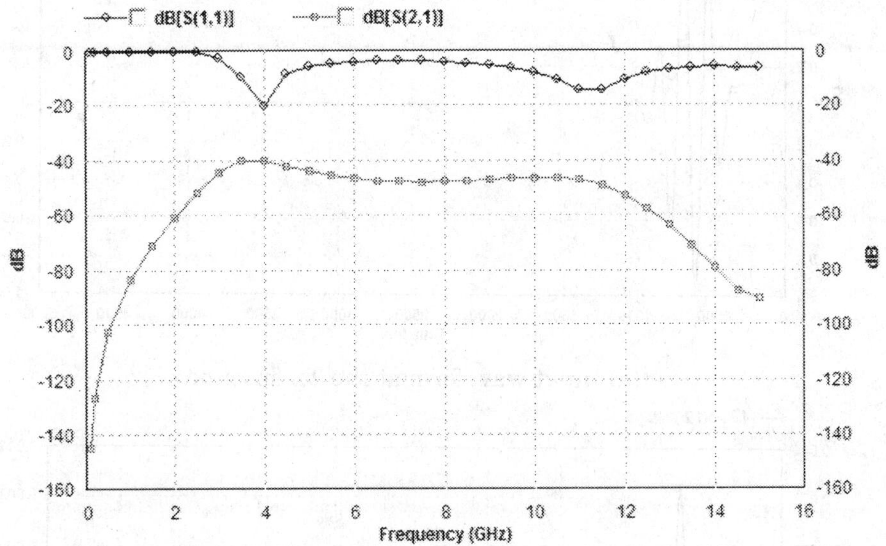

Figure 7.49 The 2-port s-parameters between the Tx and Rx antenna.

Please display the 2-port s-parameters saved in .\ie3d\ practice\ output\TX_RX_0.SP using MODUA. As you can see, the coupling between the 2-antennas is below -40 dB when they are separated by 3 feet. Please understand that the results are based upon the far-field radiation pattern. It should be reliable since the 2 antennas are far enough at the pass band which is above 3 GHz.

We can setup MDSPICE to perform a time transient analysis on the Tx Rx transmission system. We try to put a stimulus of sine modulated with Gaussian centered at 8 GHz with 3 dB bandwidth of 1 GHz at the Tx antenna and observe the waveform at the Rx antenna with a 50-ohm termination. Figure 7.50 shows the comparison between the input waveform at the Tx antenna and the output waveform at the Rx antenna. We certainly should see some difference in the waveform because the transmission properties of the system are not flat over the frequency range of interest. We also see some time delay and ringing in the output waveform. Figure 7.51 shows the comparison in the spectrum between the source waveform and the output voltage waveform. As you can see, the difference between the 2 waveforms is at about 3.8 GHz and the result is consistent with the 2-port s-parameters.

IE3D is the most powerful and versatile antenna design tool. You can use it to do much more that we demonstrate here.

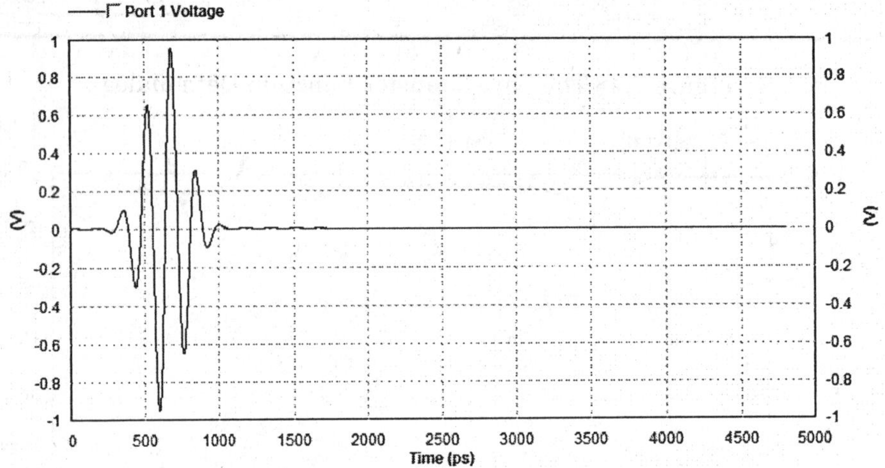

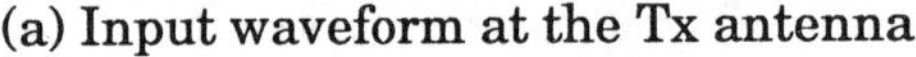

(a) Input waveform at the Tx antenna

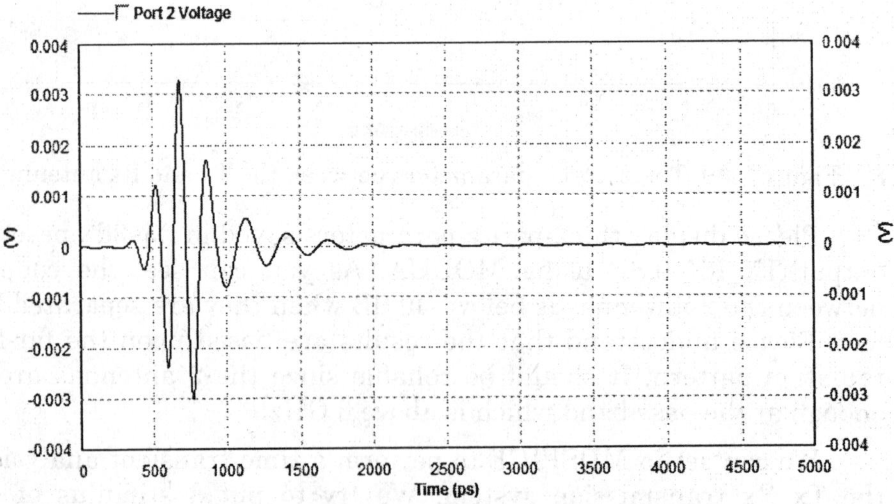

(b) Output waveform at the Rx antenna.

Figure 7.50 The Input Waveform at Tx antenna and Output Waveform at Rx antenna.

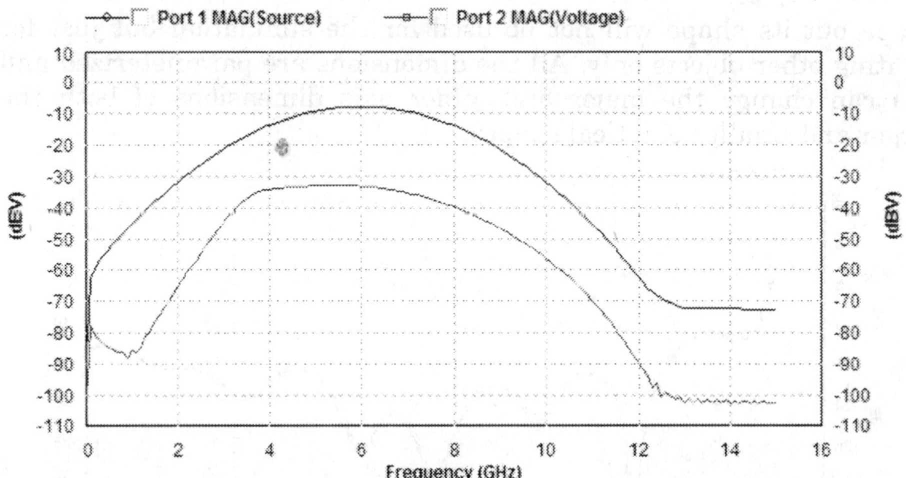

Figure 7.51 The spectrums of the source waveform and the output waveform.

Section 7.19 Parameterization of Complicated Structures

The elliptically shaped modified dipole exhibits excellent wide band characteristics. One question you may have is: Can we change the dimensions elliptical shapes. On MGRID, you may have to rebuild them. However, on IE3DLIBRARY, you are able to parameterize even more complicated structures. Saved in .\ie3d\ie3dlibrary_examples\ep4.ie3d is a more complicated structure on IE3DLIBRARY. Please use IE3DLIBRARY to open it and it is illustrated in Figure 7.52.

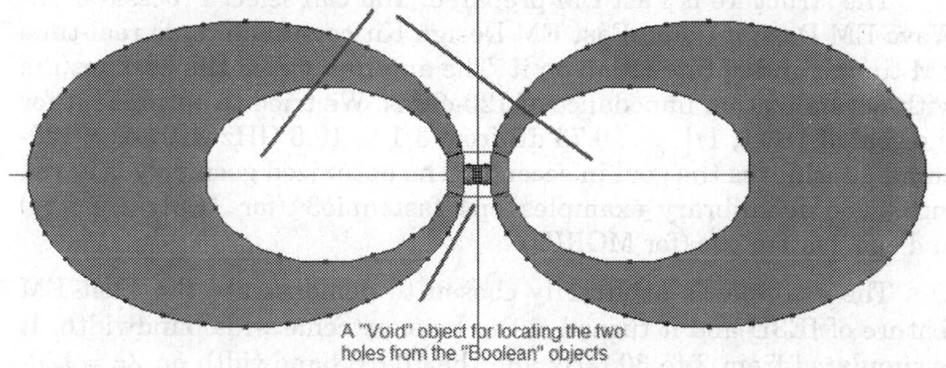

Figure 7.52 A tunable UWB antenna created using 4-elliptical shapes.

The antenna is completely parameterized. It is created mainly from 4-elliptical shapes. The two smaller elliptical shapes are "Boolean" objects. They are building the holes on the larger elliptical shapes. The symmetry of the structure is guaranteed by connecting the two "Boolean" objects to a "Void" object. The "Void" object has shape but its shape will not be used for the simulation but just for locating other objects only. All the dimensions are parameterized and you can change the major and minor axis dimensions of both the larger and smaller elliptical shapes.

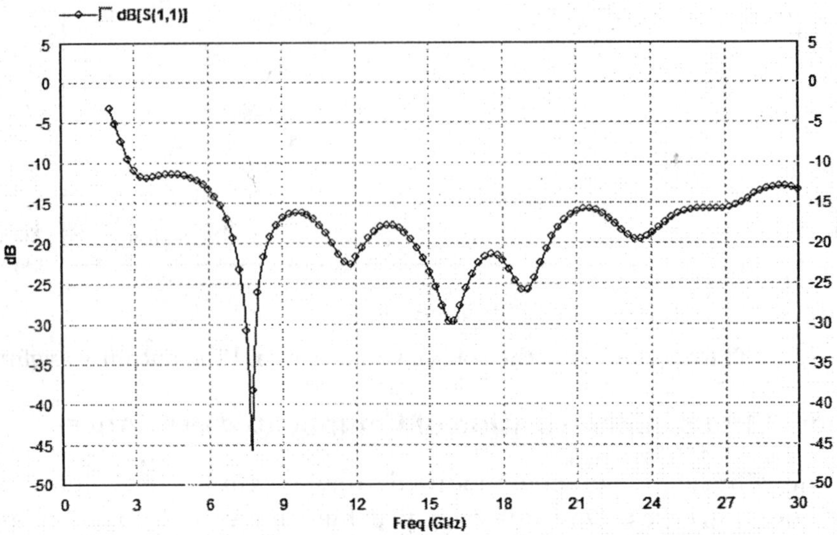

Figure 7.51 The frequency response of the UWB antenna.

Creating of the parameterized structure is documented in the IE3D Library User's Manual (see .\ie3d\manual\ie3dlibrary_manual. pdf). You can read the manual and learn how to create parameterized IE3D geometry on IE3DLIBRARY.

The structure is Fast EM prepared. You can select Process->Full-Wave EM Design Using Fast EM Design Kit command to do real-time EM tuning and optimization on it. The antenna yields the best results with normalization impedance of 120-ohms. We tried to optimize it for the goal of |S (1, 1)| < - 10.75 dB from 3.1 to 10.6 GHz with Zc = 120-ohms. It achieves the goal in seconds. The optimized geometry is saved in: .\ie3d\ie3dlibrary_examples\ep4_fastem.ie3 (for IE3D Library) and ep4_fastem.geo (for MGRID).

The example is arbitrarily chosen to demonstrate the Fast EM feature of IE3D and it turned out to have extreme wide bandwidth. It is simulated from 2 to 30 GHz and the 10-dB bandwidth on Zc = 120-ohms from below 3 GHz to more than 30 GHz (see Figure 7.51).

Section 7.20 Modeling for Complicated Wire and Patch Structures with Connections

We are able to replace the lumped elements as ports on IE3D while we can perform mixed EM and circuit co-simulation and optimization on MODUA. We will build the wire antenna with many ports (see Figure 7.54). The wire structure with many ports in Figure 7.54 can be build in one shot using the Adv Edit->Build Wire Path, possibly with the importing vertices with the command Adv Edit->Create and Edit Vertices.

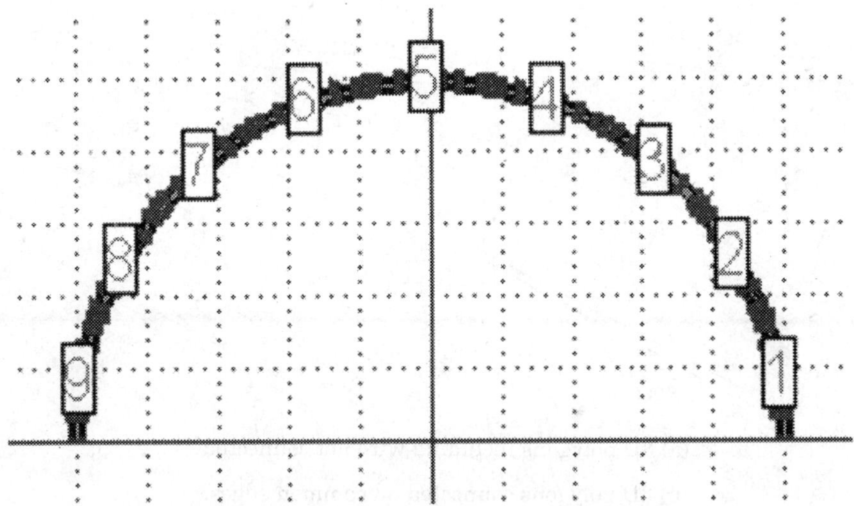

Figure 7.54 A Wire antennas with many ports for lumped element connections.

In reality, wire antennas can be more complicated. For example, a Yagi antenna may consist of multiple wire segments. We are not able to build the antenna as one single wire. We have to build them as multiple connected wires. Also, an antenna may contain plates and wires connected together. Before IE3D 14, building connections between wires and 3D polygons is not easy. We may be able to build multiple wires or 3D polygons in the same location and they may appear to be connected because they are physically overlapping each other (Figure 7.55a). However, they are not considered to be physically connected by IE3D due to the definition of connections on common edges (please refer to Chapter 4). As shown in Figure 7.55a, the polygons are not connected because there are no common edges between them. On IE3D V14, we have implemented some advance geometry modeling technology into the command Adv Edit->Connections->Adjust Geometry for Connection. The command is able to break 3D polygons and create the connections automatically for overlapped 3D polygons (see Figure 7.55). At the initial release, we would like to keep it as a manual command. You still need to select the command for the connection. If you don't select it, we will not build the connections for you. From our extensive testing, the command is quite reliable. Eventually, this command will be fully automated in IE3D's geometry modeling. Using this command, you are able to connect multiple wire segments or 3D polygons with guaranteed electrical connections easily.

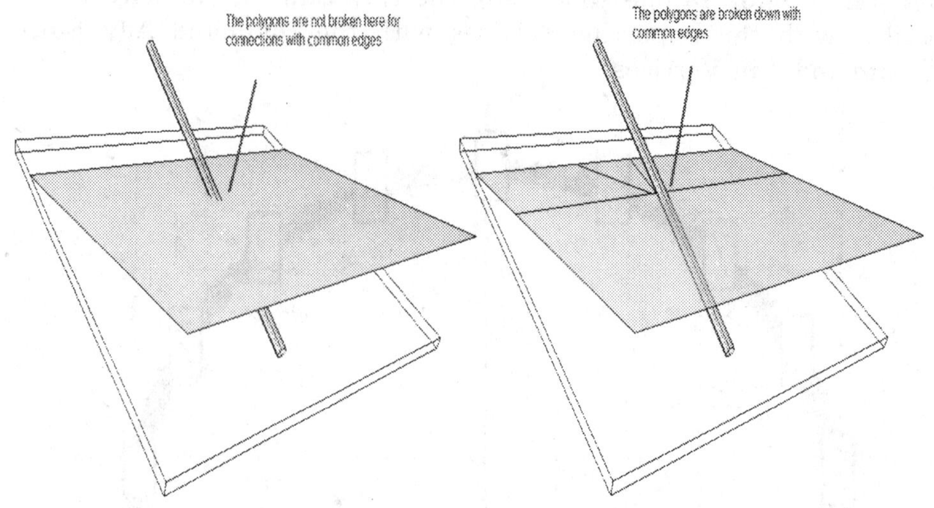

(a) 3D polygons including wire not connected

(b) 3D polygons connected on common edges.

Figure 7.55 3D polygon and wire connections on IE3D 14.

7C. ANTENNAS DESIGN, RADIATION PATTERNS AND EM OPTIMIZATION BY IE3D

One of the most important applications of the IE3D is for antenna design. We will demonstrate how we can use IE3D to perform simulation and optimization, current visualization and radiation pattern calculation.

Section 7.21 Modeling of an Edge Fed Antenna with an Inset.

The first example is an edge fed rectangular patch antenna with inset. The top view is shown in Figure 7.56. The parameters of the antenna are shown in Table 7.5. The question is how we can build the antenna. We will show you how you can use the efficient commands on MGRID to build the structure in just a few steps.

Table 7.5 The parameters of the patch antenna.

Substrate Thickness	31 mils	Dielectric Constant	4.4
Patch Length, L	1512 mils	Patch Width, W	1500 mils
Inset Width, S	115 mils	Inset Depth, D	452 mils
Strip Width, T	60 mils	Feed Line Length, F	750 mils

Step 1 Run MGRID. Select File->New command. MGRID shows you the Basic Parameters dialog. In the Length group, select the Unit as "mil". In the Meshing Parameters group, change the Highest Frequency (Fmax) to 2 GHz and the Cells per Wavelength (Ncell) to 15. Double click the "No.1 Grid Size..." in Layout and Grids group. Change it to 100 mils and select OK when you are prompted to edit the Grid Size. Select the Insert (or New) button in the Substrate Layers group. MGRID will prompt you for "Insert New Substrate". Enter "Top Surface, Ztop" as 31 (mils). Enter the Dielectric Constant, Epsr = 4.4. Select OK and the entered substrate is listed in the Substrate Layers list box. We do not need to change the Metallic Strip Types because metallic loss is normally not critical to antennas when the frequency is not very high. Select OK to continue. MGRID is ready for geometry input.

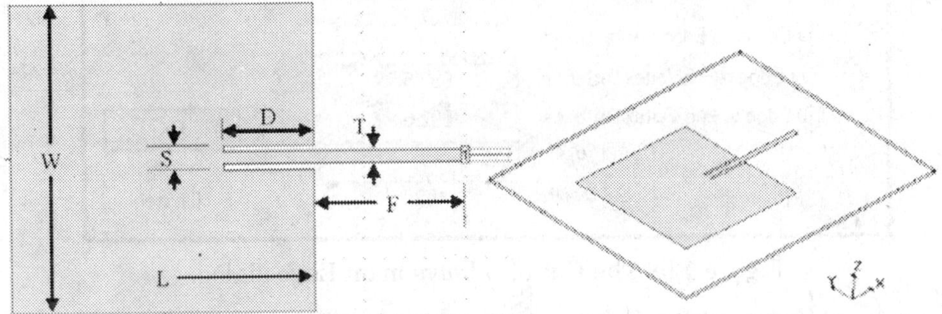

Figure 7.56 The illustration of a simple edge-fed patch antenna.

If you do not know how to define the Basic Parameters, please open .\ie3d\samples\rpatch1.geo file and select Parameters->Basic Parameters to see how we define them.

Step 2 We are going to build a rectangle without the inset first. Select Entity->Rectangle. MGRID will prompt you for the parameters of the rectangle you want to build. Enter (X, Y, Z) = (0, 0, 31). Enter Length = 1512 and Width = 1500 in the Rectangle Properties group. Select OK to continue. A rectangle is created with center at (X, Y, Z) = (0, 0, 31) as shown in Figure 5.2a.

Step 3 Press down "Shift" and window the vertices 1 and 2 in Figure 7.57a to select them. Select Adv Edit->Cut into Polygon on Edge command. MGRID prompts you for the parameters. Please enter Cut Width = 115 and the Cut Depth = 452. Select OK. MGRID will build the inset as shown in Figure 7.57b.

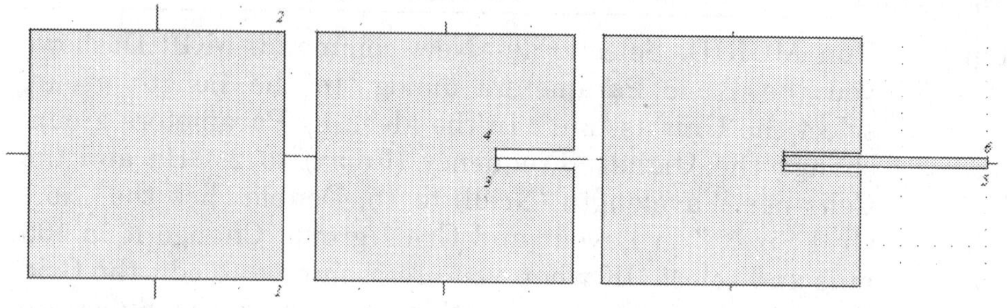

(a) Square patch (b) With inset (c) With inset and feed line

Figure 7.57 The shape of the structure at different stages.

Cut The Polygon from the Selected Edge

1st Vertex:		2nd Vertex:			
X	756	X	756	Distance:	1500
Y	-750	Y	750	Phi:	90
Z	31	Z	31	☐ Keep Cut Portion	

Cut Center to Edge Center Offset	0
Cut Edge to 1st Vertex Distance	692.5
Cut Edge to 2nd Vertex Distance	692.5
Cut Width	115
Cut Depth	452

OK Cancel

Figure 7.58 The Cut into Polygon on Edge dialog.

Step 4 Press down "Shift" and window the vertices 3 and 4 in Figure 7.57b to select them. Select Adv Edit->Continue Straight Path. Enter the Path Length = 1202 (=452+750), Path Start Width = Path End Width = 60. Select OK. MGRID will build the feed line as shown in Figure 7.57c.

Step 5 Select Port->Port for Edge Group. Choose Advanced Extension for the De-Embedding Scheme. Check the Auto Adjustment. Select OK to accept other default settings. Window the edge formed by vertices 5 and 6 in Figure 7.57c to define the port. Select Port->Exit Port to exit the state. Save the structure as: .\ie3d\practice.rpatch1.geo. The structure is ready to be simulated.

Step 6 Select Process->Display Meshing. Select AEC Layers = 1. The default AEC Ratio = 0.1 and the corresponding Width = 23.9433 mils. We know the feed line is 60 mils and the gap between the feed line and the edge of the patch is about 27.5 mils because it is 0.5*(115-60) mils. We may want to reduce the AEC Ratio to 0.05 so that the AEC Width becomes smaller compared to the gap width.

Step 7 Change the AEC Ratio from 0.1 to 0.05. The AEC Width becomes 11.9716 mils.

Please select the Meshing scheme as: Classical. Select OK to continue. MGRID will mesh the structure and create 148 cells and 256 unknowns. Select OK and MGRID displays the meshing result in Figure 7.59. The meshing result is very good. For this particular structure, the Contemporary meshing result is not as good.

Again, Process->Display Meshing is not a necessary step for a simulation. However, it is a good habit to check the meshing before some simulations in order to make sure there is no potential accuracy problem, especially when you get more experience and understand how meshing may affect simulation accuracy and efficiency.

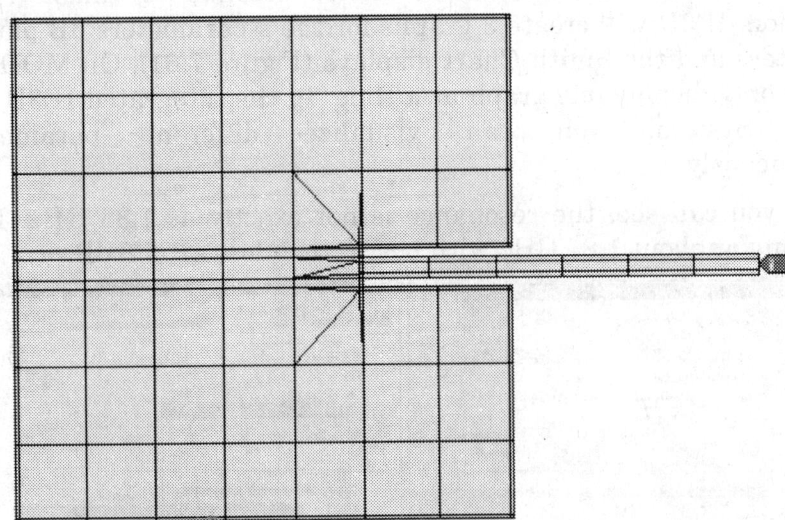

Figure 7.59 Classical meshing results on the patch antenna.

Step 8 Select Process->Simulate command. The Simulation Setup dialog comes up (Figure 7.60). We expect the resonant frequency to be about 1.88 GHz. We would like to simulate the structure from 1.7 go 2 GHz with 151 frequency points. Select the Enter button in Frequency Parameters dialog. MGRID prompts you for the frequency range. Enter Start Freq = 1.7, End Freq = 2, and Number of Freq = 151. Select

OK to add the frequency parameters into the list box. The AEC setting should be the same as before. Make sure AIF is checked.

Please select No Display for Post Processing. We don't need to use MODUA to display the s-parameters normally. We are going to use the integrated visualization on MGRID. Select the Define Graphs button (Figure 7.60). It brings up the "S-Parameters and Frequency..." dialog. Select Add Graph button. You are prompted for the Graph Type. Select S-Parameters. You will be prompted for Display Selection for the Graph. Select dB[S (1, 1)]. Select OK. The defined graph is added into the list. Select Add Graph button again. Select Z-Parameters and OK. Select Re [Z(1,1)] and Im[Z(1,1)]. Select OK to define the second graph. Select Add Graph button again. Select Smith Chart and OK. Select S (1, 1) and OK to define the third graph. Select Continue to go back to the Simulation Setup dialog. You will see "3 graphs are defined" next to the "Define Graphs" button. What we want to do is to jump into s-parameter visualization directly by pre-defining the graphs.

Select OK to continue. MGRID will invoke IE3D to perform the simulation in the background. It takes seconds to finish. After simulation, IE3D will create 3 graphs for the s-parameters dB plot, z-parameters and the Smith Chart displays (Figure 7.61). On MODUA, you can only display one graph at a time. In the integrated IE3D EM Design System, you can visualize different parameters simultaneously.

As you can see, the resonance is not exactly at 1.88 GHz. It is resonating at about 1.87 GHz with best return loss as -5.8 dB.

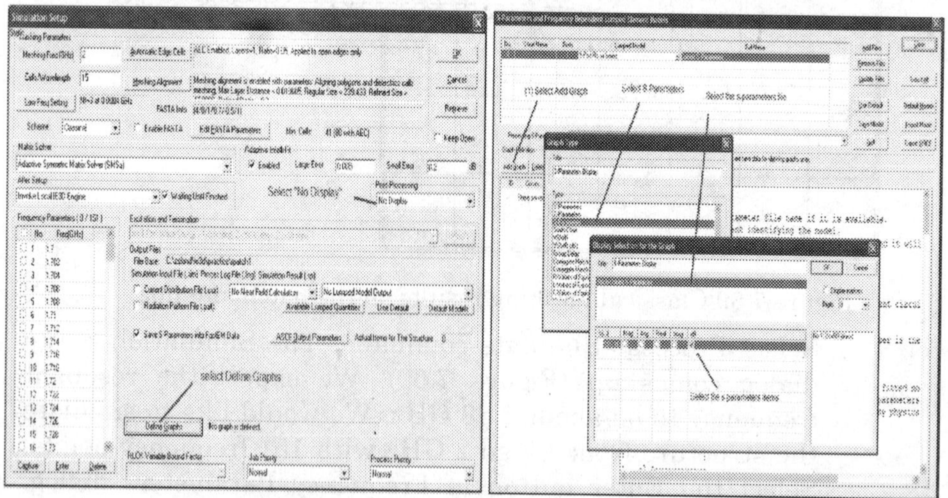

Figure 7.60. The Simulation Setup dialog and the dialogs for defining a graph.

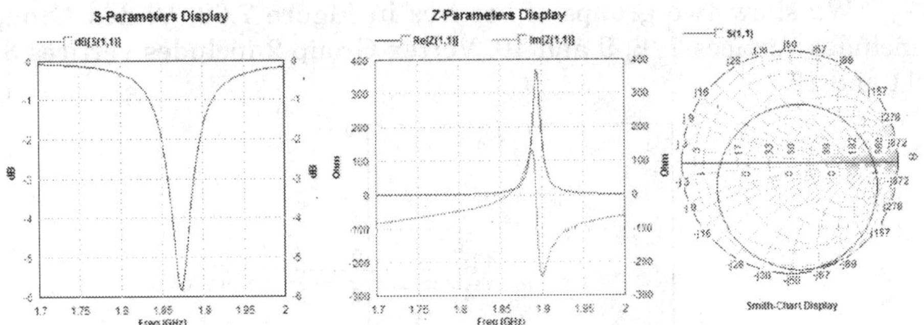

Figure 7.61. The simulation results from different meshing schemes.

Section 7.21 EM Optimization.

Assume we want to optimize this antenna to a perfect match at 1.88 GHz. We would like to study what we can do for it. EM optimization is one of the most important features of IE3D. It can help designers to achieve their goal with much less effort. Some goal may even be impossible by manual tuning. EM optimization may yield excellent results for you. However, please understand the following fact: EM optimization is not supposed to replace designers. Users should know what they want to do and what they can do before they jump into optimization. If you throw a garbage design into an EM optimizer, it is quite likely it will throw a garbage design back to you.

Before you start an optimization, you should first consider the followings: (1) What are your goals? (2) What geometry dimensions we can adjust to achieve the goals?

For the antenna we are discussing, we can adjust the length L to change the resonant frequency. We can adjust the inset depth D to tune the matching. Also, both variables are not independent or changing L may change the matching and changing D may change the resonance. If they are independent, we can optimize one dimension at a time and it may make an optimization much easier. Our goals are: Re[S(1,1)] = 0 and Im[S(1,1)] = 0 at 1.88 GHz.

On MGRID, structures are not described as parameterized objects. All structures are described by polygons and polygons are described by vertices. To change the shape of a structure, we need to change the locations of vertices. We need to identify which vertices we should adjust to change the L and D values. IE3DLIBRARY has complete parameterized structure objects with equation-based dimensions. It is extremely flexible in optimizing structures. In this chapter, we will concentrate on MGRID.

We show two groups of vertices in Figure 7.62: Vertex Group 1 includes vertices 7, 8, 9 and 10. Vertex Group 2 includes vertices 3, 4, 11 and 12.

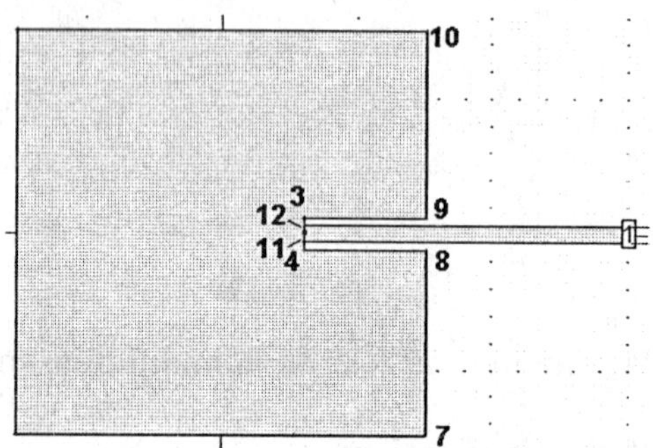

Figure 7.62 The patch and its vertices to be defined as optimization variables.

To change the patch length L, we can change the X-coordinate of Vertex Group 1. To change the inset depth D, we can change the X-coordinate of Vertex Group 2. Can we define the X-offset values of the two groups as optimization variables? Certainly, we can do it. However, it is not the best.

The reason is related to the bounds of the optimization variables. We know the difference in the X-coordinate between Vertex Group 1 and Vertex Group 2 is D = 452 mils. If we define the X-offset of Group 1 as the first variable V1, and the X-offset of Group 2 as the second variable V2, we need to make sure the high bound of V2 cannot be too big and the low bound of V1 cannot be too small. Otherwise, we may cause the vertex 3 to go to the right hand side of vertex 8. For this particular structure, such a situation may not create an invalid geometry. However, there is quite a chance we may create invalid geometry when the bounds of optimization variables are not defined properly. As you will see later, we are defining the variables and their bounds one by one. We may not know the bounds or ranges of previous variables when we are defining a specific variable.

For this particular structure, we would like to define the X-offset of the two groups as V1. Then, we define the extra X-offset of Group 2 as the V2. In this way, when we define the high bound of V2, we just need to make sure V3 will not go to the right hand side of V8 visually.

Step 1 Select Edit->Select Vertices. Window the Vertex Group 1 (or vertices 7, 8, 9 and 10) only. We can select the Vertex Group 1 and Vertex Group 2 simultaneously to define them as the No.1 variable. It is simpler. However, we would like to demonstrate how we could associate different vertex groups to one single variable by taking two steps with one step for one group.

Step 2 Select Optim->Variable for Selected Objects. MGRID will prompt you for the Optimization Variable Definition dialog. The default is "Vertices Mapped to New Variable" (Figure 7.63). Please make sure the Tuning Angle = 0 because we want to change the x-coordinate of the vertices. Select OK. MGRID is in the Low Bound definition mode. Move the mouse to the left and you can see the frame of the structure is changing. Click the mouse at appropriate location. MGRID will prompt you to Set Low Bound. Please enter the Low Bound = -100. We know the L should be close and we do not need to define very large bounds. MGRID may warn you the change of geometry will lose the results. Select YES to continue.

After you select OK, MGRID is in the mode for defining the high bound. Move the mouse to the right and you can see the frame of the structure changing. Click the mouse at appropriate location. MGRID will prompt you to Set High Bound. Please enter the High Bound = 100 and select OK. MGRID will prompt you the Defining No.1 Variable Finished dialog. It allows you to check bounds. It also allows you to make some random trials to see whether the bounds are properly defined. The random check may be able to capture possible improper settings of the bounds causing invalid structure. In such a case, MGRID allows you to find the bounds which may avoid invalid geometry problem.

Step 3 Select Continue Without Action button. MGRID will finish the definition of the first variable.

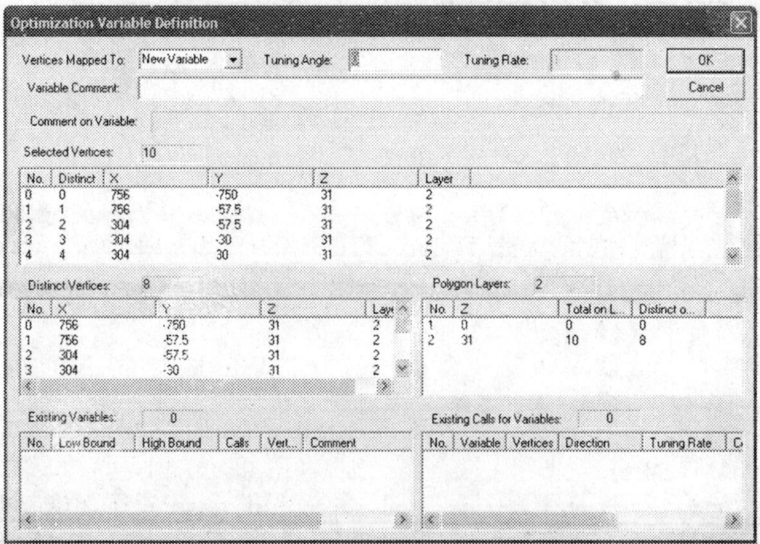

Figure 7.63. The Optimization Variable Definition dialog for the Optim->Variable for Selected Objects

Step 4 Select Edit->Select Vertices command. Window the Vertex Group 2 (or vertices 3, 4, 11 and 12) to select them. We are going to associate the vertices with the first variable.

Step 5 Select Optim->Add Selected Objects to Variable command. The Optimization Variable Definition dialog comes up again. It is very similar to Figure 7.63 except it indicates "Vertices Mapped to No.1 Variable" and the Tuning Rate is activated with a default value of 1. The default is that you want to add the selected vertices to the existing No.1 variable. If you like, you can change it to "Vertices Mapped to New Variable" to define the second variable. For this case, we will demonstrate how you define both Vertex Group 1 and Vertex Group 2 as the No.1 variable. We should accept the default.

We also want the Vertex Group 2 to change in Tuning Angle = 0. We want the Tuning Rate = 1 so that the Vertex Group 2 will change at the same speed as the Vertex Group 1. Please note that the Tuning Angle and Tuning Rate can be set different for every Optim->Add Selected Objects to Variable command. You have the flexibility for tuning very complicated structures.

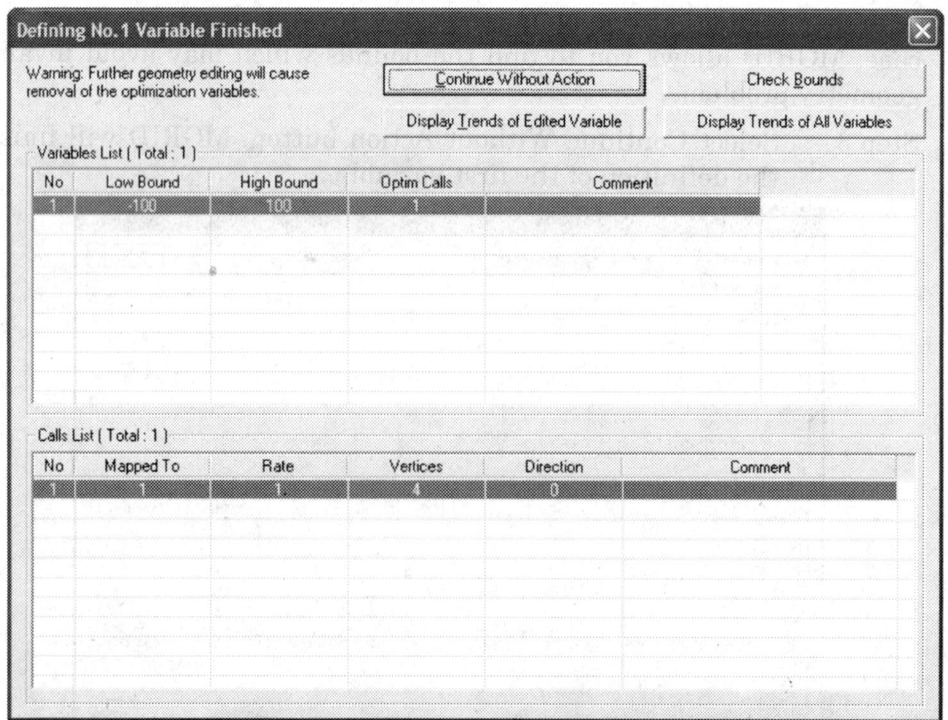

Figure 7.64. The Defining No.1 Variable Finished dialog.

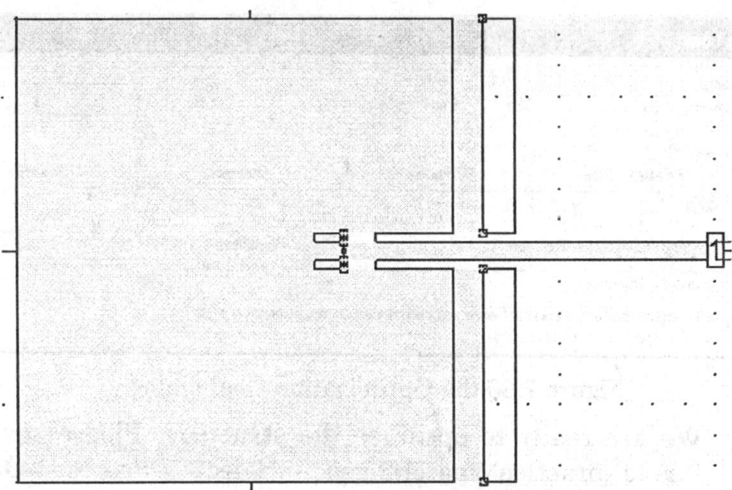

Figure 7.65 The trends and bounds of the No.1 variable.

Step 6 Select OK. MGRID will show the Defining No.1 Variable Finished dialog. Select Display Trends of Edited Variable. MGRID will close the dialog and show you the trends of the bounds and you can see the range of the vertices when it is tuned (see Figure 7.65).

Step 7 Select View->Refresh to clear the display of the bounds. Select Edit->Select Vertices. Window the Vertex Group 2 again. We will define the Vertex Group 2 alone as the second variable. Select Optim->Variable for Selected Objects. The Definition of Optimization Variable dialog comes up. By default, it is "Vertices Mapped to New Variable". The Tuning Angle = 0 and it is what we want. Select OK to continue. MGRID is in defining Low Bound mode. Move the mouse to the left and click. Enter the Low Bound as -200. Select OK. MGRID is in defining High Bound mode. Move the mouse to the right and click. Enter the High Bound as 200. Select OK. MGRID prompt you "Definition of No.2 Variable Finished" dialog. You can choose Continue Without Action or Display Trends of Edited Variable or Display Trends of All Variables. If you are not sure whether the bounds of the variables are defined correctly, you can also select Check Bounds to let MGRID to check whether the bounds are properly defined.

Step 9 Select Display Trends of All Variables. MGRID finished the definition and display the trends of the two variables for you. You can also select Optim->Geometry Tuning to see how the variables are controlling the shape of the antenna.

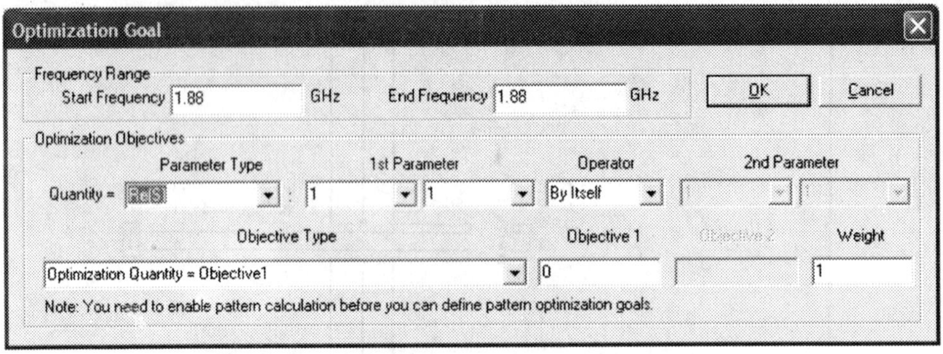

Optimization Goal

Frequency Range
Start Frequency 1.88 GHz End Frequency 1.88 GHz
OK Cancel
Optimization Objectives
Parameter Type 1st Parameter Operator 2nd Parameter
Quantity = 1 1 By Itself
Objective Type Objective 1 Objective 2 Weight
Optimization Quantity = Objective1 0 1
Note: You need to enable pattern calculation before you can define pattern optimization goals.

Figure 7.66 the Optimization Goal dialog.

Step 10 We are ready to optimize the structure. Please save it as: .\ie3d\practice\rpatch2.geo. Select Process->Optimize command. The Optimization Setup dialog comes up. It is the same as the Simulation Setup dialog except there is some minor change. Select Delete button in Frequency Parameters group. Choose Delete All and select OK to define the 151 frequency points in the list. Select Enter in the Frequency Parameters group. Enter the Start Freq = 1.88, End Freq = 1.88 and Number of Freq = 1. Select OK to add the frequency parameters into the list.

Please select the Optimization Scheme as the Powell. It is good local optimizer even though the Adaptive EM Optimizer is the best for general optimization. Please read the Appendix AU for more information on different schemes.

Select the Insert button in the Optimization Definition group. MGRID will prompt you to define an Optimization Goal. Please define Start Frequency = 1.88, End Frequency = 1.88, Quantity = Re(S), 1st Parameter = (1, 1), Operator = By Itself. Objective Type: Optimization Quantity = Objective1, Objective1 = 0, and Weight = 1. Basically, it defines Re[S(1,1)] = 0 at 1.88 GHz (see Figure7.66). Select OK to add the defined goal into the list.

Step 11 Select Add button again in the Optimization Definition group again. Change the Quantity = Im(S) and select OK to accept other settings. Basically, we define the goal as: Im[S(1,1)] = 0 at 1.88 GHz. Select OK and MGRID puts the 2nd goal into the list. We will get the Optimization Definition dialog as shown in Figure 5.12.

Step 12 Select OK to continue. MGRID will invoke IE3D to perform an optimization for the goals. The optimization dialog is shown in Figure 7.67. It is trying to perform a smart iteration until the goals are achieved.

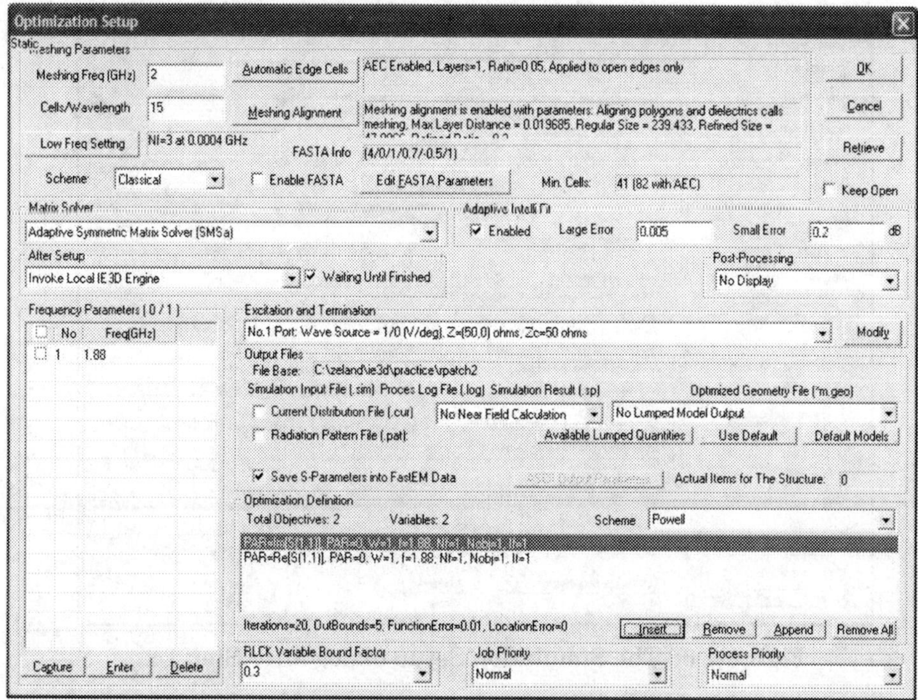

Figure 7.67 The Optimization Definition dialog after the goals are defined.

Step 13 Open .\ie3d\practice\rpatch2m.geo. Select Process->Simulate. Select Enter in Frequency Parameters dialog. Enter Start Freq = 1.7, End Freq = 2.0, Number of Freq = 151. Select Ok to add the frequency points into the list.

Step 14 Scroll the frequency list box until you see the No.91 f = 1.88 GHz. Check the check box for it. Starting from the IE3D 11, we can specify which frequency points we want the IE3D simulator to simulate it even with the AIF enabled. We want to use AIF for the 151 frequency points while we also want to make sure the IE3D engine actually simulates the 1.88 GHz. Please note that the No.1 frequency point in the list is always simulated.

Also, please check the Radiation Pattern File check box. The Current Distribution File check box is automatically checked (see Figure 7.67). Starting from IE3D 11, we can save the .cur file and .pat file even with AIF enabled. IE3D will save the .cur file and .pat file at the frequency points actually simulated. In case MGRID complains about no enough angles for the pattern, please select Modify button for the Excitation and Termination and Pattern Angles dialog. Please select Add Theta button to add 37 angles from 0 to 180 degrees and Add Phi button to add 37 angles from 0 to 360 degrees. We need to have enough angles for accurate pattern data.

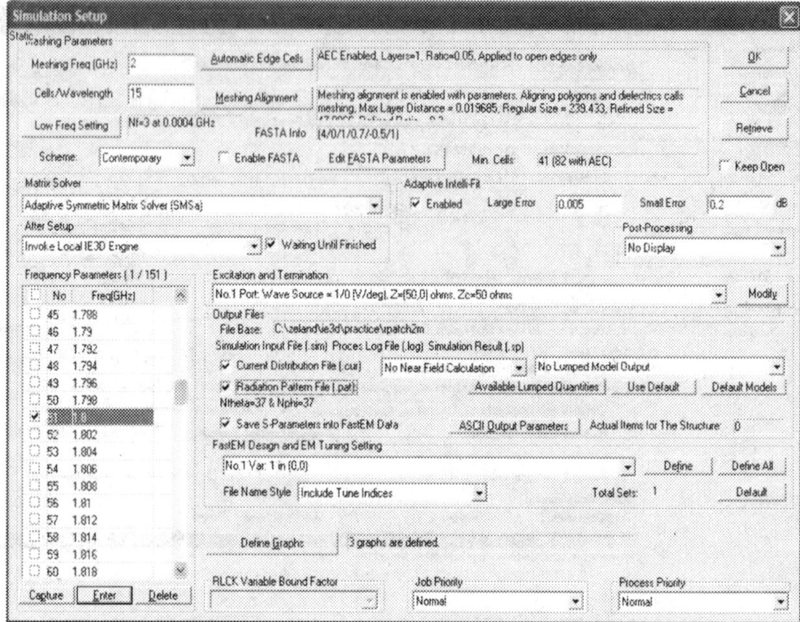

Figure 7.68 The Simulation Setup dialog after Step 14.

Starting from IE3D 11, we will save two radiation pattern related files: Radiation Pattern File (.pat) and General Pattern File (.mpa). The .pat file contains the pattern with the excitation specified and the excitation can't be changed. The .mpa file contains all the pattern data with the specified excitation while the excitation can be redefined with any possible excitation. You can always find the .pat file for a specified excitation combination from the .mpa file on MGRID or PATTERNVIEW in no time whenever necessary. More information about .pat file and .mpa file can be found in Appendix AV. Starting from IE3D 14, we can add an .mpa file into the pattern list for visualization and comparison on MGRID/PATTERNVIEW. They will treat an .mpa file as if it is a .pat file while you can change the excitation of an .mpa file in the list.

Step 15 Select OK to start the simulation for the optimized geometry. It takes seconds to finish it. The three pre-defined graphs are created automatically to display the simulation results similar to what are shown in Figure 5.14. There should not be the legend "Optimized' and the curve for the "Before" is not there. As you can see, the dB[S(1,1)] is below -35 dB and it is very close to perfect match and our goals are achieved.

Step 16 Select Window->Display S-Parameter Graphs->S-Parameters and Lumped Equivalent Circuit to bring out the dialog for s-parameter processing (Figure 5.15). Change the

Short Name to "Optimized". Select Add Files button. Select the file: .\ie3d\practice\output\rpatch1.sp". Change the Short Name to "Before". Double-click at the S-Parameter Display graph to edit it. Select the second s-parameter file in the list in the Display Selection for the Graph dialog. Check dB[S(1,1)]. Select OK to get back the s-parameter processing dialog. Select Close. MGRID will refresh the displays with the s-parameters compared in the S-Parameter Display window with the correct legends (Figure 7.69).

As it is shown in the comparison, the optimized antenna is perfectly matched at 1.88 GHz and its performance is significantly improved from the original one.

You may wonder what the L and D in the final optimized geometry are. There are multiple ways to measure them. The simplest way is to select Geometry tab in the Information Bar. Click on Z = 31 on the Layer Window to focus the mouse input on the layer. Move the mouse to an edge for the L. MGRID will show the edge's information to you. The finial L should be about 1500.72 mils and the final D should be about 615.65 mils.

We demonstrated how we could optimize the s-parameters of a structure. In fact, you can optimize many parameters such as the Y and Z-parameters, antenna gain, directivity, axial ratio, and efficiency etc. You can even optimize the shape of a pattern. When you try to optimize pattern parameters, you should first enable "Radiation Pattern File" before you define the optimization goals.

For the example demonstrated here, we defined the length and the depth of the notch as optimization variables. We can also vary other dimensions. Parameter zing a structure is much easier on IE3DLibrary. The corresponding IE3DLibrary file with all dimensions parameterized is saved in: rpatch2_tuning2.ie3. Interested users can explore it.

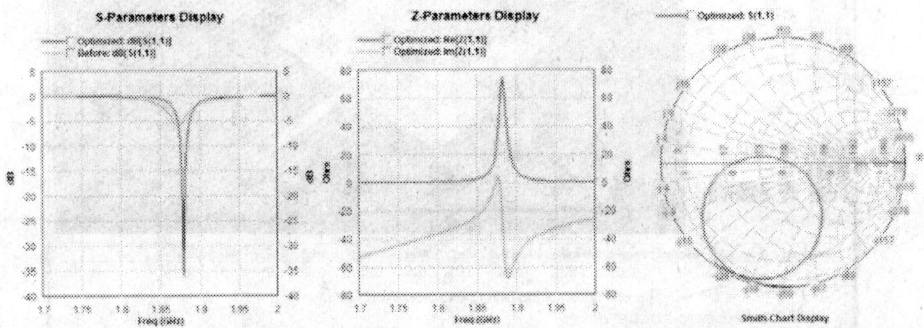

Figure 7.69 Comparison in S-parameters between original and optimized structures.

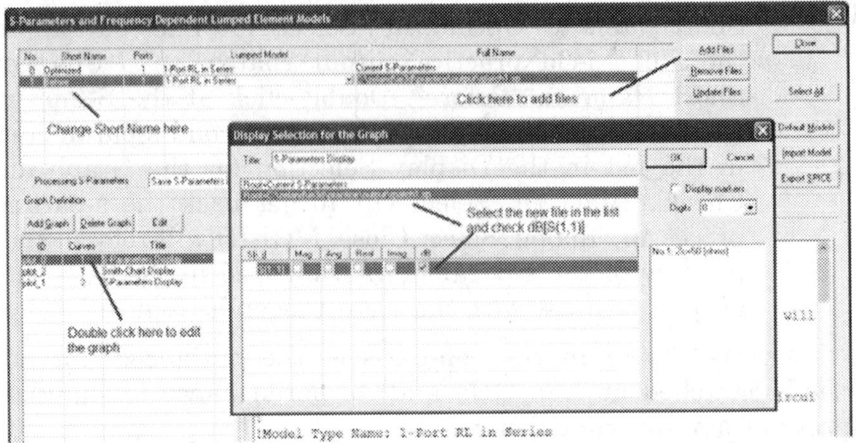

Figure 7.70 Adding more items into the graph for comparison.

Section 7.22 Current Density Distribution Visualization

After an IE3D simulation is finished, the current distribution data is saved into the file: .\ie3d\practice\output\rpatch2m.cur. Current density visualization was used to be performed on MGRID in the post-processing mode. We can still use MGRID to open the .cur file to visualize and post-processing the current distribution. However, it is not necessary for IE3D 14. After an IE3D simulation is done, the s-parameters are saved backed to the .geo file. The saved current distribution, radiation pattern data and near field data are saved into the results files (.cur, .mpa, .pat, .fld etc). However, we will automatically build a link between the .geo and the result files. When we open the .geo file again, we are able to access the current density data in the .cur file and the .mpa files through an internal link.

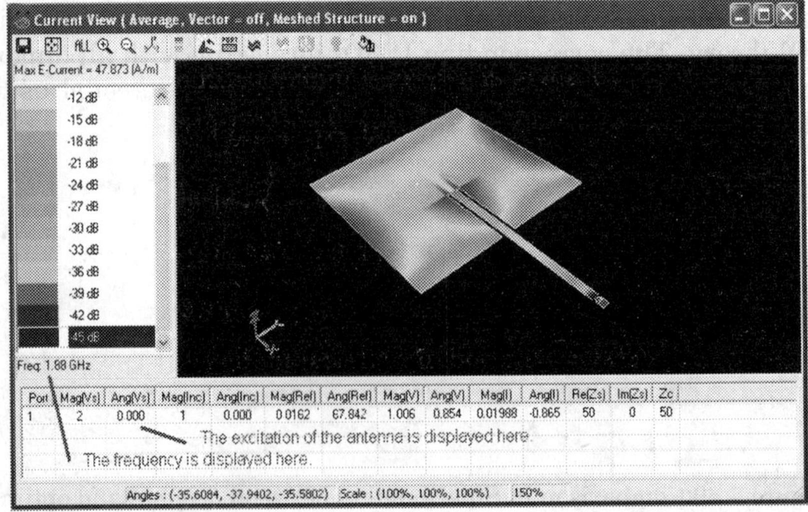

Figure 7.71 The average current distribution display at 1.88 GHz.

Step 1 While MGRID is displaying the geometry and the s-parameter results, please select Windows->3D Current Distribution Display. MGRID will show you the average current distribution similar to what is shown in Figure 5.16 except the frequency is 1.7 GHz which is the 1st one in the list. We have simulated total 151 frequency points from 1.7 to 2.0 GHz and we got accurate s-parameters over the whole band. However, we didn't actually simulate every frequency point but only selected ones. What is shown by default is the average current distribution at the 1st simulated frequency point. We can change the frequency by typing "N" for next frequency point or "P" for previous frequency point.

Please type "P" or "N" a few times to get the frequency point at 1.88 GHz or the resonant frequency. You will see the current is the densest at this point because of resonance.

There are many settings you can configure the current density display in the Options dialog (Figure 7.72a). For example, you can choose Prospective View or Orthographic View. You can choose whether you want to show the Color Palette on the left or the Port Information at the bottom. You can adjust the Light Source. In the Options->Structure Display Options, you can configure how you want to display the structure. You can show or disable the display of substrates.

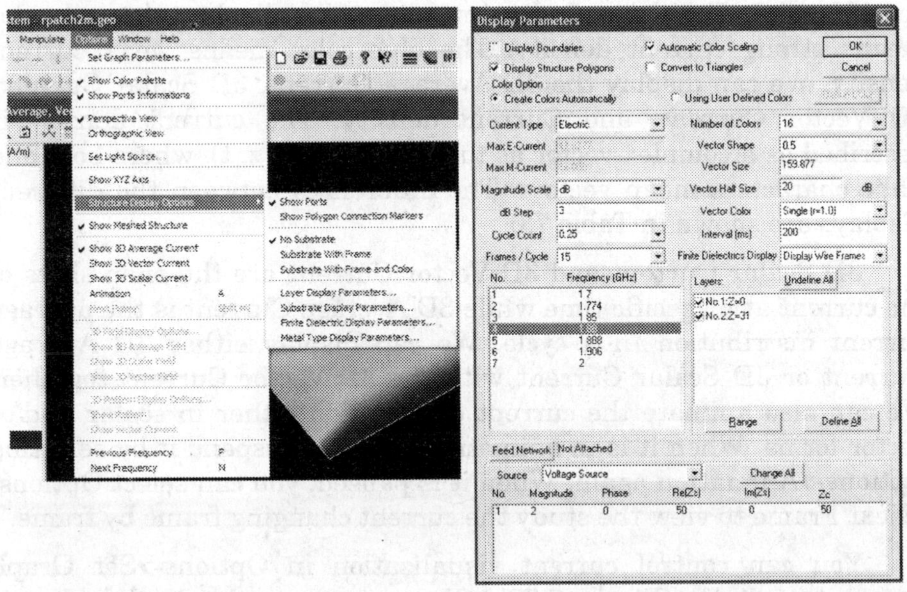

(a) The many items in the Options menu (b) The Set Graph Parameters dialog.

Figure 7.72 The Options menu and the Set Graph Parameters dialog.

$$\boldsymbol{J}(x,y,z,t) = [\, Jx(x,y,z)\,\boldsymbol{x} + Jy(x,y,z)\,\boldsymbol{y} + Jz(x,y,z)\,\boldsymbol{z}\,]\, \exp(j\omega t) \quad (5\text{-}1)$$

Table 7.6 The different current visualization schemes

Scheme	3D Average Current	3D Scalar Current and Animation	3D Vector Current and Animation
Display Quantity	It is showing the $\lvert \boldsymbol{J}(x,y,z,t) \rvert$ or the absolute value of the complex vector current in the form of colors on the structure.	It is showing $\lvert Re[\boldsymbol{J}(x,y,z,t)] \rvert$ with t defined by Cycle Count in the form of color on the structure Cycle Count = .0.25 means $\omega t = \pi/2$.	It is showing $Re[\boldsymbol{J}(x,y,z,t)]$ with t defined by Cycle Count Cycle in the form of vectors.

Step 2 To change the color scale and other settings, please select Options->Set Graph Parameters dialog and you have the dialog shown in Figure 7.72b. The color palette is automatically set by MGRID. If you want to change the corresponding 0 dB value from the default (automatically detected as 47.873 A/m), you can un-check Automatic Color Scaling and change the Max E-Current value, the Magnitude Scale, and the dB Step value.

By default, what is displayed is the average strength of the time-harmonic current density distribution at a specific frequency. The color represents the average strength of the current density at a specific point. For the default continuous tone display, the red color means strong current density. The blue color means weak current density. We can display the 3D Average Current, 3D Scalar Current, 3D Vector Current, and Current density. The current density is described as a complex vector as function of (x, y, z, t), where bold and italic characters mean vectors. The differences between the different displays are shown in Table 7.6.

3D Scalar Current and 3D Vector Current are the snap shots of the current at a specific time while 3D Average Current is the average current distribution in a cycle. We can display either 3D Average Current or 3D Scalar Current with the 3D Vector Current together. We can also animate the current distribution either in scalar and/or vector forms. When it is in animation, you can suspend it by selecting Options->Animation again. When it is paused, you can select Options->Next Frame to view the study the current changing frame by frame.

You can control current visualization in Options->Set Graph Parameters dialog (Figure 7.72b). You can choose the display electric current (E-current) or magnetic current (M-current). You can pick which layers you want to see and which layers you don't want to see. You can change the color scaling on Magnitude Scale, dB Step, and the Color Option. For scalar and vector current display, you can use the Cycle Count to change the time or phase for the snap shot. For animation, you can change the Frame/Cycle and Interval (ms) to control the updating speed. For vector display, you can change the Vector Shape, Vector Size, Vector Half Size, and Vector Color. Due to the fact that current density may change wildly at different location, we have implemented the Vector Half Size to control it. Vector Half Size = 20 dB means the following: If the length of vector A is 50% as big as vector B in the display, vector A is 20 dB smaller than vector B.

Step 3 Select Options->Show 3D Vector Current to show the vector current distribution. Select Options->Set Graph Parameters and adjust the parameters. You will get a similar display as what is shown in Figure 7.73.

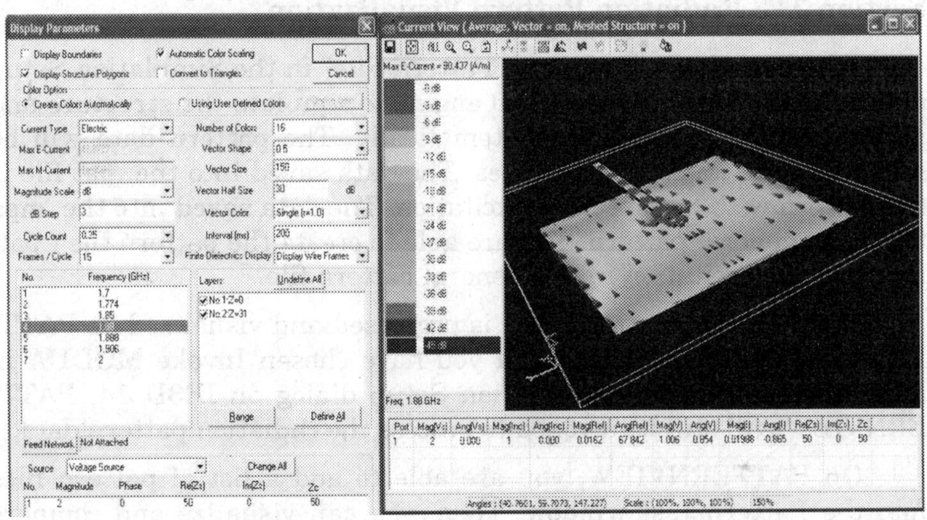

Figure 7.73. Vector current display and its setting.

Step 4 Select Options->Animation. MGRID will start animating the time-harmonic current distribution in real-time. You will see the color representing the current density distribution on the patch changes from blue to red and cycling. On the patch, it is resonating and there is little net power traveling going on. On the feed line, you will see the peak in red traveling into the patch antenna. It is indicating the antenna has reasonably good matching and power is delivered to the patch antenna at 1.88 GHz.

The animation is performed in real-time. For simple structures, you should see very smooth display. However, for complicated structures, updating each frame in animation may take some time.

You may not be able to see the animation very smooth. In such a case, you should try to perform the animation on ZDib Animator by creating the bitmap files.

Step 5 You can select File->Save to Bitmap Files. MGRID will prompt you for the Bitmap files name prefix. If you just want to save one case, you can leave it blank. Please check Invoke ZDibAnimator for animation. Select OK. MGRID will save a series of bitmap pictures recording the progress of current distribution with time. Then, it will invoke ZDibAnimator to do the animation. You can play an animation on the saved bitmap files anytime later by running the ZDibAnimator from Zeland Folder.

We have touched base with current visualization. There are many options in current visualization. Please try to explore them.

Section 7.23 Radiation Pattern Visualization

When we check "Radiation Pattern File" in the Simulation Setup dialog, MGRID will invoke IE3D engine to simulate the structure and perform pattern calculation automatically. The pattern data will be saved into the .mpa and .pat files. The data saved into the .pat file is the pattern with the specified excitation. The data saved into the .mpa file is the general pattern. You are able to create the pattern file (.pat) for a specific excitation from a general pattern file.

Traditionally, pattern data is processed and visualized on PATTERNVIEW before version 14. If you have chosen Invoke MODUA in Post-Processing in the Simulation Setup dialog on IE3D 14, PATTERNVIEW will still be invoked to process the radiation pattern data.

On PATTERNVIEW, you are able to add a list of pattern files into the Pattern List window. Then, you can visualize and compare the data. In PATTERNVIEW 14, you can add either a general pattern file (.mpa) or a specific pattern file (.pat) into the list. If it is a specific pattern file (.pat), you will not be able to change the excitation. If it is a general pattern file (.mpa), you are allowed to change the excitation anytime you want.

In fact, all the functionalities of PATTERNVIEW are implemented into MGRID on IE3D 14. You are able to post-process and visualize radiation patterns on MGRID 14 completely. On both MGRID 14 and PATTERNVIEW 14, the primary pattern files are the general pattern files (*.mpa). You can add a number of pattern files into the Pattern List. If a pattern file is a general pattern, you can even change its excitation anytime you want. We may define excitations on ports for a particular structure. It is also possible we define plane wave excitation on the structure. The radiation properties of a structure can be quite different depending upon whether it is excited by ports or by plane waves. When we add a number of files into the Pattern List for comparison, we will need to make sure all of the files are excited by either ports or plane wave simultaneously. In the Pattern List of MGRID 14 and PATTERN- VIEW 14, we will list the excitations of the patterns on the Pattern List.

Step 1 While you are opening the geometry file on MGRID, please select Windows->3D Radiation Pattern Display. The 3D Pattern Selection dialog comes up (see Figure 7.74). The rpatch2m.mpa file is automatically added into the list. In case you want to add more pattern files into the list, you can select Pattern List button to add more pattern files into the list. You can also access the pattern's properties inside Pattern List.

Step 2 For this example, we just want to display the 3D pattern, please select the only file in the list. Then, select No.3: 1.88 GHz from the frequency list. You will see the Max (dB) is updated to 4.7903 (see Figure 7.74). This is the maximum dB value of the selected field item. By default, the selected field item is the dBi(Gain) of the E-total field. It means the maximum E-total field for dBi(Gain) is 4.7903 dB. Select OK. The 3D radiation pattern is displayed (see Figure 5.20).

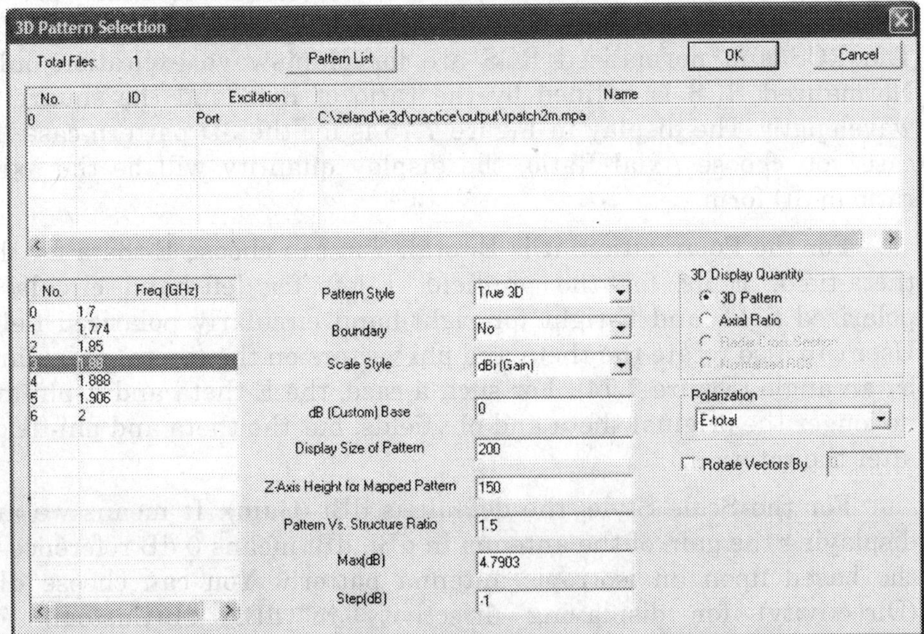

Figure 7.74 The 3D Pattern Selection dialog.

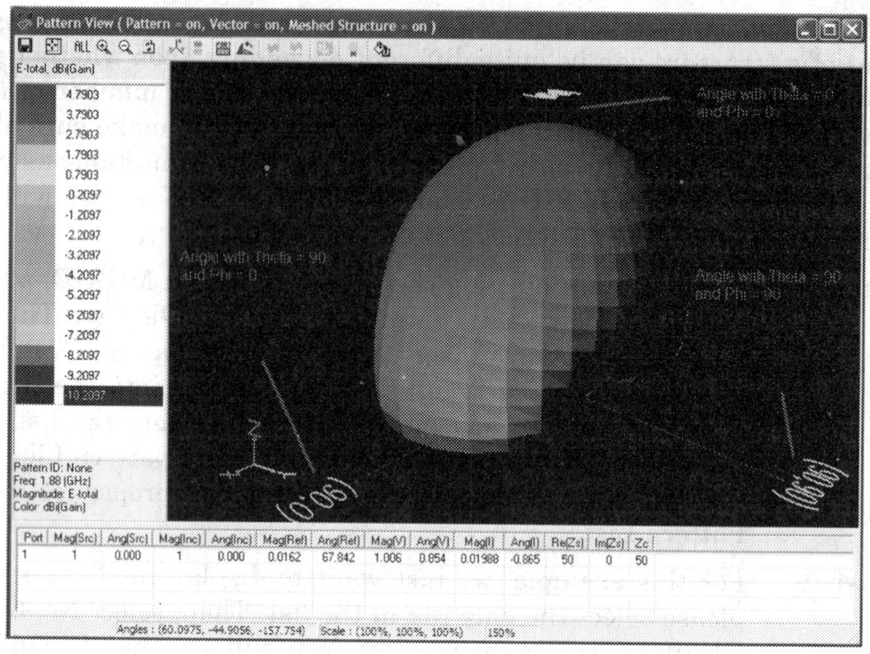

Figure 7.75 The 3D radiation pattern.

On the 3D pattern display, there are 3 axes. Those are the axis for the three main angles: (Theta, Phi) = (0, 0), (90, 0) and (90, 90).

There are many parameters you can access in the 3D pattern display (Figure 7.74). You can choose the 3D Pattern Quantity as: 3D Pattern, Axial Ratio, Radar Cross Section (RCS) and Normalized RCS. The RCS and normalized RCS are for plane-wave excitation only. Normalized RCS is defined by the ratio of RCS and the square of wavelength. The display in Figure 7.75 is for the 3D pattern case. In case you choose Axial Ratio, the display quantity will be the axial ratio in 3D form.

For the Polarization, it is E-total. You can choose E-theta for the theta-field, E-phi for the phi-field, E-left for left-hand circularly polarized field, and E-right for right-hand circularly polarized field. User can also rotate the theta and phi vectors on the (theta, phi) plane by an angle (Figure 7.74). For such a case, the E-theta and E-phi are no longer the original theta and phi fields, but the theta and phi-fields after the rotation.

For the Scale Style, the default is dBi (Gain). It means we are displaying the gain of the antenna in dBi. dBi means 0 dB reference is the based upon an isotropic antenna pattern. You can choose dBi (Directivity) for displaying directivity in dBi, dB(Custom) for displaying the pattern with use specified dB(Custom) Base, Linear scale with the maximum at 1, Phase for the phase of the field, and dBi

(Conj. Match Gain) for the gain based upon conjugate match. The standard gain definition is based upon the port impedance. For RFID applications, designers are not interested in the gain with matching to the port impedance. They are more interested in the antenna gain and efficiency based upon conjugate match. For this reason, we have defined Conjugate Match Efficiency and Conjugate Match Gain.

You may notice we do not use the terminologies of E-plane and H-plane patterns and Linear Gain. You may want to know how you can get those parameters. The answer is that we do not provide those parameters. The reason is due to the fact that IE3D is a general EM simulator. You have the complete freedom to create any antenna structure you like and simulate it. IE3D does not know what the E-plane and H-plane of the antenna are because the planes are defined by the designers. We cannot provide such information. Instead, we provide the information about Total Field, Theta-Field, Phi-Field, LH Circular Field and RH Circular Field. If your antenna is linear polarized antenna, you can consider the Total Field or the Theta Field pattern as your primary radiation pattern while the Phi Field pattern as the cross-polarization. When it is circular polarized antenna, you should use either LH Circular Field or RH Circular Field as your primary polarization and the other one as the cross-polarization. Please note the following fact: For a circular polarized antenna, the Total Field does not represent the linearly polarized field. The pattern data for E-total is assuming all the fields (both E-theta and E-phi) are received completely.

There are many items in the Window menu of MGRID (Figure 7.76a), allowing you to visualize different parameters of your structure. You can display the 2D patterns of the pattern on specified phi and/or theta angles. You can display the plot with gain vs. frequency etc.

Step 2 Select Window->Radiation Pattern Properties. MGRID will show you the Pattern List (Figure 7.76b). You can add more files into the list for display and comparison purposes.

Step 3 Select the only pattern file in the list. Select the Properties dialog. The General Pattern Properties dialog will come up (Figure 7.77a). In the dialog, you are able to change the excitations and terminations of the structure for each frequency point and post-process the pattern. In case the selected pattern file in the Pattern List dialog is a pattern with specified excitation (*.pat), you will see the specified pattern properties shown in Figure 7.77b when you select the Properties dialog in the Pattern List.

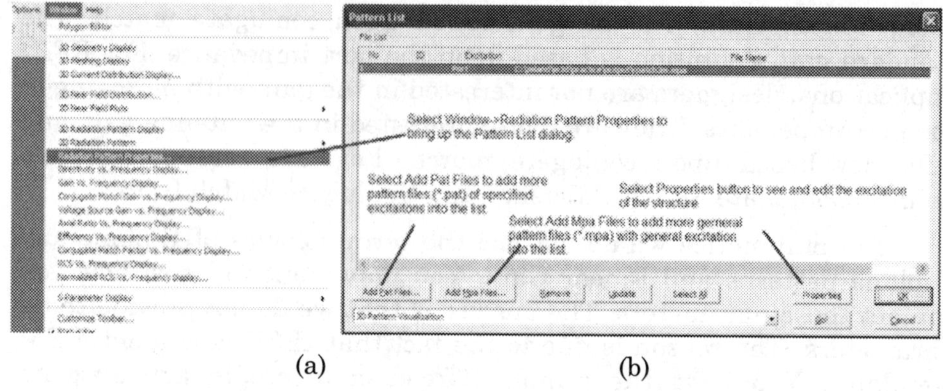

(a) (b)

Figure 7.76 The Pattern List window.

Step 4 Select "Display Detailed Pattern Properties" in the combo box (Figure 5.22a). Select GO button. MGRID will show you the detailed pattern properties of the antenna in Pattern Properties dialog (Figure 7.77b). Select View in Browser. MGRID will save the pattern properties data into an XML file and open a browser for it. There are many pattern parameters available in the dialog. The meanings of the parameters are documented in the Definitions dialog (Figure 7.77b). The selected parameter parameters at selected frequency points are shown in Table 7.7.

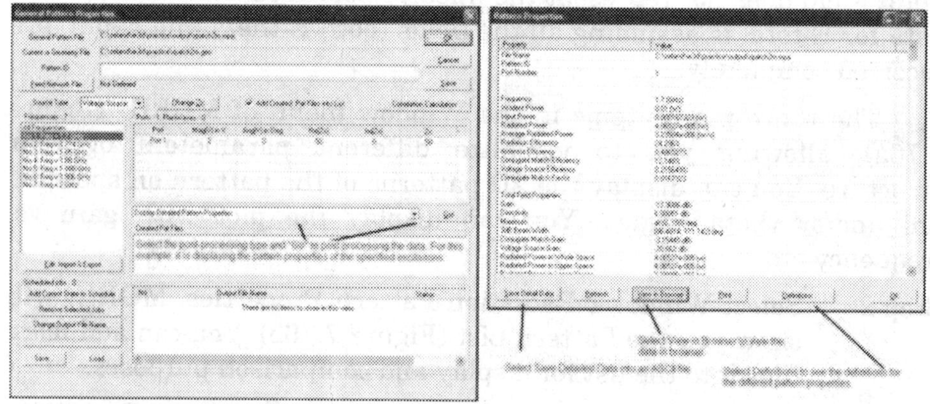

(a) General Pattern Properties dialog. (b) Pattern Properties for specified excitations.

Figure 7.77 The pattern properties dialogue for general pattern and pattern with specified excitations.

The Input Power, Radiated Power, Radiation Efficiency, Directivity, etc. are for general antenna applications. The Antenna Efficiency and Gain have included the mismatch loss to a wave source. They are normally for microwave antenna applications. RFID are widely used nowadays. For an RFID application, the source impedance is normally a voltage source with complex source impedance. For such applications, we should use the Conjugate Match Factor (CMF) and Radiation Efficiency to judge how good the RFID antenna or system is. CMF with maximum value at 1 is a criterion on how good the antenna impedance and the source impedance conjugate matching. The CMF (dB) = 20 log10(CMF) is the measure of how much loss in the delivered power to the antenna due to impedance mismatch between the source and the antenna impedances. Please read Appendix BI for more discussion on RFID designs and CMF.

Table 7.7 The pattern parameters of the optimized antenna (rpatch2m.geo) at selected frequency points.

Freq (GHz)	1.7	1.88	2.00
Incident Power (mW)	10.0	10.0	10.0
Input Power (mW)	0.167323	9.99737	0.270496
Radiated Power (mW)	0.0406527	7.19081	0.00623073
Radiation Efficiency	24.296%	71.9269%	23.0344%
Antenna Efficiency	0..406527%	71.9081%	0.623073%
Conjugate Match Efficiency	12.148%	35.9635%	11.5172%
Voltage Source Efficiency	0.215649%	36.1751%	0.197646%
Conjugate Match Factor	0.0167323	0.999737	0.0270496
Total Gain (dBi)	-17.9086	4.7903	-15.3823
Total Directivity (dBi)	6.00051	6.22253	6.67236
Conjugate Match Gain (dBi)	-3.15445	1.78114	-2.8065
Theta Gain (dBi)	-17.9086	4.7903	-15.3823
Theta Directivity (dBi)	6.00051	6.22252	6.67233
LH Circular Gain (dBi)	-20.9113	1.77843	-18.3568
LH Circular Directivity (dBi)	2.99776	3.21165	3.69786

Step 5 Select OK twice until you get back to Pattern List dialog.

Step 6 Select "3D Pattern Visualization" at the bottom of Pattern List dialog and select "Go!". MGRID will show you the 3D Pattern Selection dialog (Figure 5.19). Select the No.3 Freq = 1.88 GHz. Select Pattern Style as "True 3D". Select OK. MGRID will open a window and show you the "True 3D" pattern (Figure 7.75).

Step 7 You can select Windows->3D Radiation Pattern Display to access the dialog. Repeat Step 6 while you select the Pattern Style as "Mapped 3D". MGRID will show you the "Mapped 3D" pattern.

Step 8 You can either select Windows->2D Radiation Pattern->Define 2D Pattern Plot or select Windows->Radiation Pattern Properties->2D Pattern Visualization->Go to get into Define 2D Pattern Plots dialog (Figure 7.78).

Step 9 Select Add Plot to bring up the 2D Pattern Display dialog (Figure 7.79). You can scroll the list box to pick up any combination of theta and phi angles at any frequency. In case you have multiple files, you can also pick the items belonging to different pattern files. For this example, scroll to f = 1.88 GHz. Check the "E-Total at Phi = 0" and "E-Total at Phi = 90" (Figure 7.79). The two cuts at Phi = 0 and Phi = 90 are the two major cut planes for the pattern. Select Pattern Style as "Polar Plot". Select OK to finish defining the plot and it is added into the list of plots.

Step 10 Select Add Plot again. Select the same items "E-Total at Phi =0" and "E-Total at Phi=90" at f = 1.88 GHz. However, please choose the Pattern Style as "Cartesian Plot". Select OK to add the plot into the list.

Step 11 Select Continue and the two defined plots will be displayed on MGRID. We will have four windows showing the pattern in the forms of "True 3D" , "Mapped 3D", "Polar Plot" and "Cartesian Plot". They are representing the same data but in different forms to you.

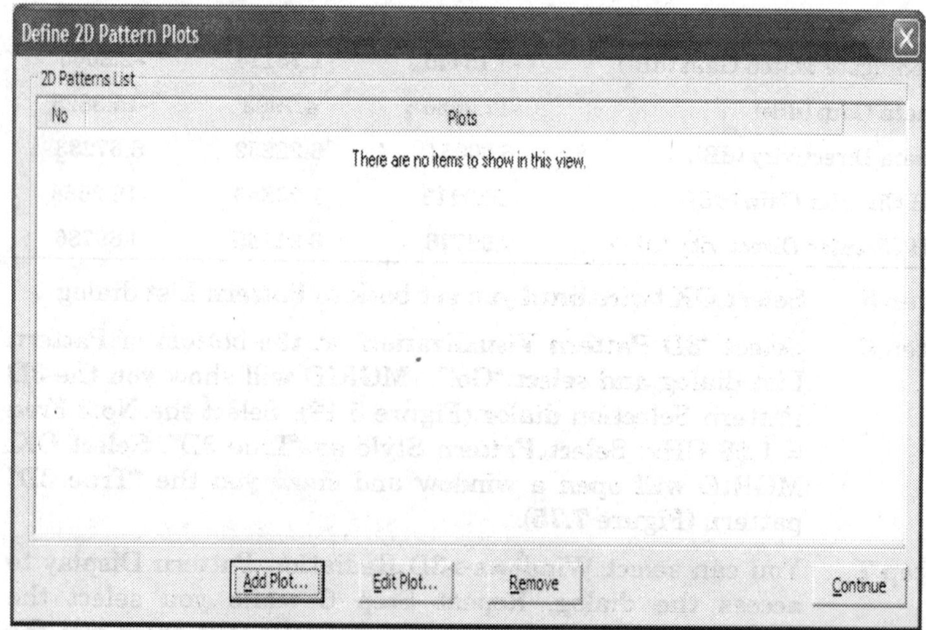

Figure 7.78 The Define 2D Pattern Plots dialog.

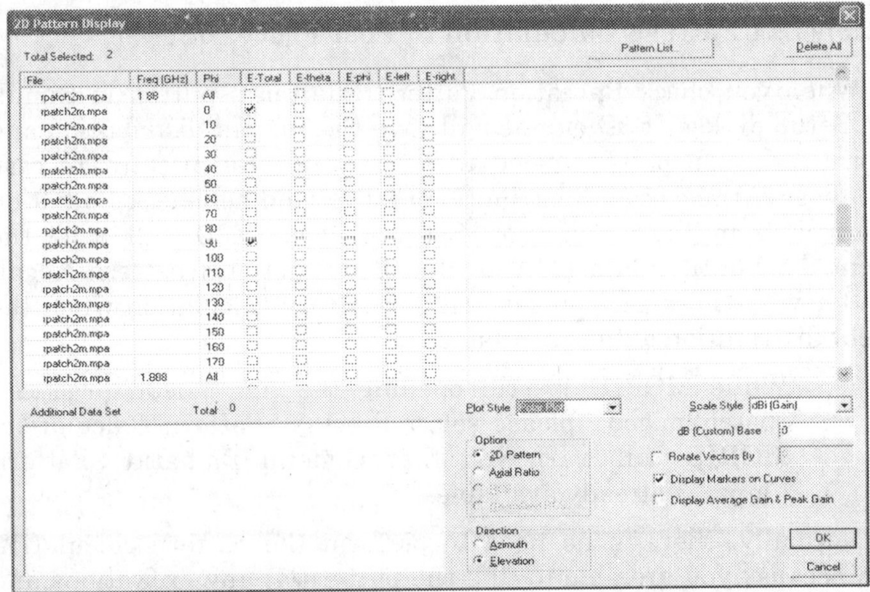

Figure 7.79 The 2D Pattern Display dialog.

Step 12 Select Windows->Gain Vs. Frequency Display. MGRID will prompt you the Frequency Respond Display dialog. Check the item: Phi = 0 and Theta(1) with Theta = 0 degree (Figure 7.80). Select OK. MGRID will open a window to display the Gain Vs. Frequency with gain at Theta = 0 and Phi = 0 (Figure 7.80). There are many other visualization capabilities. We will not explain all of them here. Please check the Windows menu.

Again, pattern visualization and processing used to be on PATTERNVIEW. For the convenience of users, we have transferred all the functionalities into MGRID. You can do the same procedures on PATTERNVIEW. You only need the general pattern files (.mpa) and/or the specific pattern files (.pat) on PATTERNVIEW.

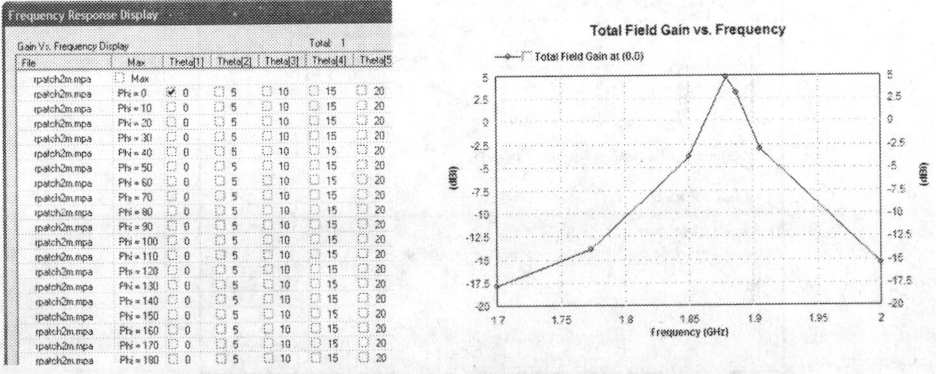

Figure 7.80 The Total Field Gain Vs. Frequency plot.

Section 7.24 Pattern Calculation in Post Processing

When you check Radiation Pattern File (.pat) button in Simulation Setup dialog, IE3D engine will save the current distribution and perform radiation pattern calculation for you automatically. In case you only save the current distribution and you did not check Radiation Pattern File in the Simulation Setup dialog, IE3D will save the current distribution data for you. It will not perform pattern calculation for you. However, you can still do pattern calculation when the current distribution data is available.

Step 1 While MGRID is still opening the file ".\ie3d\practice\rpatch2m.geo", please select Process->Pattern Calculation. MGRID will warn you that General Radiation Pattern (.mpa) is already available.

Normally, there is no need to perform the pattern calculation again because you are able to find the pattern at any excitations and terminations as you like from the .mpa file. However, there are cases you may want to do the pattern calculation again. For example, the .mpa and .pat files are lost while the .cur file is still available. We want to have finer angles for the theta and phi for better accuracy. In our example, we just want to show you that you are still able to perform pattern calculation again from the saved current distribution data.

Step 2 Select yes to continue. The Pattern Calculation Information dialog comes up (Figure 7.81). You can add more theta and phi angles for more accurate patterns. You can also change the default excitations and terminations for the general pattern.

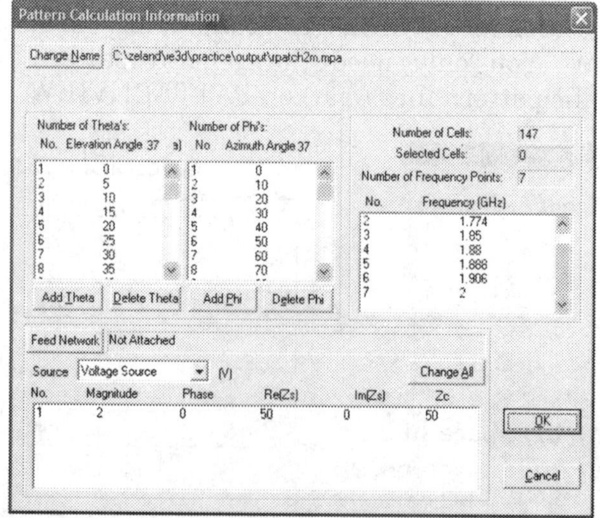

Figure 7.81 The Pattern Calculation Information dialog.

Step 3 Select OK to start the pattern calculation. For this simple structure, it takes seconds to finish the pattern calculation. After pattern calculation, Pattern List dialog comes up with the generated .mpa file updated in the list. You can continue to perform visualization and processing of the pattern data.

Section 7.25 Adding Array Factors to Radiation Patterns

To improve the directivity and gain of an antenna or change the pattern shapes, antenna designers put multiple antenna elements with appropriate spacing and excitation. Normally, there will be mutual coupling between the adjacent elements. We should simulate the coupled elements simultaneously in order to get accurate modeling of the array in case the mutual coupling between elements is strong. However, the simulation time will increase rapidly with more and more elements. We will discuss simulations of antenna arrays with mutual coupling included later. In this section, we will discuss a simple way to predict the pattern of an antenna array using the antenna array factor. When the spacing between elements is large enough, the mutual coupling between elements may not be very critical. In such a case, we can use array factor to predict the pattern of an array with reasonable accuracy.

Step 1 On MGRID, select Window->Radiation Pattern Properties to open the Pattern List dialog.

Step 2 Select Find Array Radiation Pattern in the combo box at the bottom of the dialog. Select the "Go!" button on the right of the combo box. Edit->Array Pattern Calculation. The Antenna Array Parameters dialog comes up. Select the Add button (Figure 7.82). The Add Elements dialog comes up. For X-Location, enter from = 0, to = 12400 mils. For the Y-Location, enter from = 0, to = 12400 mils. Please note that 3100 is half a wavelength in air at 1.9 GHz. We are adding a 5 by 5 antenna array with spacing of half a wavelength. We will get the dialog shown in Figure 5.26. Select OK. The Antenna Array Parameters dialog is resumed. We will see total 25 elements are added into the list.

We created a uniform 5 by 5 antenna array. In fact, we can define the Phase Step for the added elements. We can enter a set of array elements in the 3D space in each Add command. We can add more to them. For each Add command, we can designate which Element Index it is using. For our current situation, we just use one element in the list: the rpatch2m.pat. In case we have multiple elements in the pattern list window, we can choose different elements for its Element Index (starting from 0). We can mix-use different types of elements for the antenna array.

We can mix-use different types of elements for the array, for each Add command and also define the Rotation angles of all the elements to be added. For the Rotations, we need to enter the 1st Phi, 1st Theta and 2nd Phi. What do they mean? The angle or matrix transformations are like that: (1) We keep the z-axis unchanged and we rotate the ϕ angle for the 1st Phi; (2) Then, we keep the x-axis unchanged and we rotate the θ angle for the 1st Theta; (3) Finally, we keep the z-axis and we rotate the ϕ angle for the 2nd Phi.

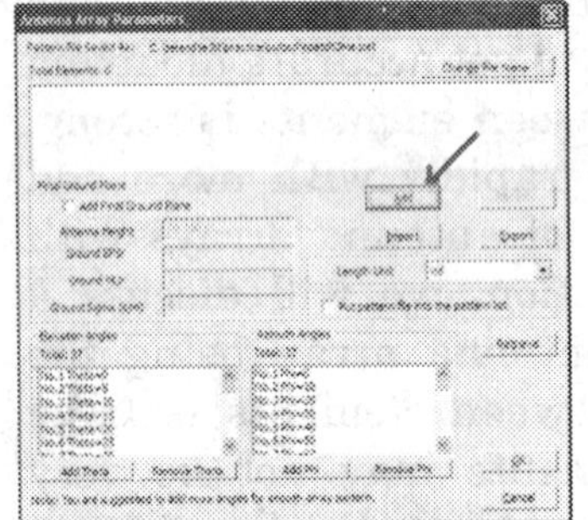

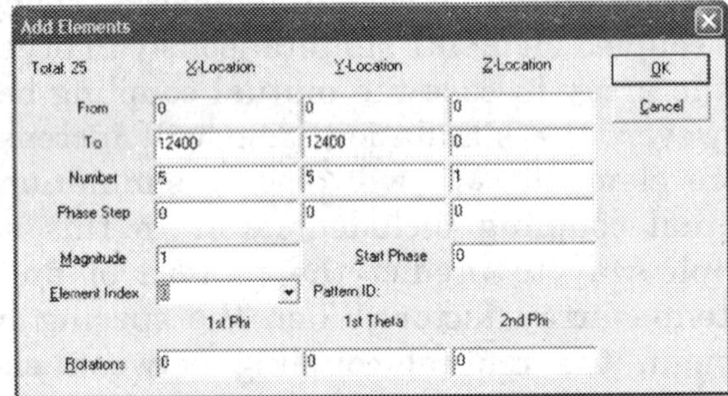

Figure 7.82 The dialogs for defining array factor.

Step 3 Select Export button and enter the file name as: rpatch2m.arr. This step is not necessary. We are saving the elements into the array file so that we can import it for other arrays of similar dimensions but different elements.

Step 4 Select Add Theta. The Add Elevation Angles dialog comes up. Enter Start = 0, End = 180, Number = 73. Select OK. The theta angles are added into the list. For more elements, we need to add more angles for smoother curves in the pattern.

Step 5 Select Add Phi. The Add Azimuth Angles dialog comes up. Enter Star t = 0, End = 360, Number = 73. Select OK. The phi angles are added into the list.

Step 6 Check the Put Pattern into the Pattern List. We will get the picture shown in Figure 7.83a. Select OK to continue.

On the MGRID in processing the patterns, it is allowed to include the effect of the Final Ground Plane (see Figure 7.83a). This is for the case we want to include the effects of the real ground plane. When we design an antenna, we always have a ground plane that is the reference of the patch. However, this ground plane may not be the final ground. For example, if we use an antenna on a base station. The final ground plane is the earth. If we want to know the effect of the earth, we can add it as the Final Ground Plane. In this example, we do not need it.

MGRID will start the calculation of the array pattern based upon the element pattern and the array factor. After it finishes the calculation, it will save the result of the array pattern into: .\ie3d\practice\output\rpatch2ma.pat. Then, it will put it into the Pattern List window. It is the No.1 pattern file in the list.

Step 7 Select "3D Pattern Visualization" in the list box at the bottom. Select "Go!". The 3D Pattern Selection dialog comes up. Select the No.1 pattern file, or the one just created with array factor, in the list. Select No.3 Frequency = 1.88 GHz. Select Pattern Style as "Mapped 3D". Select OK. The 3D Mapped pattern is displayed in Figure 7.83. How to interpret 3D mapped pattern? The (180, 0) means the point at the edge is Theta = 180 and Phi = 0. From the mid-point of the radius to the full radius, the color is blue meaning very low field value in the region. The mid-point of the radius is where the Theta = 90 angle corresponds. It means that the field is low in theta from 90 to 180 degrees. In fact, due to the infinite ground plane, the field is 0 for the angle range. The (180, 90) means that the point at the edge is Theta = 180and Phi = 90.

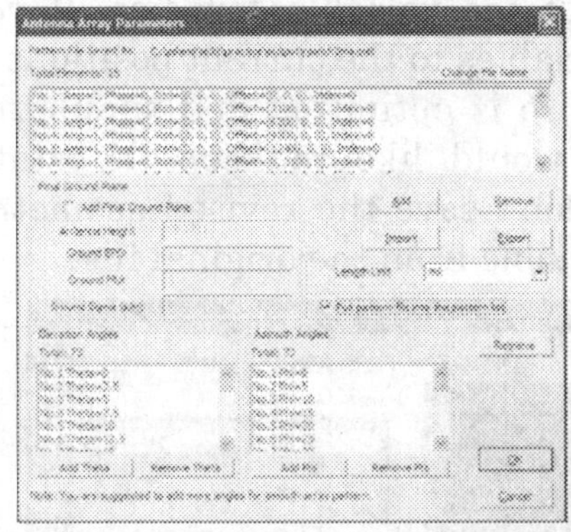

(a) The settings for array

(b) The 3D mapped pattern for the array.

Figure 7.83 The dialog for the array factor and the 3D mapped pattern.

Section 7.26 Geometry Tuning

We have demonstrated EM optimization capability of the IE3D in antenna design. The optimized structure achieves the designed performance. It is suggested that users should try to define the variables very carefully so that the geometry will be kept in good shape when the variables change. Users should also try to define the variables in reasonable ranges so that it will not become invalid. On IE3D 11 and earlier versions, users can use the command Optim>Display Trends to display the rough bounds and trends of the variables. On IE3D 12, we have implemented the geometry-tuning feature. It allows the users to examine how the geometry is tuned by the variables before any EM tuning and optimization.

Step 1 Run MGRID and open the optimized geometry file .\ie3d\practice\rpatch2m.geo file. The original optimization variables are still defined on the structure while the default shape is the optimized shape.

Step 2 Select Optim->Geometry Tuning. The IE3D Geometry Tuning dialog comes up. The left side is the top view of the structure. The color may be different due to the difference in some default setting. The right hand side has 2 tabs. One tab allows you to change the variables values in a form format. The other tab allows you to slide the bars to change the variables continuously.

Step 3 Select the Variables Sliders tab. Slide the bars and you will see how the variables are controlling the shape of the structure (Figure 7.84)

Step 4 After you finishing testing the sliders, select CANCEL to exit the dialog. If you select CANCEL, MGRID will recover the original offset values of the variables on exit. If you select OK, it will set the variables to the current positions of the bars. Since this structure is optimized and simulation results are available, we would like to keep the state. Certainly, as long as you don't save the revised geometry, the results are always accessible from the original file.

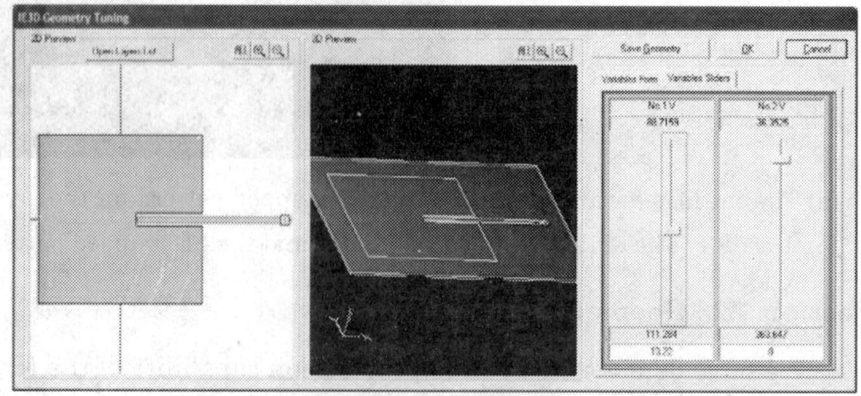

Figure 7.84 The IE3D Geometry Tuning dialog.

The geometry-tuning feature certainly provides much convenience to you. You can use it to check the ranges of variables in the structure before you do any simulation and optimization. In case there is any problem, you can fix it before you do simulation and optimization, which normally take longer time.

7D. SIMULATION OF ANTENNA ARRAYS BY IE3D

We have demonstrated how to use IE3D to simulate antenna elements. Certainly, we can use the IE3D to simulate antenna array. In this chapter, we will concentrate on antenna array simulation and design. IE3D can accept arbitrarily shaped structures with any number of ports. There is no difficulty for a user to enter an antenna array layout and simulate it. However, there are some points we may need to take into consideration. Depending upon the nature of an array or how accurate we want to model an array, we can use different approaches. Table 7.8 lists the 4 most commonly used approaches. We have successfully implemented the periodic Green's function formulation for precise modeling of planar phase array. We will also use it to simulate the following linear phase array.

This chapter is virtually not updated on IE3D 12 and IE3D 14. The newly implemented technology such as the GEMS solvers, IE3D FASTA and multiple CPU support may make the simulations of larger arrays become simple jobs. Also, many new commands will make construction of the structures much simpler. We would like to have you informed about the above fact.

Table 7.8 The common 4 approaches in array simulation and design in IE3D.

Approach	Features
Array Factor (or No Coupling)	An array is simplified as an element plus an array factor. No element coupling and feed network radiation is included. This is the fastest scheme. There is no limitation on the size of the structure. However, it is the least accurate model. The radiation from the feed network is normally not included. It is suitable for loosely coupled and equally spaced array elements with well-shielded feed network.

Periodic Wall	An array is considered as an element with duplications from the periodic walls. The major couplings between elements are included from the periodic wall condition. The radiation from the feed network is normally not included. This is also a fast scheme. It is suitable for large phased arrays with identical and equally spaced elements. The excitation phase can be tempered. One-dimensional periodic array is an approximation using finite number of images. More discussion can be found from Appendix H. Starting from the IE3D 8.0, we are able to model the 2D periodic array precisely. We can also use it to model 1D periodic array accurately because we can always define much larger element-to-element distance in the other direction. It seems that the 2d periodic array modeling is a better approach than the finite images one.
Separation	The whole structure is divided into 2 portions: all the antenna elements and the feed network. All the antenna elements are modeled in 1 simulation. The feed network is simulated in another simulation. The radiation from the elements and the feed network will be merged (you can optionally neglect the radiation from the feed network). This approach includes the coupling among elements, and the coupling within feed network. The simulation time and resource requirements are moderate. One advantage using this scheme is that you can access the power, voltage and current distribution in the whole array. Such a feature may provide very important information in array design. This new feature is also available to the previous 2 schemes. The new feature in IE3D 7.0 allows you to check the radiation from different part.
Complete Array	The complete array including elements and feed network is modeled in one single simulation. The coupling and radiation from all components are included. This certainly is the most accurate scheme. However, it requires the most computational resources. The power, voltage and current information at ports is not available for this scheme. For relative large arrays, the iterative matrix solver (IMS) or the partial matrix solver (PMS) is required to reduce the simulation time and RAM requirement.

In this chapter, we will concentrate on a 16-element linear array design. The same procedures can be applied to planar array design. The layout of the 16-element linear array is shown in Figure 7.85. We would like to optimize it for the resonant frequency at 1.88 GHz. We also would like the array to be uniformly distributed.

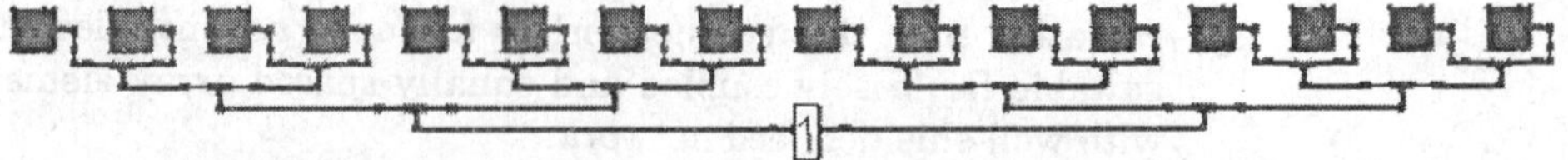

Figure 7.85 The 16-element linear array to be modeled.

Section 7.27 Element Analysis and Optimization

An array design starts from element analysis. The initial array element is saved in **.\ie3d\samples\ar_elem.geo**. The initial array element can be from some simpler tool or formula.

Step 1 Run MGRID and open geometry file **.\ie3d\samples\ar_elem.geo**. The element layout is shown in Figure 7.86. The feed line of the element was optimized for 50-ohms. Please save it into **.\ie3d\practice\ar_elem.geo**. Simulate the structure from 1.85 to 1.9 GHz without edge cells. The resonant frequency may not be predicted very precisely without the edge cells. We disable **AEC** for simplicity here. In practical design, you may consider using it for higher accuracy. The simulation result is displayed in Figure 7.87.

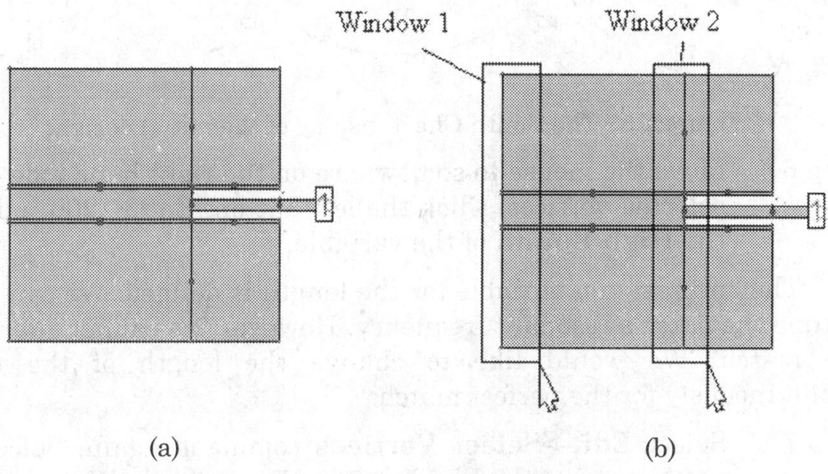

Figure 7.86 The initial array element.

As you can see, the resonance is slightly off. We would like to optimize the resonant frequency at 1.88 GHz with perfect match.

Step 2 Select **Edit->Select Vertices** command. Select the vertices using Windows 1 in Figure 7.86b.

We would like to control the length of the patch by moving its vertices at the left end in the φ = 0 angle.

Step 3 Select **Optim->Variable for Selected Objects** command. MGRID prompts you to define the optimization variable.

Step 4 Select **Vertices Mapped to: New Variable**. Enter the **Tuning Angle = 0** degree and select **OK**. We would like to change the x-coordinates of the selected vertices.

Step 5 Move the mouse to somewhere on the left hand side of the selected vertices. Click the left button. Enter "-200" mils for the **Low Bound** of the variable.

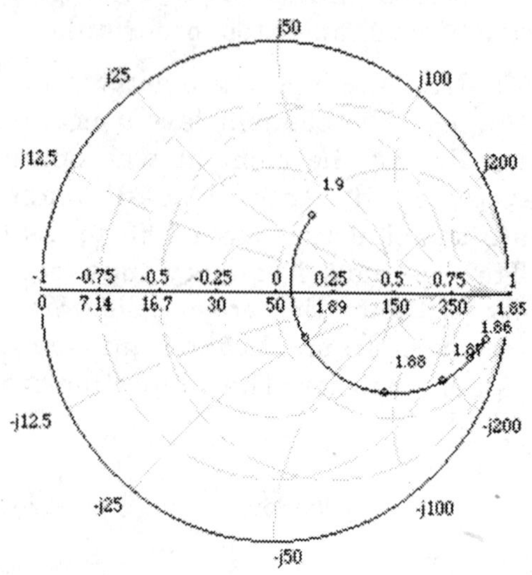

Figure 7.87 The Smith Chart display of the initial design.

Step 6 Move the mouse to somewhere on the right hand side of the selected vertices. Click the left button. Enter "200" mils for the **High Bound** of the variable.

The optimization variable for the length is defined. We can use it to tune the patch's resonant frequency. However, we cannot guarantee the match. We would like to change the length of the insets simultaneously for the perfect match.

Step 7 Select **Edit->Select Vertices** command again. Select the vertices using the Window 2 in Figure 7.86b. We would like to control the length of the insets by moving the vertices at the middle.

Step 8 Select **Optim->Variable for Selected Objects** command. MGIRD will prompt you there are total 22 vertices and 8 distinctive vertices selected.

Step 9 Select **Vertices Mapped to: New Variable**. Enter the **Tuning Angle = 0** and select **OK**. We would like to change the x-coordinates of the selected vertices.

Step 10 Move the mouse to somewhere on the left hand side of the selected vertices. Click the left button. Enter "-200" mils for the **Low Bound** of the variable.

Step 11 Move the mouse to somewhere on the right hand side of the selected vertices. Click the left button. Enter "200" mils for the **High Bound** of the variable. The optimization variable for the length of the insets is also defined. We can use it to tune the patch's matching.

Step 12 Save the geometry as **.\ie3d\practice\ar_ele1.geo**. We will optimize Re[S(1,1)] = Im[S(1,1)] = 0 at 1.88 GHz. Select Set Optimization in Process menu. Enter 1 frequency at 1.88. Select Add button. Define the **Parameter Type** as Re(S), **1st Parameter** as (1, 1). **Objective Type** as **Optimization Quantity** = Objective1, **Objective1** = 0. Select OK to add the 1st optimization goal. Select Add button again. Define the **Parameter Type** as Im(S), **1st Parameter** as (1, 1). **Objective Type** as **Optimization Quantity** = Objective1, **Objective1** = 0. Select OK to add the 2nd optimization goal. Select **Powell** for the optimization scheme. Select OK to start the optimization.

The default generation number is 565 times. Interestingly, the optimizer achieves the goals in only tens of simulations (You can check the **.\ie3d\practice\output\ar_ele1.log** file for the optimization information). The best result is saved **in .\ie3d\practice\ ar_ele1m.geo**.

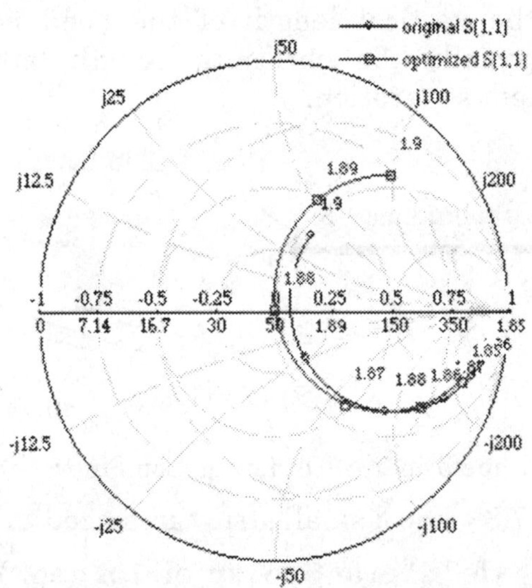

Figure 7.88 The comparison between results before and after optimization.

Step 13 Open the optimized geometry **.\ie3d\practice\ ar_ele1m. geo**. Simulate it from 1.85 to 1.9 GHz. We will use it later. Display the result before and after the optimization using MODUA. You will see the optimizer really did the job (see Figure 7.88).

Section 7.28 The "Complete Array" Approach

For convenience reason, we will discuss the "Complete Array" approach in the array analysis. We would like the element-to-element distance from center to center to be 3,500 mils. This distance is a little bit more than half of a wavelength in free space (about 3,140 mils). No grating lobe is created because the element pattern will suppress it. The input impedance for each element is 50-ohms. We will combine each pair of the elements. Before we combine them, we need to transform the 50-ohms to 100-ohms. Therefore, we need a transformer from 50-ohms to 100-ohms and we need a power combiner from two 100-ohm lines back to 50-ohm line. The geometry file for the transformer is saved in **.\ie3d\samples\ar_trans.geo** (see Figure 13.5a). The power combiner is saved in **.\ie3d\samples\ar_comb. geo** (see Figure 7.89b). The **ar_trans.geo** is a quarter wavelength power transformer with the middle section of Zc = 70-ohms and quarter of a wavelength. The total length for **ar_trans.geo** is 1,200 mils. The total length for the **ar_comb.geo** is 300 mils. We can use the Symmetrical T-Junction and Symmetrical Step in Entity of MGRID 4.16 to build the 2 structures easily. Another small geometry we want to be handy is a 50-ohm bend. The geometry is saved in **.\ie3d\samples\ar_bend.geo**. The horizontal length of the bend is 400 mils and the vertical length of the bend is 600 mils. The 2 dimensions, especially the horizontal length, are planned for the element-to-element separation.

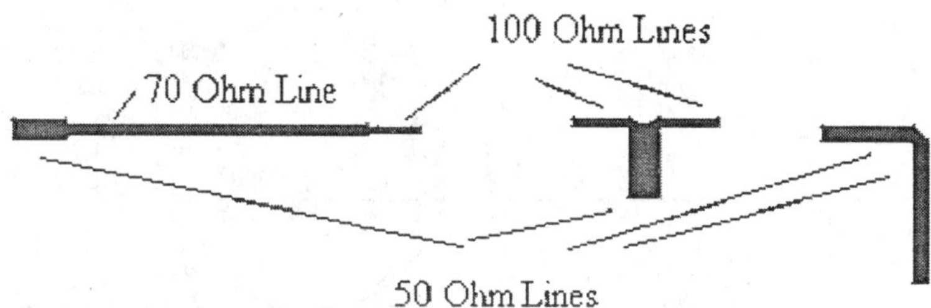

(a) Transformer (b) Power Combiner (c) Bend

Figure 7.89 The 3 small structures need to be handy.

Step 1 Open **.\ie3d\practice\ar_elelm.geo**. Drop the optimization variables in the geometry file. Save it into **.\ie3d\practice\ar_comp.geo**. Select **Port->Delete All Ports** command to delete all the ports.

Step 2 Run another MGRID. This MGRID will be referenced as the 2nd MGRID. Open **.\ie3d\samples\ar_bend.geo**. Press down "shift" and window all the polygons in **ar_bend.geo**.

Select Copy in Edit menu. We are going to copy the bend from ar_bend.geo to ar_ele1m.geo.

Step 3 Go the 1st MGRID. Select **Edit->Paste** command. The bend is moving with the cursor. Move the cursor to the feed line. The bend will be snapped to the feed line. Click the left button to locate the pasted bend. MGRID will prompt you for the offset values for the coordinates. Accidentally, they are 0's. Normally, you do not need to change them. MGRID automatically snap vertices. Select OK to continue. The bend will be copied and connected to the feed of the patch (Figure 7.90).

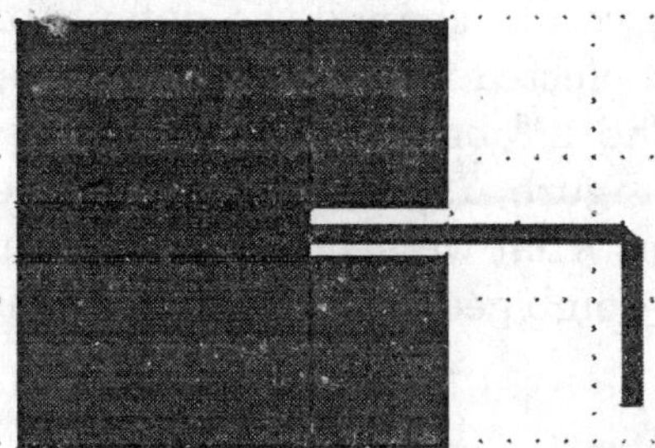

Figure 7.90 The patch with the bend connected.

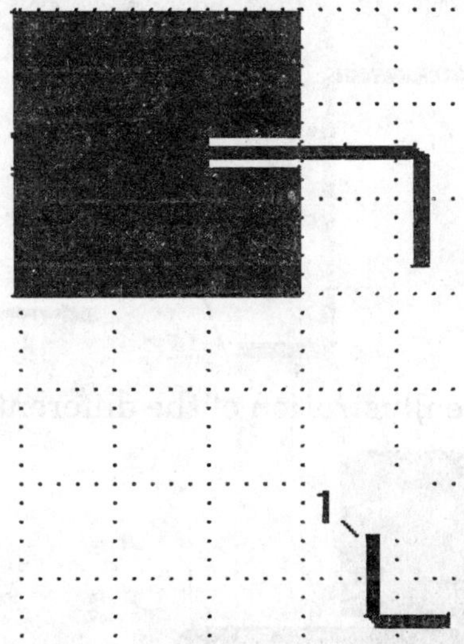

Figure 7.91 The patch, the original bend and the copied and rotated bend.

Step 4 Press down "Shift" button and window the copied bend to select it. Select **Edit->Copy and Rotate** command. Enter Object Rotation Angle = 180 degrees. Enter the Reference Point: X-Coordinate = -4000 and Y-Coordinate = 4000. Select Copied Objects Selected for After Rotation. Select OK. The bend is copied with 180-degree rotation (see Figure

7.91). The copied and rotated bend is still being selected. Please note that the Reference Point coordinates are arbitrarily chosen.

Step 5 Select **Edit->Move Objects** command. The copied and rotated bend is moving with the cursor. One of the vertices is the moving reference and it is at the location of the cursor. It may not be the vertex 1 in Figure 7.91. Click the right mouse button multiple times until the vertex 1 becomes the moving reference. Move the bend toward the end of the original bend until it snaps to it and connect. Click the left mouse button to locate the 2nd bend. MGRID will prompt you to change the default offset values. Select Objects De-Selected for the After Move option. Select OK to continue. The 2nd bend is connected to the end of the 1st bend (see Figure 7.92). We will save the change into ar_comb.geo. What we need to do is to copy the ar_trans.geo and the ar_comb.geo to connect them as shown in Figure 7.92.

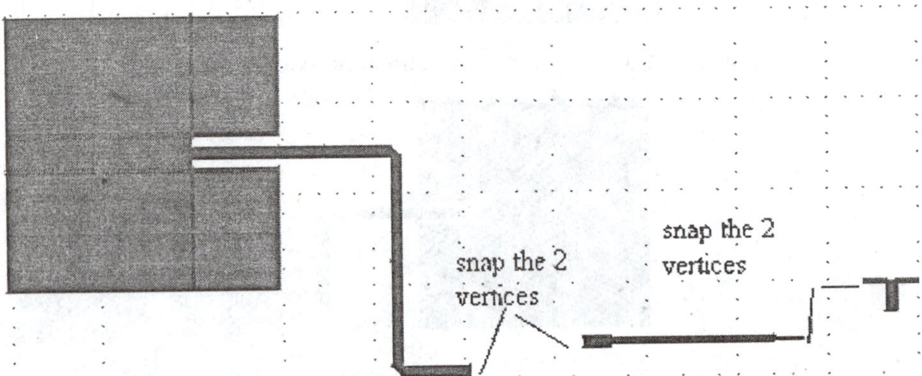

Figure 7.92 The illustration of the different components.

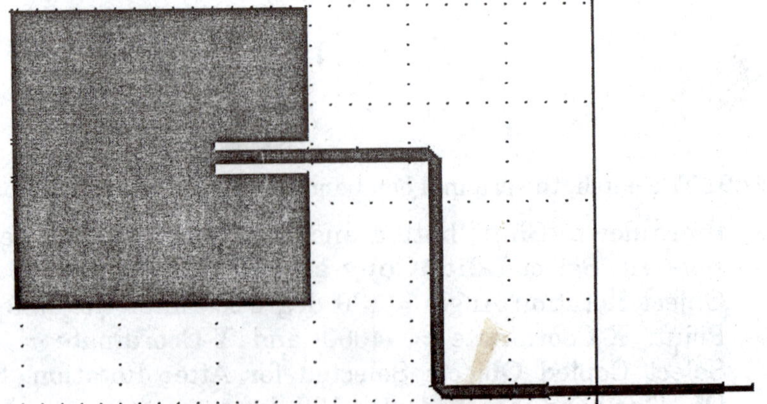

Figure 7.93 The ar_trans.geo is connected to the feed line.

Step 6 On the 2nd MGRID window containing the ar_bend.geo. Select **File->Open** command and open ar_trans.geo. Select **Port->Delete All Ports** command to delete the two ports. Press down "Shift" and window all the polygons. Select Copy in Edit menu.

Step 7 On the 1st MGRID window containing ar comp.geo, select Paste in Edit menu. The ar_trans.geo is following the mouse. One of the vertices is chosen as the moving reference. Click the left t mouse multiple times until one of the two vertices on the left end of the ar_trans.geo becomes the moving reference. Move it toward the end of the feed-line of the patch antenna, until it snaps. Then, click the left mouse button to locate it. The x-offset should be –3500 and the y-offset should be 757.5. Select OK to continue. The ar_trans.geo is connected to the end of the feed line of the patch.

Step 7 On the 2nd MGRID window, open .\ie3d\samples\ar_comb.geo. Press down "Shift" button and window all the polygons to select them. Select **Edit->Copy** command.

Step 8 Go to the 1st MGRID. Select **Edit->Paste** command. The polygons from ar_comb.geo are following the cursor. Click the right mouse button multiple times until one of the two vertices on the left edge of ar_comb.geo becomes the moving reference. Move it until the ar_comb snaps to the end of the transformer. Click the left mouse button. MGRID will prompt you to change the offset values. Accidentally, they are 0's. Select OK to continue. We will get the picture in Figure 7.94. The next step is to copy and reflect the 2nd bend and the transformer to the right hand side to get the picture in Figure 7.95.

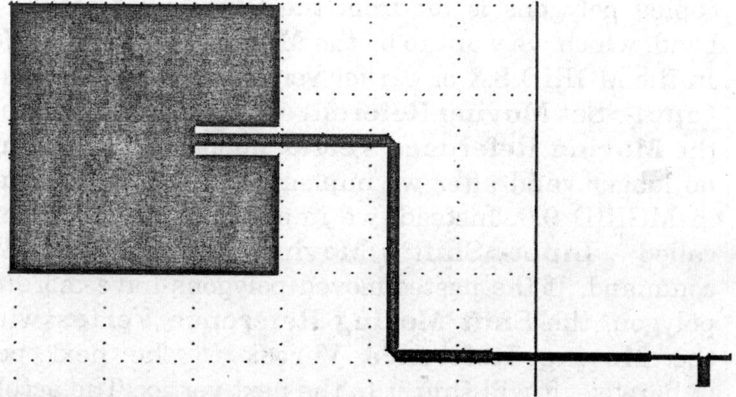

Figure 7.94 The ar_comb.geo is connected to the transformer.

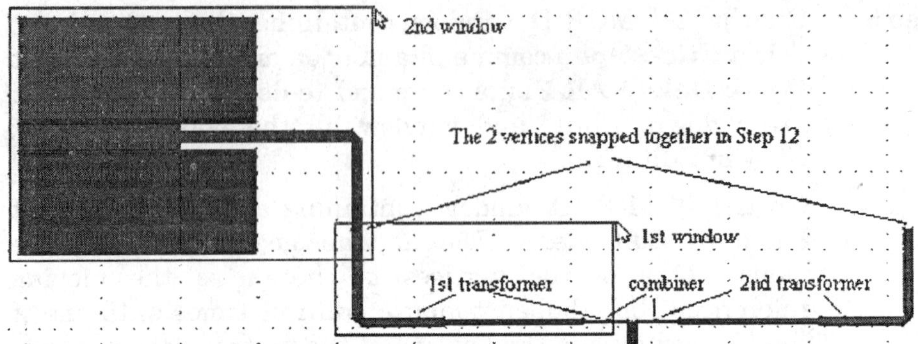

Figure 7.95 The second transformer is copied and connected to the combiner.

Step 9 Press down "Shift" and windows the 2nd bend and the transformer to select them. Select **Edit->Copy and Reflect** command. MGRID will prompt you for the Object Reflection Angle. Enter 0 for the angle. Select OK to continue. The copied and reflected polygons are following the mouse cursor. Click the left mouse somewhere. MGRID will prompt you for the distance. Enter the distance as 300 mils, which is the length of the power combiner. Check All Objects De-Selected. A transformer and a bend are created on the right hand side of the power combiner. The connection between the power combiner and the second transformer is guaranteed, because we have entered the exact length value of 300 mils. The picture is shown in Figure 7.95. Save the file as: .\ie3d\practice\ar_compl.geo.

Step 10 Press down "Shift" button and window the patch and the 1st bend (see the 2nd window in Figure 7.95) to select them. We will duplicate them for the second patch. Select **Edit->Copy** command. The **Moving Reference Vertex** for the copied polygons is far from the lower right corner of the bend, which we want to be the **Moving Reference Vertex**. In the MGRID 8.X or earlier version, we had the command **Input->Set Moving Reference Vertex** command to select the **Moving Reference Vertex** quickly. This command is no longer valid after we implemented the Paste command in MGRID 9.0. Instead, we implemented a new command called **Input->Shift Moving Reference Vertex** command. If the pasted/moved polygons have more than 1 polygon, the **Shift Moving Reference Vertex** will shift the **Moving Reference Vertex** to the next polygon. Otherwise, it will shift it to the next vertex. The accelerator for the **Input->Shift Moving Reference Vertex** command is the "tab" key.

Step 11 Click the "tab" key several times until you see the 1st vertex in the polygon of the lower right of the 2nd bend is the Moving Reference Vertex. Then, click the right mouse button until one of the two vertices at the lower right end becomes the Moving Reference Vertex. Move the pasted polygons to snap to the right end of the structure until they snap. Click the left mouse button to locate it. MGRID should prompt for the X-offset and Y-offset values. Make sure the X-offset value is 3500 mils and the Y-offset value is 0. Select OK to continue. We will get the structure in Figure 7.96. Save the file as: .\ie3d\practice\ar_comp2.geo.

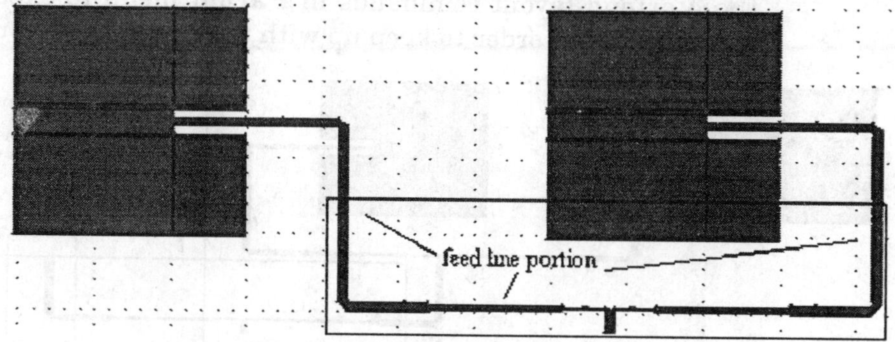

Figure 7.96 The 2 patches in the linear array.

Step 12 From here to the 16-element array is easy. Please select **Edit->Select Polygon Group** command. Window all the feed line portion of the structure, as shown in Figure 7.96. The feed line portion of the structure includes the power divider, the 2 transformers and the 2 bends, each connected to the transformer. In fact, it is the whole structure without the 2 patches and the 2 bends. Copy, Paste and the "tab" button to connect them as shown in Figure 13.13. The X-offset = 1750 mils and the Y-offset = -757.5 mils.

The copied feed line, which spans about 3,500 mils, is too short. We need to double its length, in order to accommodate the 4 patches.

Step 13 Select **Edit->Select Vertices** command. Window the vertices as shown in Figure 13.13 (the window for the 1st selection). Then, select **Edit->Move Objects** command. Click somewhere and enter the X-offset = 1,750 mils = 3,500 / 2 mils, and Y-offset = 0 mil. Uncheck the **Object De-Selected** before selecting **OK** to exit the selection mode. The left arm of the second feed network is stretched.

Step 14 Select **Edit->Select Vertices** command. Define the second window in Figure 13.13 to select the vertices. Then, select **Move Objects** in **Edit** menu. Click somewhere and enter the X-offset = 1,750 mils = 3,500 / 2 mils, and Y-offset = 0 mil. Uncheck the **Object De-Selected** before selecting **OK** to exit the selection mode. The right arm of the second feed network is stretched. (see Figure 13.14). Save the file as: .\ie3d\practice\ar_comp3.geo.

Step 15 Select the 2 patches and the feed network in Figure 13.12 and use the Copy, Paste and "tab" commands to create the 4-patch array in Figure 13.15. The X-offset = 7000 and the Y-offset = 0. Save the file as: .\ie3d\practice\ar_comp4.geo. You should also use the **Whole Circuit** in **View** menu or the 4 extend layout commands in **Param** menu to expand the view size in order to keep up with the expanded layout.

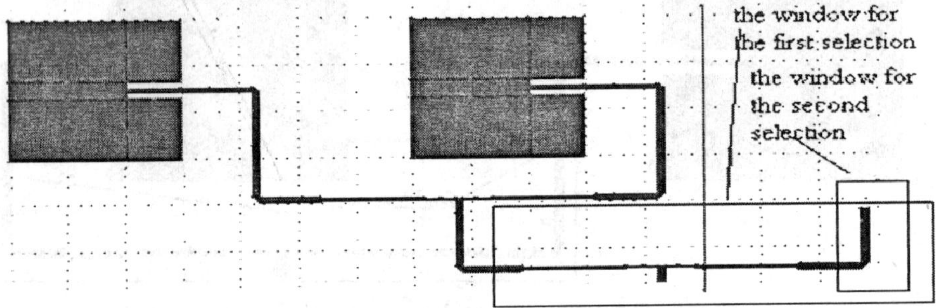

Figure 7.97 The copied feed line connected to the original feed line.

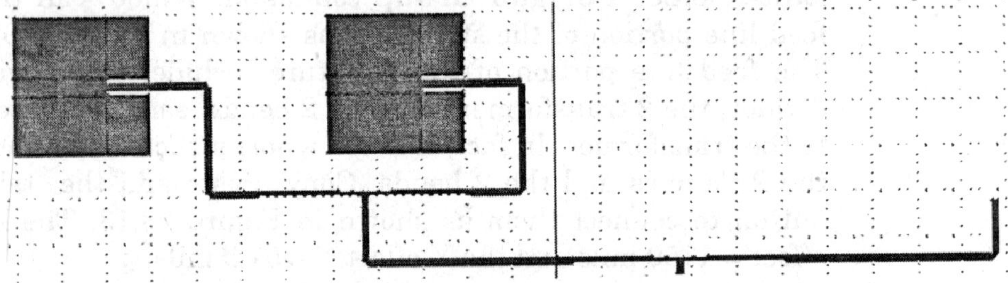

Figure 7.98 The second feed line is stretched for the 4 patches.

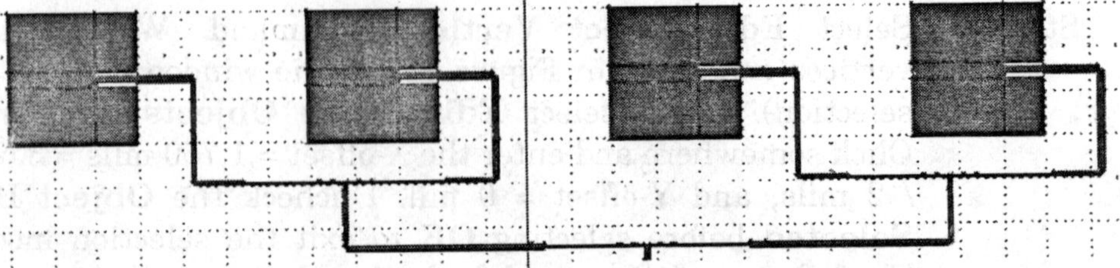

Figure 7.99 The 4 patch array is created.

Step 16 For a few more steps, you should be able to create the 16-element linear array shown in Figure 7.85. The steps will be similar to those from 2-patch array to 4-patch array. The 8-element array is saved as **.\ie3d\practice\ar_comp4.geo** and the 16-element array with 1-port defined is saved as: **.\ie3d\practice\ar_comp5.geo**.

By this step, we have created the complete antenna array. Simulation of the complete structure takes 2857 cells and 4041 unknowns. It requires about 136 MB RAM using the SMSi solver. The simulation time is about 185 seconds per frequency point on Pentium 4 2.8 GHz. Only about 33 seconds is used to mesh the geometry and fill the matrix. Using the AIMS III with Separation Distance = 10 Cells and Pre-Conditioner = "2 0.0007", the simulation time is reduced to 97 seconds. The matrix fullness is about 2.32% and the RAM requirement is about 33 MB only. You should understand that the AIMS II and AIMS III yield the accurate result without neglecting the far coupling.

In the simulation, you should enable **Radiation Pattern File** in the simulation setup. IE3D will simulate the structure and find the radiation pattern automatically for you. When you enable **Radiation Pattern File**, MGRID will prompt you for the pattern calculation parameters. Please enter total 73 data point from 0 to 180 degrees for the Theta angle, and total 73 data points from 0 to 360 degrees for the Phi angle. The radiation pattern calculation may take couple extra minutes.

Section 7.29 The "Separation" Approach

We cannot get any information about the power distribution on the feed network. It would be nice if we can know the power, voltage and current distribution on the feed network. We certainly can do it using the **"Separation"** approach as we discuss in the following.

Step 1 Open .\ie3d**practice\ar_comp5.geo** (or **.\ie3d\samples\ar_comp5.geo**). Delete the 16-patches and the 16-bends connected to the patches. Define the ports on the terminals. The 17-port feed network is created and shown in Figure 13.16. Please save it as **.\ie3d\practice\ar_feed.geo.**

Some people may be curious about the return loss, the power distribution and the radiation of the feed network. We will show how we can get the information.

Step 2 Open .\ie3d**practice\ar_comp5.geo**. Delete the feed network except the 16-patches and the 16-bends connected to the patches. Define the ports on the terminals. The 16-port feed network is created as shown in Figure 7.101. Please save the geometry as **.\ie3d\practice\ar_ps.geo**.

We are going to perform separate electromagnetic simulations on both the **ar_feed.geo** and the **ar_ps.geo**. Certainly, the simulations cannot include the coupling between the portion of elements and the feed network, even though the coupling between elements and the coupling inside the feed network will be included.

On the IE3D 6.0, we can simulate the two structures separately. Then, we connect the s-parameters of the two separated structures using a similar design setup shown in Figure 7.102. We can use the **Process->Simulate and Find Excitation** command in MODUA to find and save the voltage and current at each port when the elements are fed by the feed network. Finally, we can use MGRID or CURVIEW to calculate the radiation pattern from the elements using the excitation file saved from MODUA. The radiation from the feed network will be neglected. It is a valuable feature of the IE3D because it provides a sophisticated tool to fine tune the array. However, the process is a multi-step process and it neglects the radiation from the feed network. Also, the final radiation pattern created does not include the feed-network as part of the structure. Information such as return loss, gain and efficiency does not include the effect of the feed network.

Starting from the IE3D7.0, we are able to automate the process. We are able to merge the patterns of the elements and the feed network automatically, in a one step process. Every element in the whole design will be considered as part of the system.

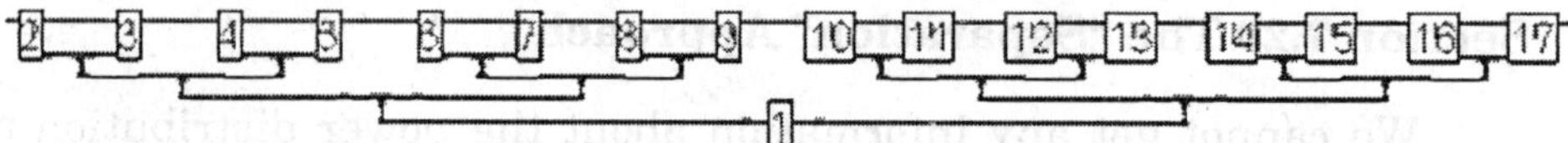

Figure 7.100 The 17-port feed network without the elements.

Figure 7.101 The 16-port structure of array elements without feed network.

Section 7.30 Pattern Merging and Simulation of Large Arrays

Step 1 Run MODUA. Select **Add Geometry Module** in **File** menu. Select the file **.\ie3d\practice\ar_feed.geo**. Select OK twice and click at a spot to drop it. Click an empty spot to de-select it. Select **Add Geometry Module in File** menu again. Select the file **.\ie3d\practice\ar_ps.geo**. Select OK twice and click at a spot to drop it. Click an empty spot to de-select it. Select **Connection** in **Element** menu to define the connections one by one shown in Figure 3.18.

Select **Exit Element** in **Element** menu to exit from the mode. Select **Define All Ports** in **Element** menu to define the port 1 in Figure 7.102. Save the design file as: **.\ie3d\practice\ar_fd_ps.dsg**.

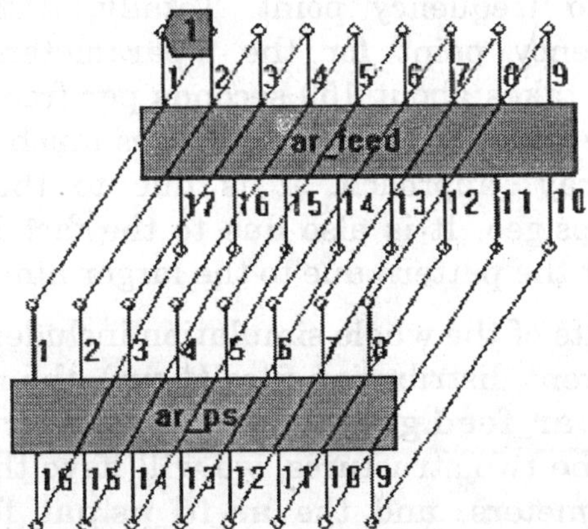

Figure 7.102 The connection between the elements and the feed network.

Step 2 Select **Simulate** in **Process** menu of MODUA. Enter 1 frequency point at 1.88 GHz. Select OK to continue. The simulation setup dialog similar to the one in MGRID comes up.

Step 3 Check the **Radiation Pattern File** checkbox. The **Current Distribution File** checkbox will be automatically checked after you check the **Radiation Pattern File**. The **Radiation and Excitation Parameters** dialog comes up.

Step 4 Enter **Start Angle** = 0, **End Angle** = 180, **Number of Angles** = 73. Select **Add Theta** to add the angles to the elevation angles. Enter **Start Angle** = 0, **End Angle** = 180, **Number of Angles** = 73. Select **Add Phi** to add the angles to the azimuth angles. Select OK to finish defining the radiation and excitation parameters.

Step 5 Since this structure has many ports, it might be better to use the AIMS II instead of the AIMS III. Select the AIMS II matrix solver. MGRID will prompt you for the parameters. Select the Separation Distance = 10 cells. However, the actual SD = 2950 mils (different from the 1796 mils in **ar_comp.geo**). It is due to the fact that MODUA use the cell size in free space as the criteria. We can change the Separation Distance in Cells to 7 and it is about 2000 mils.

Step 6 Select OK to invoke the IE3D simulation on the complete design.

The IE3D will simulate the 2 structures sequentially. The feed line takes about 1127 cells and 1143 unknowns. It takes about 22 seconds (2-iterations) per frequency point for the feed line. The patch creates 1856 cells and 3040 unknowns. It takes about 247 seconds (5-iterations) for each frequency point. Totally, it takes about 270 seconds per frequency point for the s-parameters. However, the pattern calculation takes about 150 seconds per frequency point. You may notice that the "Separate" approach takes much longer time than the "Complete Array" approach. It is due to the 5-iterations in simulating the ar_ps.geo. It is also due to the fact it requires many more calculation for the pattern due to the larger number of ports.

The final results of the whole simulation include: the s-parameter files (*.sp), the current distribution files (*.cur), the radiation pattern files (*.pat) of the **ar_feed.geo** and **ar_ps.geo** structures. For the design combining the two structures, we will have the **ar_fd_ps.spm** file for the s-parameters, and the **ar_fd_ps.pat** for the radiation pattern, and the **ar_fd_ps.ect** for the excitation information for the complete structures. The intermediate result files from **ar_feed**, **ar_ps** and **ar_fd_ps** can be used for more detail information for the complete structures. For example, you can open **ar_ps.cur** file and attach the **ar_fd_ps.ect** file to calculate the radiation pattern of the elements assuming the radiation from the feed network (**ar_feed.geo**) is negligible. You can also open **ar_feed.cur** file and attach the **ar_fd_ps.ect** file to investigate the radiation from the feed network. You should understand that such an approach on pattern post-processing is an approximation. The radiated power of the complete structure will be different from the sum of powers of the two structures when they exist individually.

Step 7 Use an ASCII editor to open the **ar_fd_ps.ect** file. We will get all the information about the excitation and termination at the ports of the feed network and the antenna elements.

As you can see from the above procedure, we can use this approach to model very large arrays with non-identical elements. As we will see in the next section, we can use array factor to model arrays with uniformly spaced identical elements. However, the array factor approach cannot be applied to any of the following situations: (1) The elements are not identical, (2) The rotation or orientation of each element is not the same, and (3) The elements are not uniformly spaced. Instead, the "separation" approach allows more flexibility in modeling general arrays. Both approaches do not include coupling between elements (or simulation blocks).

Section 7.31 The "Array Factor" Approach

We have not created the models for the "Array Factor" and "Periodical Walls" approaches. Creating the single element from **ar_ps.geo** is simple. You can just delete all but 1-element. The remaining element will be the single element. The single element file is saved as **.\ie3d\samples\ar_p1.geo**. Simulating **ar_p1.geo** takes almost no time. To get the array radiation pattern is simple. We can calculate the element pattern on either IE3D (invoked by MGRID) or by the Pattern Calculation on MGRID as post-processor. Then, we can use PATTERNVIEW to add the array factor.

Run MGRID. Open **ar_p1.geo**. Simulate it again at 1.88 GHz. Please remember to enable Radiation Pattern File. The IE3D will be invoked to simulate it and find the single element pattern in .\ie3d\practice\ar_p1.pat file. The PATTERNVIEW will be invoked to display the pattern. The ar_p1.pat file is added into the pattern list automatically.

Add Elements

Total: 16	X-Location	Y-Location	Z-Location
From	0	0	0
To	52500	0	0
Number	16	1	1
Phase Step	0	0	0

OK Cancel

Magnitude 1 Start Phase 0

Element Index 0

	1st Phi	1st Theta	2nd Phi
Rotations	0	0	0

Figure 7.103 The Add Elements dialog after we entered the parameters.

On the PATTERNVIEW, select **Edit->Array Pattern Calculation** command. The **Antenna Array Parameters** dialog comes up (Figure 7.104). Please select **Add Theta** button to define 73 angles from 0 to 180 degrees. Please select **Add Phi** button to define 73 angles from 0 to 360 degrees. Then, select the **Add** button. The Add Elements dialog comes up. On the X-Location column, enter From = 0, To = 52500, Number = 16, Phase Step = 0. Please do not change the default values in Y-Location column and Z-Location column (From =0, To = 1, Number = 1, Phase Step = 0). Please enter the common parameters as: Magnitude = 1, Start Phase = 0, Element Index = 0, 1st Phi = 0, 1st Theta = 0 and 2nd Phi = 0.

What do the parameters mean? For the X-Location, we want to add 16-elements in the X-direction. The 1st element will be located at X = 0 and the last element will be located at X = 52500. The distance between adjacent elements is 52500/(16-1)= 3500 mils. The Phase Step = 0 means that there is no tapered phase in the elements. The meanings of the Y-Location and Z-Location are similar to the X-Location. All the elements have the magnitude of 1. The Start Phase defined the No. (1, 1, 1) element's phase. The phase of any element (ix, iy, iz) can be calculated as: Phase = Start_Phase + X_Phase_Step * (ix-1) + Y_Phase_Step * (iy-1) + Z_Phase_Step * (iz-1). The Rotations parameters define how the elements are rotated.

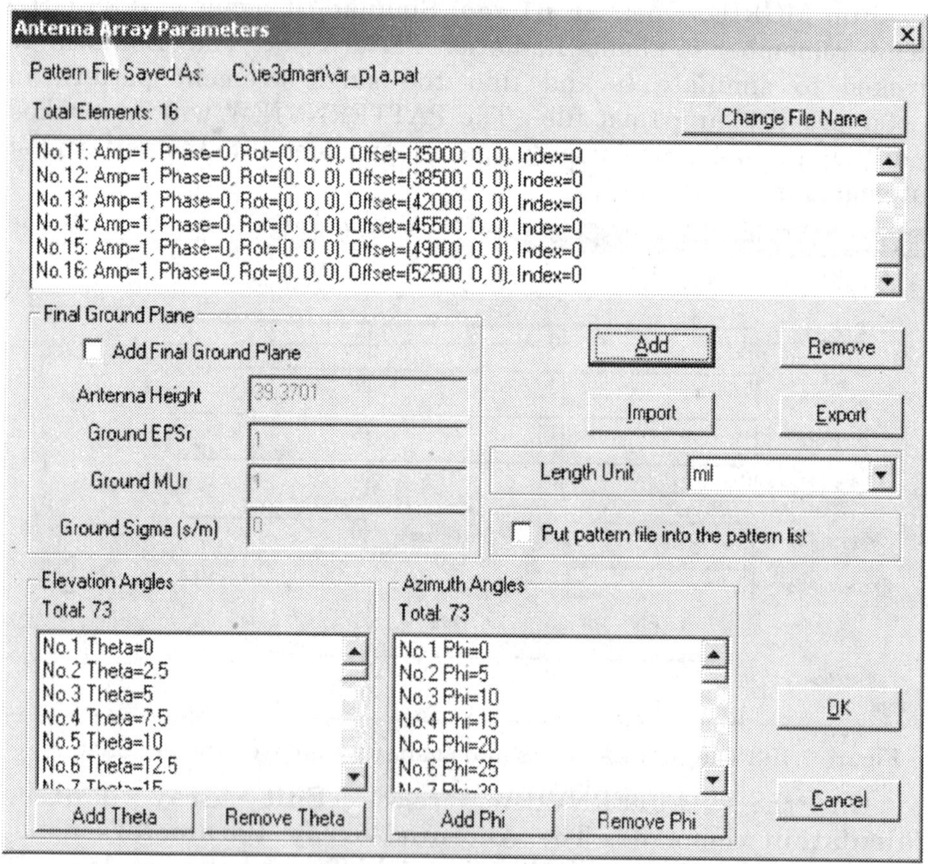

Figure 7.104 The Antenna Array Parameters dialog after the parameters are entered

There is an additional parameter: Element Index. It is for the index of a specific pattern in the pattern list window. Basically, PATTERNVIEW allows a user to combine the radiation patterns of different elements with different rotations. It is extremely flexible. In our example, we only added one element into the pattern list. We want to find the pattern of a uniform array of the only element. However, the user should understand that you could achieve much more on the PATTERNVIEW.

After you select OK in the Add Elements dialog. The 16-elements are added into the list in the Antenna Array Parameters dialog (see Figure 7.104). Each element is assigned with amplitude, phase, rotation parameters, offset parameters and element index. Old IE3D users should be very careful on the number of elements. In the CURVIEW program, which is out phased and still available on IE3D 9.0, the number of elements added is not the total number of elements. The total elements are the added elements plus the original one. On the PATTERNVIEW, the number of elements added is the total number of elements of the array.

If you need to add more elements, you can select Add button again. Not only that, a user also has the option to Add Final Ground Plane. This option allows the user to add the true ground plane's effect into the pattern when the ground plane is in the far-field zone.

Please check the Put Pattern File into the Pattern List. It allows PATTERNVIEW to save the array pattern file into the default name: ar_p1a.pat file and add the file into the pattern list automatically. Select OK to continue. PATTERNVIEW will perform the pattern calculation for you. It takes a second to finish it. Then, the resulting ar_p1a.pat file will be added into the list. You can display the pattern parameters on ar_p1.pat and ar_p1a.pat. We will delay the discussion on them to later sections.

Section 7.32 The "Periodic Walls" Approach

There are two ways to model the periodic walls: (1) Model it as a 1-dimensional infinite array using 1 pair of periodic walls; (2) Model it as a 2-dimensional infinite array using 2 pairs of periodic walls. For a large planar phased array, we should undoubtedly choose the 2-dimensional infinite array approach. For a 1-dimensional linear array, the 1-dimensional infinite array is supposed to be the appropriate one. However, it seems to us that it is better to choose the 2-dimensional infinite array approach on the IE3D 8.0 even for 1-dimensional linear array.

On the IE3D 8.0, we have implemented the periodic Green's functions for the 2D planar array case. Its simulation result is precise. For the 1D linear array, we use a pair of parallel periodic walls to approximate it. Unfortunately, we have not implemented the exact Green's functions for the one pair of parallel periodic walls for it. We also do not have the intention to implement it in the near future. We are using finite number of images to approximate it. On the other hand, it might be equally well or even be better if we use the 2D periodic Green's function to approximate it. The 2D periodic Green's function uses two pairs of parallel periodic walls for the modeling. An infinite number of images are included and the results are quite stable. If we make the 2nd pair of parallel walls to be far enough, the results should approach the one for 1 pair of parallel walls.

Creating the files for the "Periodic Walls" approaches is quite simple. We open the single element file **.\ie3d\samples\ar_p1.geo**. Then, we will define the periodic walls. We select the **Param->Basic Parameters** command. Double click at the **No. 0: No Side Walls** in the Enclosure section to edit it (see Figure 7.105). Select **Xmin Wall** as **Periodical** at –6,600 mils. Select **Xmax Wall** as **Periodical** at –3100 mils. For the X-Direction, please enter the **Pattern Images** = 15 (the number of images used in the pattern calculation), the **Spatial Images** = 4 (the coupling included in the simulation), the Phase Factor = 0 degree (the phase difference between adjacent cells), the **Image Index = 7**. This is a finite array. The Image Index = 7 means that the element will be considered as the No.7 element in the 16-element array. We will get the dialog in Figure 7.105. Totally, we have 16-elements in consideration. Please save the file as: **ar_pp.geo**.

Edit Enclosure

Comment: Editing No.0 Enclosure   OK   Cancel

X-Direction				Y-Direction			
Xmin Wall	Periodical	at	-6600	Ymin Wall	None	at	
Xmax Wall	Periodical	at	-3100	Ymax Wall	None	at	
Spectral Terms	N/A	Pattern Images	15	Spectral Terms	N/A	Pattern Images	N/A
Spatial Images	4			Spatial Images	4		
Phase Factor (deg)	0	Image Index	7	Phase Factor (deg)	N/A	Image Index	N/A

Figure 7.105 The Edit Enclosure dialog for ar_pp.geo.

Please select **Param->Basic Parameters** command and edit the **No.0: No Side Walls** in the Enclosure section again. Please select the **Ymin Wall** as **Periodical** at 0 and the **Ymax Wall** as **Periodical** at 10000 mils. Please define the Y-Direction Parameters as: Pattern Images = 0 (No image in the y-direction).

After we define all the 4 side walls as **Periodical** walls, the Spectral Terms in both X-Direction and Y-Direction are active. They are controlling the accuracy for the matrix elements in the simulation. Normally, their default values are 15. Please define X-Direction: Spectral Terms = 15 (default) and the Y-Direction Spectral Terms = 30 (larger value may improve accuracy for larger span between the walls) (see Figure 7.106). Select OK and we will get the 2D periodic walls defined as shown in Figure 7.107. Save the geometry file as:

ar_pp2.geo. The **ar_pp2.geo** is the 2D periodic wall approximation to the 1D array.

Please simulate the **ar_pp.geo** and **ar_pp2.geo** at 1.88 GHz. Please enable the **Radiation Pattern File** in the simulation setup. IE3D will calculate the pattern and save the pattern into **ar_pp.pat** and **ar_pp2.pat** files, respectively.

Section 7.33 The Comparison of Different Approaches

The comparison on the radiation parameters between the 5 schemes is shown in Table 7.9. We have listed three efficiencies: Element Efficiency (dB), the Antenna Efficiency with Feed Network (dB) and the Antenna Efficiency without Feed Network (dB). For the "Array Factor" and the 2 "Periodical Walls" approaches, the efficiency in the radiation parameters is the Element Efficiency. It is the efficiency of the elements without the feed network. The Efficiency (dB)'s for the two cases are estimated values based upon the results of the "Separation" case.

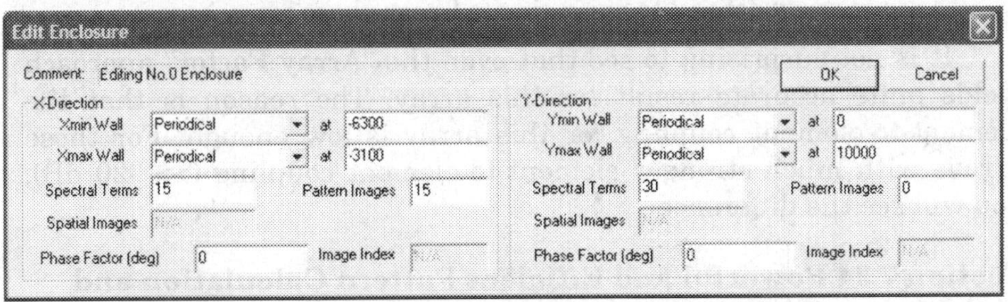

Figure 7.106 The Edit Enclosure dialog for ar_pp2.geo.

For the "Separation" case, the feed network loss is displayed as –2.02 dB. It is calculated based upon the input power at port 1 and the output power at port 2 to port 17 of the ar_feed.geo (listed in Table 7.9). We also use it for the feed loss of the 1st three cases. For the 1st four schemes, we can use the IE3D to simulate the feed network. In fact, we can also use the **Process->Create TLN S-Parameter File** command on MODUA to create the s-parameters for the feed network. For the "Separation" case, the Antenna Efficiency with Feed Network (dB) is the radiation efficiency in the **ar_fd_ps.pat** file.

Certainly, the most accurate method is the "Complete Array" approach. The "Separation" should be the 2nd most accurate approach. The "Periodical Walls" approaches are the 3rd. The "Array Factor" is the simplest approach and the fastest approach. The rule is the more accurate approach normally requires more computational effort. For this case, the "Separation" approach takes more time. However, "Complete Array" approach will be slower for bigger arrays.

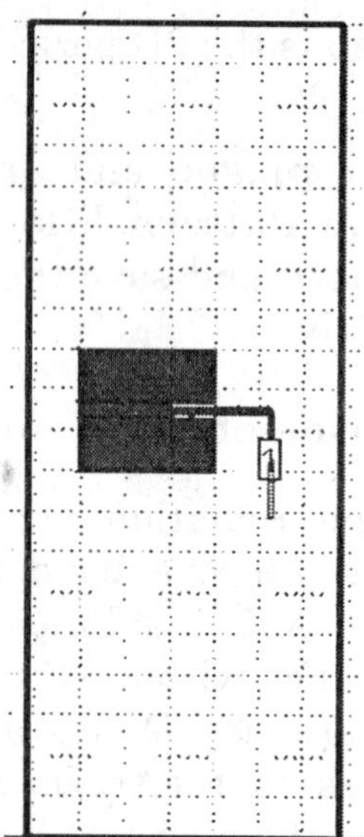

Figure 7.107. The 2 pairs of parallel periodic walls for precise modeling of planar phase array. Large distance between the top and bottom walls is purposely defined to approximate linear phase array here.

It is not surprising to see that even the "Array Factor" approach yields quite accurate result for this array. The reason is that the element-to-element coupling for this array is low enough. For those arrays with much stronger element-to-element coupling (>> -20 dB), you will see the difference.

Section 7.34 Powerful and Efficient Pattern Calculation and Handling on PATTERNVIEW.

We have demonstrated different ways to analyze antenna array. We can use the PATTERNVIEW to find the radiation pattern of an array based upon the element pattern. The antenna array is not limited to single type of basic elements. It can have different types of basic elements. Each element of the array can be of different shape, different orientation and different excitation. We can also add the final ground' effects.

On the PATTERNVIEW, we can also estimate the field propagation and distribution in some specific locations in the far field zone based upon the far field pattern.

A user can also merge different patterns from different radiators or antennas. We will not demonstrate all the features here. However, interested users can explore the power of the PATTERNVIEW by himself. We will conclude this chapter here.

Table 7.10 The comparison between the 4 approaches at 1.88 GHz.

	"Array Factor" (ar_p1. geo)	"Periodic Walls" (ar_pp. geo)	"Periodic Walls" (ar_pp2. geo)	"Separation" (ar_ps.geo/ ar_feed. geo)	"Complete Array" (ar_comp. geo)
Directivity (dB)	18.95	18.52	18.38	18.73	18.48
Antenna Gain with Feed Network (dB)	N/A	13.08*	13.03*	12.43	11.39
Antenna Gain without Feed Network (dB)	13.55	N/A	N/A	15.10	N/A
Mismatch (dB)	N/A	N/A	N/A	-0.095*	-0.068
Element Efficiency (dB)	47.1%	N/A	N/A	N/A	N/A
Loss in Feed Network (dB)	-2.02*	-2.02*	-2.02*	-2.02*	N/A
Antenna Efficiency with Feed Network (dB)	N/A	N/A	28.2%	23.47%	19.57%
Antenna Efficiency without Feed Network (dB)	40.0%	44.3%	44.8%	39.1%	N/A
3 dB Beam Width (deg)	5.55	5.88	6.41	5.87	5.93
Unknowns	190	190	190	3040	4039
Seconds/Freq (Structure)	0.1	4	21	247 (Aims II)	97 (Aims III)
Seconds/Freq (Feed Network)	22	22	22	22	N/A
Seconds/Freq (Pattern Calc)	1	5	2	150	18
RAM Requirement (MB) (with Feed Network)	1 (N/A)	1 (10)	1 (10)	30 (Aims II)	20 (Aims III)

The "*"s indicate the values are not direct output from the software. It needs calculation based upon information in the .pat files and .ect files.

APPENDIX

CONTENTS

APPENDIX

1. ANTENNA DIVERSITY

Using antenna arrays for **diversity** response is one of the most uncomplicated uses of antenna arrays. Because the power level of a received signal can vary significantly with small changes in distance, a diversity array simply uses a set of antennas and combines the signals to obtain the maximum signal. Consider the example of Figure 1. Someone is talking on their cell phone, and a hypothetical (though reasonable) power is shown in the areas in the region of the user.

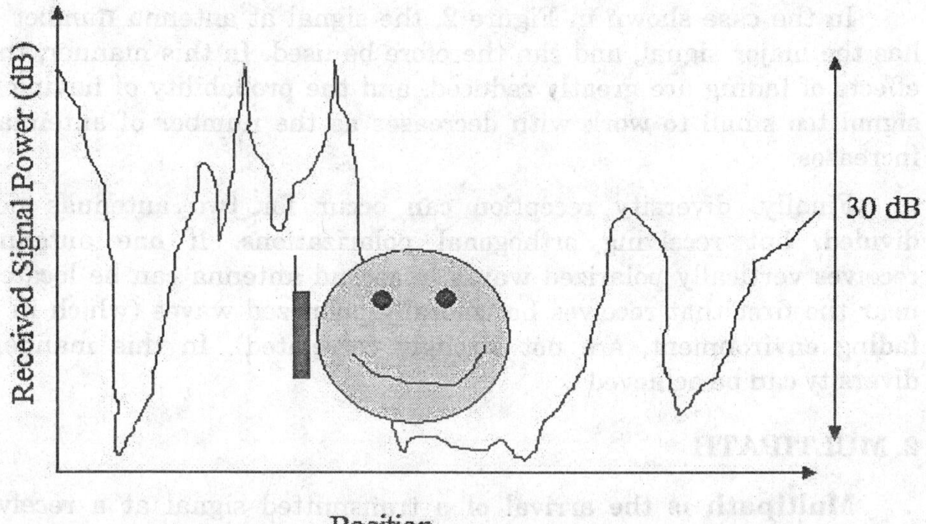

Figure 1. Example of signal level versus position

As Figure 1 illustrates, the received power can change by 30 dB (a factor of 1000) by moving the cell-phone a moderately small distance. This strong variation in signal power is caused by motion of surrounding objects (the user, cars, windy trees, etc) and multipath.

To contest this effect (known as fading), an array of antennas can be used. For instance, if 3 antennas are placed in the above situation, as shown in Figure 2, the antenna with the maximum signal can be selected and used.

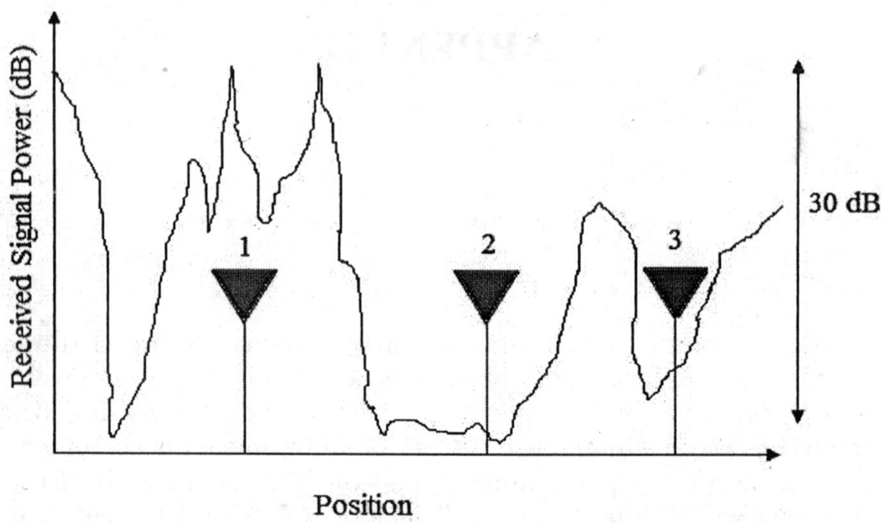

Figure 2. Three antennas used in a diversity array

In the case shown in Figure 2, the signal at antenna number 1 has the major signal, and can therefore be used. In this manner, the effects of fading are greatly reduced, and the probability of having a signal too small to work with decreases as the number of antennas increases.

Finally, diversity reception can occur for two antennas not divided, but receiving orthogonal polarizations. If one antenna receives vertically polarized waves, a second antenna can be located near the first that receives horizontally polarized waves (which in a fading environment, are not strongly correlated). In this manner, diversity can be achieved

2. MULTIPATH

Multipath is the arrival of a transmitted signal at a receive antenna from multiple paths due to reflections. The reflections can crop up from the ground, trees, buildings, people, or whatever is around. As an example, see Figure 3. The dominant signal is often the **Line of Sight (LOS)** signal, which is the direct path from the receiver to the transmitter. The other arriving signals (due to reflections) are known as multipath. In some situations (particularly in urban environments), there is no line of sight signal, and all received rays are from multipath. Multipath causes significant variation in received signal-sometimes the multipath reinforces the received signal, but often it subtracts from it (a phenomenon known as fading). See also diversity reception.

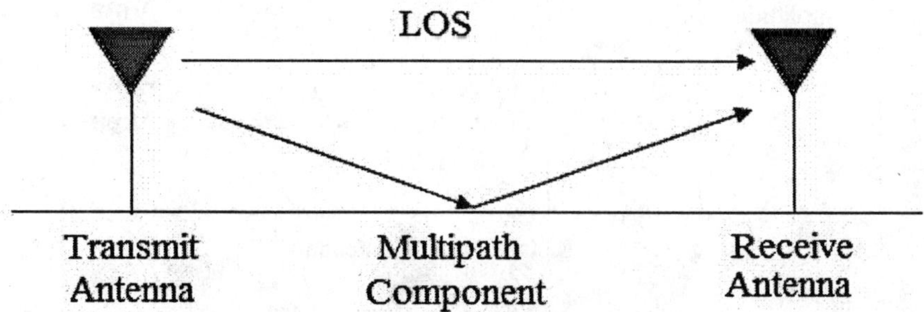

Figure 3. Example of multipath

3. VSWR (VOLTAGE STANDING WAVE RATIO)

Voltage Standing Wave Ratio (VSWR) or sometimes just Standing Wave Ratio (SWR) is a measure of how well coordinated an antenna is (in terms of impedance) to the transmission line it connects to. If the reflection coefficient is given by Γ, then the VSWR is defined as:

$$VSWR = \frac{1+|\Gamma|}{1-|\Gamma|}$$

The VSWR is always a real and positive number for passive (non-powered) loads, as in generally the case. The smaller the VSWR is, the better the antenna is matched to the transmission line and the more power is delivered to the antenna. The minimum VSWR=1.0, in which case none of the power is reflected, which is the ideal case.

Often antennas must satisfy a bandwidth requirement that is given in terms of VSWR. For instance, an antenna might maintained to operate from 100-200 MHz with VSWR<3. This implies that the VSWR is <3 over the specified frequency range; further it implies the power reflected is less than half (Γ<0.5) over the quoted frequency range.

Physically, the VSWR is determined from the voltage measured along a transmission line leading to an antenna. It is the ratio of the peak amplitude of a standing wave to the minimum amplitude of a standing wave (Figure 4).

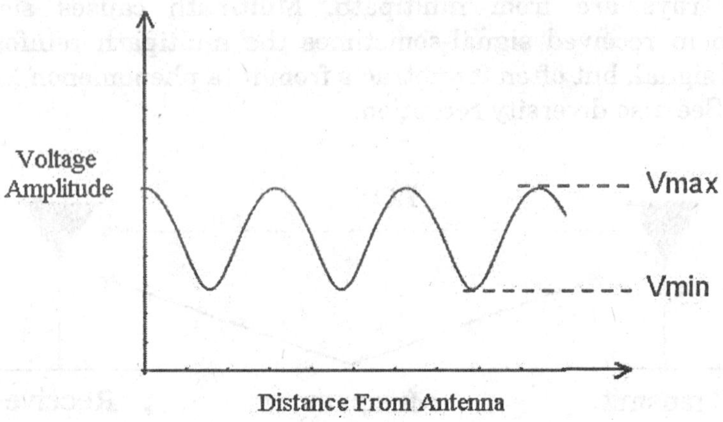

Figure 4.

In industry, VSWR is sometimes pronounced "viz-wer".

4. PERMEABILITY

Permeability is a property of a medium or a region of space. On this page, I'm going to give its meaning as it relates to antenna theory, with only a small explanation on the physics behind it. The permeability, as we saw relates the magnetic flux density to the electric field. The permeability is given in units of Henries/meter; since Henries relates to inductance, a material with a higher permeability can be thought of as storing more magnetic energy.

The permeability of a vacuum (or free space) is given by:

$$\mu_0 = 4\pi * 10^{-7} \text{ [H/m]}$$

This is approximately the permeability of air on Earth. Suppose now we are looking at a magnetic material (for instance, iron). This material will affect the magnetic field because the magnetic moments of the molecules that make up the material align themselves in the direction of an external magnetic field. In addition, these magnetic moments tend to remain even after the external field is turned off, leading to a permanent magnetization of some materials. The permeability of a medium is typically précised by a relative permeability:

$$\mu_r = \frac{\mu}{\mu_0}$$

Materials are characterized as *diamagnetic* (relative permeability slightly less than 1.0), *paramagnetic* (relative permeability slightly more than 1.0), and *ferromagnetic* (high value for relative permeability, that also exhibits hysteresis). A table of common materials and their relative permeabilities is presented below-

Material	Relative Permeability
Gold	0.999996
Mercury	0.999997
Water	0.99999
Air	1.000004
Aluminum	1.00002
Cobalt	255
Nickel	600
Steel	2000
Iron	4500

Finally, as mentioned on the page on permittivity, the permeability also affects the speed of propagation of an electromagnetic wave in a medium with a relative permeability given as μ_r, and also its corresponding wavelength:

$$c = \frac{c_0}{\sqrt{\mu_r}}$$

$$\lambda = \frac{\lambda_0}{\sqrt{\mu_r}}$$

In the above, c is the speed of wave propagation in the medium and c_0 is the speed of light in free space (or a vacuum); also λ is the wavelength of a wave at frequency f in the medium, the wavelength at the same frequency in a vacuum would be λ_0.

5. PERMITTIVITY

Permittivity is a property of a medium or a region of space. On this page, I'm going to give its meaning as it relates to antenna theory, with only a small explanation on the physics behind it. The permittivity, as we saw relates the electric flux density to the electric field. The permittivity is given in units of Farads/meter; since Farads relates to capacitance, a material with a higher permittivity can be thought of as storing more electrical energy.

The permittivity of a vacuum (often called free space) is given by:

$$\varepsilon_0 = 8.854 * 10^{-12}\ [F/m]$$

The above is roughly the permittivity of air on earth. Suppose we have some other dielectric medium (glass, rubber, wax or other non-conducting material). The permittivity describes how an electric field is affected within a medium. The E-field tends to polarize the molecules within the material, which make up a net electric field that opposes the applied E-field. As a result, the total E-field is less than it would be in a vacuum. This effect is quantized in terms of the permittivity - which can be anisotropic (direction dependent) and frequency dependent.

As an antenna engineer, the permittivity affects the speed of propagation of a wave through a medium and also its wavelength. The permittivity of a medium is most often given as a relative permittivity:

$$\varepsilon_r = \frac{\varepsilon}{\varepsilon_0}$$

The speed of wave propagation in a given medium is given as:

$$c = \frac{1}{\sqrt{\varepsilon\mu}} = \frac{1}{\sqrt{\varepsilon_0\varepsilon_r\mu_0\mu_r}} = \frac{2.99*10^8}{\sqrt{\varepsilon_r\mu_r}} = \frac{c_0}{\sqrt{\varepsilon_r\mu_r}}\left[\frac{m}{s}\right]$$

Hence, if a medium has $\varepsilon_r = 4$, then the speed of wave propagation in the medium will be half as fast as in free space (1.5*10^8 m/s). As a result of the speed being slowed, the wavelength of a plane wave decreases in size as well (the frequency remains constant):

$$\lambda = \frac{c}{f} = \frac{c_o}{f\sqrt{\varepsilon_r}} = \frac{\lambda_o}{\sqrt{\varepsilon_r}}$$

The above fact is often used in antenna-miniaturization: since resonant antennas are often a half-wavelength in size, if they are placed in a medium with a higher permittivity the required length decreases and hence the antenna will be smaller.

Permittivity alters the direction of travel of a wave incident upon a medium through Snell's law. In addition, the reflection and transmission coefficients of a wave travelling from one medium to another are influenced by the permittivity.

6. ANTENNA Q

The **Q** of an antenna is a gauge of the bandwidth of an antenna relative to the center frequency of the bandwidth. If the antenna operates over a band between *f1* and *f2* with center frequency *fc=(f1+f2)/2*, then the Q is given by:

$$Q = \frac{f_c}{f_2 - f_1}$$

Antennas with a high Q are narrowband; antennas with a low Q are wideband. The higher the value of Q, the more sensitive the input impedance is to small changes in frequency.

7. S-PARAMETERS

S-parameters describe the input-output relationship between ports (or terminals) in an electrical system. For instance, if we have 2 ports (intelligently called Port 1 and Port 2), then S12 represents the power transferred from Port 1 to Port 2. S21 represents the power transferred from Port 2 to Port 1. In general, SNM represents the power transferred from Port N to Port M in a multi-port network.

As an example, consider the following two-port network:

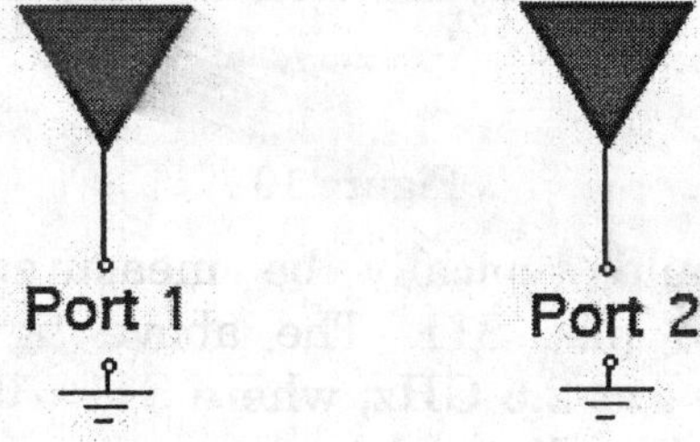

Figure 9.

In the above Figure, S21 represents the power received at antenna 2 relative to the power input to antenna 1. For instance, S21=0 dB implies that all the power delivered to antenna 1 ends up at the terminals of antenna 2 (which isn't physically possible). If S21=-10 dB, this implies that 1 Watt delivered to antenna 1 (or 0 dB), ends up as -10 dB at antenna 2, or 0.1 Watts.

In practice, the most commonly quoted parameter in regards to antennas is S11. S11 represents how much power is reflected from the antenna. If S11=0 dB, then all the power is reflected from the antenna and nothing is radiated. If S11=-10 dB, this implies that if 3 dB of power is delivered to the antenna, -7 dB is the reflected power. The rest was "accepted" by the antenna. This accepted power is any radiated or absorbed as losses within an antenna. Since antennas are typically designed to be low loss, the majority of the power delivered to the antenna is radiated.

As an example, consider the plot of S11 in the following figure:

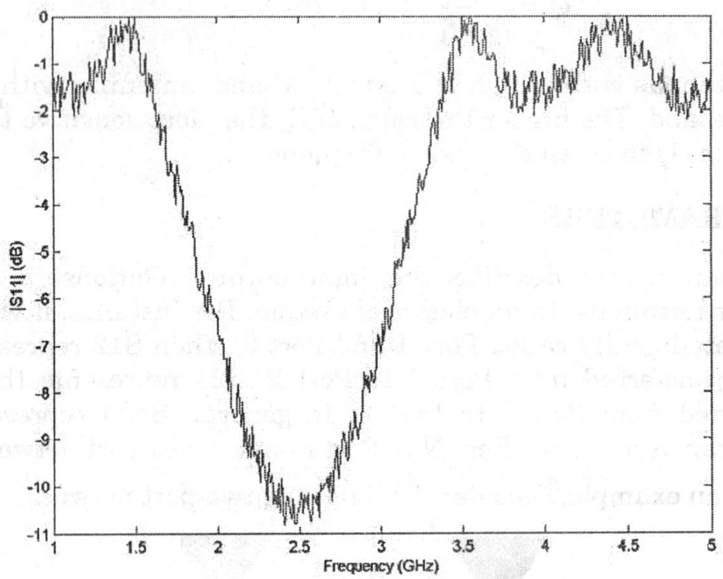

Figure 10 .

The above would typically be measured using a network analyzer, which can plot S11. The above figure shows that the antenna radiates best at 2.5 GHz, where S11=-10 dB. Further, at 1.5 GHz the antenna will radiate virtually nothing, as S11 is close to 0 dB. The bandwidth can also be resolute from the above figure. If S11 is to be <-6 dB, then the bandwidth would be roughly 1 GHz, with 3 GHz the high end and 2 GHz the low end of the band.

8. SPHERICAL COORDINATES

Understanding Spherical Coordinates is a must for the practicing antenna engineer. You are probably familiar with Cartesian Coordinates - a position (point **P**) can be précised by a triplet like (x, y, z) where x is the distance from the origin to the point along the X-axis, and so on (see Figure 1). Spherical coordinates use a different coordinate system, one with spherical symmetry, which makes it very useful in engineering and physics in certain problems.

The point **P** could be specified relative to the same origin in a different coordinate system. Spherical coordinates make use of three distinct coordinates:

R - The magnitude of the distance between the origin and the point (always positive)

θ - Angle between the z-axis and the vector from the origin to the point (ranges from 0 to 180 degrees)

ϕ - Angle between the x-axis and the projection of the point onto the x-y plane (ranges from 0 to 360 degrees)

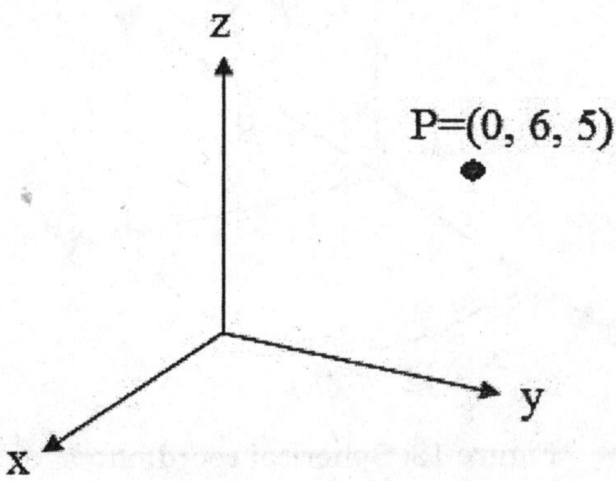

Figure 11 A point **P** defined in the Cartesian coordinate system

Any point specified in Cartesian coordinates as (x, y, z) can be re-expressed in spherical coordinates via the following transformation:

$$R=\sqrt{x^2+y^2+z^2}$$

$$\theta = \arccos\left(\frac{z}{\sqrt{x^2+y^2}}\right)$$

$$\phi = \arctan\left(\frac{y}{x}\right), \textit{if } x \geq 0$$

$$\phi = \arctan\left(\frac{y}{x}\right) + \pi \text{ , } \textit{if } x < 0$$

The above might look complex, but after you've worked with it for a while it makes a lot of sense. The point **P** = (0, 6, 5) can be evaluated in spherical coordinates as:

$$R=\sqrt{10^2+6^2+5^2} = \sqrt{61} \approx 7.81$$

$$\theta = \arccos\left(\frac{5}{\sqrt{0^2+6^2}}\right) \approx 0.568 \textit{ radians} = 33.6^\circ$$

$$\phi = \arctan\left(\frac{6}{0}\right) = \frac{\pi}{2} \textit{ radians} = 90^\circ$$

The coordinates are illustrated in Figure 12:

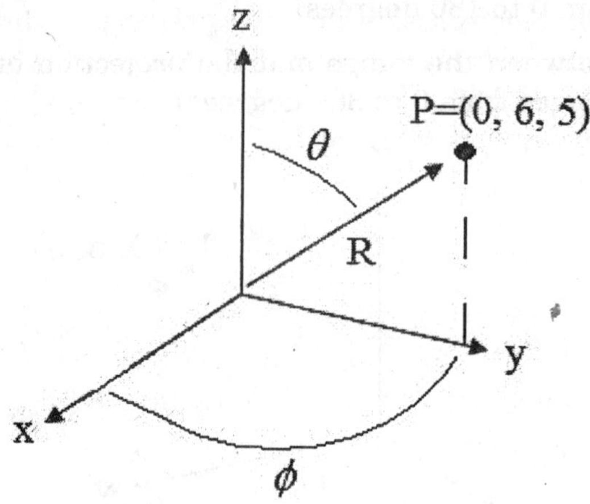

Figure 12. Spherical coordinates

Spherical coordinates are popular for antennas, since we often are only interested in the antennas response in a particular direction, not how far away something is (radiation patterns die off as 1/R^2 for all antennas in the far field). In Cartesian coordinates, 3 variables require specified determining the direction from the origin, and it is not intuitive. For spherical coordinates, once it is understood, the polar angle θ and the azimuth angle ϕ can be readily used.

For practice, make sure the following table makes sense. I give a set of rectangular coordinates on the left, and the corresponding spherical coordinates on the right.

Table 1. Conversion of Cartesian Coordinates to Spherical (these should all make sense)

Cartesian (X,Y,Z)	**Spherical (R, θ, ϕ) [angles in degrees]**
(1, 0, 0)	(1, 90, 0)
(0, 1, 0)	(1, 90, 90)
(-1, -1, 0)	(1.414, 90, 225)
(0, 0, 1)	(1, 0, 0) [Note: ϕ not unique here - could be anything. Why is that?]
(1, 1, 1)	(1.73, 45, 45)
(0, 0, -1)	(1, 180, 0) [ϕ not unique again]
(0, 0, 0)	(0, 0, 0) [ϕ and θ could be any number!]
(-5, 0, 0)	(5, 90, 180)
(2, -4, -4)	(6, 153, -135)

That is an overview of spherical coordinates. They come up a lot in the study of electromagnetic and physics, and can aid in understanding and solving of certain problems. They happen in antenna engineering most often in regards to radiation patterns, which specify how an antenna radiates versus angle (or direction) away from the antenna

9. E-PLANE

The E-plane is any plane that contains the E-field and the direction of maximum radiation from the antenna. As an example, consider a short dipole, the electric fields are given by:

$$E = \hat{\theta} \sin \theta$$

This states that in the x-y plane, the E-fields are vertically oriented (negative z-direction). See Figure 13 below.

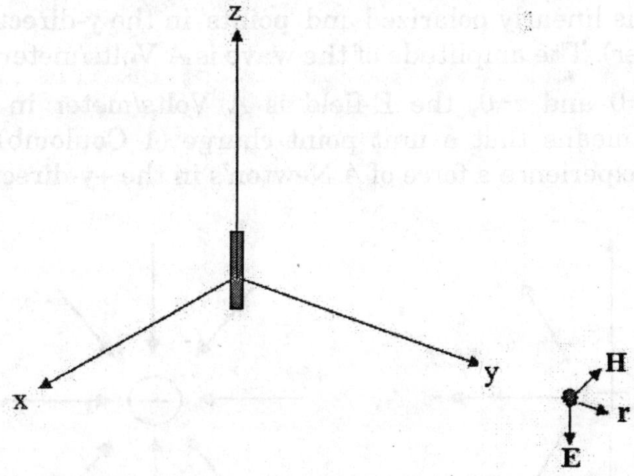

Figure 13. Fields at a point along the y-axis from a short dipole

The fields at a point along the y-axis are shown in Figure 13. The E-field is oriented in the negative z-direction, the route of propagation at the point is the +y-direction (also the direction of maximum radiation in this case), and the H-field is oriented in the negative x-direction.

The E-plane is any plane that contains the E-field and the direction of maximum radiation. Hence, the E-plane in this case is any plane that contains the z-axis (x-z plane, y-z plane, etc). The E-plane cut is not unique in this case, but a plot of the radiation pattern given in the E-plane would be a function of the polar angle (θ), which is measured off the z-axis.

10. ELECTRIC FIELD (E-FIELD)

Electromagnetic waves are made up of **Electric Fields** (often called the E-field) and magnetic fields. What is an E-field?

Technically, the E-field at a point in space is a measure of how strong the force would be on a unit point charge (a small sphere with an electric charge of 1 Coulomb on it). Hence, the units of the E-field are Newton's/Coulomb [N/C]. These units are equivalent to Volts/meter [V/m], which is what the E-field is commonly quoted in (for instance, 10 V/m).

The E-field is a vector quantity - this means at every point in space it has a magnitude and a direction. For instance, let's say an E-field exists in space given by:

$$E = A \cos(\alpha x - kz)\, \hat{y}$$

This is the E-field of a plane wave travelling in the +z-direction, and the E-field is linearly polarized and 'points' in the y-direction (k is the wave number). The amplitude of the wave is A Volts/meter.

At time t=0 and z=0, the E-field is A Volts/meter in the +y-direction. This means that a unit point charge (1 Coulomb) at this location would experience a force of A Newton's in the +y-direction.

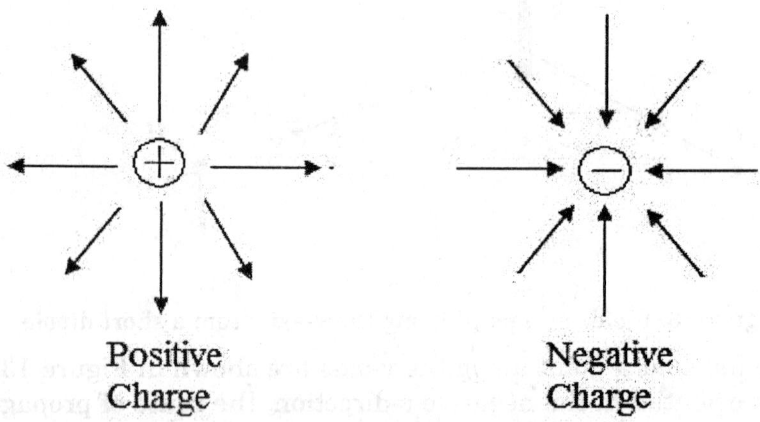

Figure 14

The electric field also relates to voltage - a stronger E-field incident upon an antenna will induce a larger voltage difference transversely the antenna's terminals. However, except for low frequencies, the relationship between E-fields and Voltage is not simple (the voltage is a potential which is subject to different definitions). At d.c., when the fields are static (no variation with time), the E-field and voltage V is linked to each other by:

$$\mathbf{E} = -\nabla V$$

$$V = -\oint E.dl$$

The electric field associated with a point charge with a positive charge point away from it at every location; the fields linked with a point charge with a negative charge point towards it.

11. ELECTRIC FLUX DENSITY (D-FIELD)

The **Electric Flux Density** (usually written as the vector quantity **D**) is often used in electromagnetic. While we won't give it a rigorous definition here, it can be sufficiently understood for the purposes of antenna theory as being proportional to the Electric Field. The proportionality constant depends on the medium being analyzed, and is recognized as the permittivity:

$$\mathbf{D} = \varepsilon\,\mathbf{E}$$

The electric flux density has units of charge/area (Coulombs/ meter-squared or [C/m^2]). The permittivity is often frequency-dependent, and is sometimes anisotropic (implying the permittivity depends on which direction the fields are in):

$$\begin{bmatrix} D_x \\ D_y \\ D_z \end{bmatrix} = \begin{bmatrix} \varepsilon_x\, E_x \\ \varepsilon_y\, E_y \\ \varepsilon_z\, E_z \end{bmatrix}$$

In the most general case the E-field is related to the electric flux density by a 3 x 3 matrix, which is also frequency-dependent.

The only other major result in relation to the electric flux density involves the continuity of fields across boundaries. Consider two materials as shown in the figure below.

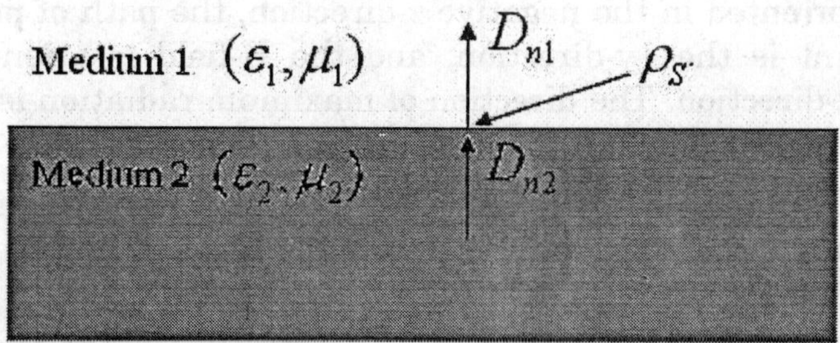

The normal component of the electric flux density will be discontinuous at the boundary by an amount equal to the surface charge density ρs, which has units of charge/area (Coulombs/meter-squared). This can be written mathematically as:

$$D_{n1} - D_{n2} = \rho_s$$

This relation holds at every point down the interface of a material discontinuity (medium change).

12. H-PLANE

The H-plane is a plane that contains the H-field and the direction of maximum radiation from the antenna. As an example, consider a short dipole, the magnetic fields are given by:

$$H = \hat{\phi} \sin \theta$$

This states that in the x-y plane, the H-fields are horizontally oriented (in the x-y plane). See Figure 15 below.

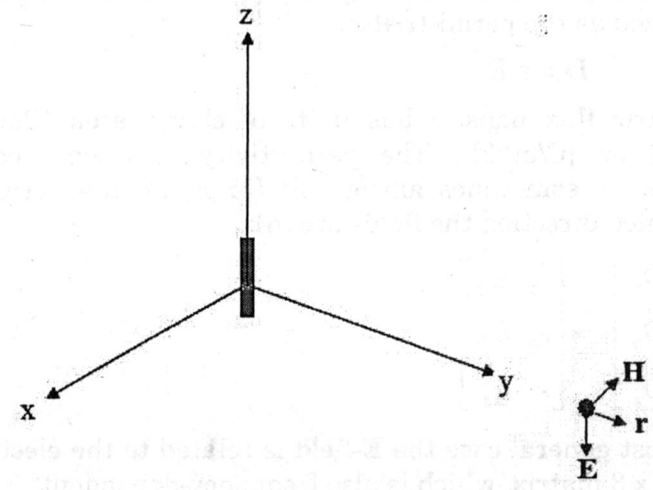

Figure 15. Fields at a point along the y-axis from a short dipole

The fields at a point along the y-axis are shown in Figure 15. The E-field is oriented in the negative z-direction, the path of propagation at the point is the +y-direction, and the H-field is oriented in the negative x-direction. The direction of maximum radiation is broadside from the antenna.

The H-plane is any plane that contains the H-field and the direction of maximum radiation. Hence, the H-plane in this case is the x-y plane.

13. MAGNETIC FIELD (H-FIELD)

Electromagnetic waves are made up of **Electric Fields** (often called the E-field) and **Magnetic fields** (also known as H-fields). What is an H-field?

The H-field is a vector quantity (has a magnitude and direction) and is measured in Amps/Meter [A/m]. Recall that the E-field points away from a positive point charge. An H-field curls (or wraps) around a wire of moving charge, as shown in Figure 16. Hence, H-fields are linked with moving electric charges.

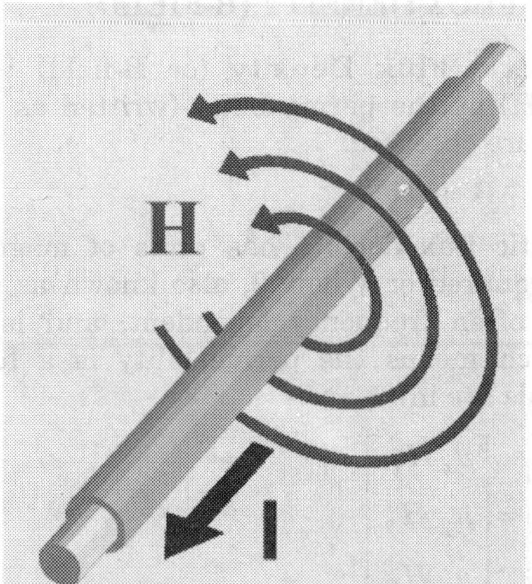

Figure 16. H-field associated with a static current.

There are no isolated magnetic charges as of 2008, so an H-field can't be distinct as a force per unit magnetic charge in the way an E-field can be distinct. However, magnetic dipoles do exist (magnets) which have a positive and negative end (or North and South). The magnetic field lines travel away from the North side and terminate on the south side.

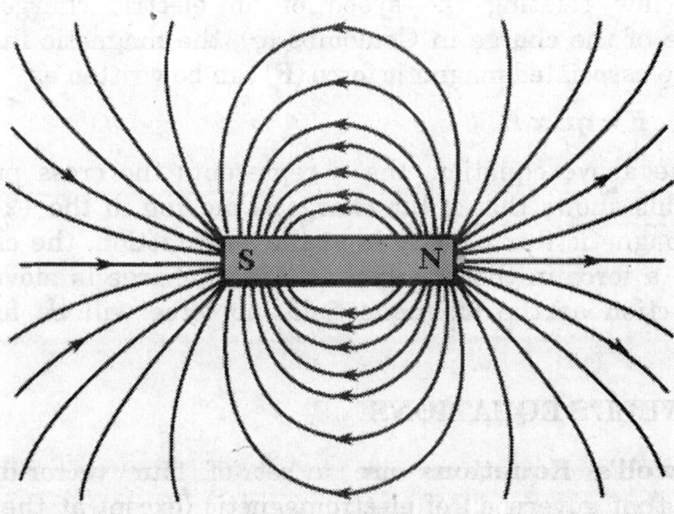

Figure 17. H-field lines associated with magnetic dipole

The H-field is orthogonal to the direction of propagation in a plane wave, as well as perpendicular to the E-field. It is the interaction of the E-field with the H-field in space that allows for wave propagation.

14. MAGNETIC FLUX DENSITY (B-FIELD)

The **Magnetic Flux Density** (or **B**-field) is related to the magnetic field (**H**) by the permeability (written as μ of the medium where the fields are observed:

$$\mathbf{B} = \mu\mathbf{H}$$

The magnetic flux density has units of magnetic charge/area (Weber's/meter-squared or Wb/m^2, also known as a Tesla (T)). The permeability is often frequency-dependent, and is also sometimes anisotropic (which means the permeability is a function of which direction the fields are in):

$$\begin{bmatrix} B_x \\ B_y \\ B_z \end{bmatrix} = \begin{bmatrix} \mu_x H_x \\ \mu_y H_y \\ \mu_z H_z \end{bmatrix}$$

In the most general case the H-field is related to the magnetic flux density by a 3x3 matrix, which is also frequency-dependent.

Recall that the electric field can be defined in terms of a "force per unit electric charge". Further, note that magnetic fields are linked with moving electric charges. The magnetic flux density can be defined in terms of a force per unit charge moving at unit velocity. An empirical law relating the speed of an electric charge (**u**), the magnitude of the charge in Coulombs (q), the magnetic flux density **B**), and the associated magnetic force (**F**) can be written as:

$$F = qu \times B$$

In the above equation, the × represents the cross product for vectors. This shows that if the charge is moving in the +x-direction, and the magnetic flux density is in the +y-direction, the charge will know-how a force in the +z-direction. If the charge is moving in the same direction as the magnetic field, no force will be felt by the charge.

15. MAXWELL'S EQUATIONS

Maxwell's Equations are a set of four vector-differential equations that govern all of electromagnetic (except at the quantum level, in which case we as antenna people don't care so much). They were first offered in a complete form by James Clerk Maxwell back in the 1800s. He didn't come up with them all on his own, but did add the displacement current term to Ampere's law which made them complete.

The four equations (written only in terms of **E** and **H**, the electric field and the magnetic field), are given below.

$$\nabla * E = \frac{\rho_v}{\varepsilon} \quad \text{(Gauss' Law)}$$

$$\nabla * H = 0 \quad \text{(Gauss' Law for Magnetism)}$$

$$\nabla \times E = -\mu \frac{\partial H}{\partial t} \quad \text{(Faraday's Law)}$$

$$\nabla \times H = J + \varepsilon \frac{\partial E}{\partial t} \quad \text{(Ampere's Law)}$$

In Gauss' law, ρ is the volume electric charge density, **J** is the electric current density (in Amps/meter-squared), ε is the permittivity and μ is the permeability.

The good news about this is that all of electromagnetic is summed up in these 4 equations. The bad news is that no matter how good at math you are, these can only be solved with an analytical solution in tremendously simple cases. Antennas don't present a very simple case, so these equations aren't used a whole lot in antenna theory (except for numerical methods, which numerically solve these approximately using a whole lot of computer power).

The last two equations (Faraday's law and Ampere's law) are responsible for electromagnetic radiation. The curl operator represents the spatial variation of the fields, which are coupled to the time variation. When the E-field travels, it is altered in space, which gives rise to a time-varying magnetic field. A time-varying magnetic field then varies as a function of location (space), which gives rise to a time varying electric field. These equations enfold around each other in a sense, and give rise to a wave equation. These equations predict electromagnetic radiation as we understand it.

16. DECIBELS (dB)

Many parameters related to antennas are calculated in decibels; for instance, gain is often specified in decibels, written as 10 dB. Or maybe the minimum received power for an antenna system to work is specified as -70 dBm (decibels relative to a mili Watt). Or the transmit power of your cell phone is given as -3 dB.

Why is this? Why don't they just tell you how much power is transmitted in Watts or the gain of an antenna in regular (linear) units?

The decibel system is used when quantities can vary by massive amounts. The Richter scale for earthquakes is also on a scale similar to decibels, because the magnitude of earthquakes can vary from barely audible with sophisticated equipment to earthquakes that demolish vast amounts of land. As an example, let's look at power received by an antenna relative to power transmitted.

Suppose we have an antenna transmitting 50 Watts of power, with a gain of 2 (=3 dB) in the direction of a receive antenna with a gain of 2 (=3 dB). Suppose the antennas are separated by 20,000 kilometers (as is the typical orbit of a GPS satellite). The GPS satellites operate at a frequency of 1.575 GHz. How much power is received?

Using the Friis Transmission Equation and the fact that wavelength equals c/f, we can calculate the received power to be:

$$P_R = \frac{P_T G_T G_R \lambda^2}{(4\pi R)^2}$$

$$P_r = \frac{(50)(2)(2)\left(\dfrac{3.10^8}{1.575.10^9}\right)^2}{(4\pi * 20 * 10^6)^2} = 1.149 * 10^{-16} \ [Watts]$$

$$= 0.0000000000000001149 \ [\mathrm{W}]$$

Of the 50 Watts transmitted, about 2.3e-16% of that power gets through. Incidentally, this might seem terrifyingly small, but yes, antenna systems can operate with this much power. So be patient with the GPS unit in your car, it is doing a good job.

Antenna engineers (all engineers actually) don't like to use linear units when the quantities can vary by such large amounts. Its real tough to picture in your head the difference between 1e17 and 1e18. To work around this, we use the decibel system. It's a simple logarithmic transformation - units in linear are easily converted via:

$$P_{dB} = 10 * \log_{10} P$$

Hence, the above quoted value of 1.149e-16 Watts is equal to -159.4 decibels, written -159.4 dB.

To give a cemented idea on the conversions, a table of linear to decibel levels is presented.

Table 2. Conversion between Linear Units and Decibels.

Linear Value	dB Value
1e-10	-100 dB
1e-9	-90 dB
1e-6	-60 dB
1e-3	-30 dB
1e-1	-10 dB
0.5	-3 dB
1	0 dB
10	10 dB
100	20 dB
1e3	30 dB
1e6	60 dB

Another common unit is dBm. This means "decibels relative to a mili Watt". In this case, we are talking about power, and the power is just specified in mili Watts instead of Watts. So:

1 W = 1000 mW = 0 dB = 30 dBm

Or

0 dBm = -30 dB = 0.001 W = 1 mW

Another common difference on dB in antenna theory is dB_i which means "decibels relative to an isotropic antenna". This just specifies the gain of an antenna relative to the isotropic gain, which is 1. So really nothing changes...:

Gain of 10 dB = Gain of 10 dB_i

The reason people specify dB_i sometimes is that other relative values are sometimes specified, like dB_d. This means "decibels of gain relative to a standard half-wave dipole antenna". The gain of a half-wave dipole is 2.15 dB_i. So:

Antenna with a gain of 10 dB_i = 10 dB = 7.85 dB_d

Gain of half-wave dipole antenna = 2.15 dB_i = 0 dB_d

16.1 Decibel Math

The decibel system (and the properties of logarithms) makes the mathematics concerned in some equations much simpler. For instance, recall the Friis Transmission Formula:

$$P_R = \frac{P_T G_T G_R \lambda^2}{(4\pi R)^2} \qquad (*)$$

To convert this equation from linear units in Watts to decibels, we take the logarithm of both sides and multiply by 10:

$$10 - \log_{10} P_R = 10 . \log_{10}\left(\frac{P_T G_T G_R \lambda^2}{(4\pi R)^2}\right)$$

A nice property of logarithms is that for two numbers *A* and *B* (both positive), the following result is always true:

$$\log_{10} (AB) = \log_{10} (A) + \log_{10} (B)$$

Equation (*) then becomes:

$$10 * \log_{10} P_R = 10 * \log_{10} (P_T) + 10 * \log_{10} (G_T) + 10 * \log_{10} (G_R) + 10 * \log_{10}\left(\frac{\lambda}{4\pi R}\right)^2$$

Using the definition of decibels, the above equation becomes a simple addition equation in dB:

$$[P_R]_{dB} = [P_T]_{dB} + [G_T]_{dB} + [G_R]_{dB} + \left[\left(\frac{\lambda}{4\pi R}\right)^2\right]_{dB}$$

The above representation is easier to work with, which is kind of cool.

17. STEERING VECTOR

A **steering vector** represents the set of phase delays a plane wave experiences, evaluated at a set of array elements (antennas). The phases are individual with respect to an arbitrary origin.

For instance, suppose a plane wave is described by a wave vector **k**. If there are *N* elements in an antenna array, with element *i* having location given by:

$$r_i = (x_i, y_i, z_i)$$

Then the steering vector is an *N*x1 complex vector representing the relative phases at each antenna, and is given by:

$$v(k) = \begin{bmatrix} e^{-jk.r_1} \\ e^{-jk.r_2} \\ \cdot \\ \cdot \\ \cdot \\ e^{-jk.r_N} \end{bmatrix}$$

The steering vector is written as **v(k)**, to make explicit that it depends on the frequency and propagation route (which determine **k**) of the plane wave.